ADVANCES IN CHEMICAL PHYSICS

VOLUME 131

Advances in CHEMICAL PHYSICS

Edited by

STUART A. RICE

Department of Chemistry
and
The James Franck Institute
The University of Chicago
Chicago, Illinois

VOLUME 131

AN INTERSCIENCE PUBLICATION
JOHN WILEY & SONS, INC.

Published by John Wiley & Sons, Inc., Hoboken, New Jersey.
Published simultaneously in Canada.

For general information on our other products and services or for technical support, please contact our Customer Care Department within the United States at (800) 762-2974, outside the United States at (317) 572-3993 or fax (317) 572-4002.

Wiley also publishes its books in a variety of electronic formats. Some content that appears in print may not be available in electronic formats. For more information about Wiley products, visits our web site at www.wiley.com.

Library of Congress Cataloging-in-Publication Data: 58:9935

ISBN-13 978-0-471-44526-6
ISBN-10 0-471-44526-6

Printed in the United States of America

10 9 8 7 6 5 4 3 2 1

CONTRIBUTORS

LUDWIK ADAMOWICZ, Department of Chemistry and Department of Physics, University of Arizona, Tucson, AZ 85721

ROBERTO BINI, LENS, European Laboratory for Non-linear Spectroscopy and INFM, I-50019 Sesto Fiorentino, Firenze, Italy; and Dipartimento di Chimica dell'Università di Firenze, I-50019 Sesto Fiorentino, Firenze, Italy

SERGIY BUBIN, Department of Physics and Department of Chemisty, University of Arizona Tucson, AZ 85721

MAURICIO CAFIERO, Department of Chemistry, Rhodes College, Memphis, TN 381120. Present address: National Institute of Standards and Technology, Gaithersburg, MD 20899

MATTEO CEPPATELLI, LENS, European Laboratory for Non-linear Spectroscopy and INFM, I-50019 Sesto Fiorentino, Firenze, Italy; and Dipartimento di Chimica dell'Università di Firenze, I-50019 Sesto Fiorentino, Firenze, Italy

LUCIA CIABINI, LENS, European Laboratory for Non-linear Spectroscopy and INFM, I-50019 Sesto Fiorentino, Firenze, Italy; and Dipartimento di Chimica dell'Università di Firenze, I-50019 Sesto Fiorentino, Firenze, Italy

MARGHERITA CITRONI, LENS, European Laboratory for Non-linear Spectroscopy and INFM, I-50019 Sesto Fiorentino, Firenze, Italy; and Dipartimento di Chimica dell'Università di Firenze, I-50019 Sesto Fiorentino, Firenze, Italy

JONATHAN S. ELLIS, Institute of Biomaterials and Biomedical Engineering, University of Toronto, Toronto M5S, 3G9, Canada

HIRO-O HAMAGUCHI, Department of Chemistry, School of Science, The University of Tokyo, Tokyo 113-0033, Japan

M. MUTHUKUMAR, Department of Polymer Science and Engineering, Materials Research Science and Engineering Center, University of Massachusetts, Amherst, MA 01003-4530

RYOSUKE OZAWA, Department of Chemistry, School of Science, The University of Tokyo, Tokyo 113-0033, Japan

VINCENZO SCHETTINO, LENS, European Laboratory for Non-linear Spectroscopy and INFM, I-50019 Sesto Fiorentino, Firenze, Italy; and Dipartimento di Chimica dell'Università di Firenze, I-50019 Sesto Fiorentino, Firenze, Italy

GERHARD STOCK, Institute of Physical and Theoretical Chemistry, D-60439, Frankfurt, Germany

MICHAEL THOMPSON, Department of Chemistry and the Institute of Biomaterials and Biomedical Engineering, University of Toronto, Toronto M5S, 2HG, Canada

MICHAEL THOSS, Department of Chemistry, Technical University of Munich, D-85748, Garching, Germany

INTRODUCTION

Few of us can any longer keep up with the flood of scientific literature, even in specialized subfields. Any attempt to do more and be broadly educated with respect to a large domain of science has the appearance of tilting at windmills. Yet the synthesis of ideas drawn from different subjects into new, powerful, general concepts is as valuable as ever, and the desire to remain educated persists in all scientists. This series, *Advances in Chemical Physics*, is devoted to helping the reader obtain general information about a wide variety of topics in chemical physics, a field that we interpret very broadly. Our intent is to have experts present comprehensive analyses of subjects of interest and to encourage the expression of individual points of view. We hope that this approach to the presentation of an overview of a subject will both stimulate new research and serve as a personalized learning text for beginners in a field.

Stuart A. Rice

CONTENTS

POLYELECTROLYTE DYNAMICS

M. MUTHUKUMAR

Department of Polymer Science and Engineering, Materials Research Science and Engineering, Center University of Massachusetts, Amherst, MA

CONTENTS

Advances in Chemical Physics, Volume 131, edited by Stuart A. Rice

I. INTRODUCTION

Considerable progress has been made during the last five decades in a molecular understanding of the behavior of solutions of uncharged polymers. In general, the polymer dynamics can be classified into the Kirkwood–Riseman–Zimm–Rouse dynamics, entropic barrier model, and reptation dynamics as the polymer concentration is progressively increased [1–6]. However, such an understanding is lacking for solutions of polyelectrolytes. The general features of the dynamical properties of polyelectrolyte solutions are qualitatively different from those of neutral polymers. Experimental observations of these differences as manifest in the occurrence of electrophoretic mobility, diffusion of polyelectrolytes, and viscosity of polyelectrolyte solutions are richly documented in the literature [7–35], although a systematic experimental investigation on a given system does not yet exist. An example of an apparently surprising behavior is that the velocity of a polyelectrolyte molecule in a dilute solution under an electric field is the same for two chemically identical polymers, differing in molecular weight by several orders of magnitude. The wisdom, with support by extensive experimental measurements on uncharged polymers, that bigger molecules move slower than smaller molecules does not hold for polyelectrolytes. Another surprising result is that in polyelectrolyte solutions without any added salt, measured diffusion coefficients using dynamic light scattering are independent of both molar mass and polymer concentration over three decades in each variable. These results and other experimental data are amply documented in the literature [7–35]. For each of the dynamical quantities, there are many theoretical attempts [36–49]. A successful attempt to provide a conceptual framework to explain all of the major features of dynamics of polyelectrolyte solutions was recently presented by the author [50]. The present review is strongly influenced by the author's theory on polyelectrolyte dynamics. There are still several questions to be answered, which are remarked at the end of the review. Before we outline the ingredients of the theory, we first review three key dynamical properties unique to polyelectrolytes.

A. Electrophoretic Mobility in an Infinitely Dilute Solution

The electrophoretic mobility μ of a polyelectrolyte chain in an infinitely dilute solution containing an added salt at concentration c_s under a constant external electric field **E**, as defined through

$$\mathbf{v}_p = \mu \mathbf{E} \tag{1}$$

where $\mathbf{v}_p$ is the velocity of the polyelectrolyte, is independent [35] of molecular weight M at all salt concentrations:

$$\mu \sim M^0 \tag{2}$$

This is a surprising result in view of the following argument [51], due to Hückel. Defining Q and D to be respectively the net charge and the translational diffusion coefficient of the polyelectrolyte, the balance of frictional force $(k_BT/D)\mathbf{v}_p$, and electrical force $Q\mathbf{E}$ gives

$$\frac{k_BT}{D}\mathbf{v}_p = Q\mathbf{E} \tag{3}$$

so that

$$\mu \sim QD \tag{4}$$

where k_BT is Boltzmann's constant times the absolute temperature. Q is proportional to M. In infinitely dilute solutions, full hydrodynamic interactions are present so that D is inversely proportional to the radius of gyration R_g of the polyelectrolyte so that

$$\mu \sim \frac{M}{R_g} \tag{5}$$

Since R_g scales with M with an effective exponent ν, $R_g \sim M^{\nu}$, we have

$$\mu \sim M^{1-\nu} \tag{6}$$

In the limit of low salt concentrations, the conformation of polyelectrolytes is expected to be rod-like so that $\nu = 1$. In this limit, Eq. (2) is recovered. However, as the concentration of the added salt increases, the effective exponent ν decreases toward $3/5$. Therefore we expect from Eq. (6) that μ will increase with M at higher salt concentrations, in contradiction with the experimentally observed result of Eq. (2). There are several erroneous claims in the literature that D is given by the Rouse law, proportional to $\frac{1}{M}$, in infinitely dilute solutions at all salt concentrations. This conjecture is produced to agree with experimental observations but without any derivation.

B. Diffusion Coefficients

The light scattering data [24–34] of polyelectrolyte solutions are characteristic of two diffusion coefficients D_f (fast) and D_s (slow). D_f for a solution with a given c_s increases with monomer concentration c and above a certain concentration c_1, D_f is independent [31, 33, 34] of both c and M.

$$D_f \sim c^0 M^0, \qquad c \geq c_1 \tag{7}$$

The slow diffusion coefficient is measurable only at high enough polyelectrolyte concentrations. The value of c at which the slow mode appears is higher if c_s is higher. When the ratio $\lambda \equiv c/c_s$ is about 1, the onset of the slow mode and the crossover between the smaller D_f for $\lambda < 1$ and higher D_f for $\lambda > 1$ occur. D_s depends [33] on c strongly,

$$D_s \sim c^{-\beta(M)} \tag{8}$$

where β is an apparent exponent depending on molecular weight. The absolute scattering intensity associated with the slow mode increases [34] upon decreasing c_s.

C. Viscosity

The reduced viscosity η_r of a polyelectrolyte solution at the monomer concentration c is defined by

$$\eta_r = \frac{\eta - \eta_0}{\eta_0 c} \tag{9}$$

where η and η_0 are the viscosities of the solution and the solvent, respectively. η_r depends on c monotonically for polyelectrolyte solutions at sufficiently high salt concentrations, as in the case of neutral polymers, and at low polyelectrolyte concentrations it can be expanded as a virial series,

$$\eta_r = [\eta][1 + k_H[\eta]c + \cdots] \tag{10}$$

where $[\eta]$ is the intrinsic viscosity and k_H is the Huggins coefficient. On the other hand, at low salt concentrations, η_r is nonmonotonic [11–23] in the dependence on c. At very low concentrations, η_r increases rapidly with c and then decreases with c at low but higher concentrations after reaching a maximum $\eta_{r\max}$ at $c_{\max}$.

At much higher concentrations (c greater than say c_e) η_r increases again with c. In the intermediate concentration region where η_r decreases with c, the experimental data at low salt concentrations are fitted empirically to the Fuoss law,

$$\eta_r \sim \frac{1}{\sqrt{c}} \tag{11}$$

$\eta_{r\max}$ increases [19] linearly with M. An increase in the salt concentration moves $c_{\max}$ toward higher c so that $c_{\max} \sim c_s$, and it drastically lowers the value of $\eta_{r\max}$. Analogous to the viscosity behavior, the dynamic storage and loss moduli also show [22] a peak with c. The unusual behavior at low c_s where the reduced viscosity increases with dilution in the polyelectrolyte concentration range between $c_{\max}$ and c_e, along with the occurrence of a peak in the reduced viscosity versus c, has remained as one of the most perplexing properties of polyelectrolytes over many decades.

In order to resolve these challenges, it is essential to account for chain connectivity, hydrodynamic interactions, electrostatic interactions, and distribution of counterions and their dynamics. It is possible to identify three distinct scenarios: (a) polyelectrolyte solutions with high concentrations of added salt, (b) dilute polyelectrolyte solutions without added salt, and (c) polyelectrolyte solutions above overlap concentration and without added salt. If the salt concentration is high and if there is no macrophase separation, the polyelectrolyte solution behaves as a solution of neutral polymers in a good solvent, due to the screening of electrostatic interaction. Therefore for scenario (a), all concepts developed for solutions of uncharged polymers can be implemented, except for the additional role played by counterions. In scenario (b), chains place themselves as far away as possible from each other due to the repulsive interactions between them. Under these conditions, the average distance Λ between the chains scales with polyelectrolyte concentration c according to

$$\Lambda \sim c^{-1/3} \tag{12}$$

as expected from the geometrical consideration. In this dilute regime, different segments in the same chain interact with screened Coloumb interaction where the electrostatic screening arises from the counterions and salt ions if present. There can also be condensation of counterions on the polyelectrolyte chain. For the salt-free solutions, the electrostatic screening length κ^{-1} is very large. In scenario (b), chain dynamics, hydrodynamics, intersegment electrostatics, and counterion dynamics describe the dynamical properties.

In scenario (c) corresponding to semidilute solutions, polyelectrolyte chains interpenetrate. Under these circumstances, there are three kinds of screening. The electrostatic interaction, excluded volume interaction, and the hydrodynamic interaction between any two segments of a labeled polyelectrolyte chain are all screened by interpenetrating chains. Each of these three interactions is associated with a screening length, namely, κ^{-1}, ξ, and ξ_H. These screening lengths are coupled to each other. By considering a thermodynamic description of the double screening of electrostatic and excluded volume interactions, it has recently been shown [48] that for semidilute conditions we have

$$\xi \sim \begin{cases} c^{-1/2}, & \text{low salt} \\ c^{-3/4}, & \text{high salt} \end{cases} \tag{13}$$

At the same time R_g of a labeled polyelectrolyte chain is derived to be

$$R_g \sim \begin{cases} \sqrt{N}c^{-1/4}, & \text{low salt} \\ \sqrt{N}c^{-1/8}, & \text{high salt} \end{cases} \tag{14}$$

These calculated results have been quantitatively verified by experiments [52–54].

The nature of screening of hydrodynamics, its coupling with the screening of electrostatic and excluded volume interactions, and its role in various dynamical properties are discussed in the following sections. In addition to the above-mentioned contributing factors, entanglements between chains play a role in high polyelectrolyte solutions. Only scaling considerations are given for this effect. Except for this effect, formulas for crossover functions in terms of chain length, salt concentration, and polymer concentration are presented. After presenting the theoretical model, along with the theory for dilute and semidilute solutions, we conclude by comparing with the qualitative features of experiments.

II. THEORETICAL MODEL

We consider a system of n polyelectrolyte chains each of N segments, n_c counterions, n_γ ions of species γ from dissolved salt, and n_s solvent molecules in volume Ω. The total charge on each polyelectrolyte is Q assumed to be proportional to N.

A. Solvent

Provided that the polyelectrolyte chains are much larger than the solvent molecules, the solvent can be treated as a homogeneous isothermal viscous continuum obeying the linearized Navier–Stokes equation (in the steady state limit)

$$-\eta_0 \nabla^2 \mathbf{v}(\mathbf{r}) + \nabla p(\mathbf{r}) = \mathbf{F}(\mathbf{r}) \tag{15}$$

Here $\mathbf{v}(\mathbf{r})$ is the velocity field at any space point $\mathbf{r}$, p is the pressure, η_0 is the shear viscosity of the solvent containing salt ions and counterions, and $\mathbf{F}(\mathbf{r})$ is the force arising from any potential field in the solution. As an example, if electrical charges are present in the solution under an externally imposed uniform electrical field $\mathbf{E}$, $\mathbf{F}(\mathbf{r})$ is given by

$$\mathbf{F}(\mathbf{r}) = \mathbf{F}_s(\mathbf{r}) + \rho(\mathbf{r})\mathbf{E} \tag{16}$$

Here $\mathbf{F}_s(\mathbf{r})$ is any force imposed by an external boundary condition such as a shear flow and $\rho(\mathbf{r})$ is the electrical charge density at $\mathbf{r}$.

The hydrodynamic property of the solvent follows from Eq. (15) as

$$\mathbf{v}(\mathbf{r}) = \int d\mathbf{r}' \mathbf{G}(\mathbf{r} - \mathbf{r}') \cdot \mathbf{F}(\mathbf{r}') \tag{17}$$

with $\mathbf{G}(\mathbf{r})$ being the Oseen tensor,

$$\mathbf{G}(\mathbf{r}) = \frac{1}{8\pi\eta_0|\mathbf{r}|}\left(\mathbf{1} + \frac{\mathbf{r}\mathbf{r}}{|\mathbf{r}|^2}\right) \tag{18}$$

Upon Fourier transform, $\mathbf{G}$ becomes

$$\mathbf{G}(\mathbf{k}) = \frac{1}{\eta_0 k^2}\left(\mathbf{1} - \frac{\mathbf{k}\mathbf{k}}{k^2}\right) \tag{19}$$

with k being the modulus of $\mathbf{k}$.

B. Polymer

The equation of motion for the position vector $\mathbf{R}_{\alpha i}$ of the ith segment of the chain α at time t is given by the Langevin equation,

$$m\ddot{\mathbf{R}}_{\alpha i} = \mathbf{F}_{\text{connectivity}} + \mathbf{F}_{\text{interactions}} + \mathbf{F}_{\text{friction}} + \mathbf{F}_{\text{ext}} + \mathbf{F}_{\text{ran}} \tag{20}$$

where the various terms on the right-hand side of the equation are described as follows.

The force arising from the chain connectivity is related to the probability of realizing particular configurations. The probability for the ith Kuhn step connecting $\mathbf{R}_i$ and $\mathbf{R}_{i-1}$ is

$$P_{\alpha i}(\mathbf{R}_{\alpha,i} - \mathbf{R}_{\alpha,i-1}) = \left(\frac{3}{2\pi\ell^2}\right)^{3/2} \exp\left[-\frac{3}{2}\frac{(\mathbf{R}_{\alpha,i} - \mathbf{R}_{\alpha,i-1})^2}{\ell^2}\right] \tag{21}$$

where ℓ is the Kuhn length. The free energy of the chain due solely to chain connectivity is

$$-k_B T \sum_{i=1}^{n} \ln\ P_i = \text{constant} + \sum_{i=1}^{N} \frac{3k_B T}{\ell^2}(\mathbf{R}_i - \mathbf{R}_{i-1}) \tag{22}$$

The connectivity force on ith segment of αth chain is

$$\begin{aligned}
\mathbf{F}_{\text{connectivity}} &= -\frac{\partial}{\partial \mathbf{R}_{\alpha,i}}\left[-\left(k_B T \sum_{i=1}^{N} \ln\ P_{\alpha,i}\right)\right] \\
&= -\frac{3k_B T}{2\ell^2}\frac{\partial}{\partial \mathbf{R}_{\alpha,i}}\left[(\mathbf{R}_{\alpha,i+1} - \mathbf{R}_{\alpha,i})^2 + (\mathbf{R}_{\alpha,i} - \mathbf{R}_{\alpha,i-1})^2\right] \\
&= -\frac{3k_B T}{\ell^2}\left[-\mathbf{R}_{\alpha,i+1} + 2\mathbf{R}_{\alpha,i} - \mathbf{R}_{\alpha,i-1}\right] \\
&= -\frac{3k_B T}{\ell^2}\sum_{i=1}^{N}(2\delta_{i,j} - \delta_{i,j+1} + \delta_{i,j-1})\mathbf{R}_{\alpha,j}
\end{aligned} \tag{23}$$

where $(2\delta_{i,j} - \delta_{i,j+1} + \delta_{i,j-1})$ is the Rouse matrix. We do not concern ourselves with the end effects, although these effects can be rigorously treated [3–4].

Assuming that the total charge Q of a polyelectrolyte chain is uniformly distributed on the chain backbone so that each segment carries a charge of ez_p where e is the electronic charge and by integrating over the positions of counterions, salt ions, and solvent molecules, the potential interaction V between any two segments i and j separated by a distance $\mathbf{R}_{ij}$ is taken to be [48]

$$\frac{V(\mathbf{R}_{ij})}{k_B T} = w\delta(\mathbf{R}_{ij}) + \frac{z_p^2 l_B}{|\mathbf{R}_{ij}|}\exp(-\kappa|\mathbf{R}_{ij}|) \tag{24}$$

where w is the short-ranged "excluded volume" denoting the solvent quality, and $l_B \equiv e^2/k_B T\epsilon$ is the Bjerrum length with ϵ being the effective dielectric constant of the solution. The Debye length κ^{-1}, within the approximations behind

Eq. (24), arises from only the counterions and the dissociated salt ions,

$$\kappa^2 = \frac{4\pi e^2}{\epsilon k_B T}\left(z_c^2\rho_c + \sum_{\gamma} Z_{\gamma}^2\rho_{\gamma}\right) \tag{25}$$

where ρ_c and ρ_γ are the number densities of counterions and the γth salt ion. Therefore the force arising from intersegment potential interactions is

$$\mathbf{F}_{\text{interactions}} = -\nabla_{\mathbf{R}_{\alpha,i}} \sum_{\beta=1}^{n}\sum_{j=1}^{N} V(\mathbf{R}_{\alpha,i} - \mathbf{R}_{\beta,j}) \tag{26}$$

The frictional force is given as

$$\mathbf{F}_{\text{friction}} = -\zeta\dot{\mathbf{R}}_{\alpha,i} \tag{27}$$

where $\dot{\mathbf{R}}_{\alpha,i}$ is the velocity of the ith segment of αth chain and ζ is a phenomenological parameter, called bead friction coefficient, representing the microscopic details of friction of a segment against the background solvent. $\mathbf{F}_{\text{ext}}$ is any externally imposed force, such as electrical or gravitational, acting at $\mathbf{R}_{\alpha,i}$. $\mathbf{F}_{\text{ran}}$ is the fluctuating noisy force arising from the solvent, acting at $\mathbf{R}_{\alpha,i}$. The correlation of this noise is stipulated by the fluctuation–dissipation theorem in terms of the temperature and ζ. Combining Eqs. (20–27), we get the equation of motion for the ith segment of αth chain as

$$\zeta\dot{\mathbf{R}}_{\alpha,i} + \Delta_\alpha \mathbf{R}_{\alpha,i} = \mathbf{F}_{\text{ext}} + \mathbf{F}_{\text{ran}} \tag{28}$$

where Δ_α denotes the terms from connectivity and interactions,

$$\Delta_\alpha = \frac{3k_BT}{\ell^2}\sum_{i=1}^{N}(2\delta_{i,j} - \delta_{i,j+1} - \delta_{i,j-1})\mathbf{R}_{\alpha,j} + \nabla_{\mathbf{R}_{\alpha,i}}\sum_{\beta=1}^{n}\sum_{j=0}^{N} V(\mathbf{R}_{\alpha,i} - \mathbf{R}_{\beta,j}) \tag{29}$$

C. Polymer and Solvent

The dynamics of the solvent and polymer are coupled in a polymer solution. Although this coupling is expressed partly through the friction and noise terms in Eq. (28), a complete description requires a boundary condition. The most common choice of this boundary condition is the no-slip condition

$$\dot{\mathbf{R}}_{\alpha i} = \mathbf{v}(\mathbf{R}_{\alpha i}) \tag{30}$$

Using this boundary condition as a constraint, the equations of motion for the combined system of polymer and solvent can be derived to be [55–59]

$$-\eta_0 \nabla^2 \mathbf{v}(\mathbf{r}) + \nabla p(\mathbf{r}) = \mathbf{F}(\mathbf{r}) + \sum_{\alpha=1}^{n} \sum_{j=0}^{N} \delta(\mathbf{r} - \mathbf{R}_{\alpha j}) \boldsymbol{\sigma}_{\alpha i} \tag{31}$$

$$\Delta_\alpha \mathbf{R}_{\alpha j} = -\boldsymbol{\sigma}_{\alpha j} + \mathbf{F}_{\text{ext}} + \mathbf{F}_{\text{ran}} \tag{32}$$

Here $\boldsymbol{\sigma}_{\alpha j}$ is a Lagrange multiplier and is to be eliminted by using the boundary condition of Eq. (30). The coupled equations are solved using the multiple scattering formalism, the details of which are provided in Refs. 55–59. The objective is to determine $\sigma_{\alpha i}$ using Eq. (30) and then calculate the microscopic velocity field from Eq. (31), which upon averaging over the distributions of polyelectrolyte chain configurations gives the viscosity of the solution and the self diffusion coefficient of the polyelectrolyte. Within the multiple scattering formalism, the effective equation of motion for the polyelectrolyte solution becomes [55–59]

$$-\eta_0 \nabla^2 \langle \mathbf{v}(\mathbf{r}) \rangle + \boldsymbol{\nabla} \langle p(\mathbf{r}) \rangle + \int d\mathbf{r}' \Sigma(\mathbf{r} - \mathbf{r}') \cdot \langle \mathbf{v}(\mathbf{r}') \rangle = \mathbf{F}(\mathbf{r}) \tag{33}$$

where the angular brackets indicate the above-mentioned average and the kernel Σ contains information about the modified viscosity and the extent of hydrodynamic screening. For the polyelectrolyte solution, the average velocity field is given by

$$\langle \mathbf{v}(\mathbf{r}) \rangle = \int d\mathbf{r}' \mathscr{G}(\mathbf{r} - \mathbf{r}') \cdot \mathbf{F}(\mathbf{r}') \tag{34}$$

whereas the hydrodynamic property of the solvent follows from Eq. (17). In the presence of polyelectrolyte chains, $\mathbf{G}(\mathbf{k})$ is replaced by $\mathscr{G}(\mathbf{k})$, where

$$\mathscr{G}^{-1}(\mathbf{k}) = \mathbf{G}^{-1}(\mathbf{k}) + \Sigma(\mathbf{k}) \tag{35}$$

with $\Sigma(k)$ being the Fourier transform of $\Sigma(\mathbf{r})$,

$$\Sigma(\mathbf{k}) = \int d\mathbf{r} \Sigma(\mathbf{r}) \exp(i\mathbf{k} \cdot \mathbf{r}) \tag{36}$$

and $\mathscr{G}^{-1}(\mathbf{k})$ being the Fourier transform of $\mathscr{G}^{-1}(\mathbf{r})$ defined by

$$\int d\mathbf{r}' \mathscr{G}^{-1}(\mathbf{r} - \mathbf{r}') \cdot \mathscr{G}(\mathbf{r}' - \mathbf{r}'') = \mathbf{1}\delta(\mathbf{r} - \mathbf{r}'') \tag{37}$$

The change $\eta - \eta_0$ in shear viscosity of the solvent due to the polyelectrolyte chains is given by

$$\eta - \eta_0 = k \xrightarrow{\lim} ok^{-2}(\mathbf{1} - \hat{k}\hat{k}) \cdot \Sigma(\mathbf{k}) \tag{38}$$

where η is the shear viscosity of the solution. As will be shown below, $\Sigma(\mathbf{k})$ becomes independent of $\mathbf{k}$ for length scales shorter than average radius of gyration of the polyelectrolyte chains at higher concentrations. We write the transverse component of the $\mathbf{k}$-independent $\Sigma(\mathbf{k})$ as $\eta_0 \xi_H^{-2}$, where ξ_H is the hydrodynamic screening length. This will be determined explicitly in the next section.

III. DILUTE SOLUTIONS

In this section, we ignore all interchain interactions and address the situation of polyelectrolyte solutions at infinitely dilute conditions. Since isolated chains are now being considered, the chain label is dropped in this section.

A. Effective Chain Dynamics

Combining Eqs. (17) and (31), the velocity field at space location $\mathbf{r}$ follows as

$$\mathbf{v}(\mathbf{r}) = \int \mathbf{dr}' \mathbf{G}(\mathbf{r} - \mathbf{r}') \cdot \left[\mathbf{F}(\mathbf{r}') + \sum_{J=0}^{N} \delta(\mathbf{r}' - \mathbf{R}_j) \boldsymbol{\sigma}_j \right] \tag{39}$$

Using the boundary condition of Eq. (30), the velocity of the ith segment is given by

$$\dot{\mathbf{R}}_i = \mathbf{v}(\mathbf{R}_i) = \int \mathbf{dr}' \mathbf{G}(\mathbf{R}_i - \mathbf{r}') \cdot \left[\mathbf{F}(\mathbf{r}') + \sum_{J=0}^{N} \delta(\mathbf{r}' - \mathbf{R}_j) \boldsymbol{\sigma}_j \right] \tag{40}$$

$$= \int d\mathbf{r}' \; G(\mathbf{R}_i - \mathbf{r}') \cdot \mathbf{F}(\mathbf{r}') + \sum_{J=0}^{N} \; G(\mathbf{R}_i - \mathbf{R}_j) \boldsymbol{\sigma}_j \tag{41}$$

For $i = j$, $G(\mathbf{R}_i - \mathbf{R}_j)$ given by Eq. (18) is not valid and it is simply the reciprocal of the bead friction coefficient. Hence

$$\dot{\mathbf{R}}_i = \int d\mathbf{r}' \mathbf{G}(\mathbf{R}_i - \mathbf{r}') \cdot \mathbf{F}(\mathbf{r}') + \frac{1}{\zeta} \boldsymbol{\sigma}_i + \sum_{J \neq i} \mathbf{G}(\mathbf{R}_i - \mathbf{R}_j) \cdot \boldsymbol{\sigma}_j \tag{42}$$

so that

$$\boldsymbol{\sigma}_i = \zeta \left[\dot{\mathbf{R}}_i - \int d\mathbf{r}' \mathbf{G}(\mathbf{R}_i - \mathbf{r}') \cdot \mathbf{F}(\mathbf{r}') - \sum_{j \neq i} \mathbf{G}(\mathbf{R}_i - \mathbf{R}_j) \cdot \boldsymbol{\sigma}_j \right] \tag{43}$$

The physical interpretation of this equation is readily apparent. $\boldsymbol{\sigma}_i$ is the force exerted on the fluid by the ith segment given by

$$\boldsymbol{\sigma}_i = \zeta(\mathbf{R}_i - \mathbf{v}'_i) \tag{44}$$

where $\mathbf{v}'_i$ is the velocity the fluid would possess at the location of the ith segment if that segment were absent. $\mathbf{v}'_i$ is the sum of velocity in the absence of the polymer chain

$$\mathbf{v}_i^0 = \int d\mathbf{r}' \mathbf{G}(\mathbf{R}_i - \mathbf{r}') \cdot \mathbf{F}(\mathbf{r}') \tag{45}$$

and the sum of velocities arising from hydrodynamic coupling with forces of all other segments of the same chain,

$$\sum_{j \neq i} \mathbf{G}(\mathbf{R}_i - \mathbf{R}_j) \cdot \boldsymbol{\sigma}_j \tag{46}$$

Returning to Eq. (43), it can be rewritten as

$$\sum_{j=0}^{N} \mathbf{D}_{ij} \cdot \boldsymbol{\sigma}_j = \dot{\mathbf{R}}_i - \int d\mathbf{r}' \mathbf{G}(\mathbf{R}_i - \mathbf{r}') \cdot \mathbf{F}(\mathbf{r}') \tag{47}$$

where the diffusion tensor is defined by

$$\mathbf{D}_{ij} = \left[\frac{1}{\zeta} \mathbf{1} \delta_{ij} + (1 - \delta_{ij}) \mathbf{G}(\mathbf{R}_i - \mathbf{R}_j) \right] \tag{48}$$

with $\mathbf{1}$ being the unit tensor. By multiplying Eq. (32) by $\mathbf{D}_{ij}$ and summing over j, it follows from Eqs. (29) and (47) that

$$\dot{\mathbf{R}}_i + \sum_j \left[\frac{1}{\zeta} \mathbf{1} \delta_{ij} + (1 - \delta_{ij}) \mathbf{G}(\mathbf{R}_i - \mathbf{R}_j) \right] \cdot \left[\frac{3k_B T}{\ell^2} \sum_p (2\delta_{j,p} - \delta_{j,p+1} \right.$$
$$\left. + \delta_{j,p-1}) \mathbf{R}_p + \nabla_{\mathbf{R}_j} \sum_p V(\mathbf{R}_j - \mathbf{R}_p) \right] = \mathbf{v}_{\text{ran}}(i) \tag{49}$$

where $\mathbf{F}$ is now taken to be zero and $\mathbf{v}_{\text{ran}}(i)$ is a noise from the background velocity proportional to $\mathbf{F}_{\text{ran}}(i)$, acting on the ith segment. Since we are interested in relatively long polymer chains, summations over segment indices can be replaced by integrals over arclength variables. Letting $s' = j\ell$,

$$\frac{1}{\ell^2}\sum_p (2\delta_{j,p} - \delta_{j,p+1} + \delta_{j,p-1})\mathbf{R}_p = -\frac{\partial^2 \mathbf{R}(s')}{\partial s'^2} \tag{50}$$

so that Eq. (49) becomes

$$\begin{aligned}\dot{\mathbf{R}}(s) + \int_0^L ds' \mathbf{D}[\mathbf{R}(s) - \mathbf{R}(s')] \cdot \Bigg[&-\frac{3k_BT}{\ell^2}\frac{\partial^2 \mathbf{R}(s')}{\partial s'^2} \\ &+\nabla_{\mathbf{R}(s')} \int_0^L \frac{ds''}{\ell} V[\mathbf{R}(s) - \mathbf{R}(s')]\Bigg] = \mathbf{v}_{\text{ran}}(s)\end{aligned} \tag{51}$$

where

$$\mathbf{D}[\mathbf{R}(s) - \mathbf{R}(s')] = \mathbf{1}\frac{\ell}{\zeta}\delta(s - s') + (1 - \ell\delta(s - s'))\mathbf{G}[\mathbf{R}(s) - \mathbf{R}(s')] \tag{52}$$

Equation (51) is highly nonlinear. One of the key approximations made to enable analytical calculations is the preaveraging approximation, whereby the diffusion tensor $\mathbf{D}[\mathbf{R}(s) - \mathbf{R}(s')]$ is replaced by its average over chain configurations,

$$\mathbf{D}[\mathbf{R}(s) - \mathbf{R}(s')] \simeq \langle \mathbf{D}[\mathbf{R}(s) - \mathbf{R}(s')]\rangle \tag{53}$$

$$= \frac{\ell}{\zeta}\delta(s - s') + [1 - \delta(s - s')]\left\langle \frac{1}{6\pi\eta_0}\frac{1}{|\mathbf{R}(s) - \mathbf{R}(s')|}\right\rangle \tag{54}$$

$$\equiv \mathbf{1}D(s - s') \tag{55}$$

which depends only on $s - s'$. We now introduce the Rouse modes by defining the Fourier transform,

$$\hat{\mathbf{R}}(q) = \int_0^L ds\, \mathbf{R}(s) \exp(iqs) \tag{56}$$

The variable q, which is conjugate to the arc length variable s, labels the normal mode. If we work with the discretized chain model (i variable, instead of s

variable), q is actually $2\pi p/L$, where $p = 0, 1, 2, \ldots, N$. The integer variable p is called the Rouse mode variable. q is proportional to L^{-1} and $p = 0$ represents the motion of the center of mass of the chain. Writing Eq. (51) in terms of its Fourier transform, we obtain

$$\dot{\hat{\mathbf{R}}}(q) + \hat{D}(q)\left[\frac{3k_BT}{\ell}q^2\hat{\mathbf{R}}(q) + \int_0^L ds' e^{-iqs'}\mathbf{F}_{\text{interaction}}[\mathbf{R}(s')]\right] = \hat{\mathbf{v}}_{\text{ran}}(q) \qquad (57)$$

where $\hat{D}(q)$ is the Fourier transform of $D(s-s')$. The second term inside the square brackets in Eq. (57) depicting potential interactions can be expressed in terms of Rouse modes. In general qth mode is coupled to all other modes. As shown in Ref. 58, the assumption of full decoupling among the Rouse modes is equivalent to replacing the factor ℓ (in the first term of square brackets) by an effective Kuhn length ℓ_1 which is q-dependent. For $q \to 0$, ℓ_1/ℓ is exactly the expansion factor of the mean square end-to-end distance $\langle R^2 \rangle$ of a chain due to excluded volume interactions (both short-ranged and electrostatic),

$$\langle R^2 \rangle = L\ell_1(q \to 0) \qquad (58)$$

The mode–mode decoupling calculation for an isolated uncharged chain in a good solution shows that [50, 58]

$$\left(\frac{\ell_1}{\ell}\right)^{5/2} - \left(\frac{\ell_1}{\ell}\right)^{3/2} = \frac{2\sqrt{3}}{\pi}\frac{w}{\sqrt{q\ell}} \qquad (59)$$

where w is the excluded volume parameter defined in Eq. (24). For small q and large w (i.e., for large values of the Fixman parameter $w\sqrt{L}$), ℓ_1 approaches

$$\ell_1 \sim q^{-1/5} \qquad (60)$$

which is consistent with

$$\ell_1 \sim q^{1-2\nu} \qquad (61)$$

where ν is the size exponent.

By making the above approximations (preaveraging and uniform expansion), the Langevin equation for the polymer segment can be written as

$$\dot{\hat{R}}(q) + \frac{3k_BT}{\ell_1}\hat{D}(q)\hat{R}(q) = \hat{v}_{\text{ran}}(q) \qquad (62)$$

The q-dependent diffusion coefficient is obtained as follows:

$$\hat{D}(q) = \int_0^L d(s-s')e^{iq(s-s')}D(s-s') \tag{63}$$

$$= \frac{\ell}{\zeta} + \frac{1}{6\pi\eta_0}\int_0^\infty d(s-s')e^{iq(s-s')}\left\langle \frac{1}{|\mathbf{R}(s)-\mathbf{R}(s')|}\right\rangle \tag{64}$$

For Gaussian chains,

$$\left\langle \frac{1}{|\mathbf{R}(s)-\mathbf{R}(s')|}\right\rangle = \left(\frac{6}{\pi\ell|s-s'|}\right)^{1/2} \tag{65}$$

With the approximation of uniform expansion by potential interactions, we can assume that

$$\left\langle \frac{1}{|\mathbf{R}(s)-\mathbf{R}(s')|}\right\rangle = \left(\frac{6}{\pi\ell_1|s-s'|}\right)^{1/2} \tag{66}$$

where ℓ_1/ℓ is the expansion factor for the mean-square end-to-end distance and is proportional to $(L/\ell)^{2\nu-1}$. Inserting Eq. (66) into Eq. (64), we obtain

$$\hat{D}(q) = \frac{\ell}{\zeta} + \frac{1}{\eta_0}\frac{1}{\sqrt{\ell_1 q}} \tag{67}$$

Since ℓ_1 is proportional to $L^{2\nu-1}$ and q is proportional to $1/L$, ℓ_1 is proportional to $q^{1-2\nu}$. Substitution of Eq. (67) into Eq.(62) gives the Langevin equation for the Rouse modes of the chain within the approximations of preaveraging for hydrodynamic interactions and mode–mode decoupling for intersegment potential interactions. Equation (62) yields the following results for relaxation times and various dynamical correlation functions.

1. *Relaxation Times*

An inspection of Eq. (62) on dimensional grounds gives the characteristic time τ_q as

$$\tau_q = \frac{\ell_1}{3k_BT\hat{D}q^2} \tag{68}$$

Substituting Eq. (67) into Eq. (68), we get

$$\tau_q = \frac{\ell_1}{3k_BTq^2\left[\frac{\ell}{\zeta} + \frac{1}{\eta_0}\frac{1}{\sqrt{\ell_1 q}}\right]} \tag{69}$$

a. Free-Draining Limit. If hydrodynamic interaction is absent ($\eta_0 \to \infty$), called the free-draining limit, τ_q becomes

$$\tau_{q,\text{Rouse}} = \frac{\zeta \ell_1}{3k_B T \ell q^2} \tag{70}$$

These are the Rouse relaxation times. For Gaussian chains ($\ell_1 = \ell$), they become

$$\tau_{q,\text{Rouse}} = \frac{\zeta}{3k_B T q^2} \tag{71}$$

$$= \frac{\zeta L^2}{12\pi^2 k_B T p^2} \tag{72}$$

where p is the Rouse mode index. For a chain with size exponent ν, Rouse relaxation times are

$$\tau_{q,\text{Rouse}} \sim \frac{\zeta}{T}\left(\frac{L}{p}\right)^{2\nu+1} \tag{73}$$

and the longest relaxation time, called the Rouse time τ_{Rouse} is proportional to $L^{2\nu+1}$,

$$\tau_{\text{Rouse}} \sim L^{2\nu+1} \tag{74}$$

b. Non-Free-Draining Limit. In dilute solutions, the hydrodynamic interactions dominate. Under this condition, τ_q becomes from Eq. (69)

$$\tau_q \sim \frac{\eta_0}{3k_B T}\left(\frac{\ell_1}{q}\right)^{3/2} \tag{75}$$

These are the Zimm relaxation times and are proportional to $L^{3\nu}$,

$$\tau_{q,\text{Zimm}} \sim \frac{\eta_0}{T} q^{-3\nu} \tag{76}$$

For Gaussian chains,

$$\tau_{q,\text{Zimm}} = \frac{\eta_0}{3(2\pi)^{3/2} k_B T}\left(\frac{L\ell}{p}\right)^{3/2} \tag{77}$$

In general, the longest relaxation time is

$$\tau_{q,\text{Zimm}} \sim \frac{\eta_0}{T} R_g^3 \tag{78}$$

Therefore we expect

$$\tau_{\text{Zimm}} \sim \begin{cases} L^3, & \text{low salt} \\ L^{9/5}, & \text{high salt.} \end{cases} \tag{79}$$

2. *Segment-to-Segment Correlation Function*

In this section, we present derivation of the correlation function, $\langle [\mathbf{R}(s,t) - \mathbf{R}(s',t')]^2 \rangle$. By defining the Fourier transform

$$\tilde{R}(q,\omega) = \int_{-\infty}^{\infty} dt \hat{R}(q,t) \exp(i\omega t) \tag{80}$$

Eq. (62) becomes

$$i\omega \tilde{R}(q,\omega) + \frac{3k_B T \hat{D} q^2}{\ell_1} \tilde{R}(q,\omega) = \tilde{v}_{\text{ran}}(q,\omega) \tag{81}$$

By formally solving this equation, we obtain

$$\tilde{R}(q,\omega) = \frac{\tilde{v}_{\text{ran}}(q,\omega)}{(i\omega + \frac{3k_B T \hat{D} q^2}{\ell_1})} \tag{82}$$

Since the stochastic term $\tilde{v}_{\text{ran}}$ is defined only in terms of its correlation function, we construct the correlation function,

$$\langle \tilde{R}(q,\omega) \cdot \tilde{R}^*(q',\omega') \rangle = \frac{\langle \tilde{v}_{\text{ran}}(q,\omega) \cdot \tilde{v}^*_{\text{ran}}(q',\omega') \rangle}{\left(i\omega + \frac{3k_B T \hat{D}(q) q^2}{\ell_1(q)}\right)\left(-i\omega' + \frac{3k_B T \hat{D}(q') q'^2}{\ell_1(q')}\right)} \tag{83}$$

where the angular brackets indicate the average over the thermal Brownian forces, and the superscript $*$ indicates the complex conjugate. For simplicity we assume, as in the case of the Brownian motion of a particle, that $\tilde{v}_{\text{ran}}(q,\omega)$ is a white noise. Now $\langle \tilde{v}_{\text{ran}}(q,\omega) \cdot \tilde{v}^*_{\text{ran}}(q',\omega') \rangle$ is proportional to $\delta(q-q')\delta(\omega-\omega')$,

$$\langle \tilde{v}_{\text{ran}}(q,\omega) \cdot \tilde{v}^*_{\text{ran}}(q',\omega') \rangle = \Theta \delta(q-q') \delta(\omega-\omega') \tag{84}$$

where the proportionality factor Θ is to be determined below from a consideration of equilibrium.

The correlation function $\langle[\mathbf{R}(s,t)-\mathbf{R}(s',t')]^2\rangle$ follows from its Fourier transform as

$$\langle[\mathbf{R}(s,t)-\mathbf{R}(s',t')]^2\rangle = \int_{-\infty}^{\infty}\frac{dq}{2\pi}\int_{-\infty}^{\infty}\frac{dq'}{2\pi}\int_{-\infty}^{\infty}\frac{d\omega}{2\pi}$$
$$\int_{-\infty}^{\infty}\frac{d\omega'}{2\pi}[e^{-i(q-q')s-i(\omega-\omega')t} - 2e^{-iqs-i\omega t+iq's+i\omega't'}$$
$$+ e^{-i(q-q')s'-i(\omega-\omega')t}]\langle\tilde{R}(q,\omega)\cdot\tilde{R}^*(q',\omega')\rangle \tag{85}$$

Substituting Eq. (83) in Eq. (85) and by performing q' and ω' integrals, we obtain

$$\langle[\mathbf{R}(s,t)-\mathbf{R}(s',t')]^2\rangle = 2\int_{-\infty}^{\infty}\frac{dq}{2\pi}\int_{-\infty}^{\infty}\frac{d\omega}{2\pi}$$
$$\frac{\Theta[1-\cos[q(s-s')]e^{i\omega(t-t')}]}{(i\omega+\frac{3k_BT\hat{D}(q)q^2}{\ell_1(q)})(-i\omega+\frac{3k_BT\hat{D}(q)q^2}{\ell_1(q)})}$$
$$= \int_{-\infty}^{\infty}\frac{dq}{2\pi}\frac{\Theta\ell_1(q)}{3k_BT\hat{D}(q)q^2}$$
$$\left[1-\cos[q(s-s')]e^{\frac{-|t-t'|}{\tau_q}}\right] \tag{86}$$

The equal time correlation function,

$$\langle[\mathbf{R}(L,t)-\mathbf{R}(0,t)]^2\rangle = \int_0^{\infty}\frac{dq}{\pi}\frac{\Theta\ell_1(q)}{3k_BT\hat{D}(q)q^2}[1-\cos(qL)] \tag{87}$$

is the equilibrium result

$$\langle[\mathbf{R}(L,t)-\mathbf{R}(0,t)]^2\rangle = L\ell_1 \tag{88}$$

By comparing Eqs. (87) and (88), we get the following expression for Θ:

$$\Theta = 6k_BT\hat{D}(q) \tag{89}$$

Therefore, $\langle[\mathbf{R}(s,t)-\mathbf{R}(s',t')]^2\rangle$ follows as

$$\langle[\mathbf{R}(s,t)-\mathbf{R}(s',t')]^2\rangle = 2\int_0^{\infty}\frac{dq}{\pi}\frac{\ell_1(q)}{q^2}\left[1-\cos[q(s-s')]e^{-\frac{|t-t'|}{\tau_q}}\right] \tag{90}$$

The mean-square displacement of a labeled monomer is

$$\langle [\mathbf{R}(s,t) - \mathbf{R}(s,t')]^2 \rangle = 2 \int_0^\infty \frac{dq}{\pi} \frac{\ell_1(q)}{q^2} \left(1 - e^{-\frac{|t-t'|}{\tau_q}}\right) \tag{91}$$

When hydrodynamic interaction dominates, $\tau_q \sim q^{-3\nu}$, so that

$$\langle [\mathbf{R}(s,t) - \mathbf{R}(s,t')]^2 \rangle \sim \int_0^\infty dq\; q^{-(2\nu+1)} (1 - e^{-q^{3\nu}|t-t'|}) \tag{92}$$

$$\sim |t - t'|^{2/3} \tag{93}$$

The incoherent dynamical structure factor is given by

$$\begin{aligned} S_{inc}(\mathbf{k}, \mathbf{t}) &= \exp(-k^2 \langle [\mathbf{R}(s,t) - \mathbf{R}(s,t')]^2 \rangle / 6) \\ &\sim \exp(-k^2 |t - t'|^{2/3}) \end{aligned} \tag{94}$$

where $\mathbf{k}$ is the scattering wave vector. $S_{inc}(\mathbf{k}, \mathbf{t})$ is therefore of the form

$$\exp[-(k^3 |t - t'|)^{2/3}] \tag{95}$$

so that the decay rate is proportional to k^3. On the other hand, if the hydrodynamic interaction is absent, $\tau_q \sim q^{-(1+2\nu)}$, so that

$$\langle [\mathbf{R}(s,t) - \mathbf{R}(s',t')]^2 \rangle \sim \int_0^\infty dq\; q^{-(1+2\nu)} (1 - e^{-q^{1+2\nu}|t-t'|}) \tag{96}$$

$$\sim |t - t'|^{\frac{2\nu}{1+2\nu}} \tag{97}$$

Now, the decay rate of the incoherent dynamic structure factor is proportional to $k^{(1+2\nu)/\nu}$. Therefore, the k-dependence of the decay rate for salt-free solutions is independent of whether the hydrodynamic interaction is present or not.

B. Translational Friction Coefficient

The translational friction coefficient f_t of the polymer chain at infinitely dilute solutions is calculated by equating the net force acting on the chain, $-\sum_i \langle \boldsymbol{\sigma}_i \rangle$, and $-f_t \dot{\mathbf{R}}^0$, where $\dot{\mathbf{R}}^0$ is the net drift velocity of the center of mass of the chain

$$\sum_i \langle \boldsymbol{\sigma}_i \rangle = f_t \dot{\mathbf{R}}^0 \tag{98}$$

If there is no velocity field in the absence of polymer ($\mathbf{F} = 0$), $\boldsymbol{\sigma}_i$ follows from Eq. (43) as

$$\langle \boldsymbol{\sigma}_i \rangle = \zeta \langle \dot{\mathbf{R}}_i \rangle - \zeta \sum_{j \neq i} \langle \mathbf{G}(\mathbf{R}_i - \mathbf{R}_j) \cdot \boldsymbol{\sigma}_j \rangle \tag{99}$$

Assuming that $\dot{\mathbf{R}}_i$ is the same as the center-of-mass velocity $\dot{\mathbf{R}}^0$ for all i and invoking the preaveraging approximation, we get

$$\langle \boldsymbol{\sigma}_i \rangle = \zeta \dot{\mathbf{R}}^0 - \zeta \sum_{j \neq i} \langle \mathbf{G}(\mathbf{R}_i - \mathbf{R}_j) \rangle \langle \boldsymbol{\sigma}_j \rangle \tag{100}$$

With the assumption of uniform chain expansion due to potential interactions, we have

$$\langle \mathbf{G}(\mathbf{R}_i - \mathbf{R}_j) \rangle = \frac{1}{6\pi\eta_0} \left\langle \frac{1}{|\mathbf{R}_i - \mathbf{R}_j|} \right\rangle \tag{101}$$

$$= \frac{1}{\eta_0} \left(\frac{1}{6\pi^3 \ell \ell_1 |i - j|} \right)^{1/2} \tag{102}$$

Defining

$$\langle \boldsymbol{\sigma}_j \rangle \equiv \phi_i \zeta \dot{\mathbf{R}}^0 \tag{103}$$

Eq. (100) yields

$$\phi_i = 1 - \frac{\zeta}{\eta_0} \frac{1}{(6\pi^3 \ell \ell_1)^{1/2}} \sum_{j \neq i} \frac{1}{|i - j|^{1/2}} \phi_j \tag{104}$$

By going to continuous variables

$$\frac{2i}{N} - 1 = x, \qquad \frac{2j}{N} - 1 = y \tag{105}$$

Eq. (104) becomes

$$\phi(x) = 1 - h \int_{-1}^{1} dy \frac{1}{|x - y|^{1/2}} \phi(y) \tag{106}$$

where

$$h = \frac{\zeta}{\eta_0} \frac{\sqrt{N}}{(12\pi^3 \ell \ell_1)^{1/2}} \tag{107}$$

The integral equation for $\phi(x)$ was originally derived by Kirkwood and Riseman [1]. By defining the Fourier transforms

$$\phi(x) = \sum_{\nu=-\infty}^{\infty} \hat{\phi}_\nu e^{i\pi\nu x} \tag{108}$$

$$\hat{\phi}_\nu = \frac{1}{2}\int_{-1}^{1} dx \phi(x) e^{-i\pi\nu x} \tag{109}$$

we get from Eq. (106)

$$\hat{\phi}_\nu = \delta_{\nu,0} - h \sum_{\nu'} a_{\nu\nu'} \hat{\phi}_{\nu'} \tag{110}$$

where

$$a_{\nu\nu'} = \frac{1}{2}\int_{-1}^{1} dx \int_{-1}^{1} dy\ e^{-i\pi\nu x + i\pi\nu' y} \frac{1}{|x-y|^{1/2}} \tag{111}$$

For $\nu = 0 = \nu'$, we obtain

$$a_{00} = \frac{8\sqrt{2}}{3} \tag{112}$$

Although $a_{\nu\nu'}$ for $\nu \neq 0$ can be readily computed from Eq. (110), we use the Kirkwood–Riseman approximation of replacing $a_{\nu\nu'}$ by the values asymptotically valid for large $|\nu|$ and $|\nu'|$,

$$a_{\nu\nu'} \sim \begin{cases} \frac{8\sqrt{2}}{3} & \text{for } \nu = 0 = \nu' \\ \left(\frac{2}{|\nu|}\right)^{1/2} \delta_{\nu\nu'} & \text{for } \nu \neq 0 \end{cases} \tag{113}$$

It follows from Eqs. (103) and (108) that

$$\begin{aligned} \sum_i \langle \boldsymbol{\sigma}_j \rangle &= \zeta \dot{\mathbf{R}}^0 \sum_i \phi_i \\ &= \frac{\zeta \dot{\mathbf{R}}^0 N}{2} \int_{-1}^{1} dx \phi(x) \\ &= \zeta \dot{\mathbf{R}}^0 N \hat{\phi}_0 \end{aligned} \tag{114}$$

$\hat{\phi}_0$ is given by Eq. (110) as

$$\hat{\phi}_0 = \frac{1}{\left(1 + \frac{8\sqrt{2}}{3} h\right)} \tag{115}$$

By combining Eqs. (98) and (114), the translational friction coefficient is given by

$$f_t = \frac{\zeta N}{\left[1 + \frac{8\sqrt{2}}{3} h\right]} \tag{116}$$

$$f_t = \frac{\zeta N}{\left[1 + \frac{8\sqrt{2}}{3\eta_0} \frac{\zeta\sqrt{N}}{(12\pi^3 \ell\ell_1)^{1/2}}\right]} \tag{117}$$

In the free-draining limit of no hydrodynamic interactions, we have

$$f_t = \zeta N \tag{118}$$

When hydrodynamic interactions are present, we obtain

$$f_t = \frac{3\eta_0}{8\sqrt{2}} (12\pi^3 N \ell\ell_1)^{1/2} \tag{119}$$

Recalling that $\ell_1/\ell (= \langle R^2 \rangle / N\ell^2)$ is the expansion factor for the mean-square end-to-end distance and the radius of gyration R_g is $\sqrt{\langle R^2 \rangle / 6}$ within the uniform expansion approximation, we have

$$f_t = \frac{9}{4} \pi^{3/2} \eta_0 R_g = \frac{3\sqrt{\pi}}{8} (6\pi\eta_0 R_g) \tag{120}$$

$$\sim \begin{cases} N, & \text{low salt} \\ N^{3/5}, & \text{high salt} \end{cases} \tag{121}$$

Although $f_t \sim N$ in both the non-free-draining limit for low salt solutions and the free-draining limit, the terms appearing as prefactors are qualitatively different.

C. Electrophoretic Mobility

In this section we consider the motion of a uniformly charged flexible polyelectrolyte in an infinitely dilute solution under an externally imposed uniform electric field **E**. The objective is to calculate the electrophoretic mobility μ defined by

$$\dot{\mathbf{R}}^0 = \mu \mathbf{E} \tag{122}$$

This is accomplished by balancing the total induced force on the polymer and the electrical force,

$$\sum_i \langle \boldsymbol{\sigma}_i \rangle = Q\mathbf{E} \tag{123}$$

where $Q = Nz_pe$, with z_pe being the charge of each of the segments of the polymer molecule. The polymer chain is surrounded by $|Nz_p/z_c|$ counterions, each with a charge of z_ce. Let $\rho_i(\mathbf{r})$ be the charge density at $\mathbf{r}$ due to the counterion surrounding the ith segment of the chain. The counterion charge density is related to the charge of the ith segment by the condition of electroneutrality,

$$z_pe = -\int d\mathbf{r}\rho_i(\mathbf{r}) \tag{124}$$

It follows from Eq. (16) that

$$\mathbf{F}(\mathbf{r}) = \mathbf{F}_s(\mathbf{r}) + \sum_{i=0}^{N} \rho_i(\mathbf{r})\mathbf{E} \tag{125}$$

In the equations below, we take the velocity field to be zero ($\mathbf{F}_s = 0$), in the absence of the polymer. Substituting Eq. (125) in Eq. (43), we obtain

$$\boldsymbol{\sigma}_i = \zeta\left[\dot{\mathbf{R}}_i - \int d\mathbf{r}'\mathbf{G}(\mathbf{R}_i - \mathbf{r}') \cdot \sum_{i=0}^{N} \rho_i(\mathbf{r}')\mathbf{E} - \sum_{j\neq i} \mathbf{G}(\mathbf{R}_i - \mathbf{R}_j) \cdot \boldsymbol{\sigma}_j\right] \tag{126}$$

By performing the average over chain configurations and using the preaveraging approximation, we get

$$\langle \boldsymbol{\sigma}_i \rangle = \zeta\dot{\mathbf{R}}^0 - \zeta\sum_{\mathrm{J}=0}^{N} A_{ij}z_pe\mathbf{E} - \zeta\sum_{j\neq i} \langle \mathbf{G}(\mathbf{R}_i - \mathbf{R}_j)\rangle\langle \boldsymbol{\sigma}_i \rangle \tag{127}$$

where

$$A_{ij}z_pe = \left\langle \int d\mathbf{r}'\mathbf{G}(\mathbf{R}_i - \mathbf{r}') \cdot \rho_i(\mathbf{r}') \right\rangle \tag{128}$$

When we use the Fourier representation of the Oseen tensor $\mathbf{G}$, A_{ij} is given by

$$A_{ij}z_pe = \int d\mathbf{r}' \int \frac{d\mathbf{k}}{(2\pi)^3} \frac{(\mathbf{1} - \hat{k}\hat{k})}{\eta_0 k^2} \langle e^{i\mathbf{k}\cdot(\mathbf{R}_i - \mathbf{r}')} \rho_j(\mathbf{r}')\rangle \tag{129}$$

$$= \int d\mathbf{r} \int \frac{d\mathbf{k}}{(2\pi)^3} \frac{(\mathbf{1} - \hat{k}\hat{k})}{\eta_0 k^2} \langle e^{i\mathbf{k}\cdot(\mathbf{R} - \mathbf{r})} \rho_j(\mathbf{r})\rangle \tag{130}$$

where $\mathbf{R} = \mathbf{R}_i - \mathbf{R}_j$ and $\mathbf{r} = \mathbf{r}' - \mathbf{R}_j$. If we assume that $\rho_j(\mathbf{r})$ is radially symmetric, Eq. (130) becomes

$$A_{ij} z_p e = \frac{2}{3\eta_0} \left\langle \left[\int_0^R dr \, \frac{r^2}{R} \rho_j(r) + \int_R^\infty dr \, r \rho_j(r) \right] \right\rangle \tag{131}$$

where

$$\mathbf{R} = \mathbf{R}_i - \mathbf{R}_j \tag{132}$$

The Poisson equation

$$\rho_j(\mathbf{r}) = -\frac{\epsilon}{4\pi} \nabla^2 \psi_j(\mathbf{r}) \tag{133}$$

reduces to

$$\rho_j(r) = -\frac{\epsilon}{4\pi} \frac{1}{r^2} \frac{d}{dr} \left(\frac{r^2 d\psi_j(r)}{dr^2} \right) \tag{134}$$

$$= -\frac{\epsilon}{4\pi} \, \frac{1}{r} \frac{d^2(r\psi_j(r))}{dr} \tag{135}$$

for the radially symmetric $\rho_j(\mathbf{r})$, and $\psi_j(r)$ is the electrostatic potential from the jth segment of charge $z_p e$. The first integral in Eq. (131) simplifies to

$$\frac{1}{R} \int_0^R dr \; r^2 \rho_j(r) = -\frac{\epsilon}{4\pi R} \int_0^R dr \, \frac{1}{dr} \left(r^2 \frac{d\psi_j(r)}{dr} \right) \tag{136}$$

$$= -\frac{\epsilon}{4\pi} R \frac{d\psi_j(r)}{dr} \bigg|_R + \frac{\epsilon}{4\pi R} a^2 \frac{d\psi_j(r)}{dr} \bigg|_a \tag{137}$$

where a is the effective radius of the polymer segment. However, by the condition of electroneutrality, we have

$$-z_p e = \int d\mathbf{r} \, \rho_j(\mathbf{r}) \tag{138}$$

$$= 4\pi \int_a^\infty dr \, r^2 \rho_j(r) \tag{139}$$

$$= \epsilon a^2 \frac{d\psi_j}{dr} \bigg|_a \tag{140}$$

because the potential vanishes at $r \to \infty$. Combining Eqs. (137) and (140), we obtain

$$\frac{1}{R}\int_0^R dr\, r^2 \rho_j(r) = -\frac{\epsilon}{4\pi} R \frac{d\psi_j}{dr}\bigg|_R - \frac{z_p e}{4\pi R} \tag{141}$$

The second integral in Eq. (131) similarly simplifies to

$$\begin{aligned}\int_R^\infty dr\, r\rho_j(r) &= -\frac{\epsilon}{4\pi}\int_R^\infty \frac{d^2(r\psi_j)}{dr^2} \\ &= -\frac{\epsilon}{4\pi}\frac{d(r\psi_j)}{dr}\bigg|_R^\infty \\ &= \frac{\epsilon}{4\pi}\frac{d(r\psi_j)}{dr}\bigg|_R \\ &= \frac{\epsilon}{4\pi} R \frac{d\psi_j}{dr}\bigg|_R + \frac{\epsilon}{4\pi}\psi_j(R)\end{aligned} \tag{142}$$

By combining Eqs. (141) and (142), we get

$$A_{ij} z_p e = \left\langle \frac{2}{3\eta_0}\left[-\frac{z_p e}{4\pi R} + \frac{\epsilon}{4\pi}\psi_j(R)\right]\right\rangle \tag{143}$$

Here R is actually $|\mathbf{R}_i - \mathbf{R}_j|$. We take the potential ψ_j as the Debye–Hückel potential,

$$\psi_j(R) = \frac{e z_p e^{-\kappa R}}{R} \tag{144}$$

Therefore A_{ij} becomes

$$A_{ij} = \frac{1}{6\pi\eta_0}\left\langle\left[-\frac{1}{|\mathbf{R}_i - \mathbf{R}_j|} + \frac{e^{-\kappa|\mathbf{R}_i - \mathbf{R}_j|}}{|\mathbf{R}_i - \mathbf{R}_j|}\right]\right\rangle \tag{145}$$

which is a function of $|i - j|$. We now repeat the calculation of the translational friction coefficient with the additional term (A_{ij}) to derive the desired expression for the electrophoretic mobility. Defining $\sum_i \langle \boldsymbol{\sigma}_i \rangle$ as in Eq. (103) in terms of ϕ_i and going to the continuous variables for i and j, we obtain from Eq. (127)

$$\zeta \dot{\mathbf{R}}^0 \phi(x) = \zeta \dot{\mathbf{R}}^0 - \frac{\zeta N}{2} z_p e \mathbf{E}\int_{-1}^{1} dy A(x,y) - h\zeta \dot{\mathbf{R}}^0 \int_{-1}^{1} dy \frac{1}{|x-y|^{1/2}}\phi(y) \tag{146}$$

Analogous to the Fourier transforms of Eq. (108), we expand $A(x, y)$ as

$$A(x, y) = \sum_{\nu} \sum_{\nu'} \hat{A}_{\nu\nu'} e^{i\pi\nu x - i\pi\nu' y} \tag{147}$$

$$\hat{A}_{\nu\nu'} = \frac{1}{4} \int_{-1}^{1} dx \int_{-1}^{1} dy A(x, y) e^{-i\pi\nu x + i\pi\nu' y} \tag{148}$$

In terms of the Fourier transforms, Eq. (146) becomes

$$\zeta \dot{\mathbf{R}}^0 \hat{\phi}_\nu = \zeta \dot{\mathbf{R}}^0 - \zeta N z_p e \mathbf{E} \hat{A}_{\nu o} - h \zeta \dot{\mathbf{R}}^0 \sum_{\nu'} a_{\nu\nu'} \hat{\phi_{\nu'}} \tag{149}$$

The total hydrodynamic frictional force $\sum_i \langle \boldsymbol{\sigma}_i \rangle$ is given by

$$\sum_i \langle \boldsymbol{\sigma}_i \rangle = \frac{N}{2} \int_{-1}^{1} dx \, \zeta \dot{\mathbf{R}}^0 \phi(x) \tag{150}$$

$$= N \zeta \dot{\mathbf{R}}^0 \hat{\phi}_0 \tag{151}$$

By invoking the Kirkwood–Riseman approximation of diagonality in $a_{\nu\nu'}$, we obtain

$$\left(1 + \frac{8\sqrt{2}}{3} h\right) \hat{\phi_0} \zeta \dot{\mathbf{R}}^0 = \zeta \dot{\mathbf{R}}^0 - \zeta N z_p e \hat{A}_{00} \mathbf{E} \tag{152}$$

so that

$$\zeta \dot{\mathbf{R}}^0 \hat{\phi}_0 = \frac{\zeta \dot{\mathbf{R}}^0}{\left(1 + \frac{8\sqrt{2}}{3} h\right)} - \frac{\zeta \hat{A}_{00} Q \mathbf{E}}{\left(1 + \frac{8\sqrt{2}}{3} h\right)} \tag{153}$$

Inserting this expression in Eq. (123), we obtain

$$Q\mathbf{E} \left[1 + \frac{N \zeta \hat{A}_{00}}{\left(1 + \frac{8\sqrt{2}}{3} h\right)}\right] = \frac{N \zeta \dot{\mathbf{R}}^0}{\left(1 + \frac{8\sqrt{2}}{3} h\right)} \tag{154}$$

providing the electrophoretic mobility as

$$\mu = Q \left[\frac{\left(1 + \frac{8\sqrt{2}}{3} h\right)}{\zeta N} + \hat{A}_{00}\right] \tag{155}$$

$$= Q \left(\frac{1}{f_t} + \hat{A}_{00}\right) \tag{156}$$

where f_t is the translational friction coefficient of the chain. $\hat{A}_{00}$ is the contribution from the counterion cloud. If this contribution is ignored, then the result is equivalent to the balancing of frictional force $f_t\dot{\mathbf{R}}^0$ and the electrical force $Q\mathbf{E}$. However $\hat{A}_{00}$ plays a crucial role in determining the electrophoretic mobility. $\hat{A}_{00}$ is given by Eqs. (145) and (148) as

$$\hat{A}_{00} = \frac{1}{4}\int_{-1}^{1} dx \int_{-1}^{1} dy\, A(x,y) \tag{157}$$

$$= \frac{1}{6\pi\eta_0}\int_0^L ds \int_0^L ds' \left\langle \left[-\frac{1}{|\mathbf{R}(s)-\mathbf{R}(s')|} + \frac{e^{-\kappa|\mathbf{R}(s)-\mathbf{R}(s')|}}{|\mathbf{R}(s)-\mathbf{R}(s')|} \right] \right\rangle \tag{158}$$

Since the first term is the negative of the second term with $\kappa = 0$, we rewrite $\hat{A}_{00}$ as

$$\hat{A}_{00} = B(\kappa) - B(0) \tag{159}$$

where

$$B = \frac{1}{6\pi\eta_0 L^2}\int_0^L ds \int_0^L ds' \left\langle \frac{e^{-\kappa|\mathbf{R}(s)-\mathbf{R}(s')|}}{|\mathbf{R}(s)-\mathbf{R}(s')|} \right\rangle$$

$$= \frac{1}{6\pi\eta_0 L^2}\int_0^L ds \int_0^L ds' \int \frac{d\mathbf{k}}{(2\pi)^3}\frac{1}{k^2+\kappa^2} \langle e^{i\mathbf{k}\cdot[\mathbf{R}(s)-\mathbf{R}(s')]} \rangle$$

$$= \frac{1}{6\pi\eta_0 N}\int \frac{d\mathbf{k}}{(2\pi)^3}\frac{S(k)}{k^2+\kappa^2} \tag{160}$$

where $S(k)$ is the static structure factor of the polyelectrolyte chain. For a Gaussian chain, $S(k)$ is the Debye structure factor,

$$S_D(k) = \frac{2N}{k^4 R_{g_0}^4}\left(e^{-k^2 R_{g0}^2} - 1 + k^2 R_{g_0}^2\right) \tag{161}$$

where $R_{g0}^2 = L\ell/6$. Substituting this result in Eq. (160) yields

$$B(\kappa) = \frac{1}{6\pi\eta_0}\frac{2}{\kappa R_{g_0}^2}\left[1 - \frac{2}{\sqrt{\pi}\kappa R_{g_0}} + \frac{1}{\kappa^2 R_{g_0}^2}(1 - e^{k^2 R_{g0}^2} erfc(\kappa R_{g_0}))\right] \tag{162}$$

The $\kappa \to 0$ result is exactly the same as the result in Eq. (102).

For a polyelectrolyte chain that has non-Gaussian statistics, exact analytical expression for B is not feasible. To get some insight, we notice that the static structure factor has the limiting behavior,

$$S(k) = \begin{cases} N(1 - \frac{k^2 R_g^2}{3} + \cdots), & \kappa R_g \ll 1 \\ \sim \frac{1}{(k\ell)^{1/\nu}}, & \kappa R_g \gg 1 \end{cases} \tag{163}$$

Therefore we construct an approximate interpolation for $S(k)$ as

$$S(k) = \frac{N}{\left(1 + \frac{2\nu}{3} k^2 R_g^2\right)^{1/2\nu}} \tag{164}$$

Substituting this expression for $S(k)$, Eq. (160) yields

$$B = \frac{1}{3\pi^2 \eta_0} \int_0^\infty dk \frac{k^2}{k^2 + \kappa^2} \frac{1}{\left(1 + \frac{2\nu}{3} k^2 R_g^2\right)^{1/2\nu}} \tag{165}$$

$$\sim \begin{cases} \frac{1}{\eta R_g}, & \kappa \to 0 \\ \frac{1}{\eta R_g^{1/\nu} \kappa^{1/\nu - 1}}, & \kappa R_g \gg 1 \end{cases} \tag{166}$$

This result reduces to Eq. (162) for Gaussian chains.

For $\kappa \to 0$, it follows from Eq. (159) that $\hat{A}_{00} = 0$, so that

$$\mu = \frac{Q}{f_t} \tag{167}$$

Since $f_t \sim R_g \sim N$ for $\kappa \to 0$, and $Q = N z_p e$,

$$\mu \sim N^0 \tag{168}$$

For $\kappa R_g \gg 1$, we get from Eqs. (156) and (166),

$$\mu = \frac{Q}{N} \left[\frac{1}{\zeta} + \frac{\beta}{\eta_0 \ell (\kappa \ell)^{1/\nu - 1}} \right] \tag{169}$$

where β is a known numerical coefficient [50]. Since $Q = N z_p e$, μ is independent of N in the high salt limit also. Furthermore, since $\nu \simeq 3/5$ for high salt solutions, we expect

$$\mu \sim \frac{1}{\eta_0 \kappa^{2/3}} \tag{170}$$

when hydrodynamic interaction dominates.

Summarizing, the electrophoretic mobility of a flexible polyelectrolyte chain in infinitely dilute solutions is given by Eq. (156):

$$\mu = Q\left(\frac{1}{f_t} + \hat{A}_{00}\right) \tag{171}$$

where the first and second terms are due to hydrodynamic interactions of polymer segments with other segments and counterion cloud, respectively. For $\kappa = 0$, $\hat{A}_{00}$ is identically zero so that

$$\mu = \frac{Q}{f_t} \sim \frac{N^0}{\eta_0} \tag{172}$$

For $\kappa \neq 0$, a part of $\hat{A}_{00}$ cancels $1/f_t$ exactly in the non-free-draining limit and the remainder is dependent on the structure factor of the polymer and the size exponent ν. For large values of κR_g, μ becomes

$$\mu \sim \frac{N^0}{\eta_0 \kappa^{\frac{1}{\nu}-1}}, \kappa R_g \gg 1 \tag{173}$$

While μ is independent of N at all salt concentrations, it decreases with the salt concentration with an apparent power law,

$$\mu \sim N^0 \kappa^{-\alpha} \tag{174}$$

where α changes smoothly from zero at low salt concentrations to $2/3$ at high salt concentrations. In reaching these conclusions the role of hydrophobic effect is ignored.

D. Coupled Diffusion Coefficient

As seen in the preceding section, the counterions play a crucial role in the mobility of the polyelectrolyte molecules. Even in the absence of an external electric field, the counterions exert an induced electric field in the immediate environment of a charged segment which in turn significantly modifies the collective diffusion coefficient of the polymer. This additional contribution is absent for uncharged polymers, where the cooperative diffusion coefficient D_c is given by the Stokes–Einstein law in dilute solutions,

$$D_c = \frac{k_B T}{f_t} \tag{175}$$

We discuss below how this law is modified by counterions for the case of polyelectrolyte solutions.

The local concentration of the polyelectrolyte molecules, in number of chains per unit volume, as defined by

$$c(\mathbf{r}, t) = \sum_{\alpha=1}^{n} \delta(\mathbf{r} - \mathbf{R}_{\alpha}^{0}(\mathbf{t})) \tag{176}$$

obeys the continuity equation

$$\frac{\partial c(\mathbf{r}, t)}{\partial t} = -\nabla \cdot \mathbf{J} \tag{177}$$

where the flux is given by

$$\mathbf{J} = -D_c \nabla c(\mathbf{r}, t) + c(\mathbf{r}, t) \mu \mathbf{E}_{\text{ind}} \tag{178}$$

Here the first term is the usual diffusive current, with D_c being the usual cooperative diffusion constant of the polymer molecule. The second term is a convective current due to the presence of induced electric field arising from all charged species in the system. μ is the electrophoretic mobility of the polymer molecule derived in the preceding section. From the Poisson equation, we obtain

$$\nabla \cdot \mathbf{E}_{\text{ind}} = \frac{4\pi}{\epsilon} \sum_{\beta} z_{\beta} c_{\beta}. \tag{179}$$

Focusing on the polyelectrolyte and the counterions by ignoring temporarily the presence of salt ions, we get

$$\nabla \cdot \mathbf{E}_{\text{ind}} = \frac{4\pi}{\epsilon} [Qc(\mathbf{r}, t) + z_c \rho_c(\mathbf{r}, t)] \tag{180}$$

where Q is the polymer charge and ρ_c is the local concentration of counterions. It is evident from Eqs. (177)–(180) that the polymer concentration is coupled to the counterion concentration. Analogous to the continuity equation for the polymer concentration, the continuity equation for the counterion concentration is

$$\frac{\partial \rho_c(\mathbf{r}, t)}{\partial t} = D' \nabla^2 \rho_c(\mathbf{r}, t) - \rho_c(\mathbf{r}, t) \mu_c \nabla \cdot \mathbf{E}_{\text{ind}} \tag{181}$$

where D' and μ_c are, respectively, the cooperative diffusion coefficient and electrophoretic mobility of the counterion, given by

$$D' = \frac{k_B T}{f_c} \tag{182}$$

$$\mu_c = \frac{z_c}{f_c} = \frac{z_c D'}{k_B T} \tag{183}$$

where f_c is the translational friction coefficient of the counterion. Defining the fluctuations in polymer concentration and counterion concentration,

$$c(\mathbf{r}, t) = \bar{c} + \delta c(\mathbf{r}, t) \tag{184}$$

$$\rho_c(\mathbf{r}, t) = \bar{\rho}_c + \delta\rho_c(\mathbf{r}, t) \tag{185}$$

where $\bar{c}$ and $\bar{\rho}_c$ are, respectively, the average concentrations of the polymer and counterions, and performing Fourier transforms, we get

$$\frac{\partial \delta c_k}{\partial t} = -D_c k^2 \delta c_k - \frac{4\pi}{\epsilon} \bar{c}\mu(Q\delta c_k + z_c \delta\rho_{c,k}) \tag{186}$$

$$\frac{\partial \delta\rho_{c,k}}{\partial t} = -D' k^2 \delta\rho_{c,k} - \frac{4\pi}{\epsilon} \bar{\rho}_c \mu_c (Q\delta c_k + z_c \delta\rho_{c,k}) \tag{187}$$

If we define

$$\kappa_c^2 = \frac{4\pi}{\epsilon k_B T} z_c^2 \bar{\rho}_c \tag{188}$$

and make the assumption that the relaxation of the counterions is much faster than that of the polymer, Eq. (187) simplifies to

$$\delta\rho_{c,k} = -\frac{\kappa_c^2}{(k^2 + \kappa_c^2)} \frac{Q}{z_c} \delta c_k. \tag{189}$$

Therefore the terms in the parentheses of Eq. (186) reduce to

$$Q\delta c_k + z_c \delta\rho_{c,k} = \frac{k^2}{(k^2 + \kappa_c^2)} Q\delta c_k \tag{190}$$

Substituting this result in Eq. (186), we obtain

$$\frac{\partial \delta c_k}{\partial t} = -\left[D_c + \frac{4\pi}{\epsilon} \frac{\bar{c}\mu Q}{(k^2 + \kappa_c^2)}\right] \delta c_k \tag{191}$$

Therefore, we can write the equation for the time evolution of fluctuations of polymer concentrations as

$$\frac{\partial \delta c_k}{\partial t} = -D_f \delta c_k \tag{192}$$

where D_f is the effective diffusion constant. Let us call D_f as the coupled diffusion coefficient to contrast with the cooperative diffusion coefficient D_c. In general, D_f depends on k.

$$D_f = D_c + \frac{4\pi}{\epsilon} \frac{\bar{c}\mu Q}{(k^2 + \kappa_c^2)} \tag{193}$$

In the limit of $k \to 0$, we get

$$D_f = D_c + \frac{4\pi}{\epsilon} \frac{\bar{c}\mu Q}{\kappa_c^2} \tag{194}$$

Noticing that the average counterion concentration is $N\bar{c}$ for fully ionized polyelectrolytes, we obtain

$$\kappa_c^2 = \frac{4\pi}{\epsilon k_B T} z_c^2 N\bar{c} \tag{195}$$

In addition, by recognizing that (for salt-free case)

$$\mu = \frac{QD_c}{k_B T} \tag{196}$$

the coupled diffusion coefficient in dilute solutions becomes

$$D_f = D_c \left[1 + \frac{1}{N} \left(\frac{Q}{z_c} \right)^2 \right] \tag{197}$$

For monovalent ions, $|z_p| = 1 = |z_c|$, we have

$$D_f = D_c(1 + N) \tag{198}$$

Therefore, the coupling of polymer segments to the counterion cloud, which is directly responsible for the term N in the above equation, dominates the collective diffusion coefficient. Since $R_g \sim N$ for salt-free solutions, D_f is independent of N.

The above derivation can readily be extended [50] to the presence of salt ions. The formula is the same as Eq. (194) except that κ_c^2 includes terms from the dissociated salt ions and becomes the reciprocal of the square of the Debye length, and μ is now given by Eq. (156).

E. Viscosity

We now consider how the polymer dynamics in turn modifies the viscosity of the solution. Recalling from Eq. (41) that the no-slip boundary condition gives an expression for the velocity of the ith segment in terms of the induced forces, $\boldsymbol{\sigma}_i$, we obtain

$$\dot{\mathbf{R}}_i = \int d\mathbf{r}' \mathbf{G}(\mathbf{R}_i - \mathbf{r}') \cdot \mathbf{F}(\mathbf{r}') + \sum_{j=0}^{N} \mathbf{G}(\mathbf{R}_i - \mathbf{R}_j) \cdot \boldsymbol{\sigma}_j \tag{199}$$

Since intersegment hydrodynamic interaction dominates the friction coefficient, we ignore the $i = j$ part that leads to the free-draining contribution. If we define the inverse propagator G^{-1} according to

$$\sum_p \mathbf{G}^{-1}(\mathbf{R}_i - \mathbf{R}_p) \cdot \mathbf{G}(\mathbf{R}_p - \mathbf{R}_j) = \mathbf{1}\delta_{ij} \tag{200}$$

then $\boldsymbol{\sigma}_i$ is given formally by

$$\boldsymbol{\sigma}_i = \sum_j \mathbf{G}^{-1}(\mathbf{R}_i - \mathbf{R}_j) \cdot \left[\dot{\mathbf{R}}_j - \int d\mathbf{r}' \mathbf{G}(\mathbf{R}_j - \mathbf{r}') \cdot \mathbf{F}(\mathbf{r}')\right] \tag{201}$$

Summing over i and taking the average over chain configurations, we get

$$\sum_i \langle \boldsymbol{\sigma}_i \rangle = \sum_i \sum_j \left\langle \mathbf{G}^{-1}(\mathbf{R}_i - \mathbf{R}_j) \cdot \left[\dot{\mathbf{R}}_j - \int d\mathbf{r}' \mathbf{G}(\mathbf{R}_j - \mathbf{r}') \cdot \mathbf{F}(\mathbf{r}')\right]\right\rangle \tag{202}$$

Upon preaveraging, this reduces to

$$\sum_i \langle \boldsymbol{\sigma}_i \rangle = f_t \langle \dot{\mathbf{R}}^0 - \int d\mathbf{r}' \mathbf{G}(\mathbf{R}_j - \mathbf{r}') \cdot \mathbf{F}(\mathbf{r}') \rangle \tag{203}$$

where

$$\begin{aligned} f_t &= \frac{3}{8\sqrt{2}} \frac{\zeta N}{h} \\ &= \frac{3\sqrt{\pi}}{8} (6\pi\eta_0 R_g) \end{aligned} \tag{204}$$

By iterating the elimination of the microscopic velocity of the jth segment in Eq. (201), we get

$$\sum_i \delta(\mathbf{r} - \mathbf{R}_i)\boldsymbol{\sigma}_i = -\int d\mathbf{r}' \left[\delta(\mathbf{r} - \mathbf{R}_i)\mathbf{G}^{-1}(\mathbf{R}_i - R_j)\delta(\mathbf{r}' - \mathbf{R}_j)\right] \int d\mathbf{r}''\mathbf{G}(\mathbf{r}' - \mathbf{r}'') \cdot \mathbf{F}(\mathbf{r}'') + \cdots \tag{205}$$

Substituting this result in Eq. (31) and performing the average over chain configurations as in Eq. (33), we obtain

$$\Sigma(\mathbf{r} - \mathbf{r}') = \left\langle \sum_i \sum_j \delta(\mathbf{r} - \mathbf{R}_i)\mathbf{G}^{-1}(\mathbf{R}_i - R_j)\delta(\mathbf{r}' - \mathbf{R}_j) \right\rangle \tag{206}$$

The Fourier transform of $\Sigma(\mathbf{r} - \mathbf{r}')$ becomes

$$\Sigma(\mathbf{k}) = \frac{n}{V}\int_0^L ds \int_0^L ds' \left\langle e^{i\mathbf{k}\cdot[\mathbf{R}(s) - \mathbf{R}(s')]}\mathbf{G}^{-1}[\mathbf{R}(s) - \mathbf{R}(s')]\right\rangle \tag{207}$$

Using the Rouse mode variable q and employing the preaveraging approximation, we obtain

$$\Sigma(\mathbf{k}) = \frac{c\ell}{\pi}\int_{2\pi/L}^{\infty} dq\, S(k, q)\mathbf{G}^{-1}(q) \tag{208}$$

where c is the segment number density and

$$S(k, q) = \int_{-\infty}^{\infty} d(s - s')e^{iq(s-s')}\left\langle e^{i\mathbf{k}\cdot[\mathbf{R}(s) - \mathbf{R}(s')]}\right\rangle \tag{209}$$

and

$$G(q) = \frac{2}{3}\int \frac{d\mathbf{k}}{(2\pi)^3}\frac{1}{\eta_0 k^2}S(k, q) \tag{210}$$

Assuming that the chain statistics is an effective Gaussian with the uniform expansion coefficient ℓ_1, we obtain

$$S(k, q) = \frac{(k^2\ell_1)/3}{\left(\frac{k^2\ell_1}{6}\right)^2 + q^2} \tag{211}$$

Substituting this result in Eq. (210), we get

$$G(q) = \frac{1}{\pi\eta_0} \frac{1}{(3\ell_1 q)^{1/2}} \tag{212}$$

When we combine Eqs. (208) and (212), $\Sigma(k)$ becomes

$$\Sigma(k) = \frac{c\ell\eta_0}{\sqrt{3}} \int_{\frac{2\pi}{L}}^{\infty} dq \frac{k^2\ell_1}{\left[\left(\frac{k^2\ell_1}{6}\right)^2 + q^2\right]} \sqrt{\ell_1 q} \tag{213}$$

The change in viscosity therefore follows from Eq. (38) as

$$\eta - \eta_0 = \lim_{k\to 0} \frac{\Sigma(k)}{k^2} = \frac{c\ell\eta_0}{\sqrt{3}} \int_{\frac{2\pi}{L}}^{\infty} dq \left(\frac{\ell_1}{q}\right)^{3/2} \tag{214}$$

Since $\ell_1 \sim q^{1-2\nu}$, the intrinsic viscosity $[\eta]$ becomes

$$[\eta] = \frac{\eta - \eta_0}{\eta_0 c} \sim L^{3\nu-1} \tag{215}$$

If we suppress the mode dependence of ℓ_1 and consider only its lowest mode value ($\sim L^{2\nu-1}$), then the intrinsic viscosity is given by

$$[\eta] = \sqrt{\frac{2}{3\pi}} \ell\ell_1^{3/2} L^{1/2} \tag{216}$$

where ℓ_1/ℓ is the expansion factor for the mean-square end-to-end distance of the chain.

IV. SEMIDILUTE SOLUTIONS

The key ingredients in calculating the various dynamical properties of semidilute solutions are the effective Oseen tensor $\mathcal{G}$ and Σ, which contains all contributions from the chains. These quantities are calculated [50] using the effective medium approximation for interchain interactions, which is consistent with the nature of approximations utilized in obtaining the dilute solution properties. Under the effective medium theory, $\sigma(\mathbf{k})$ is calculated self-consistently by assuming that $\Sigma(\mathbf{k})$ arising from all of n polyelectrolyte chains is n times the contribution from one chain with an effective hydrodynamic interaction $\mathcal{G}$, which is a functional of $\Sigma(\mathbf{k})$. The details of this procedure are provided in Refs. 55–59; only the final result is given here. $\Sigma(\mathbf{k})$ is given as a sum over the various Rouse modes

$q = 2\pi p/L$, with p being the Rouse mode index and L being the chain contour length,

$$\Sigma(\mathbf{k}) = \frac{\mathbf{1}cL}{\pi N}\int_{2\pi/L}^{\infty} dqS(\mathbf{k},q)G^{-1}(q) \tag{217}$$

$$G(q) = \int \frac{d\mathbf{j}}{(2\pi)^3}\frac{\left(\mathbf{1}-\frac{\mathbf{jj}}{j^2}\right)}{[\mathbf{1}\eta_0 j^2 + \Sigma(\mathbf{j})]}S(\mathbf{j},q) \tag{218}$$

$$S(\mathbf{k},q) = \int_{-\infty}^{\infty} d(s-s')e^{iq(s-s')}\langle \exp(i\mathbf{k}\cdot[\mathbf{R}(s)-\mathbf{R}(s')])\rangle \tag{219}$$

$S(\mathbf{k},q)$ is the Fourier transform of the static structure factor.

In obtaining Eqs. (217)–(219), we have employed the preaveraging approximation and assumed that solvent motion is instantaneous in comparison to the motion of polyelectrolytes. For a solution of polyelectrolytes, the effective medium theory for the equilibrium properties gives

$$\langle \exp(i\mathbf{k}\cdot[\mathbf{R}(s)-\mathbf{R}(s')])\rangle \simeq \exp(-k^2 l_1|s-s'|/6) \tag{220}$$

where l_1/l is the square of the expansion factor for the mean-square end-to-end distance of a labeled polyelectrolyte chain

$$\langle [R(L)-R(0)]^2\rangle \equiv Ll_1 \tag{221}$$

l_1 is a complicated function [48] of c, L, κ, w, and $z_p^2 l_B$. The asymptotic behavior of l_1 in the various limits is summarized in Table I. Equivalently, the polyelectrolyte chains can be described with bare Kuhn length but with an effective interaction Δ. Explicit formula for Δ is derived in Ref. 48. Its Fourier transform is of the form

$$\Delta_k = \frac{V_k}{\left(1+\frac{cV_k}{k^2\lambda_k^2}\right)} \tag{222}$$

where λ_k is a functional of Δ_k and l_1, and V_k is the Fourier transform of V in Eq. (24). In the high salt limit, Δ_k becomes

$$\Delta_k = \left(w + \frac{w_c}{\kappa^2}\right)k^2\xi^2/(1+k^2\xi^2) \tag{223}$$

where $\xi^2 = \lambda^2/c\left(w+\frac{w_c}{\kappa^2}\right)$ and $w_c = 4\pi z_p^2 e^2/\epsilon k_B T$. In the low salt limit, Δ_k

TABLE I

The Dependencies of Radius of Gyration R_g, Static Correlation Length ξ, Hydrodynamic Screening Length ξ_H, Viscosity η, Self-Translational Diffusion Coefficient D, Cooperative Diffusion Coefficient D_c, Coupled Diffusion Coefficient D_f, and Electrophoretic Mobility μ on c and N for Various Regimes of Polyelectrolyte and Salt Concentrations

Quantity	Salt Level	Dilute (Zimm)	Semidilute (Rouse)	Concentrated (Entangled)
$\frac{6R_g^2}{L} = l_1$	High	$\left[\frac{4}{3l^3}\sqrt{\frac{3}{2\pi}}\left(w + \frac{4\pi l_B}{\kappa^2}\right)\right]^{2/5} N^{1/5} l$	$\frac{3^{3/4}}{\sqrt{\pi}}\left(w + \frac{4\pi l_B}{\kappa^2}\right)^{1/4} c^{-1/4} l^{-3/2}$	l
	Low	$\left(\frac{4\pi l_B}{2\sqrt{6}\pi^{5/2} l}\right)^{2/3} Nl$	$\frac{\sqrt{3}}{4}\left(\frac{6\sqrt{2}}{\pi}\right)^{2/3}\left(\frac{4\pi l_B}{l}\right)^{1/6} c^{-1/2} l^{-1/2}$	l
ξ	High	$R_g \sim N^{-3/5}$	$\frac{3^{5/4}}{4\sqrt{\pi}}\left(w + \frac{4\pi l_B}{\kappa^2}\right)^{-1/4} c^{-3/4} l^{-1/2}$	$l\left[6c\left(w + \frac{4\pi l_B}{\kappa^2}\right)\right]^{-1/2}$
	Low	$R_g \sim N$	$\left(\frac{6\sqrt{2}}{\pi}\right)^{1/3}\sqrt{\frac{3}{16}}\left(\frac{4\pi l_B}{l}\right)^{-1/6}(cl)^{-1/2}$	$\left(\frac{24\pi c l_B}{l^2}\right)^{-1/4}$
ξ_H	High	∞	$\frac{2}{\sqrt{\pi}}\frac{1}{3^{3/4}}\left(w + \frac{4\pi l_B}{\kappa^2}\right)^{-1/4} c^{-3/4} l^{-1/2}$	$c^{-1/2}$
	Low	∞	$\frac{8}{\sqrt{3}\pi}\left(\frac{\pi}{6\sqrt{2}}\right)^{2/3}\left(\frac{4\pi l_B}{l}\right)^{-1/6}(cl)^{-1/2}$	$c^{-1/2}$
$\frac{\eta-\eta_0}{\eta_0}$	High	$c\left(w + \frac{4\pi l_B}{\kappa^2}\right)^{3/5} l^{6/5} N^{4/5}$	$\frac{1}{24}c^2 l^2 l_1^3 L \sim c^{5/4} N$	$c^{4.25} N^{3.4}$
	Low	$c l_B l^2 N^2$	$\frac{1}{24}c^2 l^2 l_1^3 L \sim \sqrt{c} N$	$c^{1.7} N^{3.4}$
D	High	$\frac{8}{3\sqrt{\pi}}\frac{k_B T}{6\pi\eta_0 R_g} \sim N^{-3/5}$	$\frac{k_B T \xi_H}{3\pi\eta_0 R_g^2} \sim \frac{1}{N\sqrt{c}}$	$c^{-7/5} N^{-2}$
	Low	$\frac{8}{3\sqrt{\pi}}\frac{k_B T}{6\pi\eta_0 R_g} \sim N^{-1}$	$\frac{k_B T \xi_H}{3\pi\eta_0 R_g^2} \sim c^0 N^{-1}$	$c^{-1/2} N^{-2}$
D_c	High	D	$\frac{k_B T}{6\pi\eta_0\xi}\frac{\xi_H}{\xi+\xi_H} \sim c^{3/4} N^0$	
	Low	D	$\frac{k_B T}{6\pi\eta_0\xi}\frac{\xi_H}{\xi+\xi_H} \sim c^{1/2} N^0$	
μ	High	$\frac{N^0}{\eta_0 \kappa^{2/3}}$	$\frac{QD}{k_B T} \sim N^0 c^{-1/2}$	
	Low	$\frac{N^0}{\eta_0}$	$\frac{QD}{k_B T} \sim N^0 c^0$	
D_f	High	$N^{-3/5}\left(1 + \frac{z_p^2 Nc}{z_c^2 c + 2c_s}\frac{1.17}{(\kappa R_g)^{2/3}}\right)$	$\frac{4\pi Q c\mu}{\epsilon\kappa^2 N} \sim \frac{\sqrt{c}}{c+2c_s}$	
	Low	N^0	$\frac{4\pi Q c\mu}{\epsilon\kappa^2 N} \sim N^0 c^0$	

becomes

$$\Delta_k = w_c k^2 \xi^4 / (1 + k^4 \xi^4) \tag{224}$$

where $\xi^{-4} = w_c c/\lambda^2$. The scattering function $g(k)$ per segment $(S(k, q = 0))$ is related to Δ_k through the relation

$$g(k) = \frac{1}{cV_k}\left(1 - \frac{\Delta_k}{V_k}\right) \tag{225}$$

Due to Eqs. (222)–(225), $g(k)$ becomes in the high salt limit,

$$g(k) = \frac{1}{\left(w + \frac{w_c}{\kappa^2}\right)c\xi^2} \frac{1}{(k^2 + \xi^{-2})} \tag{226}$$

and in the low salt limit,

$$g(k) = \frac{1}{cw_c\xi^4} \frac{k^2}{(k^4 + \xi^{-4})} \tag{227}$$

In infinitely dilute silutions, l_1 is proportional to N and $N^{1/5}$, and consequently the radius of gyration R_g is proportional to N and $N^{\frac{3}{5}}$, respectively, in low-salt ($\kappa \simeq 0$) and high salt ($\kappa R_g \gg 1$) limits as already pointed out. In semidilute solutions, l_1 is proportional to $c^{-1/2}$ and $c^{-1/4}$, and consequently R_g is proportional to $c^{-1/4}$ and $c^{-1/8}$, respectively, in low salt and high salt limits. As shown in Ref. 48, l_1 is intimately related to the static correlation length ξ. In infinitely dilute solutions ξ is proportional to R_g. In semidilute solutions, ξ is proportional to $c^{-1/2}$ and $c^{-3/4}$, respectively, in low and high salt limits. In concentrated solutions, ξ is proportional to $c^{-1/4}$ and $c^{-1/2}$, respectively, in low salt and high salt limits. All of the these equilibrium results have been experimentally verified [52–54]. The hydrodynamic interaction $\mathcal{G}$ is given by Eq. (35),

$$\mathcal{G}(k) = \frac{1}{\eta_0 k^2 + \Sigma(k)} \tag{228}$$

where $\Sigma(k)$ is to be solved self-consistently according to

$$\Sigma(k) = \frac{cl}{\pi} \int_{2\pi/L}^{\infty} dq \frac{k^2 l_1/3}{\left(\frac{k^2 l_1}{6}\right)^2 + q^2} \frac{1}{J(q)} \tag{229}$$

$$G(q) = \frac{1}{3\pi^2} \int_0^{\infty} dj \frac{j^2}{(\eta_0 j^2 + \Sigma(j))} \frac{j^2 l_1/3}{\left(\frac{j^2 l_1}{6}\right)^2 + q^2} \tag{230}$$

In the limit of $k \to 0$, $\Sigma(k)$ is proportional to k^2, as expected. On the other hand for $\frac{1}{R_g} < k < \frac{1}{l}$, we expect the bare hydrodynamic interaction to be screened. Therefore we look for a solution of Σ from Eq. (229) so that Σ is independent of k to entail hydrodynamic screening. When large k behavior dominates the integrals in Eq. (229), Σ can be written as

$$\Sigma(k) = \eta_0 \xi_H^{-2} \tag{231}$$

so that $\mathscr{G}$ is screened hydrodynamic interaction

$$\mathscr{G}(k) = \frac{1}{\eta_0(k^2 + \xi_H^{-2})} \tag{232}$$

where ξ_H is the hydrodynamic screening length. It follows from Eq. (230) that

$$G(q) = \frac{2\xi_H}{\pi\eta_0 l_1} \tag{233}$$

and

$$\xi_H^{-1} = \frac{\pi}{2} c l l_1 \tag{234}$$

In view of the dependence of l_1 on c as given in Table I, the hydrodynamic screening length has the following concentration dependence

$$\xi_H = \begin{cases} \frac{2}{\sqrt{\pi}} \frac{1}{3^{3/4}} \left(w + \frac{4\pi l_B}{\kappa^2}\right)^{-1/4} c^{-3/4} l^{-1/2} \\ \kappa R_g > 1 \\ \frac{8}{\sqrt{3}\pi} \left(\frac{\pi}{6\sqrt{2}}\right)^{2/3} \left(\frac{l}{4\pi l_B}\right)^{1/6} c^{-1/2} l^{-1/2}, \qquad \kappa R_g < 1 \end{cases} \tag{235}$$

It is to be noted that ξ_H is proportional to the static correlation length ξ and the numerical prefactor is not unity. ξ_H is proportional to $c^{-1/2}$ and $c^{-3/4}$, respectively, for low salt and high salt conditions. Using the results of Eqs. (232)–(235), we now calculate the diffusion coefficients, electrophoretic mobility, and viscosity of semidilute polyelectrolyte solutions.

Using $\mathscr{G}$ and Σ, the translational friction coefficient and the electrophoretic mobility can be calculated by simply replacing G by $\mathscr{G}$ in the respective formulas. Before we proceed to do this, we present the collective dynamics of monomer density and counterions.

Following the method of Ref. 59, we write the equation of motion for the collective coordinate of local segment density $\rho(\mathbf{r})$,

$$\rho(\mathbf{r}) = \sum_{\alpha=1}^{n} \sum_{i=0}^{N-1} \delta(\mathbf{r} - \mathbf{R}_{\alpha i}) \tag{236}$$

When we define the Fourier transform as

$$\rho(\mathbf{k}) = \frac{n}{\Omega} \sum_{j=0}^{N-1} e^{i\mathbf{k}\cdot\mathbf{R}_j} \tag{237}$$

the evolution of $\rho(\mathbf{k})$ follows as

$$\frac{\partial \rho(\mathbf{k})}{\partial t} = \frac{n}{\Omega} \sum_j i\mathbf{k} \cdot \dot{\mathbf{R}}_j e^{i\mathbf{k}\cdot\mathbf{R}_j} \tag{238}$$

Analogous to the derivation of the effective Langevin equation for the chain in dilute solutions, we get in semidilute solutions

$$\dot{\mathbf{R}}_i = -\sum_{j=0}^{N-1} \mathscr{G}^{-1}(\mathbf{R}_i - \mathbf{R}_j) \cdot \frac{\partial U}{\partial \mathbf{R}_j} + \mathbf{f}_{vi} + \frac{\mu}{N}\mathbf{E} \tag{239}$$

where

$$\frac{\partial U}{\partial \mathbf{R}_i} = \sum_{j=0}^{N-1} \left[\frac{3k_B T}{l^2} (2\delta_{i,j} - \delta_{i,j+1} - \delta_{i,j-1})\mathbf{R}_j + \nabla_{\mathbf{R}_i} V(\mathbf{R}_i - \mathbf{R}_j) \right] \tag{240}$$

and

$$\mathbf{f}_{vi} = \sum_{j=0}^{N-1} \mathscr{G}(\mathbf{R}_i - \mathbf{R}_j) \cdot \mathbf{f}_j \tag{241}$$

so that

$$\langle \mathbf{f}_{vi}(t) \rangle = 0 \tag{242}$$

$$\langle \mathbf{f}_{vi}(t) \mathbf{f}_{vj}(t') \rangle = 2k_B T \mathscr{G}(\mathbf{R}_i - \mathbf{R}_j)\delta(t - t') \tag{243}$$

The angular brackets indicate the thermal average for the noise $\mathbf{f}_{vi}$.

Substitution of Eq. (239) for $\dot{\mathbf{R}}_j$ in Eq. (238) enables us to write a phenomenological equation for $\rho(\mathbf{k})$:

$$\frac{\partial}{\partial t} \rho(\mathbf{k}) = -\sum_{\mathbf{k}'} L_{\mathbf{k}\mathbf{k}'} \frac{\partial U}{\partial \rho(\mathbf{k}')} + f_\rho(\mathbf{k}) + f_E(\mathbf{k}) \tag{244}$$

where

$$f_\rho(\mathbf{k}) = \frac{n}{\Omega} \sum_j i\mathbf{k} \cdot \mathbf{f}_{vj}\, e^{i\mathbf{k}\cdot\mathbf{R}_j} \tag{245}$$

and

$$f_E(\mathbf{k}) = -\frac{\mu}{N} i\mathbf{k} \cdot \sum_{\mathbf{k}'} E(\mathbf{k}')\rho(\mathbf{k} - \mathbf{k}') \tag{246}$$

The Onsager coefficient $L_{\mathbf{kk}'}$ is given by the fluctuation–dissipation theorem:

$$\langle f_\rho(\mathbf{k},t) f_\rho(\mathbf{k}',t')\rangle = 2k_BTL_{\mathbf{kk}'}\delta(t-t') \tag{247}$$

and

$$\langle f_\rho(\mathbf{k},t)\rangle = 0 \tag{248}$$

Combining Eqs. (35), (242), (243), (245), (247), and (248), we obtain

$$L_{\mathbf{kk}'} = \int \frac{d\mathbf{j}}{(2\pi)^3} \frac{\mathbf{k}\cdot(\mathbf{1}-\hat{j}\hat{j})\cdot\mathbf{k}'}{[\eta_0\, j^2\mathbf{1}+\Sigma(\mathbf{j})]} \rho(\mathbf{k}+\mathbf{j})\rho(\mathbf{k}'-\mathbf{j}) \tag{249}$$

Upon preaveraging, we obtain

$$L_{\mathbf{kk}'} = \delta(\mathbf{k}+\mathbf{k}')L_{\mathbf{k}} \tag{250}$$

where

$$L_{\mathbf{k}} = \frac{c}{\Omega}\int \frac{d\mathbf{j}}{(2\pi)^3} \frac{k^2[\mathbf{1}-(-\hat{k}\cdot\hat{j})^2]}{[\eta_0 j^2+\Sigma(j)]} g(\mathbf{k}+\mathbf{j}) \tag{251}$$

with $g(\mathbf{k})$ being the scattering function per segment,

$$g(\mathbf{k}) = \frac{\Omega}{c}\langle\rho(\mathbf{k})\rho(-\mathbf{k})\rangle \tag{252}$$

Within the effective medium theory, the potential functional can be written as [3]

$$\frac{U}{k_BT} = \frac{\Omega}{2c}\sum_{\mathbf{k}} \frac{\rho(\mathbf{k})\rho(-\mathbf{k})}{g(\mathbf{k})} \tag{253}$$

Substituting the results of Eqs. (251)–(253) in Eq. (244), we get

$$\frac{\partial}{\partial t}\rho(\mathbf{k}) = -D_c k^2\rho(\mathbf{k}) + f_\rho(\mathbf{k}) + f_E(\mathbf{k}) \tag{254}$$

where the cooperative diffusion coefficient is given by

$$D_c = k_BT\int \frac{d\mathbf{j}}{(2\pi)^3} \frac{[\mathbf{1}-(\hat{k}\cdot\hat{j})^2]}{[\eta_0 j^2+\Sigma(\mathbf{j})]} \frac{g(\mathbf{k}+\mathbf{j})}{g(\mathbf{k})} \tag{255}$$

With the use of the preaveraging approximation, Eq. (246) becomes

$$f_E(\mathbf{k}) = -\frac{\mu c}{N} i\mathbf{k} \cdot \mathbf{E}(\mathbf{k}) \tag{256}$$

which, in combination with the Poisson equation

$$i\mathbf{k} \cdot \mathbf{E}(\mathbf{k}) = \frac{4\pi}{\epsilon} \sum_i z_i \rho_i(\mathbf{k}) \tag{257}$$

and Eq. (254), leads to

$$\frac{\partial}{\partial t} \rho(\mathbf{k}) = -D_c k^2 \rho(\mathbf{k}) + f_\rho(\mathbf{k}) - \frac{c\mu}{N} \frac{4\pi}{\epsilon} \sum_i z_i \rho_i(\mathbf{k}) \tag{258}$$

Here $i = 1$ represents the polyelectrolyte with $z_1 = z_p e$ and $\rho_1 = c$. For $i \neq 1$, ρ_i is the fluctuation in the number density of the ith charged species and z_i is its charge. The Eq. (258) for the segment density of the polyelectrolyte is coupled to the equation of motion for ρ_i,

$$\frac{\partial}{\partial t} \rho_i(\mathbf{k}) = -D_{ci} k^2 \rho_i(\mathbf{k}) - \rho_{i0} \mu_i \frac{4\pi}{\epsilon} \sum_i z_i \rho_i(\mathbf{k}) \tag{259}$$

where D_{ci} and μ_i are the diffusion coefficient and the electrophoretic mobility of the ith species. The noise term is ignored in the above equation. Assuming that small ions, $i \neq 1$, relax faster than the polyelectrolyte chains, $\frac{\partial}{\partial t} \rho_i(\mathbf{k}) = 0$, Eqs. (258) and (259) get decoupled to finally give

$$\frac{\partial}{\partial t} \rho(\mathbf{k}) = -D_c k^2 \left[1 + \frac{\kappa_1^2 M_1}{(k^2 + \kappa^2)} \right] \rho(\mathbf{k}) \tag{260}$$

where

$$\kappa_1^2 = \frac{4\pi}{\epsilon k_B T} \frac{Q^2 c}{N} \tag{261}$$

and

$$M_1 = \frac{k_B T \mu}{Q D_c} \tag{262}$$

In obtaining Eq. (260), the condition

$$\mu_i = \frac{z_i D_{ci}}{k_B T} \tag{263}$$

valid for small ions $i \neq 1$ has been used. It readily follows from Eq. (262) that the time correlation function $\langle \rho(\mathbf{k}, t)\rho(-\mathbf{k}, 0)\rangle$ decays exponentially with time with rate $D_f k^2$,

$$\langle \rho(\mathbf{k}, t)\rho(-\mathbf{k}, 0)\rangle \sim \exp(-D_f k^2 t) \tag{264}$$

where the "coupled diffusion coefficient" D_f is given by

$$D_f = D_c\left[1 + \frac{\kappa_1^2 M_1}{(k^2 + \kappa^2)}\right] \tag{265}$$

In the limit of $k \longrightarrow 0$, D_f becomes

$$D_f(k \longrightarrow 0) = D_c(k \longrightarrow 0)\left[1 + \frac{\kappa_1^2 M_1}{\kappa^2}\right] \tag{266}$$

The coupled diffusion coefficient D_f is calculated using Eqs. (255)–(263).

We now present the key results of the above equations.

A. Translational Friction Coefficient

The translational friction coefficient is given by

$$f_t = \sum_{i=0}^{N-1}\sum_{p=0}^{N-1}\langle \mathscr{G}^{-1}(\mathbf{R}_i, \mathbf{R}_p)\rangle \tag{267}$$

The self-translational diffusion coefficient D is related to f_t by the Stokes–Einstein relation and is given by

$$D = \frac{k_B T}{f_t} \equiv k_B T \mathscr{G}_{00} = \frac{k_B T}{N^2}\sum_{i=0}^{N-1}\sum_{p=0}^{N-1}\langle \mathscr{G}(\mathbf{R}_i - \mathbf{R}_p)\rangle \tag{268}$$

Using Eqs. (232) and (268), we get

$$D = \frac{k_B T}{6\pi\eta_0 R_g}\left(\frac{2\xi_H}{R_g}\right)\left[1 - \frac{2}{\sqrt{\pi}}\frac{\xi_h}{R_g} + \left(\frac{\xi_H}{R_g}\right)^2\left(1 - e^{\frac{R_g^2}{\xi_H^2}}\operatorname{erfc}\left(\frac{R_g}{\xi_H}\right)\right)\right] \tag{269}$$

This reduces to the result of Eqs. (119) and (175) for dilute solutions ($\xi_H \gg R_g$). In the Rouse regime where hydrodynamic interaction is screened, D becomes

$$D = \frac{k_B T \xi_H}{3\pi\eta_0 R_g^2}, \qquad R_G \gg \xi_H \tag{270}$$

Substituting the result of Eq. (234) and noticing that $R_g^2 = Ll_1/6$, D is given by

$$D = \frac{4k_B T}{\pi^2 \eta_0} \frac{1}{cLl l_1^2} \tag{271}$$

Using the result of l_1 (Table I), the translational self-diffusion coefficient is given by the form

$$D \sim \begin{cases} \frac{1}{N\sqrt{c}}, & \kappa R_g \gg 1 \\ \frac{c^0}{N}, & \kappa R_g \ll 1 \end{cases} \tag{272}$$

where the prefactors readily follow from Eq. (271) and Table I. While the Rouse law of $D \sim N^{-1}$ is valid at all salt conditions, the dependence of D on polyelectrolyte concentration is different for different salt conditions. At low salt conditions, D is independent of polyelectrolyte concentration. The crossover between these asymptotic results is described by Eq. (271).

B. Electrophoretic Mobility

Analogous to the derivation in dilute solutions, μ is given by

$$\mu = Q(\mathcal{G}_{00} + A_{00}) \tag{273}$$

Combining Eqs. (228) and (144), $A(r)$ is given by

$$A(r) = \frac{1}{6\pi\eta_0 r} \frac{\kappa^2 \xi_H^2}{(\kappa^2 \xi_H^2 - 1)} (e^{-\kappa r} - e^{-\frac{r}{\xi_H}}) \tag{274}$$

Substituting this result in Eq. (148), A_{00} becomes

$$A_{00} = \frac{1}{6\pi\eta_0 R_g} \frac{\kappa^2 \xi_H^2}{(\kappa^2 \xi_H^2 - 1)} \left[\mathcal{M}(\kappa R_g) - \mathcal{M}\left(\frac{R_g}{\xi_H}\right) \right] \tag{275}$$

where the function $\mathcal{M}$ is defined as

$$\mathcal{M}(xR_g) \equiv \frac{R_g}{N^2} \sum_{i=0}^{N-1} \sum_{j=0}^{N-1} \left\langle \frac{e^{-xR_{ij}}}{R_{ij}} \right\rangle = \frac{2R_g}{\pi N} \int_0^\infty dk \frac{k^2}{(k^2 + x^2)} S(kR_g) \tag{276}$$

Using Eqs. (268), (273), and (275), the electrophoretic mobility μ is given by

$$\mu = \frac{Q}{6\pi\eta_0 R_g} \frac{1}{(1 - \kappa^2 \xi_H^2)} \left[\mathcal{M}\left(\frac{R_g}{\xi_H}\right) - \kappa^2 \xi_H^2 \mathcal{M}(\kappa R_g) \right] \tag{277}$$

This reduces to Eq. (156) in the infinite dilution limit ($\xi_H \to \infty$). In the Rouse regime (ξ_H small), the electrophoretic mobility becomes

$$\mu = \frac{Q}{6\pi\eta_0 R_g}\mathcal{M}\left(\frac{R_g}{\xi_H}\right) \tag{278}$$

Substituting the asymptotic form of $M(R_g/\xi_H)$, we obtain

$$\mu = \frac{Q\xi_H}{2\pi\eta_0 R_g^2} \tag{279}$$

Noticing the expression for D in the Rouse limit, Eq. (270), we see that μ becomes

$$\mu = \frac{QD}{k_B T} \tag{280}$$

in the Rouse regime. The power laws of μ on N and polyelectrolyte concentration c follow from Eq. (272) to be

$$\mu \sim \begin{cases} N^0 c^{-\frac{1}{2}}\left(w + \frac{4\pi l_B}{\kappa^2}\right)^{-1/2}, & \kappa R_g > 1 \\ N^0 c^0, & \kappa R_g < 1 \end{cases} \tag{281}$$

C. Cooperative and Coupled Diffusion Coefficients

When we substitute Eq. (231) into Eq. (255), the cooperative coefficient D_c is given by

$$D_c = \frac{k_B T}{\eta_0}\int \frac{d\mathbf{j}}{(2\pi)^3}\frac{[\mathbf{1} - (\mathbf{k}\cdot\mathbf{j})^2]}{(j^2 + \xi_H^{-2})}\frac{g(\mathbf{k}+\mathbf{j})}{g(\mathbf{k})} \tag{282}$$

The scattering function $g(k)$ is a function of static correlation length ξ as given by Eqs. (225)–(227). For semidilute solutions at high salt concentrations, D_c follows from Eqs. (226) and (282) in the $k \to 0$ limit,

$$D_c = \frac{k_B T}{6\pi\eta_0\xi}\frac{\xi_H}{(\xi + \xi_H)} \tag{283}$$

In infinitely dilute solutions we have $\xi_H \to \infty$ and the static correlation $\xi = R_g/\sqrt{3}$ as already discussed. In this limit, D_c of Eq. (175) is therefore recovered. On the other hand, in the Rouse regime, Eq. (283) yields

$$D_c \sim \frac{T}{\eta_0\xi} \sim c^{3/4} \tag{284}$$

since $\xi_H \sim \xi \sim c^{-3/4}$. For semidilute solutions at low salt concentrations, D_c follows from Eqs. (227) and (282):

$$D_c = \frac{2k_BT}{3\eta_0} \int \frac{d\mathbf{j}}{(2\pi)^3} \frac{1}{(j^2 + \xi_H^{-2})} \frac{(k+j)^2}{[(k+j)^4\xi^4 + 1]} \frac{(k^4\xi^4 + 1)}{k^2} \tag{285}$$

The cooperative diffusion coefficient in the salt-free limit is thus strongly k-dependent. In this limit the equilibrium scattering function $g(k)$ exhibits a peak at $k^* = \xi^{-1}$. Approximating $g(k)$ in Eqs. (282) and (285) by its value at the peak position k^*, Eq. (285) yields

$$D_c = \frac{k_BT}{6\pi\eta_0\xi} \frac{\sqrt{2}}{\left[1 + \left(\frac{\xi}{\xi_h}\right)^4\right]} \left[1 - \left(\frac{\xi}{\xi_H}\right)^2 + \sqrt{2}\left(\frac{\xi}{\xi_H}\right)^3\right] \tag{286}$$

In the Rouse regime (for low salt concentrations)

$$D_c \sim \frac{T}{\eta_0\xi} \sim N^0 c^{1/2} \tag{287}$$

because $\xi_H \sim \xi \sim c^{-1/2}$.

The coupled diffusion coefficient D_f can be now obtained from D_c of Eqs. (283)–(287) and μ of Eq. (281). In the Rouse regime (where the second term on the right-hand side of Eqs. (265) and (266) dominates, D_f becomes

$$D_f = \frac{4\pi Q c \mu}{\epsilon \kappa^2 N} \tag{288}$$

Writing μ explicitly in terms of R_g and ξ_H, D_f is given by

$$D_f = \frac{k_BT}{2\pi\eta_0} \frac{Z_p^2 c}{(Z_c^2 c + 2c_s)} \frac{N\xi_H}{R_g^2} \tag{289}$$

where salt ions are taken to be monovalent. In the salt-free limit, $\xi_H \sim c^{-1/2}$ and $R_g^2 \sim Nc^{-1/2}$ so that

$$D_f \sim \frac{T}{\eta_0} N^0 c^0 \tag{290}$$

Therefore, remarkably, the coupled diffusion coefficient becomes independent of N and c in the Rouse regime of salt-free polyelectrolyte solutions. This is to be contrasted with Eq. (287) for the cooperative diffusion coefficient D_c and with Eq. (272) for the translational diffusion coefficient D. In the Rouse regime at high salt concentrations, $\xi_H \sim c^{-3/4}$ and $R_g^2 \sim Nc^{-1/4}$ so that Eq. (289) yields

$$D_f \sim \frac{T}{\eta_0} \frac{Z_p^2 \sqrt{c}}{(Z_c^2 c + 2c_s)} \tag{291}$$

The coupled diffusion coefficient D_f decreases from the value of Eq. (290) to that of Eq. (291) as the salt concentration is increased.

D. Viscosity

The virial series of viscosity in polyelectrolyte concentration c can be obtained from Eq. (229) by iterating $\Sigma(k)$ to the desired order in c and then combining with Eq. (38). To the leading order in c, Eq. (229) yields

$$G(q) = \frac{1}{\pi\eta_0\sqrt{3l_1 q}}\left[1 - \frac{\pi}{2\sqrt{3q}} c l l_1^{3/2} + \cdots\right] \tag{292}$$

so that viscosity change becomes from Eq. (38)

$$\frac{\eta - \eta_0}{\eta_0} = \left(\frac{2}{3\pi}\right)^{1/2} c l l_1^{3/2} L^{1/2}\left[1 + \frac{1}{4}\sqrt{\frac{\pi}{6}} c l l_1^{3/2} L^{1/2} + \cdots\right] \tag{293}$$

Defining the overlap concentration c^* to be

$$c^* = \frac{3}{4\pi} \frac{N}{R_g^3} \tag{294}$$

where $R_g^2 = Ll_1/6$, viscosity change can be written as

$$\frac{\eta - \eta_0}{\eta_0} = \frac{9}{\pi^{3/2}}\left(\frac{c}{c^*}\right)\left[1 + \frac{9}{2\sqrt{\pi}} \frac{c}{c^*} + \cdots\right] \tag{295}$$

Alternatively, Eq. (295) can be rewritten in terms of the intrinsic viscosity given by Eq. (216) as

$$\frac{\eta - \eta_0}{\eta_0} = c[\eta] + k_H c^2 [\eta]^2 + \cdots \tag{296}$$

where

$$[\eta] \simeq \frac{1.616}{c^*} \tag{297}$$

and the Huggins coefficient k_H is

$$k_H \simeq 1.57 \tag{298}$$

This value of k_H is actually low by an order of magnitude for dilute suspensions of charged spheres of radius R_g. This is due to the neglect of interchain correlations for $c < c^*$ in the structure factor used in the derivation of Eqs. (295)–(298). If the repulsive interaction between polyelectrolyte chains dominates, as expected in salt-free solutions, the virial expansion for viscosity may be valid over considerable range of concentrations where the average distance between chains scales as $c^{-1/3}$. This virial series may be approximated by

$$\frac{\eta - \eta_0}{\eta_0} = [\eta] c \exp(k_H [\eta] c) \tag{299}$$

Substituting Eqs. (233) and (234) into Eq. (229) and (38), the viscosity change in the Rouse regime becomes

$$\frac{\eta - \eta_0}{\eta_0} = \frac{1}{24} c^2 l l_1^3 L \tag{300}$$

Using the concentration dependence of l_1 under different salt conditions, we get

$$\frac{\eta - \eta_0}{\eta_0} \sim \begin{cases} c^{5/4} N, & \kappa R_g \gg 1 \\ c^{1/2} N, & \kappa R_g \ll 1 \end{cases} \tag{301}$$

E. Frequency Dependence

The theory given above can be readily extended [60] to the frequency ω dependence of shear viscosity $\eta(\omega)$ and shear modulus $G(\omega) = i\omega\eta(\omega)$. Generalizing the equations in the preceding sections for ω dependence, we get

$$\eta(\omega) - \eta_0 = \lim_{k \to 0} \frac{1}{k^2} (\mathbf{1} - \hat{\mathbf{k}}\hat{\mathbf{k}}) \cdot \Sigma(\mathbf{k}, \omega) \tag{302}$$

$$\Sigma(\mathbf{k}, \omega) = \mathbf{1} \frac{cl}{\pi} \int_{\frac{2\pi}{Nl}}^{\infty} dq S(\mathbf{k}, q, \omega) J^{-1}(q, \omega) \tag{303}$$

$$J(q, \omega) = \int \frac{d\mathbf{j}}{(2\pi)^3} \frac{\mathbf{1} - \hat{\mathbf{j}}\hat{\mathbf{j}}}{[\mathbf{1}(\eta_0 j^2 + i\omega) + \Sigma(\mathbf{j}, \omega)]} S(\mathbf{j}, q, \omega) \tag{304}$$

$$S(\mathbf{k}, q, \omega) = \int_{-\infty}^{\infty} dt \int_{-\infty}^{\infty} d(s - s')^{i\omega t + iq(s-s')} \langle \exp(i\mathbf{k} \cdot [\mathbf{R}(s, t) - \mathbf{R}(s', 0)]) \rangle \tag{305}$$

TABLE II
Dependencies of Viscosity and Modulus in Unentangled Dilute and Semidilute Solutions on c, N, and ω for Low and High Salt Conditions[a]

		Dilute		Semidilute	
	Salt Level	$\omega \to 0$	$\omega \to \infty$	$\omega \to 0$	$\omega \to \infty$
$\frac{\eta-\eta_0}{\eta_0}$		$cN^{3\nu-1}$	$c\omega^{-\frac{(3\nu-1)}{3\nu}}$	$c^{\frac{1}{(3\nu-1)}}N$	$\omega^{-1/2}c^{\frac{3\nu}{2(3\nu-1)}}$
	Low salt	cN^2	$c\omega^{-2/3}$	$c^{1/2}N$	$\omega^{-1/2}c^{3/4}$
	High salt	$cN^{4/5}$	$c\omega^{-4/9}$	$c^{5/4}N$	$\omega^{-1/2}c^{9/8}$
$\frac{G-G_0}{G_0}$		$\omega cN^{3\nu-1}$	$c\omega^{1/3\nu}$	$\omega c^{1/3\nu-1}N$	$\omega^{1/2}c^{3\nu/2(3\nu-1)}$
	Low salt	ωcN^2	$c\omega^{1/3}$	$\omega c^{1/2}N$	$\omega^{1/2}c^{3/4}$
	High salt	$\omega cN^{4/5}$	$c\omega^{5/9}$	$\omega c^{5/4}N$	$\omega^{\frac{1}{2}}c^{9/8}$

[a]Note that ν is the radius of gyration exponent in infinitely dilute solutions.

The key results from these equations are summarized in Table II. Naturally, in the $\omega \to 0$ limit, results of Table I are recovered. Only the high frequency limit is mentioned now.

1. Dilute

In infinite dilute solutions, the intrinsic viscosity $[\eta]$ is given by [3, 4, 61]

$$[\eta] = \frac{RT}{M\eta_0} \sum_{j=1}^{N} \frac{\tau_j}{1 + i\omega\tau_j} \tag{306}$$

where R, T, M, and τ_j are, respectively, the gas constant, absolute temperature, molecular weight (proportional to the number of Kuhn segments N per chain), and relaxation time of the jth Rouse–Zimm mode. The intrinsic storage $[G']$ and loss $[G'']$ moduli are

$$[G'] = \frac{RT}{M} \sum_{j=1}^{N} \frac{\omega^2\tau_j^2}{\left(1 + \omega^2\tau_j^2\right)}$$

$$[G''] = \frac{RT}{M} \sum_{j=1}^{N} \frac{\omega\tau_j}{\left(1 + \omega^2\tau_j^2\right)} \tag{307}$$

In the high frequency limit, the intrinsic viscosity follows from Eq. (306) as

$$[\eta] \sim \int dq \frac{\tau_q}{(1 + i\omega\tau_q)} \tag{308}$$

so that both the real and imaginary parts are proportional to

$$[\eta] \sim \omega \int dq \frac{1}{q^{6\nu} + \omega^2}$$
$$\sim \omega^{-(3\nu-1)/3\nu} \tag{309}$$

Equation (307) gives the intrinsic modulus as

$$[G] \sim \omega^{1/3\nu} \tag{310}$$

2. *Semidilute*

At high frequencies, we have

$$\frac{\eta - \eta_0}{\eta_0 c} \sim \omega \int dq \frac{\tau_q^2}{\left(1 + \omega^2 \tau_q^2\right)}$$
$$\sim \frac{1}{\sqrt{\omega}} c^{\frac{(2-3\nu)}{2(3\nu-1)}} \tag{311}$$

and

$$\frac{G - G_0}{G_0 c} \sim \sqrt{\omega}\, c^{\frac{(2-3\nu)}{2(3\nu-1)}} \tag{312}$$

Therefore in this Rouse regime of unentangled semidilute solutions where hydrodynamic interaction is screened, both the reduced viscosity and reduced modulus decrease with increase in polymer concentration in salt free solutions ($\nu \to 1$),

$$\frac{\eta - \eta_0}{\eta_0 c} \sim \begin{cases} Nc^{-1/2}, & \omega \to 0 \\ \omega^{-1/2} c^{-1/4}, & \omega \to \infty \end{cases} \tag{313}$$

and

$$\frac{G - G_0}{G_0 c} \sim \begin{cases} \omega N c^{-1/2}, & \omega \to 0 \\ \omega^{1/2} c^{-1/4}, & \omega \to \infty \end{cases} \tag{314}$$

These apparent strange results are to be contrasted with the results at high salt concentrations ($\nu \to 3/5$),

$$\frac{\eta - \eta_0}{\eta_0 c} \sim \begin{cases} Nc^{1/4}, & \omega \to 0 \\ \omega^{-1/2} c^{1/8}, & \omega \to \infty \end{cases} \tag{315}$$

and

$$\frac{G - G_0}{G_0 c} \sim \begin{cases} \omega N c^{1/4}, & \omega \to 0 \\ \omega^{1/2} c^{1/8}, & \omega \to \infty \end{cases} \tag{316}$$

The above general results and the asymptotic values for high and low salt concentrations are included in Table II.

Although several experimental studies [22, 62, 63] have been reported on η, G, and the compliance of polyelectrolyte solutions, systematic investigations on the frequency dependence is still lacking.

F. Entanglement Effects

It is nontrivial to calculate the various diffusion coefficients and viscosity of polyelectrolyte solutions at very high concentrations when entanglement effects dominate. However, because of screening of electrostatic, excluded volume, and hydrodynamic interactions as the polyelectrolyte concentration is increased, we can expect the applicability of scaling arguments as in neutral solutions. Assuming that, in the asymptotic regime of very high entanglement density, reptation dynamics is valid, D is proportional to N^{-2}. The concentration dependence of D in this limit can be obtained by assuming that D is proportional to $R_g^{-1} f_e(c/c^*)$ and that the N-dependence of D should be consistent with the reptation law,

$$D \sim \begin{cases} \frac{1}{\sqrt{c}} \frac{1}{N^2}, & \kappa R_g \ll 1 \\ \frac{1}{c^{7/5}} \frac{1}{N^2}, & \kappa R_g \gg 1 \end{cases} \tag{317}$$

For concentrations between the Rouse and reptation regimes, D can depend more sensitively on N due to the entropic barrier mechanism [6, 64]. In this crossover region, D can be written as

$$D \sim N^{-\delta} \tag{318}$$

where δ can be 3 or even higher, depending on the concentration range and the system. Similarly, the electrophoretic mobility follows from the formula $\mu \sim QD$ as

$$\mu \sim N^{1-\delta} \tag{319}$$

so that $\mu \sim N^{-2}$ if entropic barriers dominate and $\mu \sim N^{-1}$ if reptation limit is approached. There has been substantial experimental support for the presence of entropic barrier mechanism for polymer dynamics [65, 66]. Recently, there have been several electrophysiology experiments [67–71], where translocation of single polyelectrolyte molecules through protein channels is investigated. Under the conditions of these experiments, the entropic barrier mechanism is the dominant polymer dynamics [72–76].

The concentration dependence of η at very high concentrations of polyelectrolytes can be obtained by assuming that $\eta - \eta_0$ is a function of c/c^* and that this function is consistent with $\eta \sim N^{3.4}$,

$$\eta \sim \begin{cases} c^{1.7} N^{3.4}, & \kappa R_g \ll 1 \\ c^{4.25} N^{3.4}, & \kappa R_g \gg 1 \end{cases} \tag{320}$$

V. CONCLUSIONS

The main results derived from the theory of polyelectrolyte dynamics are as follows.

A. Electrophoretic Mobility

In infinitely dilute solutions where full hydrodynamic interaction between segments is present as in the Kirkwood–Riseman–Zimm model, the electrophoretic mobility is independent of N in both low ($\kappa R_g \ll 1$) and high ($\kappa R_g \gg 1$) salt concentrations:

$$\mu \sim \begin{cases} N^0, & \kappa R_g \ll 1 \\ N^0 \kappa^{-2/3}, & \kappa R_g \gg 1 \end{cases} \tag{321}$$

But μ decreases with salt concentration with an apparent exponent of κ which changes from 0 at low salt concentration to $-\frac{2}{3}$ at high salt concentrations. The N-independence of μ arises from a cancellation between hydrodynamic interaction and electrostatic coupling between the polyelectrolyte and other ions in the solution. It is to be noted that the self-translational diffusion coefficient D is proportional to $\left(\frac{T}{\eta_0 R_g}\right)$ as in the Zimm model with full hydrodynamic interaction. The results of Eq. (321) are in agreement with experimental data [35]. In semidilute solutions, the hydrodynamic interaction is screened. But the electrostatic coupling between various ions are always present. When entanglement effects are unimportant, the solution is in the Rouse regime. Even in semidilute solutions, Eqs. (278)–(281) show that μ is independent of N:

$$\mu \sim \begin{cases} c^0 N^0, & \kappa R_g \ll 1 \\ \frac{N^0}{\sqrt{c}}, & \kappa R_g \gg 1 \end{cases} \tag{322}$$

In addition, we predict that μ is independent of polyelectrolyte concentration c at low salt concentrations and decreases with c as the salt concentration is increased. There is also an additional salt concentration dependence of μ as shown in Eq. (281). It is to be remarked that in the Rouse regime, the Hückel law of electrophoretic mobility is valid,

$$\mu = \frac{QD}{k_B T} \tag{323}$$

whereas in dilute solutions this law is significantly modified,

$$\mu = \frac{QD}{k_B T} \mathscr{M}(\kappa R_g) \tag{324}$$

where $\mathscr{M}(\kappa R_g)$ is defined in Eq. (276). The predictions of Eq. (323) need to be verified in order to identify the range of concentrations where Rouse regime exists.

B. Diffusion Coefficients

We have identified three diffusion coefficients. These are the self-translational diffusion coefficient D, cooperative diffusion coefficient D_c, and the coupled diffussion coefficient D_f. D_c is the cooperative diffusion coefficient in the absence of any electrostatic coupling between polyelectrolyte and other ions in the system. D_f is the cooperative diffusion coefficient accounting for the coupling between various ions. For neutral polymers, D_f and D_c are identical. Furthermore, we identify D_f as the fast diffusion coefficient as measured in dynamic light scattering experiments. The fourth diffusion coefficient is the slow diffusion coefficient D_s discussed in the Introduction. A satisfactory theory of D_s is not yet available.

1. *Self-Translational Diffusion Coefficient*

Equations (116) and (271) give D in dilute and semidilute limits. For low polyelectrolyte concentrations, we have

$$D \sim \begin{cases} \frac{1}{N}, & \kappa R_g \ll 1 \\ \frac{1}{N^{3/5}}, & \kappa R_g \gg 1 \end{cases} \tag{325}$$

In the Rouse regime we have derived

$$D \sim \begin{cases} \frac{c^0}{N}, & \kappa R_g \ll 1 \\ \frac{1}{N\sqrt{c}}, & \kappa R_g \gg 1 \end{cases} \tag{326}$$

The numerical prefactors for the relations in the above two equations are given in Eqs. (116) and (271).

2. *Cooperative Diffusion Coefficient*

In dilute solutions, $D_c = D$. In the Rouse regime, D_c depends on both static and hydrodynamic screening lengths ξ and ξ_H,

$$D_c = \frac{k_B T}{6\pi\eta_0\xi} f_d\left(\frac{\xi}{\xi_H}\right) \tag{327}$$

where f_d is given in Eqs. (283)–(287). ξ_H is given in Eqs. (234) and (235) and ξ was already reported in Ref. (48) and summarized in Table 1. ξ_H is proportional to ξ. As discussed in the previous section, D_c can be strongly k-dependent. The asymptotic results for D_c follow from Eqs. (283)–(287) as

$$D_c \sim \begin{cases} N^0 c^{1/2}, & \kappa R_g \ll 1 \\ N^0 c^{3/4}, & \kappa R_g \gg 1 \end{cases} \tag{328}$$

These results can readily be obtained by a scaling argument, $D_c \sim R_g^{-1} f_c(c/c^*)$, where we insist on the behavior of f_c on c/c^* so as to produce independence of D_c an N. The details of the prefactors are provided by formulas used in the derivation of Eq. (328).

3. *Coupled Diffusion Coefficient*

By accounting for the coupling between the dynamics of polyelectrolyte chains and their counterions and salt ions and assuming that small ions relax faster than polyelectrolyte chains, we have derived D_f to be

$$D_f = D_c + \frac{4\pi}{\epsilon\kappa^2}\frac{Qc\mu}{N} \tag{329}$$

In dilute polyelectrolyte solutions containing monovalent salt ions, D_f is given by

$$D_f \sim \begin{cases} \frac{T}{\eta_0}\left(1 + \frac{1}{N}\right), & \kappa R_g \ll 1 \\ \frac{T}{\eta_0}\frac{1}{\left(w + \frac{4\pi l_B}{\kappa^2}\right)^{1/5} N^{3/5}}\left[1 + \frac{z_p^2 Nc}{(z_c^2 c + 2c_s)}\frac{1.17}{(\kappa R_g)^{2/3}}\right], & \kappa R_g \gg 1 \end{cases} \tag{330}$$

Therefore we expect D_f, identified as the fast diffusion coefficient measured in dynamic light-scattering experiments, in infinitely dilute polyelectrolyte solutions to be very high at low salt concentrations and to decrease to self-diffusion coefficient $D(\kappa R_g \gg 1)$ as the salt concentration is increased. The above result for $\kappa R_g \ll 1$ limit is analogous to the Nernst–Hartley equation reported in Ref. 33. The theory described here accounts for structural correlations inside polyelectrolyte chains.

In the Rouse regime ($c > c^*$), the result is

$$D_f \sim \begin{cases} \frac{T}{\eta_0} N^0 c^0, & \kappa R_g \ll 1 \\ \frac{T}{\eta_0} \frac{z_p^2 \sqrt{c}}{(z_c^2 c + 2c_s)}, & \kappa R_g \gg 1 \end{cases} \tag{331}$$

Thus in salt-free semidilute solutions, the fast diffusion coefficient is expected to be independent of both N and c, although the polyelectrolyte concentration is higher than the overlap concentration. This remarkable result is in agreement with experimental data [31, 33, 34] discussed in the Introduction. Upon addition of salt, D_f decreases from this value as given by the above formulas.

C. Viscosity

In dilute solutions, the viscosity is given by

$$\frac{\eta - \eta_0}{\eta_0} = c[\eta][1 + k_H c[\eta] + \cdots] \tag{332}$$

where the intrinsic viscosity $[\eta]$ follows the Kirkwood–Riseman–Zimm formula as given by Eq. (216):

$$[\eta] \sim \begin{cases} l_B l^2 N^2, & \kappa R_g \ll 1 \\ \left(w + \frac{4\pi l_B}{\kappa^2}\right)^{3/5} l^{6/5} N^{4/5}, & \kappa R_g \gg 1 \end{cases} \tag{333}$$

The Huggins coefficient k_H is of order unity for neutral chains and for polyelectrolyte chains at high salt concentrations. In low salt concentrations, the value of k_H is expected to be an order of magnitude larger, due to the strong Coulomb repulsion between two polyelectrolyte chains, as seen in the case of colloidal solutions of charged spheres. While it is in principle possible to calculate the leading virial coefficients in Eq. (332) for different salt concentrations, the essential feature of the concentration dependence of η can be approximated by

$$\frac{\eta - \eta_0}{\eta_0} = c[\eta] \exp(k_H [\eta] c), \qquad c < c^* \tag{334}$$

in dilute solutions. Therefore, in dilute solutions where Zimm dynamics is valid, the specific viscosity increases roughly with concentration according to

$$\frac{\eta - \eta_0}{\eta_0} = 1.616\left(\frac{c}{c^*}\right) \exp\left[1.616 k_H \left(\frac{c}{c^*}\right)\right], \qquad c < c^* \tag{335}$$

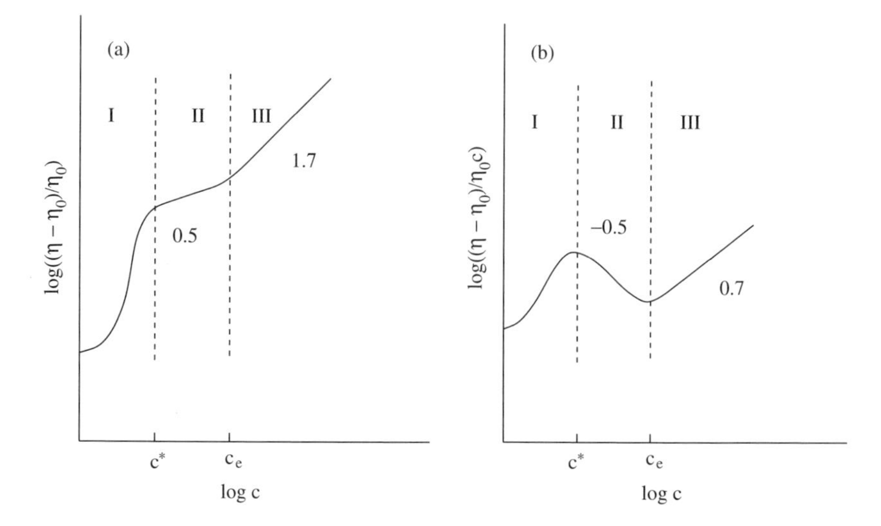

Figure 1. Sketch of double logarithmic plot of $(\eta - \eta_0)/\eta_0$ and $(\eta - \eta_0)/\eta_0 c$ versus c in (*a*) and (*b*), respectively, in salt-free solutions. I, II, and III represent Zimm, Rouse, and entanglement regimes, respectively. The expected slopes are also indicated.

where the relation of Eq. (297) is used. The increase of η with c in dilute solutions is steeper at low salt concentrations due to the higher value of k_H. This behavior is sketched as regime I in Figs. 1 and 2. The concentration dependencies of $(\eta - \eta_0)/\eta_0$ and $(\eta - \eta_0)/\eta_0 c$ are sketched in Figs. 1a and 1b, respectively,

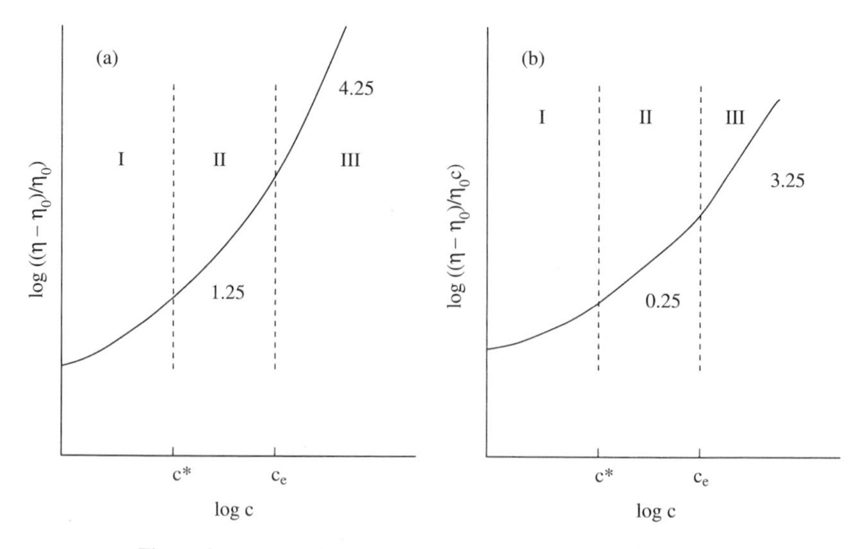

Figure 2. Same as in Fig. 1, except now the salt concentration is high.

for polyelectrolyte solutions in salt-free conditions. Figures 2a and 2b are similar sketches for high salt concentrations. In the Rouse regime where hydrodynamic interaction is screened, the viscosity is given by Eqs. (300) and (301) to be

$$\frac{\eta - \eta_0}{\eta_0} \sim \begin{cases} c^{1/2}N, & \kappa R_g \ll 1 \\ c^{5/4}\, N, & \kappa R_g \gg 1 \end{cases} \tag{336}$$

This behavior is sketched as regime II in Figs. 1 and 2. It is to be remarked that the viscosity change is monotonic with concentration in both salt-free and salty polyelectrolyte solutions. However, a plot of $(\eta - \eta_0)/\eta_0 c$ against c will show a peak in salt-free solutions, because $(\eta - \eta_0)/\eta_0$ is proportional to $\sqrt{c}$ in the Rouse regime. In such a plot, the peak will be absent at high salt concentrations as shown in Fig. 2, because now $(\eta - \eta_0)/\eta_0$ is proportional to $c^{5/4}$(with $\frac{5}{4}$ being greater than unity) in the Rouse regime. Thus we identify the empirical Fuoss law observed for salt-free polyelectrolyte solutions to be the same as the Rouse regime (with screened hydrodynamic interaction and no entanglements).

When entanglement effects dominate, scaling arguments suggest

$$\eta \sim \begin{cases} c^{1.7}N^{3.4}, & \kappa R_g \ll 1 \\ c^{4.25}N^{3.4}, & \kappa R_g \gg 1 \end{cases} \tag{337}$$

These expectations are indicated as regime III in Figs. 1 and 2 for polyelectrolyte concentrations greater than the entanglement concentration c_e.

D. Unresolved Issues

Although the theory of polyelectrolyte dynamics reviewed here provides approximate crossover formulas for the experimentally measured diffusion coefficients, electrophoretic mobility, and viscosity, the validity of the formulas remains to be established. In spite of the success of one unifying conceptual framework to provide valid asymptotic results, in qualitative agreement with experimental facts, it is desirable to establish quantitative validity. This requires (a) gathering of experimental data on well-characterized polyelectrolyte solutions and (b) obtaining the relationships between the various transport coefficients. Such data are not currently available, and experiments of this type are out of fashion. In addition to these experimental challenges, there are many theoretical issues that need further elaboration. A few of these are the following:

1. Although the coupling of counterion dynamics and polyelectrolyte dynamics has been accounted for at the mean field level, the relaxation of counterion cloud needs to be included in comparing with experimental data.

2. The crossover between the Kirkwood–Riseman–Zimm behavior and the Rouse behavior requires a better understanding, in terms of the contributing factors for the occurrence of a maximum in the plot of reduced viscosity against polyelectrolyte concentration at low salt concentrations. A firm understanding of the structure factor of polyelectrolyte solutions at concentrations comparable to the overlap concentration is necessary.
3. The role of semiflexibility of the polymer backbone on the dynamical properties is necessary to explore the dynamical properties of biologically relevant polyelectrolytes. Orientational correlations among molecules are expected to enhance the richness of the dynamical properties.
4. As discussed extensively in this chapter, most of the surprising properties of polyelectrolyte dynamics are due to the coupling of counterion dynamics with polymer dynamics. But, there is no adequate understanding of how much of the counterions are mobile and how much are effectively condensed on polymer chain backbone. Theoretical attempts [77, 78] on counterion condensation need to be extended to concentrated polyelectrolyte solutions.
5. The effective dielectric function of polyelectrolyte solutions remains as a mystery, demanding a better understanding of structure of solvent surrounding polyelectrolyte molecules.
6. The origins of experimentally observed slow mode need to be explored in terms of hydrophobic effects arising from the intrinsic chemical properties of polymer molecules.

Acknowledgments

Acknowledgment is made to the following: the NIH Grant No. 1RO1HG002776-01; the NSF Grant No. DMR-0209256; and the Materials Research Science and Engineering Center at the University of Massachusetts.

References

1. J. G. Kirkwood and J. Riseman, *J. Chem. Phys.* **16**, 565 (1948).
2. B. H. Zimm, *J. Chem. Phys.* **24**, 269 (1956).
3. M. Doi and S. F. Edwards, *The Theory of Polymer Dynamics*, Clarendon Press, Oxford, 1986.
4. H, Yamakawa, *Modern Theory of Polymer Solutions*, Harper and Row, New York, 1971.
5. P. G. de Gennes, *Scaling Concepts in Polymer Physics*, Cornell University Press, Ithaca, NY, 1979.
6. M. Muthukumar, *J. Non-Cryst. Solids* **131–133**, 654 (1991); also see references therein.
7. K. S. Schmitz, *Macroions in Solution and Colloid Suspension*, VCH Publishers, New York, 1993.
8. K. S. Schmitz, *An Introduction to Dynamic Light Scattering by Macromolecules*, Academic Press, San Diego, CA, 1990.

9. H. Dautzenberg, W. Jaeger, J. Kötz, B. Philipp, Ch. Seidel, and D. Stscherbina, *Polyelectrolytes*, Hanser Publishers, 1994.
10. S. Förster and M. Schmidt, *Adv. Polym. Sci.* **120**, 51 (1995).
11. R. M. Fuoss and U. P. Strauss, *J. Polym. Sci.* **3**, 246 (1948).
12. R. M. Fuoss and U. P. Strauss, *J. Polym. Sci.* **3**, 603 (1948); **4**, 96 (1949).
13. R. M. Fuoss, *Discuss. Faraday Soc.* **11**, 125 (1952).
14. H. Fujita and T. Homma, *J. Colloid Sci.* **9**, 591 (1954).
15. G. Eisenberg and J. Pouyet, *J. Polym. Sci.* **13**, 85 (1954).
16. J. A. V. Butler, A. B. Robins, and K. V. Shooter, *Proc. R. Soc. London* **A241**, 299 (1957).
17. H. Vink, *Makromol Chem.* **131**, 133 (1970).
18. H. Vink, *J. Chem. Faraday Trans.* **83**, 801 (1987).
19. J. Cohen, Z. Priel and Y. Rabin, *J. Chem. Phys.* **88**, 7111 (1988).
20. J. Cohen and Z. Priel, *Macromolecules* **22**, 2356 (1989).
21. J. Yamanaka, H. Aguchi, S. Saeki, and M. Tsuokawa, *Macromolecules* **24**, 3206, 6156 (1991).
22. D. F. Hodgson and E. J. Amis, *J. Chem. Phys.* **94**, 4581 (1991); **91**, 2635 (1989).
23. H. Vink, *Polymer* **33**, 3711 (1992).
24. A. M. Jamieson and C. T. Presley, *Macromolecules* **6** 358 (1973).
25. T. Raj and W. H. Flygare, *Biochemistry* **13**, 3336 (1974).
26. S. C. Lin, W. I. Lee, and J. M. Schurr, *Biopolymers* **17**, 1041 (1978).
27. K. S. Schmitz, M. Lu, N. Singh, and D. J. Ramsay, *Bioploymers* **23**, 1637 (1984).
28. M. Drifford and J. P. Dalbiez, *J. Phys. Lett.* **46**, L-311 (1985).
29. M. Drifford and J. P. Dalbiez, *Biopolymers* **24**, 1501 (1985).
30. M. Schmidt, *Makromol. Chem. Rapid Commun.* **10**, 89 (1989).
31. S. Förster, M. Schmidt, and M. Antonietti, *Polymer* **31**, 781 (1990).
32. R. M. Peitzsch, M. J. Burt, and W. F. Reed, *Bioploymers* **30**, 1101 (1990).
33. M. Sedlak and E. J. Amis, *J. Chem. Phys.* **96**, 817, 826 (1992).
34. M. Sedlak, *J. Chem. Phys.* **105**, 10123 (1996); also see references therein to his earlier work.
35. D. Hoagland, D. L. Smisek, and D. Y. Chen, *Electrophoresis* **17**, 1151 (1996).
36. P. G. de Gennes, P. Pincus, R. M. Velasco, and F. Brochard, *J. Phys. France* **37**, 1461 (1976).
37. T. A. Witten and P. Pincus, *Europhys. Lett.* **3**, 315 (1987).
38. R. Borsali, T. A. Vilgis, and M. Benmouna, *Macromolecules* **25**, 5513 (1992).
39. R. Borsali and M. Rinaudo, *Makromol. Chem. Theory Simul.* **2**, 179 (1993).
40. T. Odijk, *Macromolecules* **12**, 688 (1979).
41. L. Wang and V. A. Bloomfield, *Macromolecules* **22**, 2742 (1989).
42. W. I. Lee and J. M. Schurr, *J. Polym. Sci.* **13**, 873 (1975).
43. P. Tivant, P. Turq, M. Drifford, H. Magdalenat, and R. Menez, *Biopolymers* **22**, 2762 (1983).
44. M. Benmouna, T. A. Vilgis, and F. Hakem, *Macromolecules* **25**, 1144 (1992).
45. M. Muthukumar, *J. Chem. Phys.* **86**, 7230 (1987).
46. M. Muthukumar, in *Molecular Basis of Polymer Networks, A. Baumgärtner and C. E. Picot, eds., Springer Proceedings in Physics*, Vol. 42, Springer, New York, 1989, pp. 28–34.
47. A. V. Dobrynim, R. H. Colby, and M. Rubinstein, *Macromolecules* **28**, 1859 (1995).
48. M. Muthukumar, *J. Chem. Phys.* **105**, 5183 (1996).

49. M. Muthukumar, *Electrophoresis* **17**, 1167 (1996).
50. M. Muthukumar, *J. Chem. Phys.* **107**, 2619 (1997).
51. W. B. Russel, D. A. Saville, and W. R. Schowater, *Colloidal Dispersion, Cambridge University Press, New York, 1989.*
52. K. Kaji, H. Urakawa, T. Kanaya, and R. Kitamaru, *J. Phys. (France)* **49**, 993 (1988).
53. K. Nishida, K. Kaji, and T. Kanaya, *J. Chem. Phys.* **114**, 8671 (2001).
54. V. M. Prabhu, M. Muthukumar, G. D. Wignall, and Y. B. Melnichenko, *J. Chem. Phys.* **119**, 4085 (2003).
55. S. F. Edwards and K. F. Freed, *J. Chem. Phys.* **61**, 1189 (1974).
56. K. F. Freed and S. F. Edwards, *J. Chem. Phys.* **61**, 3626 (1974).
57. M. Muthukumar, *J. Phys. A: Math. Gen.* **14**, 2129 (1981).
58. M. Muthukumar and S. F. Edwards, *Polymer* **23**, 345 (1982).
59. S. F. Edwards and M. Muthukumar, *Macromolecules* **17**, 586 (1984).
60. M. Muthukumar, *Polymer* **42**, 5921 (2001).
61. J. Ferry, *Viscoelastic Properties of Polymers* Wiley, New York, 1980.
62. Y. Takahashi, H. Hase, M. Yamaguchi, and I. Noda, *J. Non-Cryst. Solids* **911**, 172 (1994).
63. I. Noda and Y. Takahashi, *Ber. Bunsenges Phys. Chem.* **100**, 696 (1996).
64. A. Baumgärtner and M. Muthukumar, *Adv. Chem. Phys.* **XCIV**, 625 (1996).
65. C. R. Calladine, C. M. Collis, H. R. Drew, and M. R. Mott, *J. Mol. Biol.* **221**, 981 (1991).
66. N. A. Rotstein and T. P. Lodge, *Macromolecules* **25**, 1316 (1992).
67. J. J. Kasianowicz, E. Brandin, D. Branton, and D. W. Deamer, *Proc. Natl. Acad. Sci. USA* **93**, 13770 (1996).
68. S. E. Henrickson, M. Misakian, B. Robertson, and J. J. Kasianowicz, *Phys. Rev. Lett.* **85**, 3057 (2000).
69. A. Meller, L. Nivon, E. Brandin, J. Golovchenko, and D. Branton, *Proc. Natl. Acad. Sci. USA* **97**, 1079–1084 (2000).
70. S. Howorka, L. Movileanu, X. Lu, M. Magnon, S. Cheley, O. Braha, and H. Bayley, *J. Am. Chem. Soc.* **122**, 2411 (2000).
71. J. J. Kasianowicz, S. E. Henrickson, H. H. Weetall, and B. Robertson, *Anal. Chem.* **73**, 2268 (2001).
72. W. Sung and P. J. Park, *Phys. Rev. Lett.* **77**, 783 (1996).
73. M. Muthukumar, *J. Chem. Phys.* **111**, 10371 (1999).
74. M. Muthukumar, *Phys. Rev. Lett.* **86**, 3188 (2001).
75. M. Muthukumar, *J. Chem. Phys.* **118**, 5174 (2003).
76. C. Y. Kong and M. Muthukumar, *J. Chem. Phys.* **120**, 3460 (2004).
77. G. S. Manning, *Q. Rev. Biophys.* **11**, 179 (1978).
78. M. Muthukumar, *J. Chem. Phys.* **120**, 9343 (2004).

HYDRODYNAMICS AND SLIP AT THE LIQUID–SOLID INTERFACE

JONATHAN S. ELLIS

Institute of Biomaterials and Biomedical Engineering
University of Toronto, Toronto, Canada

MICHAEL THOMPSON

Department of Chemistry and the Institute of Biomaterials and Biomedical Engineering, University of Toronto
Toronto, Canada

CONTENTS

I. INTRODUCTION

Recent times have seen much discussion of the choice of hydrodynamic boundary conditions that can be employed in a description of the solid–liquid interface. For some time, the no-slip approximation was deemed acceptable and has constituted something of a dogma in many fields concerned with fluid mechanics. This assumption is based on observations made at a macroscopic level, where the mean free path of the liquid being considered is much smaller

Advances in Chemical Physics, Volume 131, edited by Stuart A. Rice

than the length scale under consideration. However, in configurations where this length condition does not hold, the no-slip approximation may be invalid.

The contemporary focus on boundary conditions is, in part, connected to areas of application in which slip-based phenomena could be important, such as polymer extrusion, flow in pores and tubules, nanomaterials science, microfluidics, and biophysics. There is particular interest in the latter two categories on our part, due to efforts to combine them as strategies for chemical detection in drug discovery and clinical diagnostic technology. Certainly, when considering microfluidic devices, where length scales and channel dimensions can range from hundreds of microns to hundreds of nanometers, the mean free path of the liquid is comparable to the size of typical channels. In such cases, slippage and coupling effects may become prevalent and, accordingly, it is necessary to assign alternate boundary conditions. Biomolecule-detection technology involves flow of an analyte-containing solution over macromolecular receptors such as proteins and oligonucleotides immobilized onto sensor surfaces. With respect to biochemical species under these conditions, many factors come into play. For example, the material properties of biochemical moieties are highly dynamical and nonlinearly dependent on the particular surface chemistry. Also, these dynamics can be highly stochastic, such that the properties of the overall system will exhibit significant spatial and temporal variability. As will be seen subsequently, it is generally the case that the particular structure of a material surface is expected to have a profound influence on coupling.

The nature of the liquid in contact with a surface is also very important, with respect to boundary conditions. Although slip has long been observed for highly non-Newtonian, viscoelastic liquids such as polymer flows and extrusions, many recent studies have reported slippage of Newtonian liquids under a variety of experimental conditions. This clearly indicates that care must be taken when modeling any type of micro- or nanofluidic system, no matter which liquid is employed.

As we shall see in the present chapter, it is generally the situation that a number of specific surface properties govern boundary conditions, and these have been considered extensively in the literature. These include, but are not limited to, wettability and substrate–liquid affinity [1–15], hydrodynamic shear rate [16–20], surface topology and roughness [21–29], and surface material properties [30–36].

We begin in Section II with a review of the fundamental concepts of hydrodynamics and boundary conditions. In Section III, we present some common descriptions of coupling, followed in Section IV by a discussion of viscoelastic adsorbate films and the so-called inner slip. In Section V, we consider with the concept of stochastic boundary conditions, which we believe will be an important topic in situations where random fluctuations are strong. Finally, in Section VI, we present our concluding ideas and discuss some areas for future study.

II. FUNDAMENTALS OF HYDRODYNAMICS AND BOUNDARY CONDITIONS

When considering flow of a liquid in contact with a solid surface, a basic understanding of the hydrodynamic behavior at the interface is required. This begins with the Navier–Stokes equation for constant-viscosity, incompressible fluid flow, such that $\partial\rho/\partial t = 0$,

$$\rho_L \frac{D\mathbf{v}}{Dt} = -\nabla p + \eta_L \nabla^2 \mathbf{v} + \mathbf{F} \tag{1}$$

Here, $\mathbf{v}$ is the velocity vector field, ρ_L is the mass density of the fluid, $D/Dt = \partial/\partial t + \mathbf{v} \cdot \nabla$ is the material derivative, ∇p is the gradient of the pressure, η_L is the shear viscosity, and $\mathbf{F}$ is the external force acting on the fluid volume. The right-hand side of Eq. (1) is a momentum balance between the internal pressure and viscous stress and the external forces on the fluid body. Any excess momentum contributes to the material acceleration of thc fluid volume, on the left-hand side.

When considering a solid–liquid interface, we begin with the simplest case of steady shear flow parallel to the surface, with $D\mathbf{v}/Dt = 0$ and $\mathbf{v} = v_x\mathbf{x}$. Equation (1) reduces to Newtonian viscous flow,

$$\sigma_x = \eta_L \frac{\partial v_x}{\partial z} = \eta_L \dot{\gamma} \tag{2}$$

where σ_x is the shear stress in the x plane and the shear rate $\dot{\gamma} = \partial v_x/\partial z$ is the slope of the velocity gradient normal to the wall. At the wall, the tangential shear stress becomes

$$\sigma_w = \eta_L \dot{\gamma}_{z=0} \tag{3}$$

where the wall boundary is at $z = 0$.

Two common types of one-dimensional flow regimes examined in interfacial studies: Poiseuille and Couette flow [37]. Poiseuille flow is a pressure-driven process commonly used to model flow through pipes. It involves the flow of an incompressible fluid between two infinite stationary plates, where the pressure gradient, $\partial p/\partial x$, is constant. At steady state, ignoring gravitational effects, we have

$$\eta \frac{\partial^2 u_x}{\partial y^2} - \frac{\partial p}{\partial x} = 0 \tag{4}$$

This gives a solution of the form

$$u_x(y) = -\frac{1}{2\eta}\frac{\partial p}{\partial x}y^2 + A_1 y + A_2 \tag{5}$$

where the A_i are integration constants, which can be determined by specifying boundary conditions at an interface. In fact, a multilayered system can be examined by solving the governing differential equations (equations of motion, electromagnetic equations, etc.) that define the system, then solving the linear system defined by the boundary conditions. It is evident from the form of Eq. (5) that the velocity profile for Poiseuille flow is parabolic.

Couette flow is shear-driven flow, as opposed to pressure-driven. In this instance, two parallel plates, separated by a distances h, are sheared relative to one another. The motion induces shear in the interstitial fluid, generating a linear velocity profile that depends on the motion of the moving surface. If we assume a linear shear rate, the shear stress is given simply by

$$\sigma = \eta\dot{\gamma} = \eta\frac{\partial v_x}{\partial y} \approx \eta\frac{V}{h} \tag{6}$$

where V is the velocity of the upper plate and h is the separation between plates. Couette flow is commonly used to model boundary lubrication, where one surface is sliding relative to the other, separated by a lubricant.

When considering boundary conditions, a useful dimensionless hydrodynamic number is the Knudsen number, $Kn = \lambda/L$, the ratio of the mean free path length to the characteristic dimension of the flow. In the case of a small Knudsen number, continuum mechanics will apply, and the no-slip boundary condition assumption is valid. In this formulation of classical fluid dynamics, the fluid velocity vanishes at the wall, so fluid particles directly adjacent to the wall are stationary, with respect to the wall. This also ensures that there is a continuity of stress across the boundary (i.e., the stress at the lower surface—the wall—is equal to the stress in the surface-adjacent liquid). Although this is an approximation, it is valid in many cases, and greatly simplifies the solution of the equations of motion. Additionally, it eliminates the need to include an extra parameter, which must be determined on a theoretical or experimental basis.

However, in the case of large Kn, the no-slip approximation cannot be applied. This implies that the mean free path of the liquid is on the same length scale as the dimension of the system itself. In such a case, stress and displacement are discontinuous at the interface, so an additional parameter is required to characterize the boundary condition. A simple technique to model this is the one-dimensional slip length, which is the extrapolation length into the wall required to recover the no-slip condition, as shown in Fig. 1. If we consider

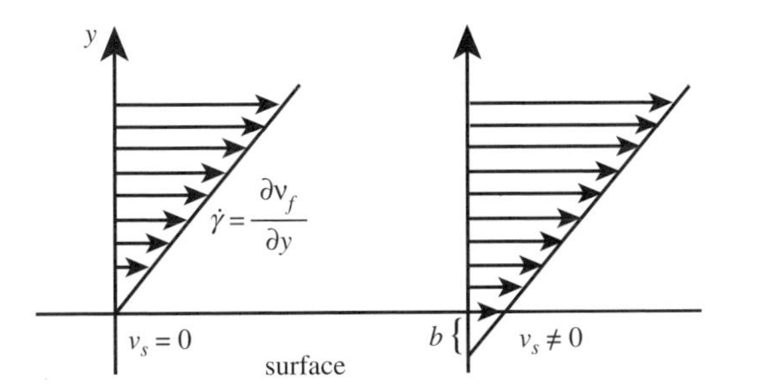

Figure 1. No-slip condition and slip condition with slip length, for one-dimensional shear flow. The slip length b is the extrapolation distance into the solid, to obtain the no-slip point. The slope of the linear velocity profile near the wall is the shear rate $\dot{\gamma}$.

frictional Stokes flow, in which the shear stress on a planar surface caused by a fluid flowing across it is proportional to the velocity of the adjacent flow, we obtain

$$\sigma_w = kv_s \tag{7}$$

where v_s is the slip velocity and k is the friction coefficient, in units of viscosity per unit length. Friction is also known as the interfacial viscosity, which is written as η_2, because it represents interfacial, two-dimensional drag. Equation (7) describes the friction or drag stress on a solid object in a flow field. Equating Eqs. (3) and (7), we find

$$\eta_L \left(\frac{\partial v_x}{\partial z} \right)_{z=0} = kv_s \tag{8}$$

To include the slip in this model, we use the slip length. In this approach, if slip occurs, the slip velocity v_s is proportional to the velocity gradient at the wall $\partial v_x / \partial z$:

$$v_s = b \left(\frac{\partial v_x}{\partial z} \right)_{z=0} \tag{9}$$

where the proportionality constant is the slip length b. Combining Eqs. (8) and (9), we obtain a relation for the slip length as the ratio of the bulk and interfacial viscosities:

$$b = \frac{\eta_L}{k} \tag{10}$$

This ratio implies that slip will increase as the bulk viscosity increases relative to the interfacial viscosity, or as the liquid self-cohesiveness increases and the liquid-surface affinity decreases.

III. DESCRIPTIONS OF SLIP

The dependence of slip on the interaction strength of the surface and liquid was studied early on by Tolstoi [3], later revisited by Blake [12], a model that is linked to interfacial viscosity. Tolstoi modeled the surface using Frenkel's model for the bulk mobility of a liquid molecule [38],

$$u = \frac{r^2}{6k_B T \tau_0} \exp(-W/k_B T) \tag{11}$$

where r is the mean center-to-center distance of adjacent molecules, τ_0 is the relaxation time for molecular displacements, and W is the molecular displacement activation energy. W is the energy required to form a cavity that an adjacent molecule will occupy. Frenkel suggested that this is equal to the product of the surface area of the cavity and the liquid–vapor surface tension, $W = A\gamma_{\mathrm{LV}}$, where A (the molecular surface area) can be approximated from the molecular radius, $A = 4\pi\sigma^2$, for a spherical molecule. Considering the layer of liquid molecules directly adjacent to the surface, we have

$$u_W = \frac{r^2}{6k_B T \tau_W} \exp\left(-\frac{W_W}{k_B T}\right) \tag{12}$$

where the subscript W indicates the molecular properties of the surface-adjacent liquid molecules. Tolstoi made the approximation that the relaxation time constants of the bulk and surface liquids are equal, $\tau_0 = \tau_W$. Using this approximation, the ratio of the surface to bulk mobilities, the fact that $W - W_W = \alpha_0 A(\gamma_{\mathrm{SL}} + \gamma_{\mathrm{LV}} - \gamma_{\mathrm{SV}}) = \alpha_0 A \gamma_{\mathrm{LV}}(1 - \cos\theta)$, where α_0 is the fraction of the cavity within the liquid, and a geometrical argument, we arrive at a simple quantitative expression for the slip length:

$$b = r \exp[\alpha_0 A \gamma_{LV}(1 - \cos\theta)/k_B T] \tag{13}$$

There are significant limitations to the above theory, many of which are discussed by Tolstoi [3] and Blake [12]. The foremost of these is the difficulty in estimating the microcavity fraction α_0. This was noted by Tolstoi, who simply assumed $\alpha_0 \approx 1/6$ and 1/9 to correspond to experimental data. In doing so, α_0 becomes another fitting parameter. Another limitation is that Frenkel's original approach is schematic, and therefore empirical. More mathematical rigor, in terms of a diffusion treatment, would yield more realistic results. In addition, Tolstoi's initial approximation, $\tau_0 = \tau_W$, is approximate and potentially invalid. Due to molecular ordering of the surface, the relaxation times of surface versus bulk velocities may be significantly different.

Another notion regarding the interfacial viscosity is as a measure of the motion of an interfacial layer. Consider an interfacial layer, a monolayer bound to the surface with interfacial density ρ_2 a mass per unit area, moving at a constant velocity. If the surface stops suddenly, the monolayer velocity will decay exponentially with a relaxation time τ. Through a quantum mechanical derivation involving the sliptime, Krim and Widom [39] determined a relationship between the interfacial viscosity, the interfacial density, and the sliptime, of the form

$$k = \eta_2 = \frac{\rho_2}{\tau_s} \tag{14}$$

τ_s is the slip relaxation time, or sliptime, which is the relaxation time for the monolayer slip velocity as it decays. For a rigidly bound monolayer, the velocity will decay very rapidly, at a rate comparable to the relaxation of the surface itself and τ_s will be near zero, yielding large interfacial friction and no slip. However, if the surface-monolayer bonds are highly dissipative, the time constant will be large and appreciable slip will occur. Rearranging Eq. (10) and equating to Eq. (14), we arrive at

$$b = \frac{\eta_L \tau_s}{\rho_2} \tag{15}$$

This result is interesting, since it gives the slip length b as a function of parameters that can be measured experimentally or *a priori*, for simple systems in a linear approximation. The bulk shear viscosity can be approximated from the literature, and the monolayer density can be determined from optical techniques. To a first approximation, for rigidly adsorbed layers, the sliptime is related to the autocorrelation function of random momentum fluctuations ΔP_x in the film, given by [40]

$$\tau_s = \int_0^{\infty} dt G(t) \cos(\omega_0 t)$$

where the autocorrelation function is

$$G(t - s) = \frac{1}{k_B T M} \langle \Delta P_x(t) \Delta P_x(s) \rangle \tag{16}$$

which is normalized to $G(0) = 1$. Momentum fluctuations determined from computer simulations can be used to predict the slip length for model systems involving adsorbed films, and then they can be correlated to experimental results if slip occurs.

A. Two-Parameter Model of Slip

Slip is not always a purely dissipative process, and some energy can be stored at the solid–liquid interface. In the case that storage and dissipation at the interface are independent processes, a two-parameter slip model can be used. This can occur for a surface oscillating in the shear direction. Such a situation involves bulk-mode acoustic wave devices operating in liquid, which is where our interest in hydrodynamic coupling effects stems from. This type of sensor, an example of which is the transverse-shear mode acoustic wave device, the oft-quoted quartz crystal microbalance (QCM), measures changes in acoustic properties, such as resonant frequency and dissipation, in response to perturbations at the surface–liquid interface of the device.

The shear-mode acoustic wave sensor, when operated in liquids, measures mass accumulation in the form of a resonant frequency shift, and it measures viscous perturbations as shifts in both frequency and dissipation. The limits of device operation are purely rigid (elastic) or purely viscous interfaces. The addition of a purely rigid layer at the solid–liquid interface will result a frequency shift with no dissipation. The addition of a purely viscous layer will result in frequency and dissipation shifts, in opposite directions, where both of these shifts will be proportional to the square root of the liquid density–viscosity product $\sqrt{\rho_L \eta_L}$.

Consider an experiment where both the frequency and dissipation change, and independent investigation reveals no mass accumulation or changes in surface or bulk material properties. If the frequency and dissipation shifts are both proportional to the density–viscosity product, then the shifts are related to a change in interfacial slip, and a single-valued slip length [Eq. (9)] can be used to describe the change in coupling. This situation characterizes purely viscous slip, where the only physical manifestation of slip is to partially decouple the liquid flow from the surface motion.

However, if the frequency and dissipation shifts are not proportional to the density–viscosity product, and there is again no mass accumulation and no change in surface or bulk properties, the change in coupling is no longer purely viscous and cannot be explained with a single-valued slip length. In this situation, a two-valued coupling parameter or boundary condition is required to account for (a) the magnitude of the boundary condition and (b) the relative contributions from storage and dissipation processes. This can be done using a complex slip parameter, where the complex magnitude gives degree of coupling, and the phase shift represents out-of-phase motion of the interface.

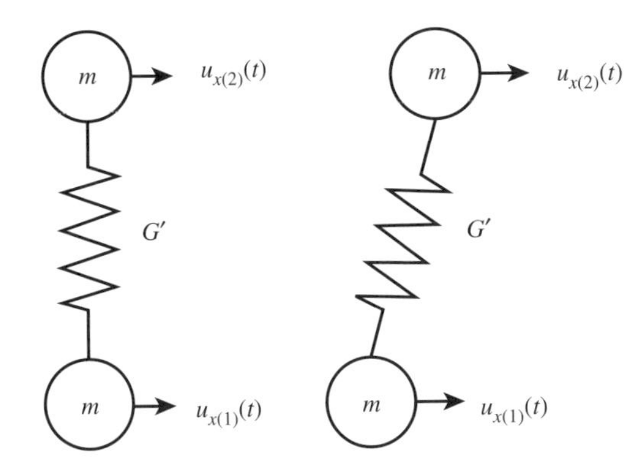

Figure 2. Mass–spring model for slip motion. The upper and lower masses are allowed to move horizontally in one dimension, based on the spring dynamics. The slip parameter is the ratio of the displacements $\alpha = u_{x(2)}/u_{x(1)}$.

This type of boundary condition has been explored extensively in the acoustics literature. Ferrante et al. [41] modeled the interface between a surface oscillating in the shear direction and a semi-infinite liquid as a spring connecting two masses (Fig. 2). The no-slip boundary condition for the displacements at the interface is replaced by a complex-valued ratio of the upper and lower displacements,

$$u_{L,x} = \alpha u_{W,x} \tag{17}$$

where the subscripts *L,x* and *W,x* represent the liquid and surface displacements tangential to the plane of motion. The slip parameter α is complex, given as

$$\alpha = \mathrm{Re}(\alpha) + i\mathrm{Im}(\alpha) \tag{18}$$

with magnitude $|\alpha| = ([\mathrm{Re}(\alpha)]^2 + [\mathrm{Im}(\alpha)]^2)^{1/2}$ and phase shift $\angle\alpha = \tan^{-1}(\mathrm{Re}(\alpha)/\mathrm{Im}(\alpha))$. In this configuration, it is assumed that the surface is driving the liquid. When $\alpha = 1 + i0$, this is the no-slip condition. If the magnitude is less than unity, then there is some degree of slip, and any imaginary component indicates some viscoelastic coupling effects. At a planar interface with a purely viscous fluid, it has been shown [42] that the complex slip parameter reduces to the slip length *b*. By comparing the boundary conditions for the two models [Eqs. (9) and (17)], we obtain an expression relating the two models:

$$\alpha = \exp\left(i\sqrt{2i}\,\frac{b}{\delta} \right) \tag{19}$$

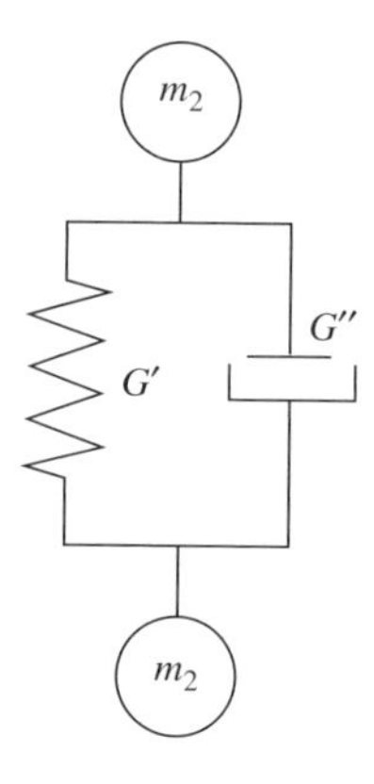

Figure 3. Mass–spring–damper model of slip, from Lu et al. [7].

where $\delta = (2\eta/\omega\rho)^{1/2}$ is the decay length of the damped shear wave into the bulk fluid. Although this does present a method for relating the two models, the slip length model involves the assumption that structure and ordering does not impart any viscoelastic properties to the interface and that the interface itself is planar.

Lu et al. [7] extended the mass–spring model of the interface to include a dashpot, modeling the interface as viscoelastic, as shown in Fig. 3. The continuous boundary conditions for displacement and shear stress were replaced by the equations of motion of contacting molecules. The interaction forces between the contacting molecules are modeled as a viscoelastic fluid, which results in a complex shear modulus for the interface, $G^* = G' + i\omega G''$, where G' is the storage modulus and G'' is the loss modulus. G^* is a continuum molecular interaction between liquid and surface particles, representing the force between particles for a unit shear displacement. The authors also determined a relationship for the slip parameter Eq. (18) in terms of bulk and molecular parameters [7, 43]:

$$\alpha = \frac{G^*/\bar{r}}{G^*/\bar{r} + iG_L k_L} \tag{20}$$

where $\bar{r}$ is the molecular separation between the surface and liquid particles at the wall, $G_L = (c_L + i\omega\eta_L)$ allows for viscoelastic behavior in the liquid by setting $c_L \neq 0$, and $k_L = \omega[\rho_L/(c_L + i\omega\eta_L)]^{1/2}$ is the acoustic wavenumber for the liquid. By setting $G^*/\bar{r}$ large compared to the other terms in the denominator, the no-slip condition is obtained. If the other terms in the denominator become significant, however, slip will become apparent. Although this model can be used to estimate deviations from the no-slip boundary condition, it approximates molecular interaction forces, such as G^* and $\bar{r}$, as bulk continuum parameters.

The authors noted that when their friction parameter $\mathscr{R} = (\rho G_L)^{1/2}\delta/G^*$ is real, it is equivalent to the real slip parameter $s = k^{-1}$ used by McHale et al. [14]. From this analysis, a real interfacial energy G^*/δ is related to the slip length b, for a purely viscous fluid, by

$$\frac{G^*}{\delta} = \frac{\alpha}{1-\alpha} iG_L k_L = iG_L k_L \frac{\exp(\sqrt{-2i}\, b/\delta)}{1 - \exp(\sqrt{-2i}\, b/\delta)} \tag{21}$$

However, in this formulation, G^*/δ is simply a fitted parameter, since there is no rigorous molecular definition for this form of the interfacial energy. In addition, this model suffers the same problem as do other continuum models, in that it contains parameters that must be determined from experimental data and that cannot be estimated *a priori* from molecular techniques.

There are a number of factors that contribute to coupling changes, such as liquid–solid interaction and wetting, shear rate, surface material properties, and surface roughness. In a recent review [44], we discussed many of these factors and outlined a number of techniques used to measure coupling. To summarize the results, a decrease in fluid–wall interactions will normally reduce the coupling at the interface and could lead to slip. Vinogradova [5, 45-47] has performed extensive work on the slippage of liquids on a hydrophobic surface, including deviations from Reynolds theory of drag in the drainage of a liquid squeezed between a surface and sphere during surface force apparatus (SFA) measurements. In this case, two spheres or cylinders, separated by a liquid, are brought into contact at a constant approach velocity dh/dt. The Reynolds drag force for two approaching cylinders is

$$F_h = -\frac{2(R_1 R_2)^{3/2}}{R_1 + R_2} \frac{6\pi\eta}{h} \frac{dh}{dt} f^* \tag{22}$$

where R_1 and R_2 are the cylinder radii, h is the separation, and f^* is the slippage factor for the configuration. In the case of no-slip, $f^* = 1$ and the classical drag force is recovered. However, in the case of slip, there is a finite slip length b and the slippage factor is

$$f^* = 2\frac{h}{6b}\left[\left(1 + \frac{h}{6b}\right)\ln\left(1 + \frac{6b}{h}\right) - 1\right] \tag{23}$$

Typically, the authors found slip lengths over a very wide range, from a few nanometers to a few centimeters, but this could be due to different slip lengths on each cylinder.

Molecular dynamics has been used extensively to explore the solid–liquid interface. In one such study, a modified Lennard-Jones potential has been used to model this interaction in the spreading of a droplet [4], of the form

$$U(r_{ij}) = 4\varepsilon\left[\left(\frac{\sigma}{r_{ij}}\right)^{12} - c_{\mathrm{AB}}\left(\frac{\sigma}{r_{ij}}\right)^{6}\right] \tag{24}$$

c_{AB} is the cohesiveness between molecules (c_{LL} is the liquid–liquid interaction and c_{LS} is the liquid–surface interaction), and r_{ij} is the separation between molecules. This potential has since been used by others [48]. Slip lengths in the nano- and micrometer range have been observed experimentally for partially and nonwetting interfaces at high shear rates [10, 13, 17, 19, 20, 22], and some studies have even reported slip for wetting surfaces [15]. Surface roughness, which we will consider in more depth below, could either induce slip or lead to the no-slip condition, depending on the nature of the roughness and the liquid–solid interaction. From molecular dynamics studies [6], it has been shown that increasing the wall stiffness can reduce coupling between the wall and liquid. However, this depends also on the strength of the liquid–wall interaction. For a strong interaction, a rigid wall reduces slip, whereas a compliant wall may increase slip, especially at high shear rates. However, for low interactions, wall stiffness can lead to increased slip.

A wide range of results has been reported in the literature for the effect of roughness on slip [21, 23, 29]. Roughness may either cause slip or lead to no-slip, depending on the surface–liquid interaction and magnitude of roughness. Again, for a consideration of some of the previous results in this area, refer to Ref. [44]. Recently, Ponomarev and Meyerovich [27] performed a rigorous derivation for roughness-driven slip, which actually turns out to be stick, where a hydrodynamic boundary condition with random surface asperities can be described by a smooth, flat surface with a stick length [a negative slip length in Eq. (9)] and an increased effective viscosity within the stick layer. They performed the analysis for an oscillating rough surface in contact with a viscous liquid, ignoring any coupling related to surface wetting or liquid–surface adhesion properties. Their derivation is valid for situations when the viscous wave penetration depth into the liquid is larger than the size of the random surface asperities, $\sqrt{2}R/\delta \ll 1$, where R is the mean radius of curvature of the surface asperities and $\delta = \sqrt{2\eta/\omega\rho}$ is the viscous wave penetration depth. The stick length can be expressed in terms of the surface inhomogeneity correlation function

$$\zeta(x) = \langle \xi(x_1)\xi(x_1 + x)\rangle \tag{25}$$

and its Fourier image

$$2\pi\zeta(k_x)\delta(k_x + k'_x) = \langle\xi(k_x)\xi(k'_x)\rangle \tag{26}$$

where $\delta(\ldots)$ is the delta function and $\xi(x)$ is an unknown dimensionless random function with zero mean, giving the surface fluctuations in the vertical direction. At low frequencies, their expression for the stick length is

$$b_{eff} = -2\frac{h^2}{\pi R}\int_0^\infty dk_x\,\zeta(k_x)k_x \tag{27}$$

where h and R are the mean height and radius of the asperities, respectively. The change in interfacial viscosity is given by

$$\eta_{eff}(y) = \eta[1 + \beta\delta(y)] \tag{28}$$

where $\delta(y)$ is the Dirac delta function and, for small Λ,

$$\beta = 2\left[\frac{b_{eff}}{R} + \frac{\Lambda}{\sqrt{2}}\left(\frac{h}{R}\right)^2\right] \tag{29}$$

is the viscosity renormalization parameter. Integration of this extra term in the viscosity for the energy dissipation equation of a smooth surface leads to a correction term that recovers the dissipation result for a rough surface. This analysis can be used to determine the characteristic size of the surface asperities directly from frequency and dissipation measurements experiments using acoustic wave devices.

Similar results were obtained by McHale and Newton [24], who treated a liquid layer trapped in surface inhomogeneities on an oscillating surface as a rigid mass layer. They found that a smooth, flat surface in contact with a liquid, with a stick length b and a surface location at half the mean roughness height, is equivalent to a no-slip condition on the same surface, along with an additional "mass" layer of rigid water of thickness $|b|$, which is half the height of the mean surface roughness. This analysis is equivalent to the rigorous theory mentioned above [27], but the interfacial layer is treated as a rigid mass instead of as having an effective viscosity. However, in both of these studies, no wettability or surface–liquid interaction effects were examined, which will have an effect in most cases and would normally result in a positive slip length. Since most surfaces are both rough and involve surface–liquid coupling effects, both of these factors will be important.

Recently, McHale et al. [25, 26] discussed the mechanisms involved in wetting behavior on rough surfaces. The authors focused particularly on superhydrophobic surfaces, defined as surfaces having a contact angle in water $\theta > 150^\circ$. On flat surfaces, the contact angle is a result of interfacial surface tensions, determined by the surface chemistry. A balance of forces at the triple line (the line where the solid, liquid, and vapor phases meet) is given by Young's equation,

$$\cos\theta_e = \frac{\gamma_{SV} - \gamma_{SL}}{\gamma_{LV}} \tag{30}$$

where θ_e is the equilibrium contact angle and γ_{IJ} are the interfacial surface tensions at interface *IJ*. The surface chemistry of the interface determines the interaction forces, and hence the flat-surface contact angle; the surface topography determines how these forces are distributed, as well as the effect this has on the observed contact angle. There are two types of topography-enhanced wetting. The first is Wenzel-type wetting, wherein the liquid is able to penetrate into the surface inhomogeneities and adhere to the contours of the surface. The observed rough contact angle is given by

$$\cos\theta_e^W = r\cos\theta_e \tag{31}$$

where θ_e is the equilibrium contact angle on a flat surface of the same materials and r is an empirical roughness factor, $r > 1$, such that the roughness always serves to amplify the wetting properties. Nonwetting ($\theta_e > 90^\circ$) surface chemistry will become more nonwetting, even superhydrophobic, whereas partially wetting surfaces ($\theta_e < 90^\circ$) could become completely wetting surfaces. This is caused by an increase in the total surface area in contact with the liquid. If the surface is partially wetting, the excess liquid in contact with the surface causes a change in the contact angle to balance Young's equation, Eq. (30), since the surface tension remains unchanged [25]. If the surface is very hydrophobic, the increase in surface area will cause the liquid to be repelled by the surface even more, resulting in a superhydrophobic-range contact angle. Any change in the surface roughness, which may be due to a shift in surface stiffness, or a conformational shift, in the case of a biological surface, would result in a change or amplification of the wetting properties. This can be derived from the Wenzel equation, Eq. (31), as

$$\Delta\theta_e^W = r\left[\frac{\sin\theta_e}{\sin\theta_e^W}\right]\Delta\theta_e \tag{32}$$

where $\Delta\theta_e^W$ is the change in observed contact angle, following a change in surface chemistry resulting in chemically induced contact angle shift $\Delta\theta_e$. As can be seen, this form of topography-induced superhyrdrophobicity [26] or superwetting [25] can amplify the surface chemistry signal. Furthermore, it is characterized by a large amount of contact angle hysteresis, wherein the advancing and receding contact angles are different.

The other type of superhydrophobicity is the Cassie–Baxter form, wherein the liquid does not fully penetrate the surface topography, but instead will sit atop surface asperities and air pockets. For nonwetting surfaces, this is a more likely scenario, since it may be more energetically favorable for the liquid to be in contact with air than a repelling surface. This type of wetting can be described by

$$\cos\theta_e^C = \varphi_s \cos\theta_e - (1 - \varphi_s) \tag{33}$$

where θ_e^W is the measured contact angle, φ_s is the fraction of the liquid in contact with the surface asperities, and $1 - \varphi_s$ is the fraction in contact with air. Due to the liquid forming a hydroplane on the tops of the solid, this type of superhydrophobicity results in slippery droplets, where the drop will roll off a tilted surface at low tilt angles. This is contrasted with the Wenzel form discussed above, which results in sticky drops. Also, there is no contact angle amplification in the case of Cassie–Baxter superhydrophobicity, since $\varphi_s < 1$, and there is little hysteresis between the advancing and receding angles.

Such a situation could occur in many instances and may prove to have strong analytic potential. Changes in surface chemistry, material properties, and surface topography could be detected by transitions between hydrophobic and super hydrophobic configurations, which can be measured by optical or hydrodynamic techniques. Superhydrophobic surfaces can lead to a monolayer cushion of a dissolved gas between the surface and the liquid, such that the liquid is not in contact with the surface [1]. This creates a low-density, low-viscosity layer above the surface, resulting theoretically in very large slip lengths in the micrometer range. While this type of situation is idealized, it may partially explain cases of large apparent slip effects, due to low-viscosity interfacial regions.

B. Interfacial Viscosity

When a liquid is in contact with the surface, adhesive forces from the surface will cause the liquid molecules adjacent to the surface to adopt an ordered structure and, depending on the structure and nature of the surface, could lead to nonlinear (elastic) or anisotropic properties in the liquid. This will, however, depend on the nature and wettability of the surface. A hydrophilic surface will induce strengthened hydrogen-bonding for up to 10 layers [49], potentially producing a layered or gradient viscosity for as much as 3 nm into the liquid. A hydrophobic surface will repel the adjacent liquid, resulting in reduced or no ordering of the liquid near the surface and an interfacial viscosity similar to the bulk viscosity. Some liquids, such as nitrobenzene, will form crystalline layers at the surface. This has, however, not been observed for water, for which properties will gradually approach those of the bulk as the distance from the solid–liquid interface increases [50].

To model this, Duncan-Hewitt and Thompson [50] developed a four-layer model for a transverse-shear mode acoustic wave sensor with one face immersed in a liquid, comprised of a solid substrate (quartz/electrode) layer, an ordered surface-adjacent layer, a thin transition layer, and the bulk liquid layer. The ordered surface-adjacent layer was assumed to be more structured than the bulk, with a greater density and viscosity. For the transition layer, based on an expansion of the analysis of Tolstoi [3] and then Blake [12], the authors developed a model based on the nucleation of vacancies in the layer caused by shear stress in the liquid. The aim of this work was to explore the concept of graded surface and liquid properties, as well as their effect on observable boundary conditions. They calculated the first-order rate of deformation, as the product of the rate constant of densities and the concentration of vacancies in the liquid,

$$R_{\mathrm{def}} = k[\ ]_{\mathrm{vac}} = \sigma \frac{V_{\mathrm{mol}}}{h} \exp([-\Delta G^{*} - \Delta H_{\mathrm{vac}}]k_B T) \qquad (34)$$

Here, σ is the shear stress in the transition liquid layer, V_{mol} is a molecular volume, h is Planck's constant, ΔG^{*} is the free energy change of the movement of a molecule into a vacancy, and ΔH_{vac} is the enthalpy of formation of a vacancy. The rate of deformation of a liquid is the strain rate, $\dot{\gamma}$ [see Eq. (2)], so the right-hand side of Eq. (34) can be used to estimate the viscosity of the transition layer.

The presence of a low-viscosity interfacial layer makes the determination of the boundary condition even more difficult because the location of a slip plane becomes blurred. Transitional layers have been discussed in the previous section, but this is an approximate picture, since it still requires the definition of boundary conditions between the interfacial layers. A more accurate picture, at least from a mesoscopic standpoint, would include a continuous gradient of material properties, in the form of a viscoelastic transition from the solid surface to the purely viscous liquid. Due to limitations of time and space, models of transitional gradient layers will be left for a future article.

We now turn our attention to the interstitial viscoelastic film between the solid and liquid, and we discuss its effect on coupling and boundary conditions.

IV. EFFECT OF VISCOELASTIC FILM ON COUPLING, AND INNER SLIP

Up until this point, we have been considering only surface–liquid interactions where the solid is purely rigid and is perfectly secured to some lower surface upon which it rests. This is usually a good approximation; however, in some circumstances, the ability of the surface to deform or shift could influence energy transfer modes within the system and may even lead to coupling changes at the surface–liquid interface. In this discussion, we now consider the surface as an adsorbate layer, supported by a solid layer beneath and in contact with a liquid above. This situation can be applied to a variety of areas, such as flow in coated pipes, fluid in confined geometries with polymeric wall, and substrates coated with biomaterials, such as proteins or nucleic acids, in a liquid environment. Of particular interest are conformational shifts and geometric changes caused by a biochemical event, potentially leading to a corresponding change in interfacial coupling. In addition, coupling effects within the film or at the adsorbate–substrate interface may influence the system response, so we consider multi-layered devices.

Energy can be transmitted between layers viscously or elastically. As discussed above [Eq. (7)], viscous energy is frictional energy, dissipated as heat into the contacting medium, and is proportional to the velocity of the interfacial particles. Because it is velocity-dependent, dissipated energy normally has a time constant associated with it, which is reflected in the sliptime, Eq. (14). In the case of an interfacial polymer adsorbate, this could become manifest through the random entanglement of particles or strands within the layer. Energy transmitted elastically is stored and returned to the material. An elastic deformation, in the case of a viscoelastic material adhered to a substrate, will occur instantaneously. In the case of an acoustic device or sensor, a standing wave generated in the substrate is transferred through interfacial bonds. Elastic energy will be stored and reflected at the interface, maintaining the standing wave, whereas energy dissipated viscously is lost either in the adsorbate layer as heat, or in the overlying liquid, which cannot support standing oscillations.

Using the no-slip assumption at the substrate–adsorbate interface, we assume that energy is transmitted perfectly elastically, without any viscous loss. This implies that the bonds at the interface act perfectly rigid and transmit energy completely, as though they were perfect covalent bonds in the classic model of self-assembled monolayers. However, this is not always the case, as shown, for instance, by Legget and co-workers [51–53] in studies of the photoinduced oxidation of SAM on gold. The thiol bonds are not perfectly rigid, and they may in fact be quite labile. At any given instant, some thiol bonds may be attached to the surface, while others may be away from an energy minimum and not chemisorbed to the surface. The importance of these thiol "jumps" would become apparent if the time constant of jumps is proportional to the time constant of the process to be measured. Such SAMs are indeed stable on gold; however, the stability may indeed rest in interfilm bonding, as opposed to the stability of the thiol bonds themselves. Any unbound thiols would not transmit energy elastically, since they would have no direct energetic contact with the surface, so any energy associated with these molecules or binding sites would be purely viscous, friction-type transmission. This situation, with slip at both interfaces, was modeled recently, and interesting dissipation patterns were observed [54]. In addition, there is a stochastic factor to this type of coupling, since the fraction of thiols bound at any given time will depend of the molecular and environmental properties of the system. Similar stochastic arguments can be made for the adsorption of biological molecules on surfaces, in terms of both coupling and topography.

Shanahan and Carre [31–36, 55, 56] have done extensive theoretical work on the coating of viscoelastic surfaces and the effect of soft surfaces on hydrodynamic forces. Again, we have considered this area in a recent review [44]. This area is important in how energy is transferred or lost at the interface. Coupling changes at an inner interface can result in either an increase or decease in the energy dissipated. This has been discussed and observed for a number of acoustic systems [40, 41, 54, 57, 58].

Inner slip has been discussed in a recent review by us [44]. In many cases, the adsorbate is rigidly or covalently bound to the substrate, so energy can be transmitted completely through the interface. One common such instance is surface-adhered self-assembled monolayers (SAMs), whereby an alkyl-chain is anchored to a substrate through some surface chemistry, such as a silane or thiol bond. The gold–thiol bond, common in SAM chemistry, is generally assumed to be a rigid interfacial structure. However, the recent work by Leggett and co-workers referred to above has shown that the thiol–gold bonds are in fact not entirely stable and that they can be photo-oxidized in vacuum. The reported

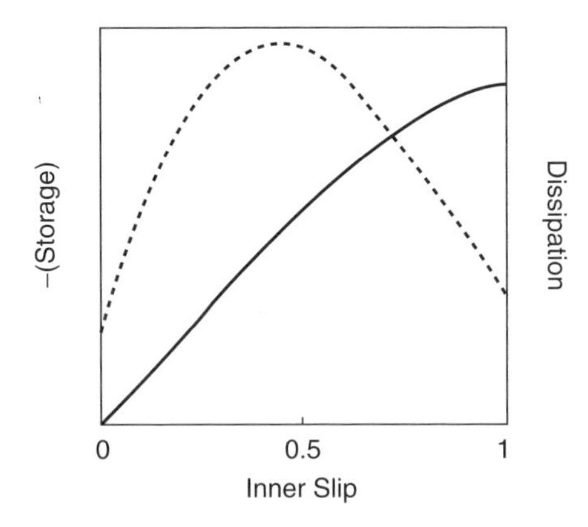

Figure 4. Theoretical trends for –(storage) and dissipation as the inner slip is varied between no slip (0) and strong slip (1) for a coated transverse shear acoustic wave device in water. The thickness of the film is 5 nm. The solid line displays the decrease in storage, and the dashed line shows the change in dissipation.

stability of these monolayers may be due to interchain bonding within the monolayer itself. Because of this, the thiol bonds can easily bend, rotate, or translate in response to stress, especially on a moving surface. As a result, there could be storage and dissipation effects within the thiol bonds themselves, leading to inner coupling effects as described above. We calculated the effect of inner slip on the storage and dissipation parameters and found, from a model for a shear mode acoustic wave device, that an increase in inner slip corresponds to the expected decrease in energy storage, but an unexpected nonmonotonic curve for dissipation, where we first see an increase, then a decrease (Fig. 4). This behavior was attributed to bending and torsion of the interfacial bond [44, 54].

The lability of the thiol bonds, and the energy dissipation processes resulting from their motion, will have very short time constants, and would likely be stochastic in nature. Even at steady-state, bond fluctuations may have an effect. For this reason, we now turn our attention to stochastic coupling.

V. STOCHASTIC COUPLING

In most problems involving boundary conditions, the boundary is assigned a specific empirical or deterministic behavior, such as the no-slip case or an empirically determined slip value. The condition is defined based on an averaged value that assumes a mean flow profile. This is convenient and simple for a macroscopic system, where random fluctuations in the interfacial properties are small enough so as to produce little noise in the system. However, random fluctuations in the interfacial conditions of microscopic systems may not be so simple to average out, due to the size of the fluctuations with respect to the size of the signal itself. To address this problem, we consider the use of stochastic boundary conditions that account for random fluctuations and focus on the statistical variability of the system. Also, this may allow for better predictions of interfacial properties and boundary conditions.

A stochastic boundary condition has been used for a convection–diffusion problem involving chemically interacting species on a flat wall [59, 60] The authors examined zero-concentration (absorbing) and zero-flux (reflecting) boundary conditions; but it is important note that this work does not describe interfacial boundary conditions for slip versus no-slip situations, but instead describes boundary conditions for the particle flux at the wall. However, the method described below may be useful for describing stochastic interfacial coupling. The approach involves the use of Green's functions to define particle distribution weighting functions. Given a linear differential equation of the form $Lu(x) = f(x)$, where L is a differential operator, $u(x)$ is an unknown function, and $f(x)$ is a known nonhomogeneous term, a solution of this equation would take the form $u(x) = L^{-1}f(x) = \int d\xi G(x;\xi)f(\xi)$, where L^{-1} is the inverse of L and an integral operator, and $G(x;\xi)$ is the Green's function of the differential operator L. For a more in-depth examination of Green's functions and solutions of boundary value problems, refer to any text in advanced calculus (for example, see Ref. 61). The discussion begins with the mirror image of the infinite-space propagator [59], which is the Green's function for a reflecting wall at $y = 0$,

$$G(y, y', \Delta t) = \frac{e^{-(y-y')^2/4D\Delta t}}{\sqrt{4\pi D\Delta t}} \tag{35}$$

For the half-space above the wall ($y > 0$), the Green's function for a reflecting wall, in the absence of flow, becomes

$$G_r(y, y', \Delta t) = G(y, y', \Delta t) + G(y, -y', \Delta t) \tag{36}$$

A particle distribution weighting function $p(\tilde{y} \mid y, y', \Delta t)$ represents the mean time a particle resides at a position $\tilde{y}$ during its move from y to y' in a time step Δt, and it is given by

$$p(\tilde{y} \mid y, y', \Delta t) = \frac{\int_0^{\Delta t} dt G(y, \tilde{y}, \Delta t - t) G(y, \tilde{y}, t)}{G(y, \tilde{y}, \Delta t)\Delta t} \tag{37}$$

Assuming that the flow above the wall is linear shear flow, the velocity field can be modeled, from Eq. (6), as $v_x = \dot{\gamma} y$, where the velocity increases linearly away from the wall, with slope $\dot{\gamma}$. Near the wall, a reflection of the weighting function, similar to the reflection of the Green's function [Eq. (36)], must be applied,

$$p_r(\tilde{y} \mid y, y', \Delta t) = p(\tilde{y} \mid y, y', \Delta t) + p(-\tilde{y} \mid y, y', \Delta t) \tag{38}$$

which, again, applies only in the region $\tilde{y} > 0$. The average convective velocity is then found by integrating the shear flow field v_x at each $\tilde{y}$, giving

$$\bar{v}_x = \dot{\gamma} \int_0^{\infty} d\tilde{y} [\tilde{y} p(\tilde{y} \mid y, y', \Delta t)] \tag{39}$$

The authors applied this model to the situation of dissolving and deposited interfaces, involving chemically interacting species, and included rate kinetics to model mass transfer as a result of chemical reactions [60]. The use of a stochastic weighting function, based on solutions of differential equations for particle motion, may be a useful method to model stochastic processes at solid–liquid interfaces, especially where chemical interactions between the surface and the liquid are involved.

An early model for slip of fluids in tubes is due to Maxwell [62], wherein the velocity distribution function parallel to the wall is a linear combination of the distribution functions for the fluid particles reflected and adsorbed at the wall,

$$f_m(v_{\|}) = \alpha f(v_{\|}) + (1 - \alpha) f(v_{\|} - \bar{u}_{\rm in}) \tag{40}$$

Here, $\bar{u}_{in}$ is the mean streaming velocity of particles approaching the wall and $(1 - \alpha)$ is the fraction of fluid particles reflected at the wall, so the first term represents the distribution of particles adsorbed. The velocity distribution functions, $f(v)$, are assumed to be Maxwellian,

$$f(v) = \sqrt{\frac{m}{2\pi k_B T}} \exp\left(-\frac{mv^2}{2k_B T}\right) \tag{41}$$

From this, the velocities of particles flowing near the wall can be characterized. However, the absorption parameter α must be determined empirically. Sokhan et al. [48, 63] used this model in nonequilibrium molecular dynamics simulations to describe boundary conditions for fluid flow in carbon nanopores and nanotubes under Poiseuille flow. The authors found slip length of 3 nm for the nanopores [48] and 4–8 nm for the nanotubes [63]. However, in the first case, a single factor [4] was used to model fluid–solid interactions, whereas in the second, a many-body potential was used, which, while it may be more accurate, is significantly more computationally intensive.

Stochastic friction has been used to explain dissipation processes in noncontact AFM studies [64], which involves a nano-tip approaching a solid surface, such that atomic forces opposing the approach can be measured. Dissipation in the approach is attributed to nonequilibrium stochastic processes, due surface atoms opposing the tip that can be modeled as a friction force. Due to high rates of approach and the constant motion of the tip, the surface atoms do not come to complete equilibrium. The lack of equilibrium results in a dissipative force proportional to the velocity of the tip, with the friction coefficient proportional to the integral of the time correlation function of the tip-surface microscopic force. This is analogous to motion of a Brownian particle in solution, where friction is caused by stochastic collisions with the liquid particles. In a particular treatment [64], the authors presented a method for calculating the friction coefficient from a general expression for the friction force, which is related to Green's function of the lattice dynamics. By increasing the interaction strength between the tip and the surface, the energy dissipated increased.

Although this treatment does not explicitly involve interactions at a solid–liquid interface, the application of Green's function to find the stochastic friction force may be an excellent opportunity for modeling interfacial friction and coupling, in the presence of liquid. An interesting note made by the authors is that the stochastic friction mechanism is proportional to the square of the frequency. This will likely be the case for interfacial friction as well.

VI. CONCLUSIONS

We have examined the many of the various factors that determine the proper boundary condition to use at the solid–liquid interface and considered many of the models associated with theses factors. The single-valued slip length model is the simplest and most convenient boundary condition, and it has been used successfully in many studies. However, it cannot describe coupling changes where there are changes in both the storage and dissipation properties. In this situation, a two-parameter complex value may be necessary.

Wettability, hydrodynamic shear rate, surface roughness, and the interface material properties all play a role in determination of the boundary condition. Slip will be more likely in nonwetting systems, but has been observed, or at least inferred, in a number of systems showing complete wetting. Most surfaces can only sustain the no-slip boundary condition up to a critical shear stress, after which the assumption fails and slip can occur. For some surfaces, however, this shear rate is quite high. Surface roughness can have different effects on coupling. Depending on the surface–liquid interaction strength and the degree of roughness, increasing the roughness could lead to either increased slip or increased stick. The material properties of the adsorbate or interface also greatly influence slip, and they will likely have important implications when considering biological or biochemical surfaces.

Inner slip, between the solid wall and an adsorbed film, will also influence the surface–liquid boundary conditions and have important effects on stress propagation from the liquid to the solid substrate. Linked to this concept, especially on a biomolecular level, is the concept of stochastic coupling. At the molecular level, small fluctuations about the ensemble average could affect the interfacial dynamics and lead to large shifts in the detectable boundary condition. One of our main interests in this area is to study the relaxation time of interfacial bonds using slip models. Stochastic boundary conditions could also prove to be all but necessary in modeling the behavior and interactions of biomolecules at surfaces, especially with the proliferation of microfluidic chemical devices and the importance of studying small scales.

References

1. P. G. de Gennes, *Langmuir* **18**, 3413–3414 (2002).
2. J. N. Israelachvili, *Intermolecular and Surface Forces*, 2nd ed., Academic Press, London, 1991.
3. D. M. Tolstoi, *Dokl. Akad. Nauk. SSSR* **85**, 1329 (1952).
4. J. L. Barrat and L. Bocquet, *Phys. Rev. Lett.* **82**, 4671–4674 (1999).
5. O. I. Vinogradova, *Int. J. Miner. Process.* **56**, 31–60 (1999).
6. A. Jabbarzadeh, J. D. Atkinson, and R. I. Tanner, *J. Chem. Phys.* **110**, 2612–2620 (1999).
7. F. Lu, H. P. Lee, and S. P. Lim, *Smart Mater. Struct.* **12**, 881–888 (2003).

8. R. G. Horn, O. I. Vinogradova, M. E. Mackay, and N. Phan-Thien, *J. Chem. Phys.* **112**, 6424–6433 (2000).
9. C. Cottin-Bizonne, S. Jurine, J. Baudry, J. Crassous, F. Restagno, and E. Charlaix, *Eur. Phys. J. E* **9**, 47–53 (2002).
10. E. Bonaccurso, M. Kappl, and H. J. Butt, *Phys. Rev. Lett.* **88**, 076103 (2002).
11. L. Bocquet and J. L. Barrat, *Phys. Rev. Lett.* **70**, 2726–2729 (1993).
12. T. D. Blake, *Colloids Surf.* **47**, 135–145 (1990).
13. J. Baudry, E. Charlaix, A. Tonck, and D. Mazuyer, *Langmuir* **17**, 5232–5236 (2001).
14. G. McHale, R. Lucklum, M. I. Newton, and J. A. Cowen, *J. Appl. Phys.* **88**, 7304–7312 (2000).
15. R. Pit, H. Hervet, and L. Leger, *Phys. Rev. Lett.* **85**, 980–983 (2000).
16. F. Brochard and P. G. de Gennes, *Langmuir* **8**, 3033–3037 (1992).
17. V. S. J. Craig, C. Neto, and D. R. M. Williams, *Phys. Rev. Lett.* **8705**, 054504-1–4 (2001).
18. L. Leger, H. Hervet, G. Massey, and E. Durliat, *J. Phys.-Condes. Matter* **9**, 7719–7740 (1997).
19. C. Neto, V. S. J. Craig, and D. R. M. Williams, *Eur. Phys. J. E* **12**, S71–S74 (2003).
20. Y. X. Zhu and S. Granick, *Phys. Rev. Lett.* **8709**, 096105-1–4 (2001).
21. E. Bonaccurso, H. J. Butt, and V. S. J. Craig, *Phys. Rev. Lett.* **90**, 144501-1–4 (2003).
22. C. Cottin-Bizonne, J. L. Barrat, L. Bocquet, and E. Charlaix, *Nature Mater.* **2**, 237–240 (2003).
23. A. Jabbarzadeh, J. D. Atkinson, and R. I. Tanner, *Phys. Rev. E* **61**, 690–699 (2000).
24. G. McHale and M. I. Newton, *J. Appl. Phys.* **95**, 373–380 (2004).
25. G. McHale, N. J. Shirtcliffe, S. Aqil, C. C. Perry, and M. I. Newton, *Phys. Rev. Lett.* **93**, 036102-1–4 (2004).
26. G. McHale, N. J. Shirtcliffe, and M. I. Newton, *Analyst* **129**, 284–287 (2004).
27. I. V. Ponomarev and A. E. Meyerovich, *Phys. Rev. E* **67**, 026302-1–12 (2003).
28. M. Urbakh and L. Daikhin *Abstr. Pap. Am. Chem. Soc.* **212**, 82-COLL (1996).
29. Y. X. Zhu and S. Granick, *Phys. Rev. Lett.* **88**, 106102-1–4 (2002).
30. F. Baldoni, *Z. Angew. Math. Mech.* **79**, 193–203 (1999).
31. A. Carre, J. C. Gastel, and M. E. R. Shanahan, *Nature* **379**, 432–434 (1996).
32. A. Carre and M. E. R. Shanahan, *Langmuir* **11**, 24–26 (1995).
33. A. Carre and M. E. R. Shanahan, *Abstr. Pap. Am. Chem. Soc.* **212**, 282-POLY (1996).
34. M. E. R. Shanahan and A. Carre, *J. Adhes.* **57**, 179–189 (1996).
35. M. E. R. Shanahan and A. Carre, *C. R. Acad. Sci. Ser. IV—Phys. Astrophys.* **1**, 263–268 (2000).
36. M. E. R. Shanahan and A. Carre, *Langmuir* **11**, 1396–1402 (1995).
37. G. Karniadakis and A. Beskok, *Micro Flows: Fundamentals and Simulation*, Springer, New York, 2002.
38. I. A. Frenkel, *Kinetic Theory of Liquids*. Clarendon Press, Oxford, 1946.
39. J. Krim and A. Widom, *Phys. Rev. B* **38**, 12184–12189 (1988).
40. A. Widom and J. Krim, *Phys. Rev. B* **34**, 1403–1404 (1986).
41. F. Ferrante, A. L. Kipling, and M. Thompson, *J. Appl. Phys.* **76**, 3448–3462 (1994).
42. J. S. Ellis and G. L. Hayward, *J. Appl. Phys.* **94**, 7856–7867 (2003).
43. F. Lu, H. P. Lee, and S. P. Lim, *J. Phys. D* **37**, 898–906 (2004).
44. J. S. Ellis and M. Thompson, *Phys. Chem. Chem. Phys.* **6**, DOI:10.1039/b409342a (2004).
45. O. I. Vinogradova, *J. Phys.-Condes. Matter* **8**, 9491–9495 (1996).

46. O. I. Vinogradova, *Langmuir* **14**, 2827–2837 (1998).
47. O. I. Vinogradova, *Abstr. Pap. Am. Chem. Soc.* **215**, 093-COLL (1998).
48. V. P. Sokhan, D. Nicholson, and N. Quirke, *J. Chem. Phys.* **115**, 3878–3887 (2001).
49. M. L. Gee, P. M. McGuiggan, J. N. Israelachvili, and A. M. Homola, *J. Chem. Phys.* **93**, 1895–1906 (1990).
50. W. C. Duncan-Hewitt and M. Thompson, *Anal. Chem.* **64**, 94–105 (1992).
51. E. Cooper and G. J. Leggett, *Langmuir* **14**, 4795–4801 (1998).
52. E. Cooper and G. J. Leggett, *Langmuir* **15**, 1024–1032 (1999).
53. D. A. Hutt and G. J. Leggett, *J. Mater. Chem.* **9**, 923–928 (1999).
54. J. S. Ellis and M. Thompson, *Chem. Comm.* 1310–1311 (2004).
55. A. Carre and M. E. R. Shanahan, *J. Colloid Interface Sci.* **191**, 141–145 (1997).
56. M. E. R. Shanahan, A. Carre, S. Moll, and J. Schultz, *J. Chim. Phys.-Chim. Biol.* **84**, 199–200 (1987).
57. E. D. Smith, M. O. Robbins, and M. Cieplak, *Phys. Rev. B* **54**, 8252–8260 (1996).
58. M. Cieplak, E. D. Smith, and M. O. Robbins, *Science* **265**, 1209–1212 (1994).
59. P. Szymczak and A. J. C. Ladd, *Phys. Rev. E* **68**, 036704 (2003).
60. P. Szymczak and A. J. C. Ladd, *Phys. Rev. E* **69**, 036704 (2004).
61. F. B. Hildebrand, *Advanced Calculus for Applications*, 2nd ed., Prentice-Hall, Englewood Cliffs, NJ, 1976.
62. J. C. Maxwell, *Philos. Trans. R. Soc.* (1879). Reprinted in *The Scientific Letters and Papers of James Clerk Maxwell*, vol. 3, P. M. Harman, ed., Cambridge University Press, Cambridge, 2002, p. 802.
63. V. P. Sokhan, D. Nicholson, and N. Quirke, *J. Chem. Phys.* **117**, 8531–8539 (2002).
64. L. N. Kantorovich, *Phys. Rev. B* **64**, 245409-1–13 (2001).

STRUCTURE OF IONIC LIQUIDS AND IONIC LIQUID COMPOUNDS: ARE IONIC LIQUIDS GENUINE LIQUIDS IN THE CONVENTIONAL SENSE?

HIRO-O HAMAGUCHI and RYOSUKE OZAWA

Department of Chemistry, School of Science, The University of Tokyo, Tokyo, Japan

CONTENTS

I. INTRODUCTION

The term *ionic liquid* (IL) refers to a class of liquids that are composed solely of ions. It is a synonym of molten salt. Although molten salt implicitly means a high-temperature liquid that is prepared by melting a crystalline salt, IL includes a new class of ionic compounds that are liquids at the ambient temperature [1]. Thus, IL in a narrow sense often stands for room-temperature ionic liquid (RIL). In the present chapter, IL is used in a broader sense and, if necessary, RIL is used to clarify that it is liquid at the ambient temperature. The history of ILs has already been reviewed [2].

Advances in Chemical Physics, Volume 131, edited by Stuart A. Rice

In contrast to ordinary molecular liquids, in which the dipolar and/or higher-order multipolar electrostatic interactions are dominating, the Coulomb interaction plays a major role in ILs. It is well known that the long-range nature of the Coulomb interaction makes the melting points of ionic crystals much higher than those of molecular crystals. In that sense, RILs are somewhat extraordinary; the melting point of a typical RIL, 1-butyl-3-methyl-imidazolium iodide, is −72°C [3], while a typical ionic crystal NaI melts only at 651°C. Here, we have the most fundamental and profound question about RIls. Why are RIls liquids at the ambient temperature, despite the fact that they are composed solely of ions? A few important questions then follow. Are there any liquid structures that are characteristic of the dominating Coulomb interactions in RILs? Will there be any novel properties that originate from those specific structures of RILs? In order to answer these questions, we need to elucidate the structure and dynamics of RILs and related IL compounds.

In spite of these basic interests of high importance, the structural studies of Ils are scant and are still in a developing stage. Most of the IL studies so far are directed toward their possible applications [4–6]. The high-temperature stability, nonvolatility, nonflammablility, amphiphilicity, and many other characteristics of RILs as solvents are potentially useful as a new class of "Green" solvents. The wide potential window (large difference in the oxidation and reduction potentials) of ILs promises applications as new electrochemical materials. New types of catalysts based on Ils have also been explored [7]. Under these circumstances, it is not possible to provide a perspective overview on the structure and dynamics of Ils in general. We therefore focus here on the structural studies on Ils having the imidazolium-based cations—in particular, the Raman spectroscopic and x-ray studies carried out in the past few years. Thus, the present chapter is more like an advances report rather than a comprehensive review. The authors hope that it would nevertheless be helpful to the many researchers who are coming into the wonder world of ionic liquids.

The 1-butyl-3-methylimidazolium cation, $bmim^+$ (Fig. 1), makes a number of ILs with varying properties, when combined with different anions [4]. BmimCl and bmimBr are crystals at room temperature, while bmimI is a RIL. By cooling down molten bmimCl and bmimBr below the melting points, their

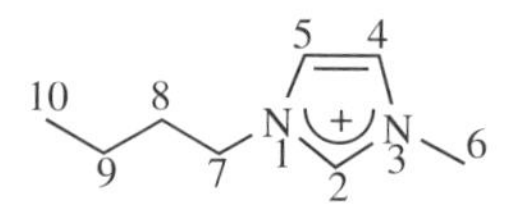

Figure 1. The 1-butyl-3-methylimidazolium cation, $bmim^+$.

supercooled liquids are easily obtained. Those halogen salts thus comprise a unique system for studying the structure of the bmim$^+$ cation in the crystalline and liquid states at room temperature. X-ray diffraction can determine the structures in crystals, while Raman spectroscopy facilitates comparative studies of the structures in crystals and liquids. The two salts, bmim[BF_4] and bmim[PF_6], are prototype RILs that are most extensively used in basic IL investigations as well as in practical applications. Therefore, the elucidation of the crystal and liquid structures of the bmim$^+$-based Ils will be an important first step for the understanding of ILs in general.

In the following, we first discuss the crystal polymorphism of bmimCl and the crystal structures of bmimCl and bmimBr. The results of single crystal x-ray analysis are given to show that the crystal polymorphism is due to the *trans–gauche* rotational isomerism of the butyl group of the bmim$^+$ cation. They also show that the cations and the halogen anions in bmimCl and bmimBr crystals separately form characteristic column structures extending along the crystal *a* axis. Then, comparison of the Raman spectra and the normal coordinate analysis lead to a conclusion that at least two rotational isomers, one having a *trans* conformation and the other having a *gauche* conformation with regard to the C_7–C_8 bond of the butyl group of the bmim$^+$ cation, coexist in the ionic liquid state. A few pieces of experimental evidence are then given which are indicative of some local structures existing in ILs. The unusual long equilibration time between the *trans* and *gauche* conformers upon melting of a small single crystal of the *trans* polymorph of bmimCl indicates that the rotational isomers do not interconvert with each other at the molecular level. The two rotational isomers are most likely to be incorporated in their specific local structures, and they can interconvert with each other only through the conversion of the local structure as a whole. The apparent enthalpy differences between the *trans* and *gauche* conformers in a group of 1-alkyl-3-methylimidazolium tetrafluoroborate ILs are much smaller than the corresponding enthalpy difference between the *trans* and *gauche* conformers of the free alkyl chain. This finding also indicates that the 1-alkyl-3-methylimidazolium cation form local structures specific to each rotational isomers. Coexistence of these local structures incorporating different rotational isomers may well hinder crystallization and hence lower the melting points of the 1-alkyl-3-methylimidazolium based ILs. The wide-angle x-ray scattering results on bmimI IL show prominent peaks in the residual radial distribution curve, indicating certain periodical arrangements of the iodide anions. These local structures, if they exist, distinguish ILs from the conventional molecular liquids. It is argued that ILs may form an entirely new material phase that exists in between liquid and crystal. These local structures may also lead to unique properties of ionic liquids. For example, if magnetic anions are aligned in ILs, novel magnetic liquids will be created.

II. CRYSTAL POLYMORPHISM OF bmimCl

Crystal polymorphism of bmimCl was reported almost simultaneously by two groups [8, 9]. We found by chance that two different types of crystals, Crystal (1) and Crystal (2), formed when liquid bmimCl was cooled down to −18°C and was kept for 48 h [8]. Orthorhombic Crystal (2) dominantly formed but monoclinic Crystal (1) also formed occasionally (see Fig. 2).

Upon leaving Crystal (2) for more than 24 h at dry-ice temperature, Crystal (2) was converted to Crystal (1). It is not clear yet whether Crystal (1) forms directly from the liquid state or not. Holbrey et al. [19] independently obtained two crystal polymorphs, orthorhombic Crystal I and monoclinic Crystal II. Crystal I was obtained by cooling down slowly the molten liquid to room temperature, while Crystal II was obtained by cooling ionic liquid mixtures, bmimCl/bmim[PF_6] and bmimCl/bmim[BF_4]. Crystal II was also obtained by crystalization from a hexane–benzene mixed solvent. The melting points of Crystals I and II measured by DSC were reported by Holbrey et al. to be 66°C and 41°C, respectively. Independent DSC measurements by us [10] showed somewhat different results. Crystal (1) melted at varying temperature in the range between 47°C and 67°C depending on individual crystals, although single crystals were used for the DSC measurements. Crystal (2) melted at 64°C. The DSC curves of the two polymorphs showed broad melting peaks as also mentioned by Holbrey et al. These extraordinary DSC behaviors are indicative of complex structure and dynamics of bmimCl Crystall (1) with regard to the temperature change. From the crystal types and the melting points,

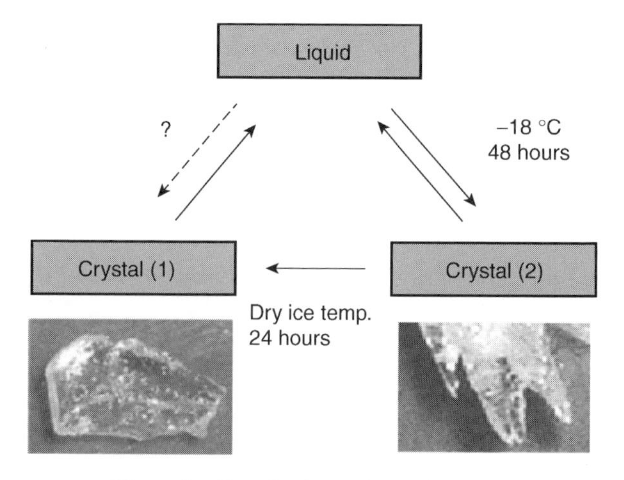

Figure 2. Crystal polymorphism of bmimCl.

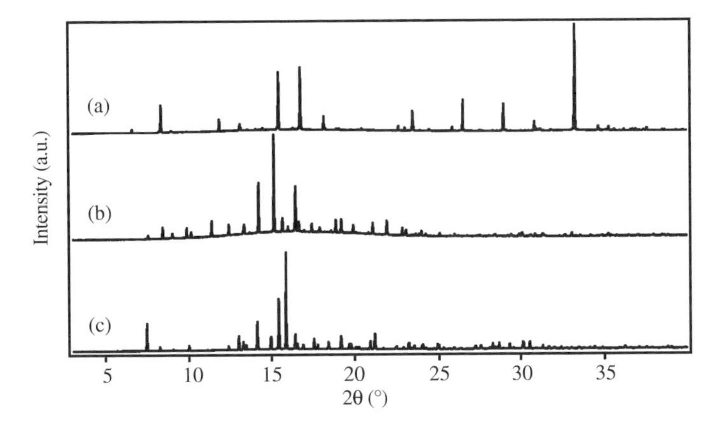

Figure 3. X-ray powder diffraction patterns of (*a*) bmimCl Crystal (1), (*b*) bmimCl Crystal (2), and (*c*) bmimBr.

it is obvious that Crystal I corresponds to Crystal (2) and Crystal II to Crystal (1). In the following, we use the notation Crystal (1) and Crystal (2).

The x-ray powder diffraction patterns of bmimCl Crystal (1) and (2) are shown in Fig. 3 [8]. The sharp peaks with distinct patterns indicate that they are different crystals and that neither of them is an amorphous solid. The continuous background notable for Crystal (2) is most likely to arise from the structural disorder existing in the crystal. The x-ray powder diffraction pattern of bmimBr is also shown in Fig. 3 for comparison. The pattern of bmimBr is more close to that of bmimCl Crystal (2) than to Crystal (1).

The Raman spectra of bmimCl Crystals (1) and (2), and that of bmimBr, are compared in Fig. 4 [8]. The polymorphs of bmimCl give two distinct Raman

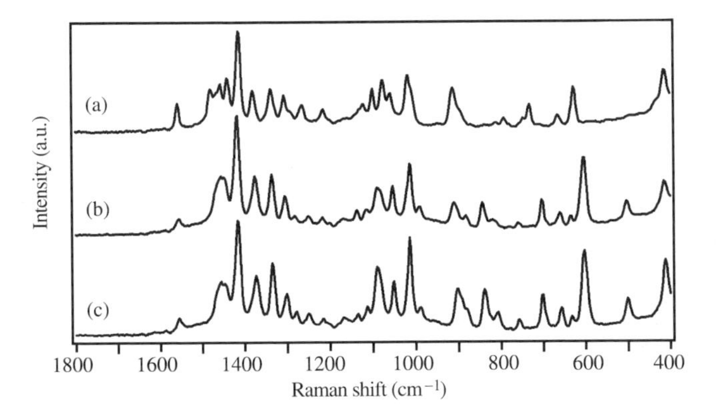

Figure 4. Raman spectra of (*a*) bmimCl Crystal (1), (*b*) bmimCl Crystal (2), and (*c*) bmimBr.

spectra, while those of bmimCl Crystal (2) and bmimBr are almost identical with each other. These findings are consistent with the results of the x-ray powder diffraction experiments. Since the halogen anions are inactive in Raman scattering except for the lattice vibrations that are expected in the wavenumber region lower than 400 cm^{-1}, all the Raman bands observed in Fig. 4 are ascribed to the $bmim^+$ cation. The Raman spectra in Fig. 4 therefore indicate that the $bmim^+$ cation takes two different structures in those salts; it takes the same molecular structure in bmimCl Crystal (2) and bmimBr, but it takes a different structure in bmimCl Crystal (1). As shown below, this structural difference originates from the rotational isomerism of the butyl chain of the $bmim^+$ cation.

III. CRYSTAL STRUCTURES OF bmimCl CRYSTAL (1) AND bmimBr

Subsequent to the discovery of the crystal polymorphism of bmimCl, the crystal structures of bmimCl and bmimBr were determined. We determined the crystal structures of bmimCl Crystal (1) and bmimBr at room temperature [11, 12]. Independently, Holbrey et al. [9] reported the crystal structures of bmimCl Crystal (1) and Crystal (2), as well as that of bmimBr at −100°C. The two sets of structures determined at different temperatures agree well with each other except for the lattice constants that vary with temperature. They also show that the molecular structure of the $bmim^+$ cation in bmimCl Crystal (2) is different from that in (1) but that it is the same as that in bmimBr, as already indicated by the Raman spectra. In the following, we discuss the crystal structures of bmimCl Crystal (1) and bmimBr as the two representative structures at room temperature.

The crystal structure of bmimCl Crystal (1) is shown in Fig 5. The detailed structural data are available from the Cambridge Crystallographic Data Centre [12] (CCDC, with deposition number 213959). BmimCl Crystal (1) belongs to the monoclinic space group $P2_1/n$ with $a = 9.982(10)$, $b = 11.590(12)$, $c = 10.077(11)$ A, $\beta = 121.80(2)^\circ$. In the crystal, both the $bmim^+$ cations and the chloride anions form separate columns extending along the *a* axis, and no specific ion pair is formed. The imidazolium rings are all planar pentagons. The *n*-butyl group of the $bmim^+$ cation takes a *trans–trans* (TT) conformation with respect to the C_7–C_8 and C_8–C_9 bonds, as shown in the inset of Fig. 5. A couple of the $bmim^+$ cations form a pair through an aliphatic interaction of the stretched *n*-butyl group. Those pairs stack together and form a column extending along the *a* axis, in which all the imidazolium ring planes are parallel with one another. Two types of cation columns with different orientations exist. The planes of the imidazolium rings in the two differently oriented columns make an angle of 69.5°. A zigzag chain of the anion Cl^- directed in the *a*

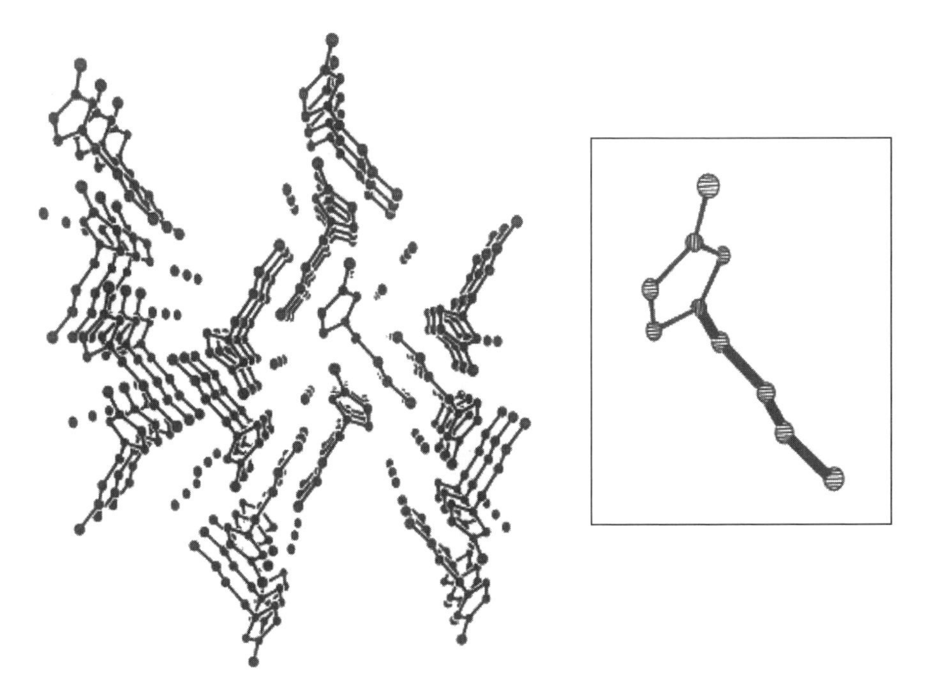

Figure 5. Crystal structure of bmimCl Crystal (1) viewed over the *a* axis. Carbon atoms, nitrogen atoms, and chloride anions are shown. The *trans–trans* conformation of the butyl group of the bmim$^+$ cation is shown in the inset with the relevant C–C bonds marked by thick bars.

direction (Fig. 6) are accommodated in a channel formed by four cation columns, of which two opposite columns have the same orientation. The shortest three distances between Cl^- anions in the zigzag chain are $r_1 = 4.84$ Å, $r_2 = 6.06$ Å, and $r_3 = 6.36$ Å. These distances are much larger than the sum of the van der Waals radii of Cl^- (3.5 Å). There seems to be no specific interactions among the Cl^- anions, and they are likely to be aligned under the

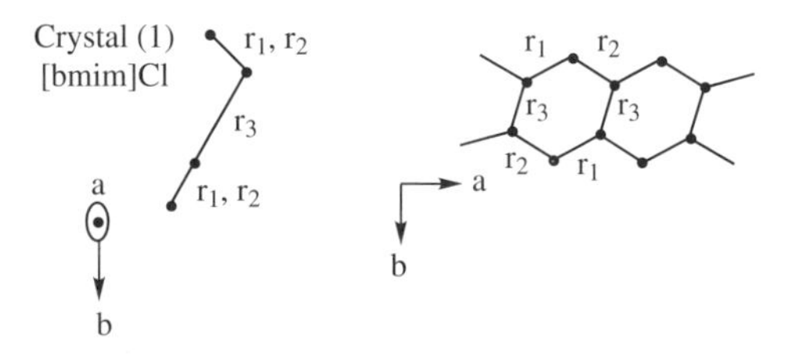

Figure 6. The zigzag structure of chloride anions in bmimCl Crystal (1).

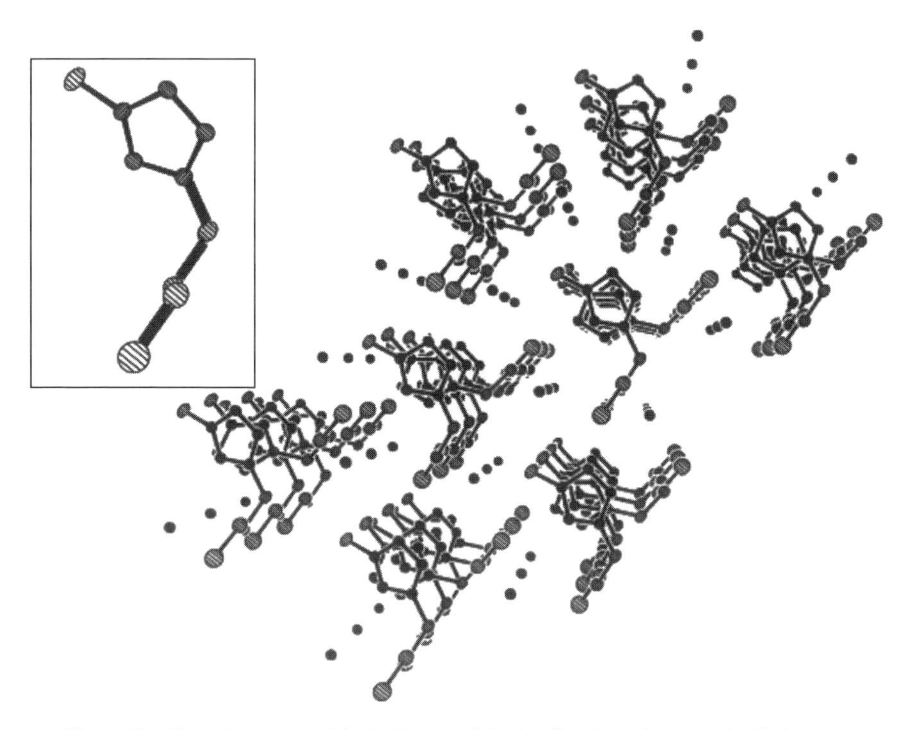

Figure 7. Crystal structure of bmimBr viewed in the direction of the *a* axis. Carbon atoms, nitrogen atoms, and bromide anions are shown. The *gauche–trans* conformation of the butyl group of the $bmim^+$ cation is shown in the inset with the relevant C–C bonds marked by thick bars.

effect of the Coulomb force. Similar crystal structures were also reported for 1-ehtyl-3-methylimidazolium chloride (emimCl) [13].

The molecular arrangements in a crystal of bmimBr are shown in Fig. 7. Detailed crystal data are registered as CCDC 213960. BmimBr belongs to the orthorhombic space group $Pna2_1$ with $a = 10.0149(14)$, $b = 12.0047(15)$, $c = 8.5319(11)$ A. As in the case of bmimCl Crystal (1), the $bmim^+$ cations and the halogen anions form separate columns extending along the *a* axis and no ion pairs exist. The *n*-butyl group takes a *gauche–trans* (*GT*) conformation with respect to the C_7–C_8 and C_8–C_9 bonds (inset of Fig. 8). Only one kind of cation column is found. The imidazolium rings stack so that the N–C–N moiety of one ring interacts with the C=C portion of the adjacent ring. The orientation of the adjacent ring is obtained by rotating a ring plane by about 73° around an axis involving the two N atoms. As in the case of [bmim]Cl Crystal (1), a zigzag chain of Br^-, directed in the *a* direction, is accommodated in a channel produced by four cation columns, which are also extending in the *a* direction. The shortest three Br^-–Br^- distances are 4.77, 6.55, and 8.30Å and are all longer than the sum of the van der Waals radii (3.7 Å). This fact again indicates that there are no specific interactions among the Br^- anions and that they are arranged in the zigzag form as a result of the Coulomb interaction.

IV. NORMAL MODE ANALYSIS AND ROTATIONAL ISOMERISM OF THE Bmim$^+$ CATION

The Raman spectral variations of the bmim$^+$ cation in bmimX crystals (Fig. 4) are interpreted very well in terms of the rotational isomerism of the butyl group. Figure 8 compares the Raman spectra of bmimCl Crystal (1) and bmimBr in the wavenumber region of 400–1000 cm^{-1}. The structures of the bmim$^+$ cation in the two crystals are also depicted in the same figure.

In the wavenumber region 600–700 cm^{-1}, where ring deformation bands are expected, two bands appear at 730 cm^{-1} and 625 cm^{-1} in bmimCl Crystal (1)

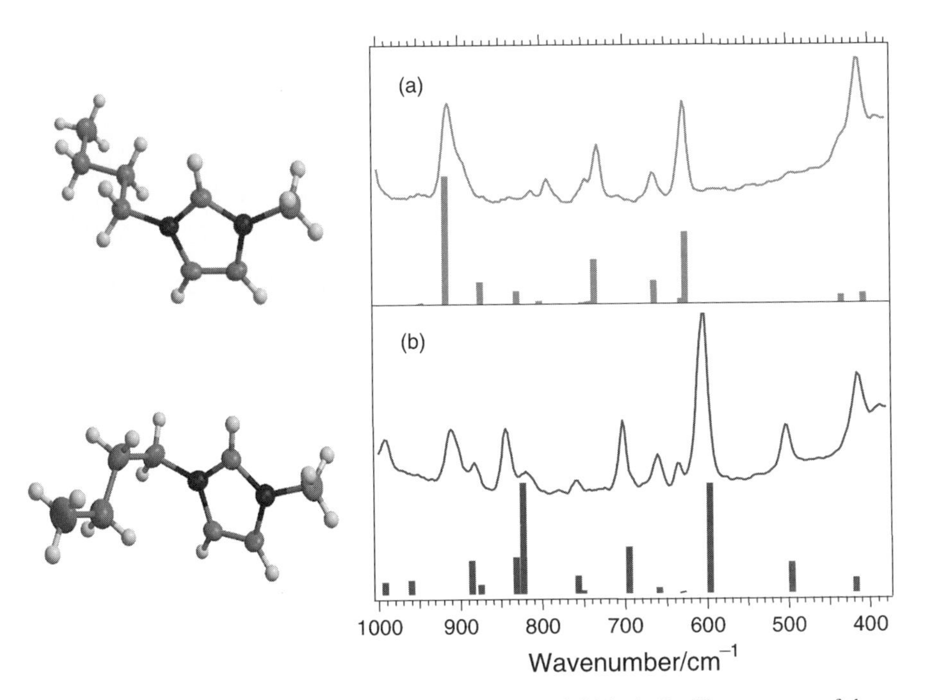

Figure 8. Raman spectra of (*a*) bmimCl Crystal (1) and (*b*) bmimBr. The structures of the bmim$^+$ cation in the two crystals are depicted on the right-hand side. The thick vertical bars indicate calculated frequencies and Raman intensities.

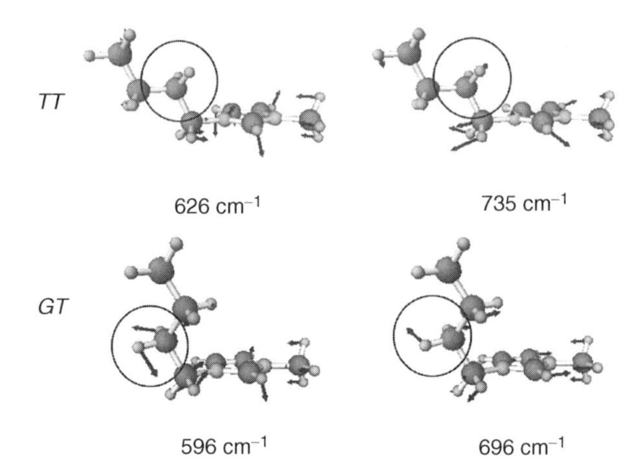

Figure 9. Calculated normal modes of the two key bands of the TT and GT forms of the $bmim^+$ cation. The arrows indicate vibrational amplitudes of atoms. The CH_2 rocking vibration of the C_8 methylene group are surrounded by a circle.

but not in bmimBr, while another couple of bands appear at 701 cm^{-1} and 603 cm^{-1} in bmimBr and not in bmimCl crystal (1). A DFT (density functional) calculation with Gaussian 98(6), B3LYP/6-31G+** level gives frequencies and intensities shown as thick vertical bars in Fig. 9 [14]. In this calculation, the structures of the *TT* and *GT* forms of $bmim^+$ were optimized in the vicinity of the structures determined, respectively, for bmimCl Crystal (1) and bmimBr by the x-ray analysis. The calculation reproduces the observed spectra very well, particularly in the wavenumber region 500–700 cm^{-1}. The calculated normal modes of vibrations are shown in Fig. 9 for the 730-cm^{-1} and 625-cm^{-1} bands of bmimCl Crystal (1) and for the 701-$cm^{-1}$and 603-cm^{-1} bands of bmimBr. It shows that the 625-cm^{-1} band of bmimCl Crystal (1) and the 603-cm^{-1} band of bmimBr originate from similar ring deformation vibrations but that they have different magnitudes of couplings with the CH_2 rocking motion of the C_8 carbon. The coupling occurs more effectively for the *gauche* conformation around the C_7–C_8 bond, resulting in a lower frequency in the *GT* form (603 cm^{-1}) than in the *TT* form (625 cm^{-1}). Note that the coupling with the CH_2 rocking motion having a higher frequency pushes down the frequency of the ring deformation vibration. The same coupling scheme holds for another ring deformation mode and the *GT* form has a lower frequency (701 cm^{-1}) than the *TT* form (730 cm^{-1}). It is made clear by the DFT calculation that the 625-cm^{-1} and 730-cm^{-1} bands are characteristic of the *trans* conformation around the C_7–C_8 bond, while the 603-cm^{-1} and 701-cm^{-1} bands are characteristic of the *gauche* conformation. In other words, we can use these bands as key bands to probe the conformation around the C_7–C_8 bond of the $bmim^+$ cation.

V. RAMAN SPECTRA AND LIQUID STRUCTURE OF BmimX

With the structural information obtained from the crystals, we are now in a position to discuss the liquid structure of bmimX ionic liquids. Raman spectra of liquid bmimX (X = Cl, Br, I, BF_4, PF_6) are shown in Fig. 10. The Raman spectra of bmimCl Crystal (1) and bmimBr are also shown as references. All Raman spectra were measured at room temperature. The Raman spectra of liquid bmimCl and bmimBr were obtained from their supercooled states.

The Raman spectra of the BF_4^- and PF_6^- anions are already well known. Except for these anion bands that are deleted in Fig. 10, the Raman spectra of liquid bmimX are surprisingly alike with one another. It seems that the structure of the $bmim^+$ cation is very similar in these liquids. Both of the two sets of key bands, the 625-cm^{-1} and 730-cm^{-1} bands for the *trans* conformation and the 603-cm^{-1} and 701-cm^{-1} bands for the *gauche* conformation, appear in all of the

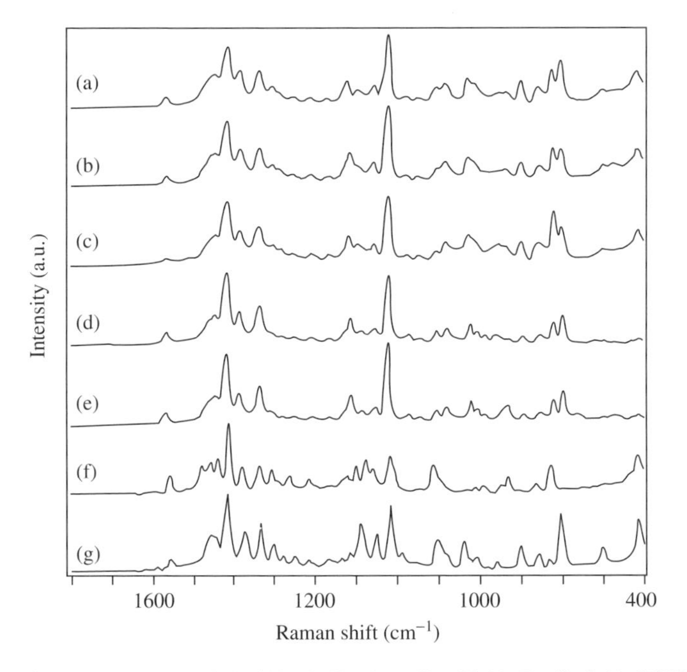

Figure 10. Raman spectra of liquid bmimX, where X = Cl (*a*), Br (*b*), I (*c*), [BF4] (*d*), and [PF6] (d). The anion bands in (d) and (*e*) are deleted. Raman spectra of bmimCl Crystal (1) and that of crystalline bmimBr are also shown as references in (*f*) and (*g*), respectively.

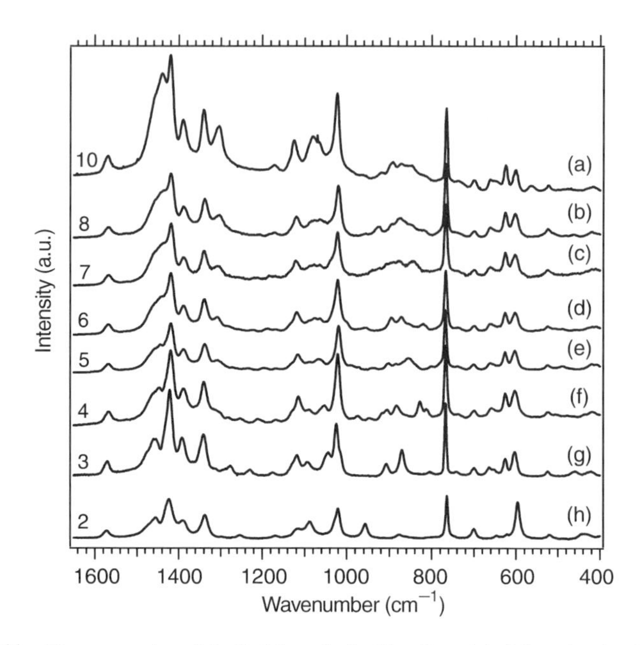

Figure 11. Raman spectra of 1-alkyl-3-methylimidazolium tetrafuluoroborate, $C_n$mim[BF_4]; $n = 10$ (*a*), 8 (*b*), 7 (*c*), 6 (*d*), 5 (*e*), 4 (*f*), 3 (*g*), 2(*h*).

liquid Raman spectra. Therefore, at least two rotational isomers, one having a *trans* conformation and the other having a *gauche* conformation around the C_7–C_8 bond, coexist in liquid [bmim]X. The relative intensity of the 625-cm^{-1} band to that of the 603-cm^{-1} band is proportional to the *trans/gauche* population ratio. According to the Raman spectra in Fig. 10, this *trans/gauche* ratio changes with the anion. For the halides, it increases in the order $Cl^- < Br^- < I^-$. It is similar for bmimCl, bmim[BF_4], and bmim[PF_6].

The *trans/gauche* ratio also changes with the cation. A series of 1-alkyl-3-methylimidazolium ($C_n$mim, where n is the number of carbon atoms in the alkyl chain) cations generate RILs with $[BF_4]^-$ [15]. Figure 11 shows the Raman spectra of $C_n$mim[BF_4] for $n = 2$ to $n = 10$ in the liquid state at room temperature. For C_2mim[BF_4], there is no rotational isomerism. Consequently, only one Raman band is observed at 596 cm^{-1}, which corresponds to the *gauche* conformation. This apparent gauche frequency is rationalized by the fact that the methyl rocking motion of the C_8 carbon couples strongly with the ring deformation vibration and pushes down the frequency as in the case of the *gauche* conformation of the C_4mim^+ ($bmim^+$) cation.

For the carbon number larger than two ($n > 2$), the 625/603-cm^{-1} Raman intensity ratio increases with increasing n. The *trans* band at 625 cm^{-1} is weaker in intensity than the *gauche* band at 603 cm^{-1} for $n = 3$, but the intensity ratio is reversed for $n = 10$. Since the vibrational modes giving rise to those bands are very similar with each other and localized within the imidazolium ring and the C_7 and C_8 carbons (see Fig. 9), their Raman cross sections are thought to be independent of the chain length. Therefore, the 625/603-cm^{-1} Raman intensity ratio can be regarded as a direct measure of the *trans/gauche* isomer ratio. The observed increase of the 625/603-cm^{-1} Raman intensity ratio then means that the *trans/gauche* isomer ratio increases as the chain becomes longer. In other words, the *trans* structure is stabilized relatively to the *gauche* for longer alkyl chains. Such stabilization of the *trans* isomer is understandable only if we assume interactions among the cations. Otherwise, the relative stability is determined by the energy difference (about 0.6 kcal/mol [16]) between the *trans* and *gauche* conformations around the C_7–C_8 bond and is more likely to be independent of the chain length. We know from the crystal structure of C_4mimCl (bmimCl) Crystal (1) that a couple of C_4mim^+ cations make a pair through an aliphatic interaction between the two alkyl groups. The chain-length dependence of the *trans/gauche* ratio is thus indicative of an interaction between two $C_n$mim cations, most probably through an aliphatic interaction between the two alkyl chains. In Fig. 11, broad Raman features are observed for longer-chain $C_n$mim[BF_4] ($n = 7$–10) in the wavenumber region of 800–950 cm^{-1}, where the rocking and the other hygrogen bending vibartion of the methylene groups are located. These broad features are also indicative of a specific interaction between the alkyl chains. The crystal structures of $C_n$mimCl and $C_n$mim[PF_6] show interdigitated structures of alkyl groups when $n > 12$ [17–19]. Similar interdigitated structures of alkyl chains are also suggested in the meso-phase of $C_n$mimCl ($n = 12$–18) [20].

VI. POSSIBLE LOCAL STRUCTURES IN BmimX IONIC LIQUIDS

The interaction through the alkyl chains is likely to operate also in bmimX ILs. In fact, there are more experimental results that strongly suggest the interaction among the cations and the formation of local structures in bmimX ILs. First, unusually long equilibration time between the *trans* and *gauche* conformers has been observed for liquid bmimCl [21]. A small piece (0.5 mm × 0.5 mm × 0.5 mm) of [bmim]Cl crystal (1) was heated abruptly with a heat gun from room temperature to 72°C, and a droplet of liquid in a nonequilibrium state was transiently formed. The sample was then kept at 72°C to get thermally equilibrated. The time-resolved Raman spectra in Fig. 12 show this melting and equilibration processes. Before melting, only the *trans* band at 625 cm^{-1} is observed in the 600- to 630-cm^{-1} region, reflecting the *trans* structure of the $bmim^+$ cation in Crystal (1). Immediately after melting, the 625-cm^{-1} band remains much stronger than the *gauche* band at 603 cm^{-1}. Then, the *gauche* band

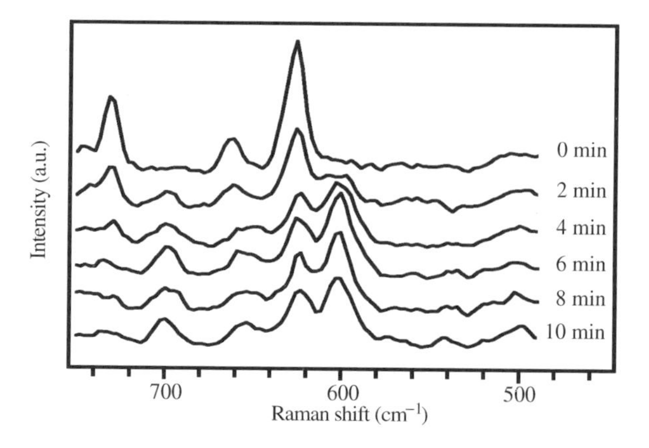

Figure 12. Time-resolved Raman spectra of the melting and thermally equilibration process of bmimCl Crystal (1).

becomes stronger as time goes on, and the *trans/gauche* intensity ratio becomes constant after 10 min. Thus, it takes about 10 min for the *trans* and *gauche* conformers in liquid bmimCl to get thermally equilibrated. If the $bmim^+$ cation undergoes the *trans/gauche* transformation at the single molecular level, as expected for a free butyl chain, the equilibration should occur instantaneously. Note that the *trans* and *gauche* isomers of alkyl chains are not separately observed but give a coalesced peak in NMR spectra, indicating that transformation between them takes place much faster than a second. The observed unusually long equilibration time, 10 min, therefore indicates that the *trans* and *gauche* conformers of $bmim^+$ in liquid bmimCl are not transformed from each other at the single molecular level but that they can be interconverted only through a slow collective transformation (analogous to phase transition) of the ensembles of $bmim^+$ cations. It is most probable that the two rotational isomers are incorporated in their specific local structures and that they can interconvert with each other only through the conversion of those local structures as a whole.

The idea of different local structures incorporating the *trans* and *gauche* conformers is also consistent with the observed enthalpy differences. Table I gives the enthalpy differences ($\Delta H = H_{gauche} - H_{trans}$) between the *gauche* and *trans* conformers in the $C_n$mim[BF_4] system determined from the temperature dependence of the Raman intensity ratio of the 625-cm^{-1} and 603-cm^{-1} bands [22]. As shown in the table, the determined enthalpy difference is very small for $n = 4, 5$, and 6, meaning that the two isomers are associated with almost the same enthalpy. It becomes the largest for $n = 10$. It is known that the ΔH value for a free alkane chain is around 0.6 kcal/mol [16]. However, all the obtained ΔH values for $C_n$mim[BF_4] are significantly smaller than 0.6 kcal/mol. This disagreement is explained well in terms of the local structure formation of

TABLE I
The Apparent Enthalpy Difference Between the *trans* and *gauche* Conformers for $C_n$mim[BF_4] [22]

n	ΔH (kcal mol^{-1})	Melting Point (°C) [15]	Glass Transition Temperature (°C) [15]
3	−0.1	—	−13.9
4	0.01	—	−71.0
5	0.02	—	−88.0
6	−0.01	—	−82.4
8	0.09	—	−78.5
10	0.19	−4.2	—

the $C_n$mim cations; the determined enthalpy differences are associated with the alkyl chains incorporated in the local structures and not with those for free alkane chains. The dependence of the enthalpy difference on the chain length is also of considerable interest. For $n = 3$–8, for which the magnitude of the ΔH values are smaller than 0.1, no liquid/crystal transitions are observed and the glass state is formed form the liquid [15]. Only for $n = 10$, for which ΔH is significantly larger than those of the others, a clear melting point is reported. This trend suggests that the coexistence of the two different local structures, which incorporate the *trans* and *gauche* conformers of the bmim$^+$ cation and have similar enthalpies, hinders crystallization and hence lowers the melting points (liquid/glass transition temperature) of $C_n$mim[BF_4] with $n = 3$–8.

The ordering of the anions in bmimX ionic liquids has also been suggested by our recent large-angle x-ray scattering experiment on liquid bmimI [23]. Figure 13 shows a differential radial distribution function obtained for liquid bmimI at room temperature. Clear peaks in the radial distribution curve are

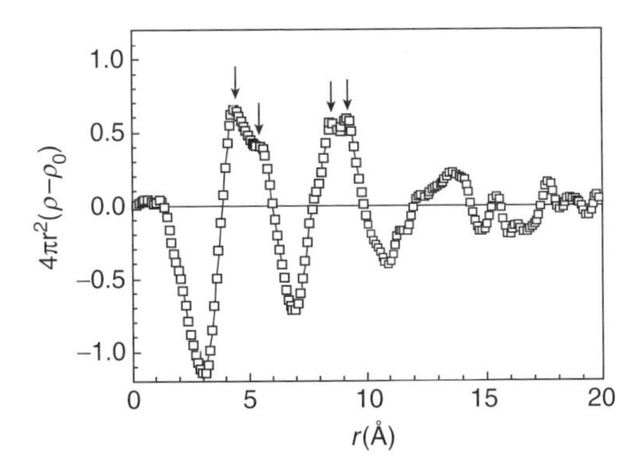

Figure 13. Differential radial distribution function of liquid bmimI by large-angle x-ray scattering [23].

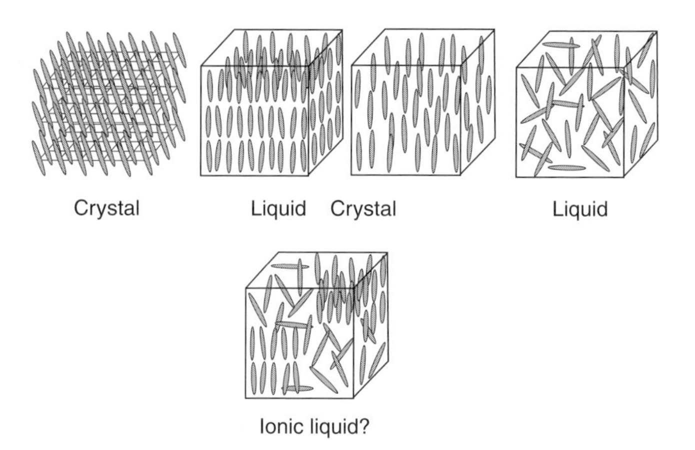

Figure 14. Conceptual structure of an ionic liquid.

observed at 4.5, 5.5, 8.5, and 9.2 Å (arrows in Fig. 14). The shortest distance, 4.5 Å, corresponds very well to the shortest halogen–halogen distance found in the crystal structures of bmimCl Crystal (1)(4.84 Å) and bmimBr (4.65 Å). The other distances can also be correlated to the other halogen–halogen distances in the zigzag chains shown in Fig. 6. It seems that the zigzag chains found in the bmimX crystals do exist, at least partially, and may be in a slightly distorted form, in the ionic liquid state as well.

In this way, by combining several pieces of experimental evidence obtained so far, we come to a thought that both the cations and anions in bmimX ILs do have local ordering or structures. The conceptual structure of a bimimX IL is shown in Fig. 14 in comparison with the structures of a crystal, liquid crystals, and a liquid. In a crystal, component molecules or ions are periodically arranged to form a lattice, and a long-range order exists. In a liquid, they take random position and orientation and there is no order. In liquid crystals, a long-range orientational order is preserved though the position is random or only partially ordered. In bmimX ILs, the supposed local structures are positioned and oriented randomly, and there is no translational and orientational order in the macroscopic level. Taking into account the fact that the bmimX ILs are all transparent, the dimension of those local structure must be much smaller than the wavelength of visible light ($<$100 Å).

As stated in the Introduction, studies on the strucutre and interactions in ILs are limitted. In the early stage of structural investingations, a focus was placed on the elucidation of hydrogen bonding between the imidazolium cations and the anions by NMR [24–28]. For emimX (X = Cl, Br, I) diluted in several solvents, the formation of hydrogen bonding has been indicated between the cation and the anion via the hydrogen atoms attached to the C_2, C_4, and C_5

carbons [25]. From the measurement of the ^{13}C dipole–dipole relaxation rate, Huang et al. [26] have suggested that the hydrogen atom attached to thc C_2 carbon participates in the hydrogen bonding with anions in neat emim[BF_4]. Similar results are also reported for bmim[PF_6] and bmim[BF_4] by Saurez et al. [27] and Lin et al. [28]. Interestingly, Saurez et al. suggest the existence of extended hydrogen-bonded networks at 279 K in bmim[BF_4]. The extended structure via hydrogen bonding is also suggested by Abdul-Sada et al. [29] for 1-alkyl-3-methylimidazolium halide by use of fast-atom bombardment mass spectroscopy. Recently, charge ordering in ILs is discissed by Hardacre et al. [30, 31]. They have obtained the radial distribution function of dimethylimidazolium chloride and dimethylimidazolium hexaflurophosphate using neutron diffraction and argued that there is significant charge ordering of ions and hence the local structures in ILs and that the local structure resembles that found in the solid state. The charge ordering is also discussed by computer simulation studies. A number of computer simulation studies have been conducted on 1-alkyl-3-methylimidazolium-based ILs in recent few years [32–41]. The radial distribution functions calculated in these articles give similar results, all suggesting the long-range charge ordering. These results, though reported fragmentally, give support to our idea that ILs are unique in that they have more ordering in structure than do conventional molecular liquids.

If local structures do exist in ILs as discussed above, many unique properties are expected to arise therefrom. In fact, unusually high viscosity of ILs is ascribable to the local structures that hinder the translational motion of the ions. Amphiphilicity of ILs is also well-explained in terms of the inhomogeneous nature of ILs having local structures; polar molecules are dissolved in a polar part composed of the cation centers and anions, while nonpolar molecules are accommodated in a nonpolar part where alkyl chains are densely located. One of the most interesting properties that are expected to arise from the local ordering of ions is the magnetism. If magnetic ions are locally aligned in a liquid and if the spin angular momenta of those ions are strongly interacting with one another, we may be able to have an unusual magnetic liquid. We recently demonstrated that bmim[$FeCl_4$] responds strongly to a magnet, as shown in Fig. 15. It turned out that this RIL was nearly paramagnetic with no strong interactions among the spins. By combining many different cations and

Figure 15. Pictures showing the response of bmim[$FeCl_4$] to a magnet [42].

magnetic anions, however, it may be possible to prepare superparamagnetic or even ferromagnetic ionic liquids in the near future.

VII. CONCLUSIONS

Raman spectroscopic and x-ray diffraction studies of bmimCl Crystal (1) and Crystal (2) and of crystalline bmimBr have revealed that two rotational isomers, the *trans–trans* and *gauche-trans* forms of the butyl group of the $bmim^+$ cation, exist in these crystals. Raman spectra and a DFT calculation show that at least two rotational isomers, one having a *trans* and the other having a *gauche* conformation around the C_7–C_8 bond of the butyl group, coexist in bmimX ionic liquids, where X = Cl, Br, I, BF_4,PF_6. Time-resolved Raman study of the melting and equilibration processes of bmimCl Crystal (1) suggests that the $bmim^+$ cations in those liquids form local structures that are specific to the *trans* and *gauche* conformers. It is likely that these local structures are similar to those found in the crystals. Coexistence of two different local structures, which incorporate the *trans* and *gauche* conformers separately, is likely to hinder crystallization of liquid bmimX and hence lower the melting points of the corresponding crystals. The size and the relaxation time of those local structures in bmimX ILs are yet to be elucidated, and they should be studied urgently in the nearest future.

If the local structure formation discussed in this chapter is not specific to bmimX ILs but applies to ILs in general, ILs may not be genuine liquids in the conventional sense. They might be better called **nano-structured fluid** or **crystal liquid**. They may form a new meso-phase that is distinct from the liquid crystal phase.

Acknowledgments

The authors are grateful to Mr. Satoshi Hayashi, Dr. Satyen Saha, Dr. Hideki Katayanagi, and Professor Keiko Nishikawa for collaboration.

References

1. K. R. Seddon, *J. Chem. Tech. Biotechnol.* **68**, 351 (1997).
2. J. S. Wilkes, *Green Chem.* **4**, 73 (2002).
3. J. G. Huddleston, A. E. Visser, W. M. Reichert, H. D. Willauer, G. A. Broker, and R. D. Rogers, *Green Chem.* **3**, 156 (2001).
4. T. Welton, *Chem. Rev.* **99**, 2071 (1999).
5. P. Wasserscheid and W. Keim, *Angew. Chem. Int. Ed.* **39**, 3772 (2000).
6. Saurez et al. *Chem. Rev.* **102**, 3667 (2002).
7. J. H. Davis, Jr, *Chem. Lett.* **33**, 1072 (2004).

8. S. Hayashi, R. Ozawa, and H. Hamaguchi, *Chem. Lett.* **32**, 498 (2003).
9. J. D. Holbrey, W. M. Reichert, M. Nieuwenhuyzen, S. Johnston, K. R. Seddon, and R. D. Rogers, *Chem. Commun.* 1636 (2003).
10. S. Wang, K. Tozaki, H. Katayanagi, H. Hayashi, H. Inaba, S. Hayashi, H. Hamaguchi, Y. Koga, and K. Nishikawa, to be published.
11. S. Saha, S. Hayashi, A. Kobayashi, and H. Hamaguchi, *Chem. Lett.* **32**, 740 (2003).
12. Cambridge Crystallographic Data Centre, http://www.ccdc.cam.ac.uk.
13. A. Elaiwi, P. B. Hitchcock, K. R. Seddon, N. Srinivasan, Y. M. Tan, T. Welton, and J. A. Zora, *J. Chem. Soc. Dalton Trans.*, 3467 (1995).
14. R. Ozawa, S. Hayashi, S. Saha, A. Kobayashi, and H. Hamaguchi, *Chem. Lett.* **32**, 948–949 (2003).
15. J. D. Holbrey and K. R. Seddon, *J. Chem. Soc., Dalton Trans.*, 2133 (1999).
16. (a) N. Shepard and G. J. Szasz, *J. Chem. Phys.* **17**, 86 (1949). (b) R. G. Snyder, *J. Chem. Phys.* **47**, 1316 (1947).
17. C. M. Gordon, J. D. Holbrey, A. R. Kennedy, and K. R. Seddon, *J. Mater. Chem.* **8**, 2627 (1998).
18. J. D. Roche, C. M. Gordon, C. T. Imrie, M. D. Ingram, A. R. Kennedy, F. L. Celso, and A. Triolo, *Chem. Mater.* **15**, 3089 (2003).
19. A. Downard, M. J. Earle, C. Hardcre, S. E. J. McMath, M. Nieuwenhuyzen, and S. J. Teat, *Chem. Mater.* **16**, 43 (2004).
20. A. E. Bladley, C. Hardcre, J. D. Holbrey, S. Johnston, S. E. J. McMath, and M. Nieuwenhuyzen, *Chem. Mater.* **14**, 629 (2002).
21. H. Hamaguchi, R. Ozawa, S. Hayashi, and S. Satyen. Abstract of Papers of the American Chemical Society, **226**, U622 (2003).
22. R. Ozawa, and H. Hamaguchi, to be published.
23. H. Katayanagi, S. Hayashi, H. Hamaguchi, and K. Nishikawa, *Chem. Phys. Lett.* **392**, 460–464 (2004).
24. W. R. Carper, J. L. Pflung, A. M. E. Elias, and J. S. Wilkes, *J. Phys. Chem.* **96**, 3828 (1992).
25. A. G. Avent, P. A. Chaloner, M. P. Day, K. R. Seddon, and T. Welton, *J. Chem. Soc. Dalton Trans.*, 3405 (1994).
26. J. F. Huang, P. Y. Chen, I. W. Sun, and S. P. Wang, *Inorg. Chim. Acta* **320**, 7 (2001).
27. P. A. Z. Suarez, S. Einloft, J. E. L. Dullius, R. F. Souza, and J. Dupont, *J. Chim. Phys.* **95**, 1626 (1998).
28. S. T. Lin, M. F. Ding, C. W. Chang, and S. S. Lue, *Tetrahedron* **60**, 9441 (2004).
29. A. K. Abdul-Sada, A. E. Elaiwi, A. M. Greenway, and K. R. Seddon, *Eur. Mass Spectrom.* **3**, 245 (1997).
30. C. Hardacre, S. E. J. McMath, M. Nieuwenhuyzen, D. T. Brown, and A. K. Soper, *J. Chem. Phys.* **118**, 273 (2003).
31. C. Hardacre, S. E. J. McMath, M. Nieuwenhuyzen, D. T. Brown, and A. K. Soper, *J. Phys. Condens. Matter.* **15**, 159 (2003).
32. C. G. Hanke, S. L. Price, and R. M. Lynden-Bell, *Mol. Phys.* **99**, 801 (2001).
33. J. K. Shah, J. F. Brennecke, and E. J. Maginn, *Green Chem.* **4**, 112 (2004).
34. T. I. Morrow and E. J. Maginn, *J. Phys. Chem. B* **106**, 12807 (2002).
35. J. D. Andrade, E. S. Böes, and H. Stassen, *J. Phys. Chem. B* **106**, 3546 (2002).

36. J. D. Andrade, E. S. Böes, and H. Stassen, *J. Phys. Chem. B* **106**, 13344 (2002).
37. C. J. Margulis, H. A. Stern, and B. J. Berne, *J. Phys. Chem. B* **106**, 12017 (2002).
38. M. G. D. Pópolo, and G. A. Voth, *J. Phys. Chem. B* **108**, 1744 (2004).
39. Z. Liu, S. Huang, and W. Wang, *J. Phys. Chem. B* **108**, 12978 (2004).
40. S. M. Urahata and M. C. C. Ribeiro, *J. Chem. Phys.* **120**, 1855 (2004).
41. J. K. Shah and E. J. Maginn, *Fluid Phase Equilibria* **195**, 222 (2004).
42. S. Hayashi and H. Hamaguchi, *Chem. Lett.* **33**, 1590 (2004).

CHEMICAL REACTIONS AT VERY HIGH PRESSURE

VINCENZO SCHETTINO, ROBERTO BINI, MATTEO CEPPATELLI, LUCIA CIABINI, and MARGHERITA CITRONI

LENS, European Laboratory for Non-linear Spectroscopy and INFM, Firenze, Italy
and
Dipartimento di Chimica dell'Università di Firenze Firenze, Italy

CONTENTS

Advances in Chemical Physics, Volume 131, edited by Stuart A. Rice

I. INTRODUCTION

In 1918 the Nobel prize in Chemistry was awarded to Fritz Haber for the method of synthesizing ammonia from molecular nitrogen and hydrogen. In the presentation speech and in the Nobel lecture it is mentioned that the synthesis could be made possible at the highest attainable pressure. It is also envisaged that in the ammonia synthesis the high pressure has the twofold effect of improving the equilibrium constant of the reaction and of increasing the reaction rate as well. Later, in 1931, the Nobel prize in chemistry was awarded jointly to Carl Bosch and Friedrich Bergius for originating and developing chemical high-pressure methods that were specifically applied to the industrial synthesis of ammonia from the elements and to the manufacture of oils and liquid fuels from solid coal. It is quite well understood that the basic motivation for the assignment of the two awards was bound to the tremendous impact on industry and economy of the production at affordable costs of ammonia, as a starting material to obtain nitrogen-based fertilizers, in essentially unlimited quantities. However, it is important to stress that these two fundamental milestones in the history of chemical industry evidenced the importance of high-pressure methods to enlarge the spectrum of attainable chemical transformations. Yet the potentialities of high-pressure methods were quite unexplored at those times. For instance, pressures up to only 200 atmospheres (~0.02 GPa) were reached in the industrial plants for the production of ammonia. It can be mentioned, however, that a considerable number of experiments had been carried over the years, starting from the beginning of the 1800s, at high pressure. These experiments were mainly concerned with experimental determination of the effect of pressure on a large variety of physical properties of materials that included the compressibility of solids, liquids, and solutions, the viscosity, the melting points and the temperatures of polymorphic transitions, the specific heat of liquids and solids, the conductivity of metals and electrolytic solutions, and many others as well. A detailed description of the early historical attempts of a foundation of the high-pressure science and technology has been outlined by P. W. Bridgman in his *The Physics of High Pressure* [1]. Among these primordial high-pressure experiments, studies of the effect of high pressure on chemical reactions are also found. Not to mention the fundamental statement of Le Chatelier on the effect of pressure on chemical equilibria, dating back to 1884, and its later formalization in terms of basic thermodynamic relations, several organic reactions were studied at the beginning of the 1900s at pressures as high as 1500 kg/cm^2 by Cohen and co-workers [2, 3], who were able to perceive the importance of the effect of pressure on the reaction rate and, above all, could first enlighten the complexity of the high-pressure chemistry involving, on the one hand, the deformation induced in the molecular edifice of the reactants and, on the other hand, the participation and rearrangement of the environment. Among the early systematic reports on high-pressure chemistry, one could also mention the study of polymerization reactions reported by Conant [4] because polymerizations, amorphization, and, in general, some kind of condensation processes will turn out to be a recurring feature of chemical reactions at ultrahigh pressures.

High-pressure science remained in its infancy until the fundamental work of Bridgman [1], who received the Nobel prize in physics in 1946 "*for his invention of an apparatus to produce extremely high pressure, and for the discoveries he made therewith in the field of high-pressure physics*." In fact, technological difficulties in assembling high-pressure equipment set severe limitations on practical methods of reaching the highest pressures compatible with the materials used in the available high-pressure apparatuses and on the accuracy of the measurement of the actual pressure applied to the samples under investigation. As a matter of fact, prior to the advent of the opposed-anvil apparatus of Bridgman and Drickamer (to be described in the next section), high-pressure experiments were confined to maximum pressures of a few thousand kg/cm^2 at most.

In Fig. 1 the variation of pressure encountered in natural environments and exceeding more than 60 orders of magnitude [5, 6], is shown and is compared with the range of static pressures that can be obtained by laboratory equipment. From the left side of Fig. 1, it can be also seen that, even if presently still limited, the range of pressure statically obtainable in available laboratory equipment can span more than 10 orders of magnitude. In Fig. 2 the profile of the pressure and of the temperature within the Earth as a functions of depth is illustrated, showing that many geophysical and geochemical processes can be imagined to have occurred or to occur in this giant natural high-pressure laboratory. A more detailed view of the P-T regions of interest in Earth and planetary sciences and of the static pressures (and temperatures) attainable by laboratory equipment, to be described in the following section, is shown in Fig. 3.

From the qualitative data represented in Figs. 1–3, it is not difficult to understand the origin of the increasing interest in high-pressure studies. In the first instance, they allow a knowledge of the behavior of materials under extreme conditions (of temperature and pressure) that is fundamental for an understanding of physical and chemical processes occurring in the Earth and planetary interiors. On the other hand, it has been found that in many instances, processing of materials at ultrahigh pressures can result in the formation of novel materials that can be recovered at ambient conditions and can exhibit

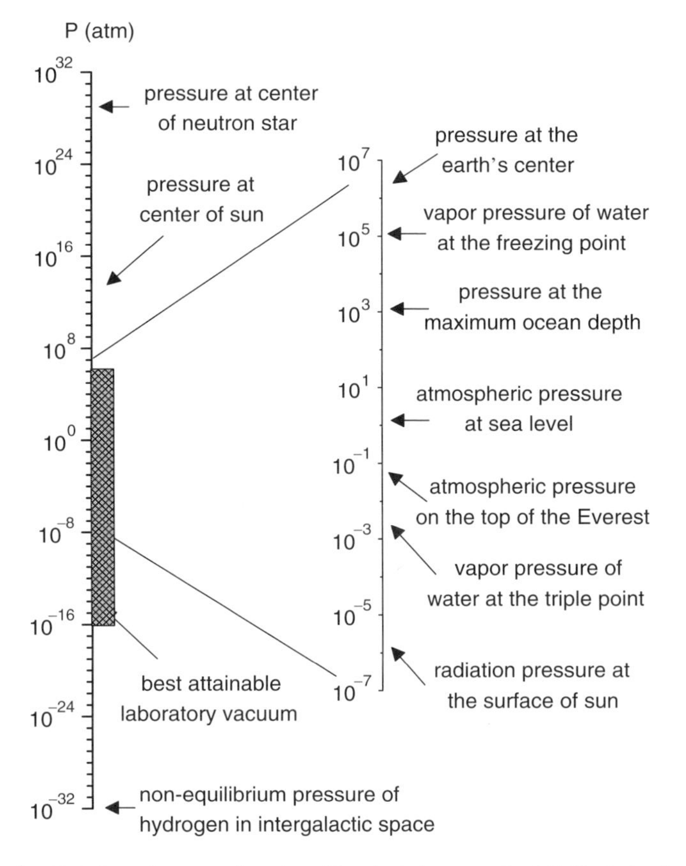

Figure 1. Variation of pressure in nature. The shaded area on the left illustrates the pressure range achievable by static experimental methods.

unique properties of high technological interest. This has been the case for the high-pressure synthesis of diamonds and more generally of superhard materials from light elements, ceramics, high-temperature superconductors, and magnetic materials [7, 8]. Of considerable interest is also the high-pressure treatment of biological materials [9, 10]. In conjunction with their practical importance, high-pressure science and technology are giving unique contributions to our basic understanding of the fundamental properties of matter. The primary effect of increasing the pressure is the reduction of volume and, hence, of the inter-atomic and intermolecular distances. As it will be discussed in the following, this allows us to explore and clarify otherwise hidden portions of the potential energy surfaces wherein the behavior of matter becomes highly unconventional.

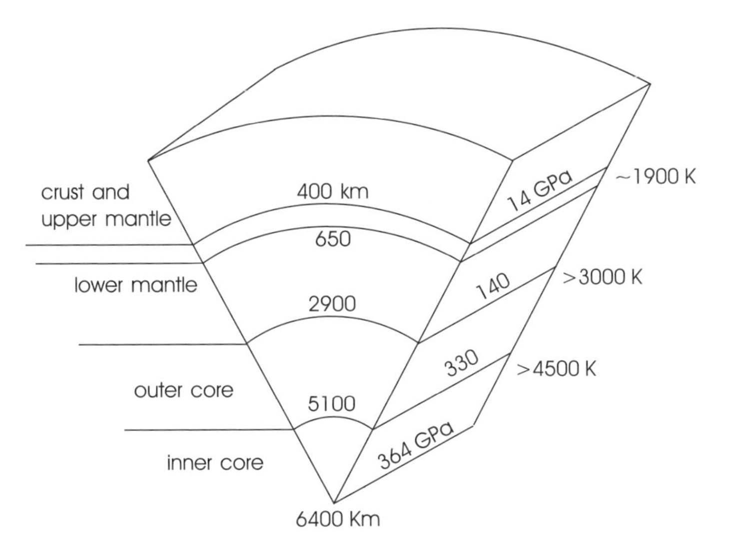

Figure 2. Schematic representation of the pressure and temperature conditions in the Earth interior.

High-pressure chemistry, although only a section of high-pressure science, has also been greatly developed in the last few decades, and the purpose of the present work is to illustrate and discuss the basic effects produced by ultrahigh pressures on chemical reactions. A classification of the processes occurring in materials subjected to high pressure has been proposed by Drickamer [11] and is

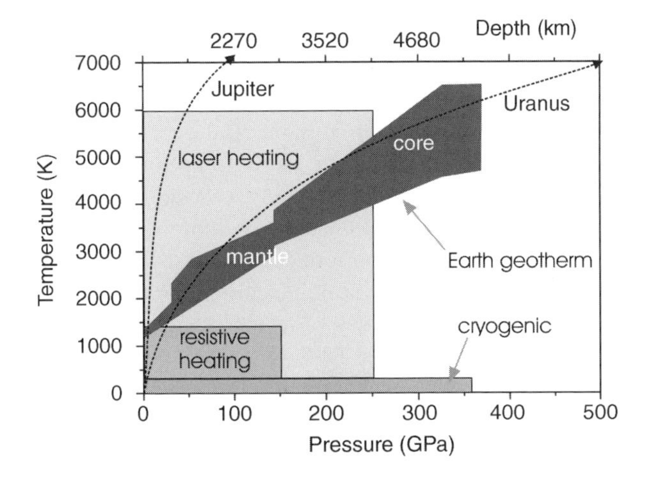

Figure 3. The regions of pressure and temperature accessible by joining static compression methods to heating and cooling techniques are compared to the Earth geotherm and to Jupiter and Uranus isentropes.

based on the assumption that the key point in discussing high-pressure effects is the evaluation of the changes induced in the electronic structure of the material. Consequently, Drickamer first classifies high-pressure transformations where changes of the electronic component is negligible. These include class I and class II of Drickamer's classification that correspond, respectively, to transformations leading to simple rearrangements of the atoms and to polymorphic transformations, with the latter involving also some changes of electric and magnetic properties. On the other hand, Drickamer identifies what he calls electronic transitions included in class III and class IV transformations. These correspond, respectively, to (a) simple electronic transitions occurring discontinuously and accompanied by a volume change and (b) electronic transitions leading to a new ground state, to cooperative phenomena, and to substantial changes of the configuration interaction. In this review we follow in essence the Drickamer's view implying a particular attention to the connection between the electronic rearrangement of the molecules induced by the high pressure and its chemical consequences. Of course it is well understood that any chemical reaction implies a rather substantial rearrangement of the electronic structure between the reactants and the products with disruption of chemical bonds and formation of new bonds, events that occur through some intermediate stage (the transition state complex). However, the attention in this work will mainly be focused on situations where the high pressure induces some kind of electronic transition on the reactant molecules prior to the chemical event and prior to the formation of a reaction intermediate. Ideally starting from the reactants at ambient conditions, one can imagine that pressurization induces a preliminary transformation of the starting material to a possibly quite different electronic structure such that the actual "high-pressure reactants" can follow reaction pathways hard to foresee for the same materials at ambient pressure. The high pressure is then essentially viewed as some kind of activation process bringing the material in the hidden region of the potential energy surface mentioned above. These effects are more likely to occur at very high pressure, even though a threshold can hardly be defined and will change for each system. At milder pressures, chemical reactions can be more "normally" affected in terms of shifts of the chemical equilibrium conditions, changes of reaction rates, and selectivity exchange in cases of simultaneously occurring reactions. As far as the effect of pressure is concerned, one can therefore loosely define two categories of chemical reactions. The first, very much resembling the ammonia synthesis from the elements, comprises reactions where the pressure is a powerful regulatory tool for the chemical equilibrium and the chemical kinetics. The second category, which is mostly the object of this review, includes reactions that follow peculiar pathways that are not predictable from a simple knowledge of the chemical structure of the reactants at ambient conditions. In various instances, and in connection with the unusual behavior of matter at high pressures, recourse has been made to the image of a new chemical frontier and even of a new periodic table where the elements occupy the same position as in the Mendeleev table but disclose completely different properties [12, 13]. These impressive and extreme statements simply summarize the results of a considerable number of high-pressure experiments where chemical reactions develop at variance with conventional laboratory practice. We shall concentrate on this particular type of chemical phenomena since they better illustrate the potentiality of a full exploitation of the pressure variable. More occasionally, reference will be made to the vast category of chemical synthesis, and we shall mainly be concerned with reactions of molecular systems in condensed phases, as it is the case at very high pressures.

The plan of this chapter is as follows. In Section II the basics of high-pressure technology and equipment are covered with particular reference to (a) the types of equipment that have actually been used to study chemical reactions and (b) the techniques in use for *in situ* and on-the-fly monitoring of chemical equilibria, products structure, reaction kinetics, and mechanism. Section III deals with fundamental concepts to treat the effect of high pressure on chemical reactions with several examples of applications, but with no claim of extensive covering of the available literature. In Section IV the results obtained in the study of molecular systems at very high pressures will be discussed, and some conclusive remarks will be presented in Section V.

II. HIGH-PRESSURE TECHNICAL SURVEY

This section contains a short overview of the technical aspects of the high-pressure generating and probing methods and is not intended as a full coverage of the field. Several excellent reviews and textbooks are available on this subject and will be quoted in the following; the interested reader should refer to them for detailed and extensive explanations. Here it is only appropriate to briefly describe the high-pressure devices that were assembled originally and that can be considered as large-volume devices. These have been used in many cases for organic synthesis at intermediate high pressure, a category of high-pressure chemical reactions that will not be dealt in too many details in this review. On the contrary, the diamond anvil cell (DAC) will be described more extensively because most of the chemical reactions discussed in this chapter have been carried with this high-pressure device.

A. The Piston–Cylinder Device

The simplest and most direct method for realizing pressure is achieved through the application of a mechanical force perpendicular and uniformly distributed over a surface. This is mostly realized by means of a piston sliding into a cylinder filled with a fluid, where the pressure is determined by the force per unit area

TABLE I
Conversion of Units of Pressure

	bar	Pa(10^5)	atm	kg/cm^2
bar	1	1	0.986923	1.019716
Pa (10^5)	1	1	0.986923	1.019716
atm	1.013249	1.013249	1	1.033226
kg/cm^2	0.980665	0.980665	0.967842	1

(conversion factors between commonly used pressure units are collected in Table I).

With this approach, final pressures ranging from few kilobars up to some gigapascals can be realized; as a consequence, the application of the piston–cylinder-based devices in studying chemical processes under moderate high-pressure conditions is incredibly wide. As a matter of fact, addition (Wittig, Michael), cycloadditions (Diels–Alder, 2 + 2, dipolar), ionogenic, substitution, and polymerization reactions have been extensively studied with these high-pressure devices [14]. A good overview of the state of art in this field is provided by Jenner [15].

The basic working geometry involves a piston sliding into a cylindrical bore machined in a containing vessel as schematically shown in Fig. 4. The sample is squeezed between the piston and the cylinder (closed-end cylinder geometry). Another geometry with two opposed pistons sliding one against the other along the cylindrical bore of the containing vessel can also be adopted (open-end

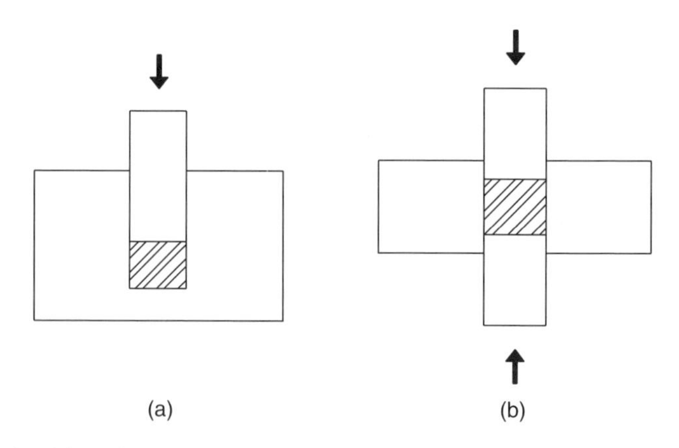

Figure 4. Schematic representation of piston–cylinder devices. (*a*) Closed-end geometry. (*b*) Open-end geometry.

cylinder geometry). With this simple concept, infinitely high pressure could in principle be generated. However, the practical realization of this experimental device presents several difficulties. If the ultimate theoretical limit is determined by the tensile strength of the material, limitations imposed by constructional geometries and size are often decisive in lowering this theoretical limit, mainly because of leakage and material distortion or rupture. Considering the components of these devices, the cylinder is not the weakest part of the apparatus. Piston, plugs, seals, and probing sections, like electrical leads and windows, are most likely prone to failure under loading. Several methods for increasing the performances, safety and reliability of piston–cylinder apparatuses have been adopted. Leak prevention in pipes and piston–cylinder devices was realized by Bridgman, who developed a leak-proof solid packing based on the "unsupported area" principle [1]. According to this principle, the sealing pressure is higher than the internal pressure; this is realized through a reduction of the area supporting the internal pressure.

Short and laterally supported pistons are stiffer, and a further reinforcement can be provided by conical shaping and additional gasketing support between the piston and the sample [16]. Sealing can be achieved with antiextrusion rings and wedges or mushroom-type seals based on the Bridgman unsupported area principle or in some cases with the employment of a teflon cell container positioned in the piston–cylinder devices [17]. Also probing gauges, like electrical leads or windows, can be introduced in the high-pressure vessel avoiding leakage, rupture and extrusion [18, 19]. Several designs can be adopted for introducing windows in high-pressure vessels in combination with indium or teflon gaskets and proper polishing of the seats, which must be perfectly flat [20, 21]. The observation of interference fringes is used to machine the seats, according to the criterion that the polishing is optimized to a single interference ring. The stress gradient between the supported and the unsupported portion of the window is greatly reduced with decreasing the aperture of the window, with respect to the portion in contact with the seat, but the unsupported principle for sealing is hardly matched in the case of very small inlet. Windows of diamonds would be the best choice in term of mechanical resistance and transparency over a large portion of the electromagnetic spectrum, but the cost is very high. Other materials like sapphire, Si, or Ge can be used, depending on the probing technique [22, 23]. Higher pressures can be reached using Drickamer windows [24, 25], which can sustain pressures as high as 15 GPa even without outer support. These windows are obtained by compressing NaCl into a small cylinder of successively increasing diameter, resulting in a clear transparent access to the high-pressure vessel. Friction forces are responsible for the high pressures sustained by these windows up to 15 GPa. Another type of optical access can be provided by the introduction of a bundle of fibers in the high-pressure container, which would be very promising for providing optical access in large-volume apparatuses. In this case, limitation in pressure up to a few gigapascals, sealing failure and fiber fragility are the main drawbacks.

Still today the piston–cylinder principle is at the basis of many large-volume hydrostatic high-pressure apparatuses dealing with pressure in the megapascal range [26]. In these devices, pressure is transmitted by a piston to the sample through a compressed fluid medium, thus granting a hydrostatic compression. The pressure transmitting fluid must be inert and stable and must remain fluid up to the maximum operating pressure. The sample itself can be used as compression medium, but normally this is not advisable, due to the risk of chemical reactions or precipitation of a solid product in the compression stages. In this case, it is mandatory to keep the sample and the pressure medium separated in different compartments. This separation may be achieved with a mechanical bellow made of an inert material that expands under the pressure exerted by the pressure medium. As a further improvement, in order to avoid unintended catalytic effects from the metallic parts of the vessel, the sample under investigation can be confined in an inner cell of quartz equipped with a movable piston, (consisting of quartz as well) perfectly fitting with the cylinder [27]. An alternative can be the employment of an internal cell made of polytetrafluorethylene (PTFE) fitting into the high-pressure autoclave in such a way that the walls of PTFE are in contact with the inner surfaces of the windows [26]. The sample is sealed inside the PTFE bag and is completely separated from the pressure medium. In transmission-type cells the changes in the optical path due to the compression or change of temperature is comparable with the thickness of the sample. In order to avoid considerable uncertainty in quantitative measurements on strongly absorbing material, an experimental device that allows to change the optical path length from the outside of the cell while under loading was developed [28].

Large-Volume Piston–Cylinder Apparatuses. Instruments able to generate pressures up to 10 kbar and working on cubic-centimeter-sized samples are in use. Their realization and operational conditions concern the maximum pressure, the sample volume, and the type of apparatus. The sample dimensions are crucial because the total dimensions cannot be increased indefinitely because of the strength limits of the materials. These devices are basically made up of two elements: one pressure container and one pressure-generating system. Pressure can be generated with a piston–cylinder or an intensifier vessel. In the first case a massive metallic pressure vessel is filled with the fluid and is compressed through a piston driven by a hydraulic ram. Sealing is provided through various types of Δ-ring and O-ring. These apparatuses operate very simply and can be combined with electric heater for working at high-temperature conditions, but their sample volume is limited in the 50- to 300 cm^3 range due to the balance of several factors related to the applied force and to the mechanical resistance and dimensions of the components. The employment of large bores requires (a) an increased force on the piston and (b) larger hydraulic rams and supporting frames. Furthermore, the length of the bore cannot be increased indefinitely due to difficulties in perfectly machining the internal walls and to the large volume variation associated with compression of liquids, which requires a longer piston and a longer displacement to achieve the desired compression. Pressure can also be generated through intensifier vessels working on the same piston–cylinder principle, but in this case the pressure is increased by gradually reducing the piston diameter through a sequence of compression stages. With this expedient, the size of the sample vessel can be increased regardless of the geometry because it can be separated from the pressure-generating unit [26]. Intensifying systems are generally more complex and expensive than piston–cylinder devices, but they can work on larger volumes. Dealing with solid samples, large volumes up to 1000 dm^3 can be pressurized in the 3 to 6 kbar range with the isostatic presses used for powder moulding. Electrical pumps able to supply oil at 10 kbar in a continuous way are now available, offering a more convenient approach with respect to intensifier techniques. High-pressure pumps offer the possibility of generating a continuous compression while providing a constant supply of reactants. Normally, reactants are not allowed to come in contact with the pressure generation unit in order to avoid corrosion of the vessel, catalytic effects, and contamination of the reaction, but this is not possible in the case of continuous flow reactors. In some cases, reactants are introduced in the high-pressure vessel in some type of flexible package compressed by a pressure transmitting medium.

Continuously operating reactors are employed for industrial purposes. These apparatuses work up to 3–4 kbar and 300°C and can be equipped with optical windows for visual observation and spectroscopic studies [29]. An example of this kind of device and of *in situ* measurements using the reflectance technique is described in Ref. 30. In general, high-pressure optical cells can be inserted in the main flow of a continuously operating reactor as a bypass separate chamber for monitoring purposes. In this way, not only the evolution of a chemical reaction can be followed, but also important parameters like the residence time distribution of the reactor can be checked—for example, by addition of a small amount of a tracer [31]—allowing optimization of the process. The free radical polymerization of ethylene [32, 33], acrylate [34], and methacrylate, as well as copolymerization [35] and terpolymerization [36], have been monitored by on line vibrational spectroscopy at high temperature and high pressure.

B. The Opposed-Anvil Devices

After the second half of the twentieth century, the performances of high-pressure devices began to be improved. The piston–cylinder device was modified by introducing external support for both the cylinder and the piston, and the concept of opposed-anvil and multianvil systems started to be successfully applied revolutionizing the static high-pressure field. The principle of opposed anvils is an evolution of the piston–cylinder principle and is based on the magnification of pressure by compression of the sample between massively supported tapered pistons. These pistons, with the flat tapered tips pointing one against the other, are named anvils. In this setup the maximum pressure experienced by the sample largely exceeds the compressive strength of the anvil material [37]. Extensive description of the most important device based on this principle, the diamond anvil cell, will be given in the following, while the most representative opposed anvil devices are schematically shown in Fig. 5.

In the Bridgman anvil device the sample is compressed between supported anvils of tungsten carbide, and lateral containment is ensured by a gasket that can be made of pyrophyllite or metallic material. A variety of cells based on the Bridgman opposed anvils has been developed [38–40]. Pressures up to 20 GPa

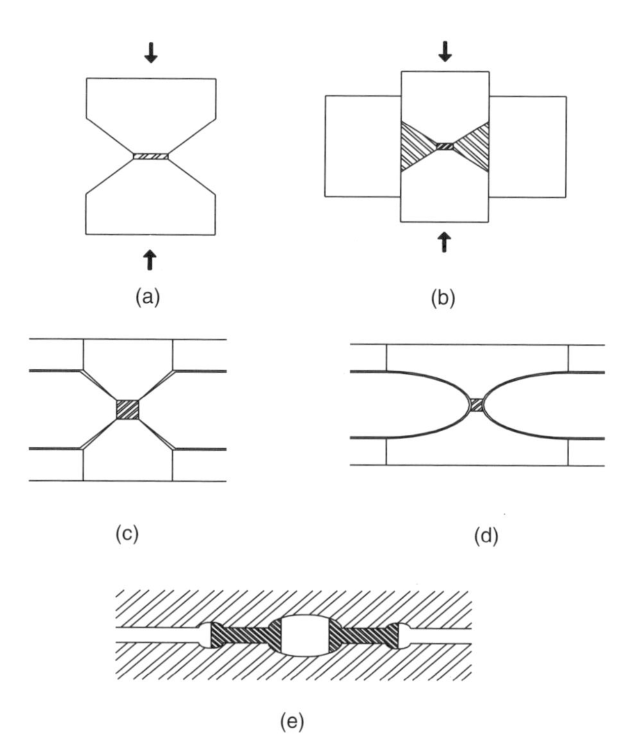

Figure 5. Different types of opposed-anvil devices: (*a*) Bridgman; (*b*) Drickamer; (*c*) girdle; (*d*) belt; (*e*) sample area of the toroidal anvil.

can be generated with this instrument, and the limit is fixed by the extrusion of the gasket. An intermediate version between the piston–cylinder and the Bridgman opposed anvil device was developed by Drickamer in his supported anvil device [41]. By inserting sintered diamond tips, pressures as high as 35 GPa could be reached. The interest in Bridgman anvils faded with the introduction of the diamond anvil cells but raised again with the availability of good-quality sintered diamonds, which allowed pressures to reach up to 40 GPa on cubic-millimeter-sized samples [42, 43]. This instrument can be also equipped with windows of transparent material to allow optical measurements up to 10 GPa [24] at ambient or low temperatures [44]. The two principles of massive support and lateral support marked a fundamental evolution of the piston–cylinder principle. Several other instruments were developed in order to reach higher static pressures and increase the sample dimensions [26]. In these devices, indicated as girdle, belt, and profiled Bridgman anvil high-pressure apparatuses, a radial support is continuously provided to the tapered pistons, and pressures up to 20 GPa can be applied on cubic-millimeter-sized samples, thus making these instruments appealing for high-pressure synthesis of superhard materials. Girdle devices use a conical piston profile, while belt apparatuses have a smoothed conical profile that has been shown to provide better support in the low-pressure region. The pistons are insulated from the belt or the girdle by a gasket assembly, and thermocouples or electrical leads can be inserted through the gasket. They are particularly suitable for synthesis of materials at high pressures and high temperature, but their assembly is quite complex with respect to the Bridgman anvil devices. The two main profiled Bridgman anvil apparatuses are the cupped and toroidal anvil cells. These instruments have a deep cavity in the center of the anvils, which can host a larger amount of sample with respect to the Bridgman and Drickamer cells ($\sim$100 cm^3); but in spite of this larger volume, the pressure limit is only slightly reduced. In the cupped type the anvils have a cup-shaped profile with a semispherical volume depression, while in the toroidal type the depression is formed by a central cavity and a circular concentric groove. In the toroidal cell the sample is confined within the cavity delimited by the profiled anvils and the gasket, which is squeezed in the volume of the toroidal groove. The role of the toroidal groove results in a favorable pressure distribution, support for the gasket, and smoothing of the pressure variation. This cell is comparable to the multianvil apparatuses in terms of volume and pressure performances but is very simple to assemble and is compact and convenient, with the main drawback being the absence of optical windows. Heater and pressure gauges can be placed in the sample cavity with the pressure transmitting medium. Due to the spheroid shape of the sample cavity, an almost perfectly hydrostatic pressure distribution is achieved on compression and high-temperature studies are allowed. The toroidal cell represents an improved version of the cupped anvil cell due to the increased stability and to the higher load [45–47]. This cell was the main device for the synthesis of superhard materials in the industry of the USSR but was not diffused in the Western countries until the early 1990s, because of doubts on the reliability of functioning of profiled anvils. Pressures up to 14 GPa can be reached with tungsten carbide anvils, and a small toroidal cell working on several hundred cubic millimeters can sustain pressures of 11 GPa and temperatures of ~200°C, while larger volumes (~800 mm^3) can resist up to 9.5 GPa and 1800°C. Pressures of 8 GPa have been applied to a ~200-cm^3 sample [48]. Several modified toroidal cells were developed and adapted to a large variety of techniques [21]. Among these, the Paris–Edinburgh cell [49], an improved version of the toroidal cell, is one of the most popular devices for high-pressure neutron diffraction [50] and x-ray [51, 52] studies, allowing pressures to reach 12 and 25 GPa when tungsten carbide and sintered diamond, respectively, are used. The possibility of *in situ* observation by x-ray diffraction may open the way to high-pressure and high-temperature synthesis, including also the polymerization of organic materials.

C. Multianvil Devices

In the case of solid samples, larger volumes can be compressed using multianvil systems. Some of these devices, which are composed of several identically shaped pistons pushed one against each other [16, 26, 53, 54], are illustrated in Fig. 6. The number of such anvils determines the shape of the sample, which is usually that of a regular polyhedron. Tetrahedral, cubic, and octahedral presses have been successfully employed, while hexagonal geometries have been abandoned [26]. Multianvil systems require a massive environment for experimental space. In the early experiments, independent hydraulic presses were used to drive each anvil. Today, thanks to the advances in machining technology, some guide blocks are compressed by a single hydraulic piston, allowing the driving of six to eight anvils. Different designs have been developed for this purpose. The Dia-type apparatus is one of the most widely used multianvil devices [55, 56]. In this device a guide block is used to compress six anvils that generate pressure on a cubic-shaped transmitting medium. The top and bottom anvils are fixed to the upper and lower guide blocks, respectively, while the four side anvils are fixed on the corresponding guide blocks. Lubrication and insulation between the guide blocks is ensured by teflon sheets and glass epoxy plates. One of the anvils, typically the top or the bottom one, does not move during compression, thus facilitating the localization of the sample. Exact knowledge of the sample position is crucial in high-pressure x-ray experiments for proper alignment of the beam and detector. Pressures up to ~10 GPa can routinely be accessed with this single-stage multianvil device, but higher pressures require a double-stage compression. In this case, pressures up to 26 GPa can be routinely reached.

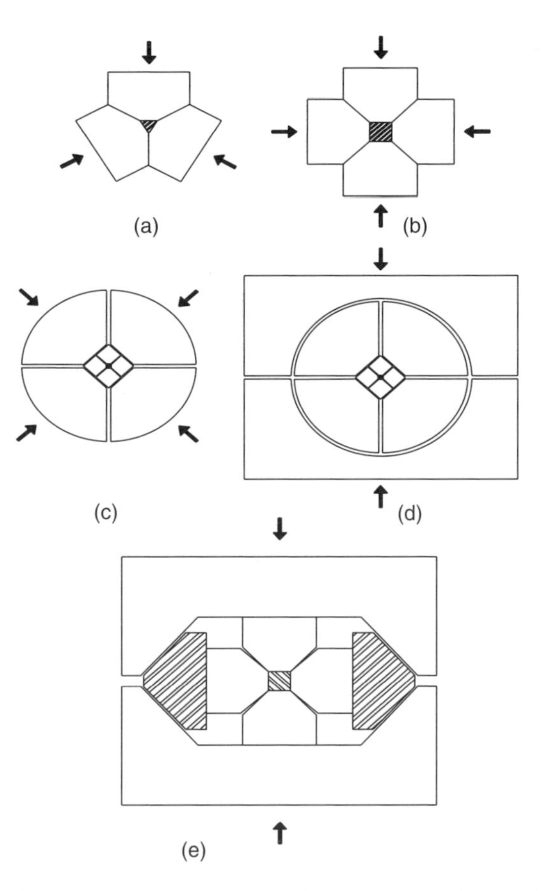

Figure 6. Most commonly employed multianvil presses: (*a*) tetrahedral, (*b*) cubic; (*c*) split-sphere; (*d*) Kawai-type; (*e*) Dia-type.

The split-spheres apparatus [57] is composed by a sphere split in six blocks. The tip of each block is truncated in such a way that assembling the cell results in a cubic-shaped space in the middle of the anvils. The second stage is realized by filling this cubic space with eight cube-shaped anvils. One vertex of each of the eight cubes is truncated so that an octahedral cavity is created. The whole assembly, composed of the six anvils of the external first stage along with the eight anvils of the internal second stage, was initially covered by two half-shells of rubber and put into oil in a high-pressure vessel. After we increase the oil pressure, a homogeneous compression is applied to the first stage and then to the second stage. Further modifications have been introduced, and compression can be mechanically achieved by means of three anvils fixed to the upper and lower blocks of the hydraulic press (Kawai-type apparatus) [58]. Another version of the Kawai-type apparatus, called Walker-type apparatus, has been widely used in petrology laboratories for reaching higher pressures with respect to piston–cylinder devices [59]. Research in this field, encouraged by recent advances in chemical vapor deposition methods for synthesis of diamonds, is looking forward to the possibility of employing large-volume diamond anvils both for increasing the accessible pressure range and allowing large-volume samples to be investigated by means of spectroscopic techniques.

D. The Diamond Anvil Cell

The use of diamond for pressure generation was first reported in 1950 [60] (in this case the diamond acted as sample container as well) and can be regarded as the beginning of a revolution in the experimental high-pressure techniques. In 1959 two different types of diamond anvil cells (DAC) were independently developed at the University of Chicago for high-pressure x-ray measurements [61] and at the National Bureau of Standards (NBS) for IR absorption measurements [62], allowing the obtainment of pressures as high as 10 GPa. Since the first versions of the DAC, many developments and improvements were introduced [63–68] as the metallic gasket technique for generation of hydrostatic pressure [69], the ruby fluorescence method for the pressure calibration [70–72], new media for hydrostatic pressure transmission at low [73], and ultrahigh (Mbar) pressure [74, 75], making the DAC the most widely used high-pressure device up to hundreds of gigapascals. The success of the DAC is essentially due to the extraordinary versatility and compatibility with many investigation techniques. The mechanical and optical properties of diamond made the DAC a suitable device for spectroscopic studies using electromagnetic radiations ranging from the x rays to the radio frequencies [12, 76] (with the only exception of UV and soft x ray, due to the absorption of the diamond between 5 and 5000 eV [26, 77–79]) and for electrical conductivity [80, 81], magnetic susceptivity, rheology, elasticity, and neutron diffraction and scattering as well [12]. The DAC has allowed the investigation of high-pressure phenomena in an astonishing and unpredictable range of combined extreme pressure and temperature conditions, ranging from 0.1 up to 500 GPa [82] and from liquid helium temperature [82, 84] (or down to a few millikelvins in some cases [85–88]) up to 5000 K [89, 90] using laser heating techniques. From a chemical point of view, the great advantage in using the DAC is represented by the possibility to maintain the required P–T conditions for the desired time allowing the kinetic, and then thermodynamic, study of a chemical reaction. The small sample dimensions did not prevent experimental measurements because many modern investigation methods can be performed even with a very reduced amount of sample. As a matter of fact, the DAC increased by more than one order of magnitude the experimentally accessible pressure range, making this variable much more effective in changing the density of matter, as handful as temperature to control. Several reviews cover technical details and evolution of the DAC [77–79, 91–95].

The basic idea of the functioning of the DAC derives from the opposed anvil concept, but in this case the anvils are made of diamonds, the hardest known material that can withstand higher pressures than any other material. As already stated, its transparency in a large region of the electromagnetic spectrum makes the optical investigation of the sample and *in situ* measurement of pressure possible through suitable sensors that can be inserted in the sample region [96–98]. The optical axis is usually perpendicular to the diamond culets. Seats, necessary for supporting the diamonds, are the limiting factor for the optical access to the sample area, thus they are usually made of very hard materials such as tungsten carbide and machined with cylindrical or cone-shaped optical access. If x-ray measurements have to be performed, the seats can be made of beryllium, which is x-ray transparent, but in this case lower pressures can be achieved.

Many types of DAC have been developed and optimized for different experiments during the years (see Fig. 7). While all are based on the opposed anvil system, they differ in the pressure-generating system exerting the force on the diamonds—that is, in the way the diamonds are pushed one against the other. All of them feature a precise guiding system for the movement of the diamonds (normally a piston–cylinder geometry or guiding rods are used) and very fine regulations for centering and aligning the diamond, in order to avoid the damaging of the diamonds at high pressure and to ensure monoaxial compression. On this basis the main schematic classification among the various types of DAC can be pictured dividing between mechanical [26, 99, 100] and pneumatic-driven systems [101]. Mechanical DACs include the National Bureau of Standard (NBS) squeezer-type cell [102, 103], the Bassett cell [100, 104, 105], the Mao-Bell cell [74, 99, 106–108], the Syassen–Holzapfel cell [109–111] and the Merrill-Bassett cell [112]. In all these cells, pressure is generated through lever arms, threads, clamps, and screws, and singular optimizations are made according to the experiments the cells are designed for [113, 114]. These types of cells have a very high mechanical stability and are suitable for constant pressure experiments; nevertheless, they present several disadvantages. As a matter of fact in the case of optical experiments, in order to regulate the applied pressure in such devices, the cell must usually be removed from the optical path of the instrument and realigned after the adjustment. This procedure can be very time-demanding, especially when dealing with vacuum chambers and heating or cooling instruments. Moreover, when keeping constant thermodynamic conditions is crucial, as in the case of kinetic studies, these devices prevent any instantaneous adjustment of the pressure. During polymerization reactions, for example, it occurs that a pressure drop due to

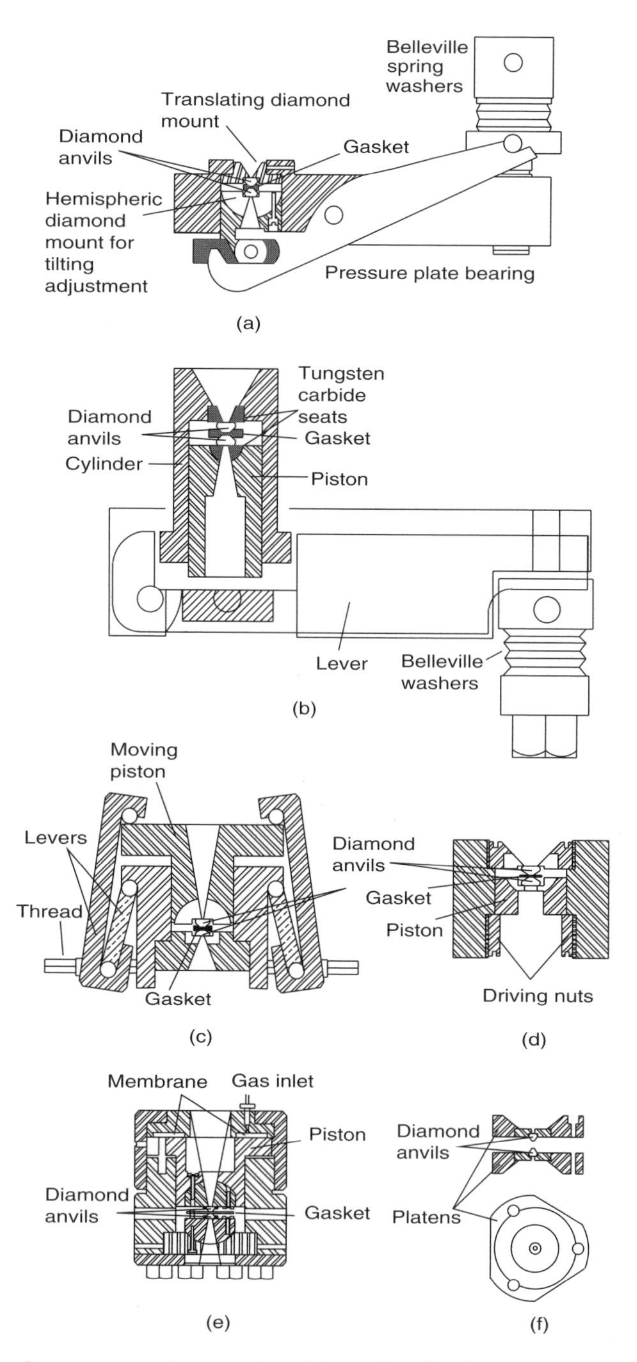

Figure 7. Some representative examples of diamond anvil cells: (*a*) NBS; (*b*) Mao–Bell; (*c*) Syassen–Holzapfel; (*d*) Bassett; (e) membrane-type; (f) Merrill–Bassett.

the contraction of the sample volume follows the polymer formation. If not compensated, this pressure variation can affect the kinetic evolution of the process or even stop the process itself [98]. Finally, pressure adjustments are not as fine as in the pneumatic systems. In the pneumatic devices the force on the diamond is applied by inflating a metallic membrane, diaphragm, or bellow by means of a fluid medium, usually helium. The membrane dilatation pushes one diamond against the other, with this mechanism typically being driven by a piston–cylinder movement or by rods. Pressure can be released with the same principle by deflating the membrane. Helium is usually used as pressurizing gas due to its low liquefaction temperature, which prevents any condensation of the gas inside the membrane down to 4 K while performing low-temperature experiments or simply in some loading procedures. When cooling or heating the cell, the helium pressure inside the membrane must be carefully monitored and adjusted in order to maintain the desired pressure inside the cell. Pneumatic DACs allow a fine and remote control of pressure, so that no realignment is necessary when adjusting the pressure. This is a significant advantage when performing low-temperature experiments or when monitoring kinetic evolutions. Several of these cells were developed for x-ray and optical spectroscopy experiments [101, 115–118]. The helium pressure inside the membrane is regulated through a thin, high-pressure stainless steel capillary ($\sim$1 mm in diameter) connected to a gas cylinder. This system is very suitable whenever fine-tuning of the pressure is required, as in the case of phase transitions, equilibrium processes, crystal growth, and processes (such as reaction kinetics) where compensation of pressure due to volume contraction is mandatory.

Diamonds. The diamonds for the anvils are polished from single-crystal gem quality and defect-free stones. Typical anvils for the DAC are 16-sided standard-cut stones with 2.7-mm height, with the large face having a diameter of 4 mm and the culet of variable diameter ranging between 50 and 700 μm, depending on the desired pressure and with the flat culet anvil set parallel to the (100) plane (see Fig. 8). The base-to-tip ratio between the diamond faces provides an indication of the pressure that can be reached. The higher this ratio, the higher the magnification of the applied force and the higher the resulting pressure on the

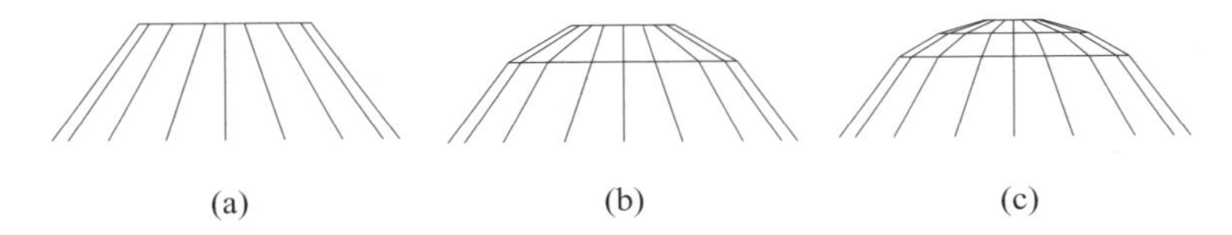

Figure 8. Diamond tips employed, in order of increasing final pressure, in experiments with the DAC: (*a*) standard, (*b*) bevel, (*c*) double-bevel.

sample squeezed between the diamonds. In order to reach high pressure and maximize the mechanical strength of diamonds, beveled and double-beveled cuts can be adopted (see Fig. 8).

Diamonds for high-pressure experiments can be selected on the basis of their optical and mechanical qualities, which vary according to the level of impurities [119]. Several reviews covering this topic are available [120–125]. The mechanical resistance of the stone is a fundamental requirement for making high-pressure experiments. Internal defects and cracks can cause a premature failure of the stones when applying pressure. A low birefringence indicates a low level of strain, cracks, and faults and hence a better mechanical resistance. Larger anvils, besides being economically unaffordable, are usually more fragile than smaller ones since the density of defects increases with dimensions. The advent of large synthetic diamonds obtained by means of chemical vapor deposition (CVD) will lift this drawback. Low luminescence is also a critical requirement concerning optical spectroscopy. Luminescence can be due to internal stress or impurities. Under laser excitation the presence of defects and impurities opens radiative decay paths that originate very broad fluorescence bands. The intensity of these bands changes in natural diamonds, so that selection must be made before planning an experiment. This is of particular importance in Raman and Brillouin spectroscopy, because the background signal can prevent any recording of signal from the sample [110, 126–129], while this is not a problem when using other techniques as in the case of infrared or x-ray spectroscopy. The position of the Raman phonon band of diamond at 1330 cm^{-1} is well known, while the far less intense two-phonon structure around 2500 cm^{-1} can be used for checking the quality of the stone. The accepted criterion, as a rule of thumb, is to compare the intensity of this band with respect to the fluorescence background [130]. However, this is not an absolute rule because a low luminescence in the two-phonon Raman region is not indicative of an overall low luminescence. Synthetic diamonds fabricated from defect-free and isotopically pure ^{12}C atoms show very low luminescence, even under compression [131, 132], providing extraordinary clarity and extending the range of optical measurements [133, 134]. A characteristic absorption region of the diamond in the IR spectrum near 2000 cm^{-1} is due to the two-phonon absorption. Depending on the amount and on the type of nitrogen impurities, several types of stones are classified according to the IR absorption spectrum. The main classification concerns type Ia and type IIa stones. Type Ia diamonds show a strong absorption in the region between 1100 and 1300 cm^{-1}, which can be very disturbing in IR spectroscopy. For this reason, type IIa diamonds, which are nitrogen-free, are more suitable in IR spectroscopy.

Even if a diamond is defect-free, failure can also be due to improper mounting and alignment. Mounting of the diamonds must be sturdy and ensure preservation of the alignment during compression. The diamonds are usually mounted by means of an epoxy glue or by means of soft copper rings set on the diamond support, which ensures a better stability. Alignment is also a critical parameter; therefore every DAC must allow a fine adjustment of the anvils concerning parallelism and centering. The procedure is normally made by direct observation of the diamonds through the optical axis. While centering is usually obtained by translation of the anvil supports [79], parallelism of the culets can be checked by tilting these supports through screws and observing the disappearance of the interference fringes when shining with white light from the rear and looking along the optical axis of the cell [135]. This method allows tuning of the parallelism in terms of the distance between the faces given by $\lambda/2$. A perfect alignment is fundamental for reaching very high pressures without damaging the anvils.

Diamonds can reach the highest pressures, but their use is limited by some severe constraints. Availability and cost of the stones are the main factors limiting the potential employment of diamonds on a large scale, forcing to use microscopic sample volumes. Despite its excellent transparency in such a wide range of the electromagnetic spectrum, the diamond signal can interfere with the sample in some specific regions covered by IR, Raman, Brillouin, UV, and soft x ray. Furthermore, high-pressure and high-temperature experiments can be performed in air only below 900 K, due to the carbon oxidation, and for higher temperatures an inert atmosphere is required. For these reasons, other materials have been sought for replacing diamonds in these conditions. Cubic zirconia [136] and sapphire [137] can reach pressures as high as 16.7 and 25.8 GPa, respectively. Sapphire is cheaper than diamond, transparent down to 140 nm, and has normally lower luminescence than diamond. For these reasons, it can be used to study wide-band-gap materials [138–140] or chemical transformations, because the transparency window allow excitation in the energy scale of chemical bonding—for example, using a 193-nm excimer laser. Moreover, it can be used in large-volume cells for neutron scattering and electrical measurements. Sapphire is very brittle, but pressures up to 16 GPa can be reached with anvil tips 1 mm in diameter and up to 8 GPa with 2 to 4-mm tips [141, 142]. A good alternative to diamonds could be represented by a synthetic gem-quality single crystal of moissanite (hexagonal SiC). Anvils of this material have been used to generate pressure up to 60 GPa [143, 144]. Moissanite is transparent to the visible light (0.4–5.5 μm) and, according to its characteristic Raman bands, the employment in Raman experiments is complementary with that of diamonds [143]. The hardness of moissanite is 3000 in the Knoop scale, compared with 2000 of sapphire and 1500 of cubic zirconia (5700–10400 of diamond). Moreover, moissanite has high thermal conductivity and has been heated up to 1100 K in air and near to 4000 K in laser heating experiments. Because moissanite exists in many polytypes, phase transitions could be a main drawback at extreme pressures and temperatures. The strong x-ray absorption can also be a problem using conventional x-ray sources, but not with synchrotron radiation. Moissanite anvils can be scaled up to three orders of magnitude larger than diamond anvils, allowing the study of cubic-millimeter-size samples. This is very important both for opening the path to neutron techniques and for improving already existing techniques.

Gasket. The introduction of the metallic gasket [69] marked a significant improvement in the history of the diamond anvil cell. The metallic gasket is a fundamental component of the diamond anvil system in order to ensure several critical functions: encapsulating sideways the sample, forming the high-pressure chamber together with the diamonds, sustaining a pressure gradient from ambient to the peak pressure, and providing lateral support for the tips of the anvils. The gasket is a metallic foil, differently shaped depending on the cell geometry and with a typical initial thickness of 100–250 μm, placed between the diamonds. Materials having simultaneously a large yield strength, a large friction coefficient on the diamond, and considerable ductility must be employed. The material must also be inert with respect to the sample in order to avoid unwanted reactivity that can cause contamination and corrosion of the gasket with loss of the sample and probable diamond failure. High-strength stainless steel has proved to be a practical material for many purposes; but in the very high pressure regime and at high temperature, rhenium is more suitable because of its higher yield strength. Boron and beryllium gaskets have also been used in order to provide optical access for x-ray measurements [145]. Gaskets with several parts optimized for different specific properties can also be realized. Inert materials like MgO or Al_2O_3 can be inserted in the metallic gasket in order to provide the electrical insulation necessary for allowing the introduction of electrical leads in the sample chamber [12]. A diamond coating can also be applied for increasing the shear strength of the central flat area of the gasket [146, 147]. Before use, the gasket must be preindented to a typical thickness of 30–50 μm by applying pressure to the diamonds. In this way the gasket deforms plastically and symmetrically on both sides, and its thickness decreases as a concentric polygonal craterlike deformation depending on the diamonds shape (see Fig. 9). The extruded material is fundamental for lateral support and for preventing failure of the anvils due to the concentration of the stress at the edge of the anvil culets [77]. A further reduction of the stress can be obtained by using beveled diamonds [148]. The sample chamber is obtained by drilling a cylindrical hole centered, within micrometric precision, with respect to the indented area of the gasket throughout the indented thickness. For this purpose, microdrilling machines or spark erosion techniques are generally used, while the alignment and the drilling process are followed by means of a suitable optical microscope. A centering as perfect as possible is important to ensure stability to the sample

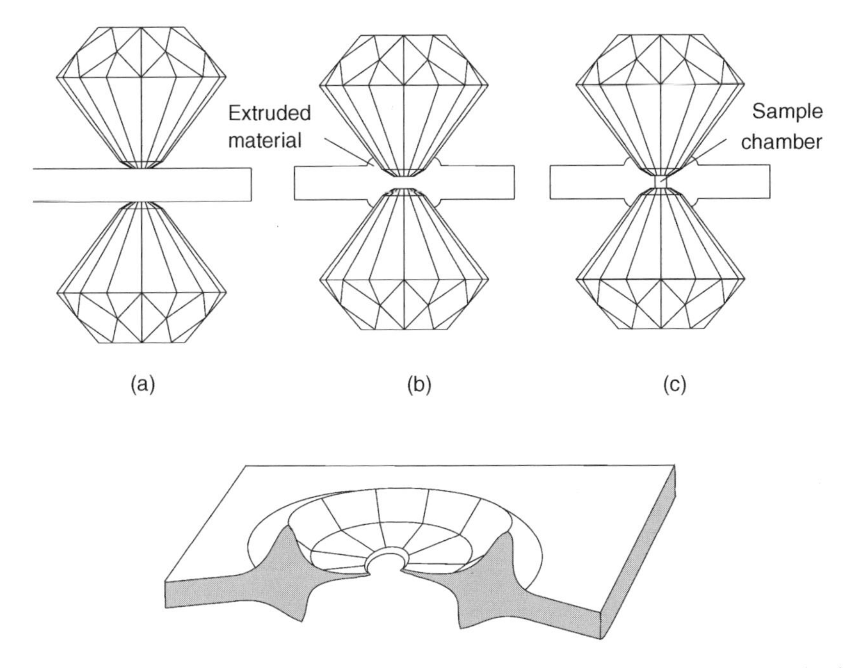

Figure 9. The gasket preparation procedure. (*a*) The metal foil employed for the gasket is inserted between the diamonds. (*b*) By applying load to the diamonds, the gasket is deformed by the anvils (indenting). (*c*) After the indented gasket has been drilled, it is properly positioned between the diamonds. In the lower section the final shape of the gasket after indenting and drilling is reported.

area when load is applied. The pressure distribution at the diamond surface is not constant but shows a pronounced peak at the center and decreases going to the edges of the gasket, as shown in Fig. 10. For this reason, it is advisable to use a gasket hole/culet area ratio as small as possible, in order to ensure the most homogeneous compression on the whole sample. Nevertheless, it must be kept in mind that the dimensions of the gasket hole will decrease with compressible sample and will increase with incompressible ones for sufficiently high pressures.

Loading and Compression Media. The loading procedures of the DAC depend on the thermodynamic properties and chemical characteristic of the sample. Liquid samples at ambient conditions are generally easy to be loaded, because a droplet can be positioned in the sample chamber to completely fill the gasket hole. Solid samples can be crumbled and cut in the desired dimensions and then positioned in the gasket hole. Powders as well can be loaded in the same way.

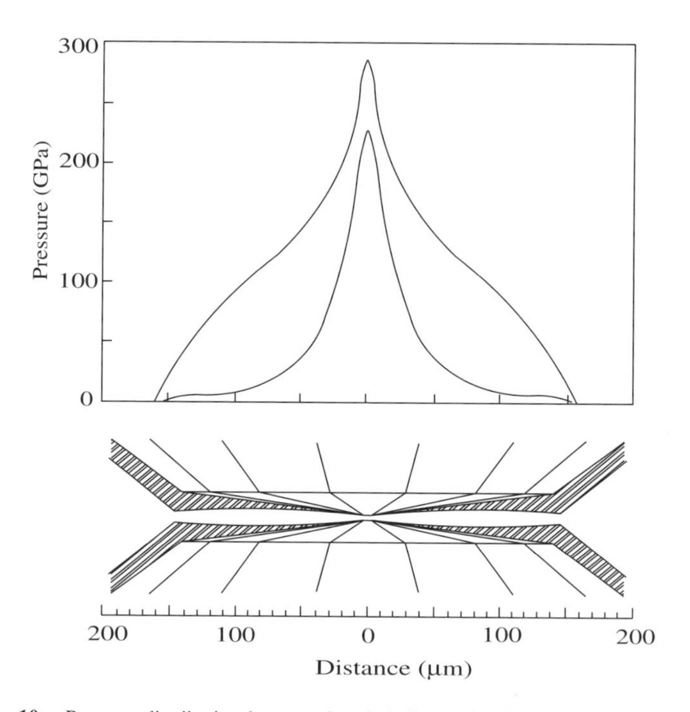

Figure 10. Pressure distribution between beveled diamond culets under two different loads. The central flat region, where the sample is confined, is 10 μm in diameter while the rest of the culet is immersed in the gasket material. The lower picture schematically shows the diamond culets corresponding to the two pressure profiles. The shaded area indicates the extent of the diamond deformation at the highest load. (Adapted from Ref. 149.)

Nevertheless, in these cases a fluid compression medium must be used to reduce the pressure anisotropy in the sample due to unfilled space in the sample chamber. Hydrostatic pressure conditions can be approached using liquids or condensed noble gases. Mixtures of methanol/ethanol (4:1) and methanol/ ethanol/water (16:3:1) can be used up to 10.4 [150] and 14.5 GPa [151] at ambient temperature, respectively. In the first case, above 10.4 GPa the mixture becomes a very hard glass with large pressure gradients (3 GPa over 100 μm at 50 GPa). The glass transition is slow and can be overcome by a fast pressurization; in this case, homogeneous nonhydrostatic conditions up to 35 GPa have been achieved [26, 152]. For pressures higher than 15 GPa, commercial silicon oil is also available. All these compression media present the remarkable disadvantage to be strongly absorbing in IR measurements and to perturb the chemical environment when a reaction occurs. Rare gases are the most appropriate hydrostatic compressing media, especially for the chemical inertia and for the employment at low temperatures [77]. Helium and neon can be loaded under moderate pressure (150 MPa) at ambient temperature enclosing the DAC in a loading vessel [75, 153, 154]. Helium solidifies at 11 GPa but a homogeneous distribution of pressure across the helium sample has been measured up to 60 GPa [155]. Evidence of a nearly hydrostatic behavior up to 120 GPa has been reported by Loubeyre et al. [156]. Neon can be considered hydrostatic up to 16 GPa [157], and no line broadening of the ruby gauge is observed in helium up to 70 GPa [155]. Argon is probably the most used gas because it can be easily loaded at liquid nitrogen temperature [158] and is a reliable quasi-hydrostatic compressing medium up to $\sim$30 GPa [159]. Xenon can be considered, according to the splitting of the ruby lines, as a hydrostatic pressure medium up to 55 GPa [160]. Hydrogen remains a very hydrostatic medium up to 177 GPa [21].

In the case of gaseous samples, several techniques are available depending on the thermodynamic and chemical properties of the sample. The general principle for loading a gaseous sample with the DAC is to condense it in order to have enough matter when applying compression to prevent the gasket closure. This can be obtained using low temperature (cryogenic loading) or high pressure (gas loading). In the cryogenic loading the low temperature is usually realized by means of cryostats or liquid nitrogen thermal baths. Condensation at liquid helium temperature is more complex and requires a vacuum chamber, a pumping system, and a helium cryostat. For higher temperatures, liquid nitrogen is more practical. The cell must be inserted in the condensation chamber with the diamond slightly spaced ($\sim$150 μm); then the container is purged, evacuated, and filled with the sample gas while continuing to supply the gas during condensation. When the condensation temperature is reached, the liquid starts to flood in the container; and as the sample chamber is filled, the DAC can be closed by sealing the sample between the diamond anvils. Different types of DAC may require different operational details for this procedure. The gas loading technique is based on the principle of filling the cell with a gas sufficiently dense for having enough condensed matter once pressure is applied. In this case, high density of the gas is achieved by means of a compressor, and typically the sample pressure should exceed 1 kbar. The sealing is achieved by applying an excess pressure on the diamonds.

When dealing with flammable, toxic, and explosives gases, the cryogenic and gas loading techniques may not be suitable. A different approach must be adopted by condensing a very small amount of sample directly on the diamond faces [98]. For increasing the probability of filling the gasket hole as the condensation occurs, advantage can be taken from the indium dam technique [161], in order to collect as much sample as possible in the sample chamber. This technique consists of the application of a thin indium ring fixed around the sample chamber. The cell is then cooled in an inert atmosphere. When the temperature of the cell is close to the condensation temperature of the sample, the sample gas is flown through a small capillary, inserted between the slightly spaced (~150 μm) anvils. In order to prevent the obstruction of the capillary due to the condensation of the sample inside it, a very weak flow of helium at room temperature is maintained through the capillary while the temperature decreases. Once the condensation has occurred, the cell can be closed by sealing the sample with applied pressure.

E. High- and Low-Temperature Techniques

One of the main advantages in using the DAC is the possibility of generating high pressure while maintaining full control over the other thermodynamic variables. The available techniques enable us to reach temperature conditions down to a few millikelvins and up to thousands of kelvins. The low temperature concerning the coupling of the DAC to commercial cryostats for optical and x-ray diffraction measurements has been extensively described [21, 26]. The main difficulties of this coupling concern the small working distance of the focusing and collecting optics that must fit to the sample dimensions and cryostat windows. Vacuum chambers including the cryostat with the cell and the necessary optics have been realized for both Raman [162] and IR [117] spectroscopies, but other solutions like the employment of fiber optics or suitably shaped cryostats are rather common.

From the high-pressure chemistry point of view, much more important is the possibility to heat, under controlled conditions, the compressed samples. Two main techniques are used for performing high-pressure and high-temperature experiments: the resistive heating and the laser heating.

Resistive Heating. In the resistive heating technique, electrical current is used to heat the sample in the DAC. Resistive heating can be further classified as internal or external. In the internal heating technique, only the sample chamber is heated by means of a resistive wire, and temperature is determined from the spectrum of the thermal radiation [163–166]. While the thermal insulation is ensured by the pressure medium, the electrical insulation can be achieved with Al_2O_3 or MgO coating of the electrical leads [165]. Despite temperature is very stable, gradients occur inside the resistive elements of the wire. The pressure is measured placing a ruby chip inside the sample chamber nearby the gasket, where temperature is close to room conditions. In this way the pressure can be measured, avoiding line broadening and frequency shift of the ruby fluorescence lines due to high temperature. The main drawbacks of this method are the large temperature gradients and the introduction of the electrical leads inside the sample chamber. Furthermore, the sample must be electroconductive. Within these limitations, the melting of tungsten has been observed, indicating a temperature as high as 3700 K.

In the external resistive heating technique the whole DAC is uniformly heated. This exposes diamond anvils to the risk of oxidation, graphitization (diamond is actually a metastable form of graphite), chemical reaction with the gasket material, and plastic deformation (above 1700°C). Note that 1500°C seems to be a limit temperature fixed by diamond failure, even in vacuum conditions and when using an inert gas flow. Two methods for external heating have been developed. One is to place the DAC inside a furnace. This method ensures a uniform temperature over the sample, but all the components of the cell are exposed to thermal stress, resulting in mismatch between the different materials composing the cell. Tungsten carbide, boron carbide, rhenium, inconel, and udimet 700 are generally used because of their low thermal expansion coefficients [167, 168]. Despite the difficulties in combining optical measurements with the use of a furnace, Raman spectra have been recorded [168, 169]. In the second method, only the cavity around the diamond anvils is heated [167, 170–172] . This allows a relatively easy employment of normal DAC for obtaining temperatures up to 1000 K at several tens of gigapascals. As a further improvement, an additional heater can be used as an external thermal shield to increase the temperature stability and improve the heating performances. Temperature as high as 1500 K and pressures up to 10 GPa were reached [167, 170], while 72 GPa at 1100 K were recently obtained [173]. Temperature can be measured by means of W–Re thermocouples, which can also operate above 2200 K [21]. The main problem is related to the positioning of these temperature sensors that must be placed as close as possible to the sample area in order to minimize errors in the temperature measurement. This drawback can in a certain extension be overcome by calibrating the thermocouple with some material having a known transition temperature [174, 175], but for very high temperature the spectroradiometry of the blackbody emission or the intensity ratio between the corresponding Stokes and anti-Stokes Raman bands should be used (see the laser heating section). Diamond and cubic boron carbide are very promising materials for this purpose because they have an intense Raman signal and they are chemically inert in most conditions. From the experimental point of view, measuring the Raman intensity in terms of absolute values in order to compare the Stokes and anti-Stokes regions is not trivial, due to different instrumental sensitivities and different luminescence of the diamonds in the two regions. In principle, also the ruby luminescence line shift could be used as a temperature gauge if pressure is known from any other technique, but practically for temperature exceeding 400°C the decrease in the signal-to-noise ratio prevents any use of the ruby for this purpose.

Laser Heating. The other main technique used for reaching high-pressure and high-temperature conditions is laser heating [176]. Geophysics is now the largest application field of laser heating at high pressure in the attempt to (a) gain experimental information on the interior of the Earth and of other planets and (b) test theories on their evolution and on their geomagnetic properties [177, 178]. Laser heating allows us to reach the highest static conditions of pressure and temperature [89] (4000 K at 200 GPa and 6000 K at lower pressures), while maintaining independent control on the two thermodynamic variables. The achievement of such high temperatures without damaging the diamonds is due to stability of the diamond with respect to the graphitization above 10 GPa [21]. Heating is obtained by irradiating the sample with high-power laser lines in the IR. Typical wavelengths are the 1.064-μm (Nd:YAG), the 1.053-μm (Nd:YLF), and the 10-μm (CO_2) laser lines, depending on the absorption of the sample. As shown in Fig. 11, the sample must be surrounded and thermally insulated by a chemically inert pressure medium having a negligible absorption of both the irradiating wavelength and the emission from the sample, which in these conditions becomes incandescent. Despite their high compressibility, which

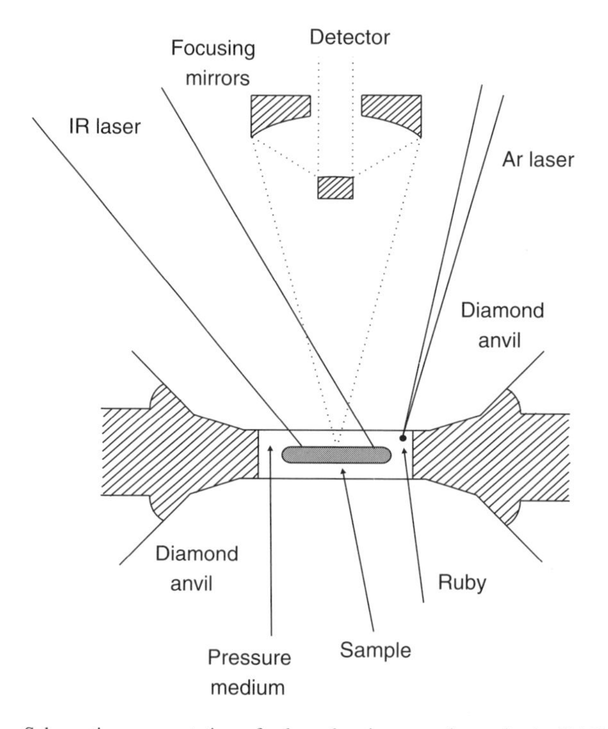

Figure 11. Schematic representation of a laser heating experiment in the DAC. The IR laser beam is directed onto the absorbing sample immersed in a compression medium acting also as thermal insulator. The thermal emission of the sample is employed for the temperature measurement, while the local pressure is obtained by the ruby fluorescence technique (see next section).

reduces the thickness of the insulating layer between the sample and the diamonds at high pressure, noble gases are generally used for this purpose, due to their low thermal conductivity.

In order to achieve a uniform temperature distribution, a homogeneous beam profile, a high power stability, and a precise and stable beam positioning are requested for the heating laser [179]. Furthermore, the beam can be defocused in order to irradiate the whole sample and minimize the thermal gradient. Very high laser power (50–150 W) is necessary because the diamond anvils absorb a significant part of the beam (~10% for a IIa-type and 30–40% for a Ia-type diamond anvil [176]). Moreover, because diamonds have a very high thermal conductivity, they act as thermal sink in conducting away from the sample a large amount of the generated heat, despite the presence of the insulating pressure medium. Of course, ordinary temperature gauges like thermocouples cannot be employed at such high temperatures. The temperature is measured by collecting the thermal emission of the incandescent sample and by fitting the emission profile with the Planck equation for the black-body emission [180–182]:

$$I(\lambda) = \frac{\epsilon c_1 \lambda^{-5}}{e^{c_2/\lambda T} - 1} \tag{1}$$

where $I(\lambda)$ is the measured intensity, ϵ is the emissivity, and c_1 and c_2 are constants. The whole procedure is not a trivial task. Several experimental difficulties have limited the accuracy of this method, by affecting both the generation of high temperature and its measurement. The main problem concerning the generation of high temperature is related to the large gradients in the heat distribution due to the localization and to the spatial profile of the laser beam. Power stability of the laser is also crucial, and fluctuations of less than 2–3% in the laser output may result in temperature fluctuations of ~200 K at about 2000 K [183, 184]. Several techniques allow us to stabilize the beam intensity and position, dramatically reducing the temperature fluctuations to a few kelvins [182, 185, 186]. Due to the large extension of the black-body emission at this temperature, reflecting optics with no chromatic aberration [187] and accurately calibrated CCD detectors are mandatory [176] to collect reliable spectra over the broadest possible emission range and to allow a correct fit of the emission profile.

The intensity versus wavelength distribution according to the Planck equation for the black-body emission is used to calculate the temperature (see Fig. 12). This calculation is based on two severe approximations. The first concerns with the assumption that the system is an ideal black body, which corresponds to assuming that the emissivity ϵ equal to 1. On the contrary, real systems are gray bodies that possess emissivity values less than 1. In addition, the ϵ dependence on the wavelength and on the pressure is generally neglected.

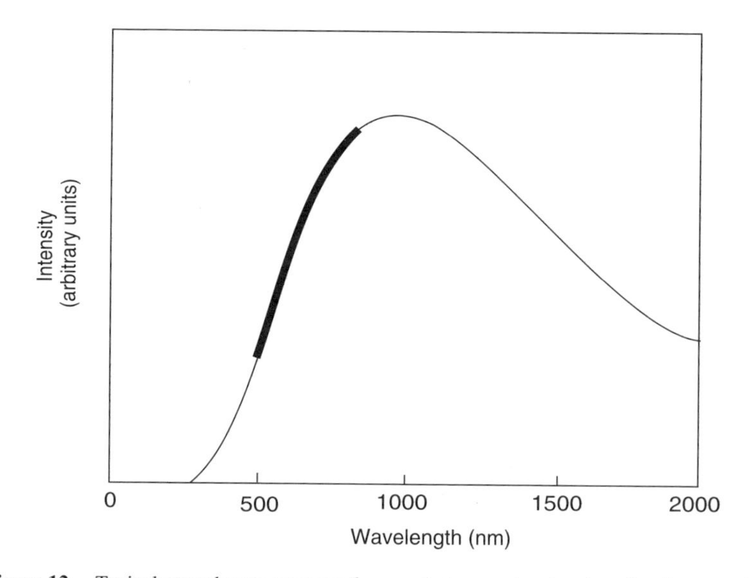

Figure 12. Typical procedure to measure the sample temperature in a laser heating experiment. The experimental emission spectrum (thick line) fitted according to Eq. (1) (thin line) gives a sample temperature of 3000 K.

The second approximation, whose results are extremely evident from Fig. 12, concerns the limited spectral region where the emission is measured, generally never exceeding the 450- to 850-nm range.

Temperature can also be measured from the intensity ratio of the corresponding Stokes and anti-Stokes Raman bands. The different intensities of these bands only depend on temperature regardless of the material and is determined by the Boltzmann distribution, which depends on the transition energy and on the temperature. This method is quite common in ambient pressure measurements, but only recently this technique has been applied to measure the temperature in laser heating Raman experiments on carbon dioxide at 67 GPa and 1660 K [188]. The employment of the Stokes/anti-Stokes intensity asymmetry in measuring temperature was also decisive to reveal important temperature gradients when transparent samples are heated by a metal foil employed as an absorber of the heating laser radiation. The temperature measured by the analysis of the emitted light, prevalently due to the metal foil, was always higher than the real sample temperature measured by the Stokes/anti-Stokes intensity ratio [189]. This difference is extremely relevant ($\sim$400 K) below 40 GPa, but it decreases continuously on further compression due to the larger thermal conductivity of both the sample and the metal foil with increasing pressure.

F. Dynamic Techniques: Shock Waves

Dynamic techniques are based on shock wave compression, which can have either mechanical or electromagnetic origin. The pressure range accessible by these techniques ranges from few gigapascals up to hundreds of megabars, but the overlap with the static (DAC) approach is extremely limited because temperatures as high as thousands of kelvins are generated together with pressure.

Shock waves represent a powerful experimental tool in probing high P–T states, such as Hugoniot equations of state, that are of both fundamental and geophysical interest. The capability of producing such high-pressure and high-temperature values also allows hot dense plasma to be produced and the corresponding properties investigated. From the chemical point of view, this kind of experimental study is rather uncommon due to the difficulty of, as well as the limited amount of usable techniques for, probing the reactive processes occurring in these conditions. Nevertheless, the chemistry that can be realized in these extreme high-temperature high-pressure conditions is very broad, covering topics of applicative interest such as the oxidation of small molecules in flames (combustion) or the detonation of energetic materials, but also fundamental arguments such as the formation and the dynamics of interstellar clouds and, more generally, the chemical processes occurring in planetary science [190, 191]. Recent applications of dynamic compression include the synthesis of materials recovered from shock compression such as nanocrystalline materials with controlled size distribution depending on the duration of the pulse [192], quenched films [193], superconductors [194, 195], hard material films [196], or fine-grained diamonds from fullerenes [197] and diamond powder for abrasives [198]. As a matter of fact, no other technique is able to provide a quenching rate up to 10^{12} bar/s and 10^{9} K/s as shock compression does. Finally, dynamic compression provides an absolutely independent measure of pressure and is of fundamental importance in static high-pressure science for calibration of the high-pressure scale.

Dynamic compression over 1 Mbar is achieved through planar shock compression by high-velocity impact obtained with light gas guns, pulsed lasers, high pulsed electrical currents, and explosives. Differently from the static compression, where virtually no restriction on time is imposed to the investigation of the sample, the dynamic compression is limited in time to the propagation of the shock wave and to the persistence of the steady-state condition after the shock front. Characteristic time scale for these processes is typically in the 10^{-7} to 10^{-9}-s range. For this reason, fast diagnostics, like VISAR [199] or transversal radiography [200, 201], are required. Measurements of sound velocities and electric conductivity, Raman spectroscopy, and x-ray measurements are available techniques under shock compression on a nanosecond time scale [21, 202–212]. Mechanical shock waves are obtained by shooting a planar impactor against a target. In this case, chemical propellents and compressed gases are used to accelerate the impactor [205, 213–215]. Higher pressures (up to 570 Mbar) can be achieved with underground nuclear explosions [211, 216, 217]. Electromagnetic shock waves are generated through the absorption of a high power (10^{15}–10^{19} W) short (1–100 ps) laser pulse focused on a thin metal plate. The ablated material, consisting of ionized plasma, expands in the vacuum, producing a shock wave in the opposite direction toward the sample [21, 128]. Shock waves can also be generated through the employment of a pulsed magnetic field used for detonating the explosive. This method can be viewed as intermediate between shock wave and static methods because the sample is compressed slowly enough to avoid the generation of shock waves and allows the process to be isentropic ($\sim 10^{-5}$ s) [219]. With this method, pressures up to 2000 GPa could be reached using a two-stage generator [21, 220]. Recently, a new instrument for producing quasi-isentropic compression of solids to high pressure using pulsed magnetic loading has been developed [211, 222]. The advantages of this technique derive from the possibility of (a) launching metallic flyer plates (typically aluminum or titanium) to velocities exceeding $\sim$20 km/s (three times higher than those allowed by traditional gas gun technology) and (b) producing a well-defined shock on the sample with constant pressure over a 30-ns pulse [223]. Furthermore, larger and thicker samples can be used.

The principle at the basis of the shock wave generation can be pictured in the following way. A disturbance traveling through a compressible medium, such as a solid object moving through the air, induces a disturbance of the medium which propagates as a pressure wave at the speed of sound. If the object is moving slower than the speed of sound, the medium is allowed to redistribute after the disturbance has passed by. Nevertheless, if the disturbance moves at supersonic speed, and then faster than the possibility of the medium to rearrange, a steep rise in the pressure and in other thermodynamic variables occurs, propagating through the medium. In static pressure conditions, like those realized in diamond anvil cells, hydrostatic conditions are achieved by isotropic compression and granted by the presence of a pressure medium. Hydrostaticity in shock compression is more complex to figure out, due to the strong anisotropy of the compression, which occurs along the direction of propagation of the shock wave. Nevertheless, quasi-hydrostatic conditions can be reached during shock pulses. Every material is characterized by a Hugoniot elastic limit (a few gigapascals for metals) related to the yield strength of the material. As long as the Hugoniot elastic limit is reached, the shock compression is uniaxial; but for stronger compression, hydrostatic isotropic components increase with respect to the uniaxial one, which remains constant [21, 96]. The velocity of propagation is called *shock velocity*, and the point of discontinuity in the thermodynamic properties of the medium is named *shock front*. The shock front is not an ideal discontinuity. More realistically, it can be pictured as a microscopic spatial region where the sample passes from the initial unshocked conditions to the compressed state in thermodynamic equilibrium of temperature and density. Rise-time values for the shock front depend on the duration of the pressure pulse and typically range from ~1 ns to ~1 μs. Within this time scale, the process is adiabatic and a temperature increase of several thousands of kelvins occurs on compression [224, 225]. The introduction of shocks generated by pulsed currents in which pressure is raised in few nanoseconds (against the picosecond rise time typical of mechanical compression), and new advances in designing shock profiles, allow to perform isentropic compression (ICE). This relatively long rise time decreases the temperature and increases the density of the sample. As a matter of fact, the higher the density achieved, the lower the temperature increase experienced by the sample [213, 222]. For this reason, multiple shocks and cryogenically and precompressed samples can be used to reduce the rise in temperature and therefore to sample different regions of the P–T diagram.

The behavior of shock waves is ruled by the Rankine–Hugoniot equations, which express the conditions for conservation of mass, momentum, and energy and can be used to design suitable shock profiles. Referring to the PV diagram (see left panel of Fig. 13), the compressed state (P_H, V_H) can be represented as

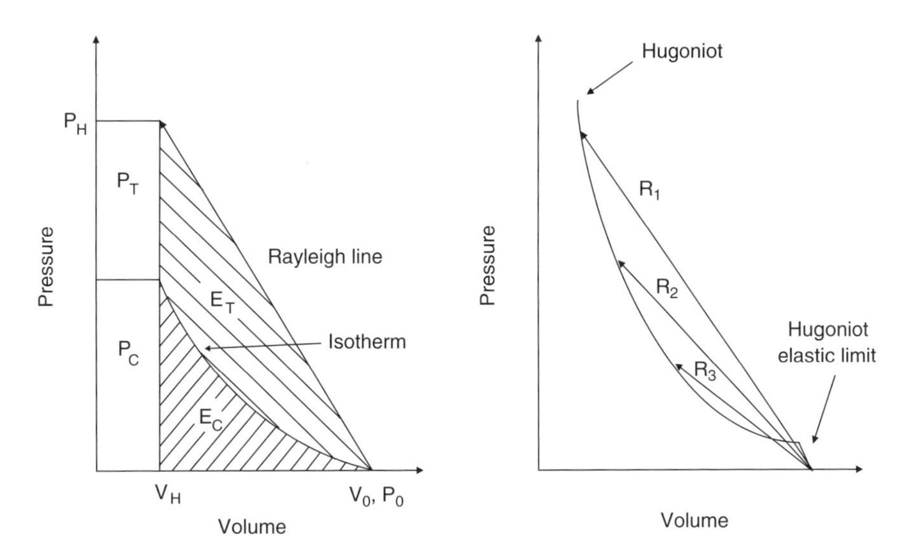

Figure 13. Left panel: Schematic representation of a shock compression in the PV diagram. **Right panel:** Building up of a Hugoniot by the Rayleigh lines (R_i) obtained from different shock experiments.

connected to the initial state through a straight line, called the Rayleigh line. This line does not represent the isotherm of compression, and a higher pressure characterizes the shocked volume with respect to the isothermal process. The increase in thermal energy is composed of two contributions: the zero temperature energy E_C, due to the isothermal process and represented by the area under the isotherm curve, and the thermal energy E_T represented by the area between the Rayleigh and the isotherm curve. This last contribution increases with the amplitude of the shock wave and accounts for the steep rise in temperature and entropy which accompanies the compression, leading to the appearance of a thermal pressure P_T. The locus of the possible states that can be reached from a given initial state upon different shocks is called Hugoniot (see right panel of Fig. 13). The Hugoniot does not represent the thermodynamic path of compression along the PV diagram. This path is given for every point of the Hugoniot by the Rayleigh line connecting it with the initial unshocked state. If the sample is in bulk thermodynamic equilibrium at both dynamic and static high-pressure–high-temperature conditions, the data from the two experiments can be related by a thermal model [213, 226].

G. Pressure Measurement

As a consequence of the reduced sample dimensions and of the complex experimental apparatuses employed to generate high pressure, the determination of the exact pressure experienced by the sample is generally one of the main experimental difficulties. Also, the realization of a pressure scale for calibration up to the highest accessible experimental conditions is a primary requirement. Different methods are employed, depending on which high-pressure apparatus is in use. The most accurate measurements of pressure are obtained with piston–cylinder devices, where pressure is derived absolutely and directly as the measurement of the force applied on a surface. This method is used for the realization of an absolute high-pressure scale and for pressure calibration of other secondary gauges. In these cases, great accuracy (10 ppm) is required for (a) the determination of the effective area and (b) corrections for friction and for buoyancy of the piston [26]. Absolute calibrations up to 5 GPa can be performed with this method, even if the accuracy is reduced above 1 GPa due to friction. Fixed-point secondary gauges play an important role in piston–cylinder, belt, and multianvil large-volume apparatuses, where no optical access is provided. They must have sharp and reproducible first-order transitions with very low hysteresis, and changes in the electrical resistance or in the volume are also commonly used [21, 26, 227]. For pressures higher than 5 GPa a pressure scale derived from a combination of theoretical and semiempirical calculations of equations of state and experimental data obtained from shock wave experiments [175, 228, 229] is used. Experimentally, the P–V relation can be obtained from the particle velocity (U_p) and the shock velocity (U_s) through the Hugoniot equations. These direct measurements provide a primary standard pressure calibration and can be used for calibration of secondary standards of more practical use. A primary pressure scale can be derived by combining measurements of elasticity and density [230], and it has been obtained from high-precision x-ray diffraction data and Brillouin spectroscopy on MgO in hydrostatic conditions up to 55 GPa [231].

The ruby fluorescence method is the most widely used technique for the local pressure determination in optical experiments [96, 103]. In this method the pressure is measured through the wavelength position of a ruby fluorescence emission, excited by absorption of suitable laser lines. A ruby chip with typical diameter ranging between 1 and 10 μm can be placed in the sample chamber, allowing *in situ* measurement of pressure. Ruby consists of Al_2O_3 corundum containing Cr^{3+} ions as impurities substituting the Al^{3+} ions. Depending on the concentration of Cr^{3+} ions pink, standard and dark rubies are obtained. The corundum structure is a hexagonal close-packed oxygen lattice with Al occupying two-thirds of the octahedral sites. When no lattice distortion is present, the symmetry of these octahedral sites is given by the O_h cubic group [232, 233], and the energy levels are well-described by the ligand field theory [234]. The energy terms deriving from the electronic configuration of the three d electron belonging to the Cr^{3+} ions are, in order of increasing energy, the $^4A_2(t_2^3)$ ground state and the $^2E(t_2^3)$, $^2T_1(t_2^3)$, $^4T_2(t_2^2e)$, $^2T_2(t_2^3)$, $^4T_1(t_2^2e)$ excited states. The electronic transitions from the excited to the ground state are defined as $^4A_2(t_2^3) \leftarrow {}^2E(t_2^3)$ (R transition), $^4A_2(t_2^3) \leftarrow {}^2T_1(t_2^3)$ (R′ transition), $^4A_2(t_2^3) \leftarrow {}^4T_2(t_2^2e)$ (U transition), $^4A_2(t_2^3) \leftarrow {}^2T_2(t_2^3)$ (B transition), $^4A_2(t_2^3) \leftarrow {}^4T_1(t_2^2e)$ (Y transition). Repulsive interactions between neighboring Al ions generate a slight trigonal distortion, lowering the aluminum site symmetry to C_3 and removing the inversion symmetry at the aluminum sites. This distortion, together with spin-orbit coupling, gives rise to a splitting of the energy levels. In particular, the $^2E(t_2^3)$ state gives rise to a two-level splitting, whose emissions are the R_1 and R_2 ruby fluorescence lines with respective wavelength of 694.25 and 692.74 nm at 300 K and ambient pressure [235, 236]. The ruby fluorescence method is based on the pressure-dependent shift to longer wavelength of the R_1 and R_2 ruby fluorescence emission lines. These emissions are very narrow and have a very high fluorescence quantum yield and a strongly pressure-dependent frequency shift. Furthermore, they maintain these characteristics with increasing pressure. All these features are shown in Fig. 14. The resolution of the R_1 and R_2 lines is an indication of the quasi-hydrostatic character of the environment.

The linear calibration of the ruby fluorescence emission, initially based on the equation of state of sodium cloride [71, 89, 96, 103, 237], is reliable up to 30 GPa for a quasi-hydrostatic environment. The quasi-hydrostatic calibration was extended up to 80 GPa [157], and a slight nonlinearity of the calibration curve at high pressure was found. Calibration of the ruby scale against primary

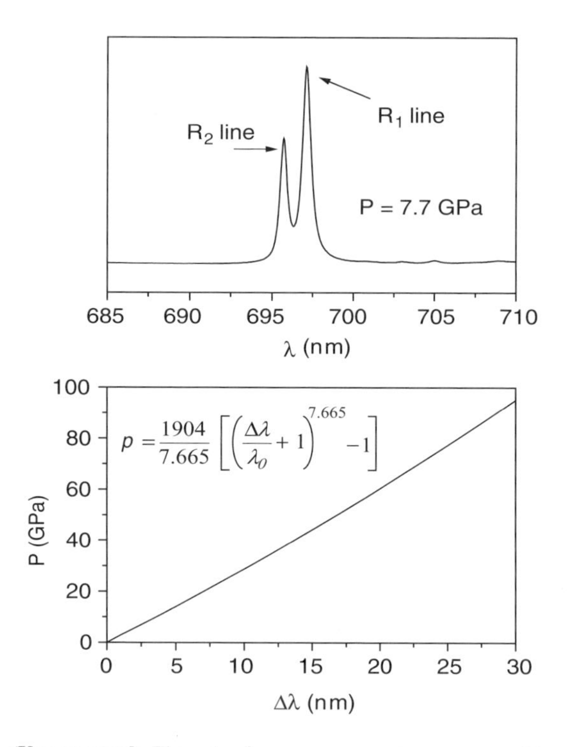

Figure 14. (**Upper panel**) The ruby fluorescence spectrum measured in quasi-hydrostatic conditions at 7.7 GPa. (**Lower panel**) The empirical law describing ruby R_1 line shift with pressure [96] is also reported.

shock-wave standards was performed up to 180 GPa [238, 239]. With improvements in spectroscopic techniques, the precision of the ruby measurements is now between ±0.5% above 20 GPa under hydrostatic conditions, but the shock-wave primary calibration still has a ±5% uncertainty and becomes the limiting accuracy for the ruby fluorescence secondary calibration [156, 240]. Experimental results have been reported in the 150 to 200-GPa range [241, 242] using extrapolations of the quasi-hydrostatic ruby pressure scale and in the 450 to 550-GPa range [243, 244] using extrapolations of the nonhydrostatic ruby pressure scale.

The empirical equation correlating the pressure and the ruby wavelength for quasi-hydrostatic conditions is

$$P = \frac{A}{B}\left[\left(\frac{\Delta\lambda}{\lambda_0}+1\right)^{B}-1\right] \tag{2}$$

where P is expressed in gigapascals, A and B are parameters whose values [157] are 1904 GPa and 7.665, respectively, and λ_0 is the wavelength (expressed in nanometers) of the R_1 ruby fluorescence line at ambient pressure and at the reference temperature, typically the room temperature. $\Delta\lambda = \lambda - \lambda_0$ is the R_1 line shift with respect to the ambient pressure wavelength. A correction for the temperature dependence of the wavelength can be applied by including a correction term $\Delta\lambda_T$ in the expression of $\Delta\lambda$. The expression of this term is given by a third-order polynomial:

$$\Delta\lambda_T = C_1(T - T_0) + C_2(T - T_0)^2 + C_3(T - T_0)^3 \tag{3}$$

where T_0 is the reference temperature and C_1, C_2, and C_3 comprise a set of coefficients. The temperature correction becomes important above 100 K, where a 6 K variation is equivalent to 1 kbar in terms of ruby fluorescence shift [245]. In spite of the very good accuracy of the ruby pressure scale, the significant temperature coefficient $C_1 \sim 0.007$ Å K^{-1}, the thermal line broadening responsible for the overlapping of the R_1 and R_2 bands above 400–500 K, and the decrease in the fluorescence intensity around 1 Mbar determine a large uncertainty in the determination of the ruby fluorescence peak. Provided that pressure is measured with another method, the isobaric frequency shift of the ruby fluorescence with temperature can be estimated [246]. In a recent study, a comparison between previously published x-ray diffraction data on diamond and Ta and low-pressure ultrasonic studies was performed, and on this basis a refinement of the ruby fluorescence scale was proposed, leading to the following equation [247]:

$$P = \frac{A}{B + C}\left(\exp\left\{\frac{B + C}{C}\left[1 - \left(\frac{\Delta\lambda}{\lambda_0}\right)^{-C}\right]\right\} - 1\right) \tag{4}$$

where $A = 1820$ GPa and $B = 14$ represent the slope and curvature at low pressure, while $C = 7.3$ is a parameter accounting for different curvature at high pressure. Equation (4) corresponds to Eq. (2) in the limit for $C \to 0$.

The ruby fluorescence emission is induced by laser excitation and can be revealed through a monochromator and a CCD detector. The wavelength and the power of the laser excitation are not restrictive at low pressure, and even few milliwatts of the 647.1-nm excitation line of a Kr ion laser can induce an easily detectable fluorescence emission. Any lower wavelength can be used as well. Typical exciting laser lines used are the 488- and 514.5-nm emissions of an Ar ion laser. Things are more complicated at pressures of ~100 GPa, where the ruby signal decreases in intensity and the two components are unresolved [235, 248–251]. Recently, it has been demonstrated by means of x-ray diffraction that ruby shows a phase transition to a *Pbcn* structure, analogous to Rh_2O_3(II), at 96 GPa and high temperatures (~1200 K) [252], and it has been observed that the ruby fluorescence spectra of the quenched samples at ambient conditions show a significant red shift and line broadening. This behavior is consistent with the different local environment determining the ligand field experienced by the Cr^{3+} ions and suggests a reexamination of the ruby pressure scale at high pressure and temperature. Another complication arises from the electronic absorption edge of the diamond shifting with pressure in the visible spectrum and absorbing the excitation wavelengths [249, 253].

Several other pressure calibrants working on the same principle of the ruby, the frequency shift with pressure of selected fluorescence lines, have been reported: Eu:YAG [254], Sm:SrB_2O_4 [255], Sm:SrB_4O_7 [256], Sm:SrFCl [257], Sm:YAG [258, 259]. Among these materials, Sm:YAG shows an intensity comparable with that of ruby, with the advantage that the pressure-induced frequency shift is temperature-independent, thus making Sm:YAG a very appealing pressure gauge eligible for high-temperature and high-pressure experiments without using any temperature correction. Unlike ruby, the Sm:YAG fluorescence intensity is strongly dependent on the excitation wavelength and excitation wavelengths lower than 500 nm are required.

The ruby fluorescence method allows us to perform pressure measurements in a short time scale (1–10^{-3} s), providing a real-time access for pressure control comparing to the time scale of many solid-state chemical processes. As a matter of fact, real-time pressure measurements are necessary when studying kinetic processes [117], but it is also important to minimize the laser power used for measuring the ruby fluorescence in order to avoid undesired photochemical effects on the sample, whenever these are possible. In the case of IR absorption studies, which are commonly used for kinetic purposes, the advantage of using the ruby fluorescence method, once photochemical effects are prevented, with respect to the employment of vibrational gauges is that no additional absorption bands are introduced in the IR spectrum.

Vibrational pressure gauges have been used both in infrared and Raman spectroscopy. In this case the pressure is determined by the frequency shift of a selected vibrational band under compression. The marker peak, besides fulfilling the requirement of stability, chemical inertness, and significant intensity and frequency shift with pressure, must be isolated and should not overlap with the sample signals. Infrared gauges, in particular, can be very useful to avoid risk of unwanted photochemical reactivity induced by the laser used for exciting the ruby fluorescence. Polyatomic ionic salts are the best candidates for matching these requirements thanks to the strong transition dipole moments of the antisymmetric vibrational modes. Small amounts of the selected salt are typically used as a dilute solid solution in a transparent matrix. The asymmetric stretching vibrations of NO_2^- and NO_3^- have been successfully used as pressure calibrants in the infrared region [97]. In particular, the antisymmetric stretching mode of the NO_2^- mode has been characterized up to 50 GPa by using a 3% solution of $NaNO_2$ in NaBr [98]. $MgCO_3$ is also reported to be a useful vibrational gauge for pressure in infrared spectroscopy due to the pressure behavior of the asymmetric C—O stretching vibration of the CO_3^{2-} ion [260].

As Raman experiments are concerned, the N_2 vibron has been employed [261], being particularly useful whenever nitrogen can be used as a compression medium. The Raman signal from the diamond is reported to be used as a pressure gauge, taking advantage of both (a) ^{13}C isotopic enrichment [262] and (b) the splitting and frequency shift of the diamond peak upon uniaxial stress along the [100] direction [263]. Indeed, in these conditions the threefold degenerate Raman band of the diamond splits in a singlet and a doublet. The stress-induced shift and the relative intensities of the singlet and doublet depend on the distance from the center of the anvil culet in the sample–anvil interface, and specifically the singlet intensity increases with respect to the doublet one moving away from the center. Also, cubic boron nitride (c-BN) [264] and quartz [265] are reported to have a useful Raman signal for pressure calibration.

Finally, in x-ray diffraction measurements the knowledge of the equation of state (EOS) of specific materials is used to calculate the pressure by the measurement of selected reflections. For this purpose, Au [266], NaCl [267], Re [268], Pt [269], and MgO [231] have been used in x-ray scattering measurements when no optical access was available.

III. FUNDAMENTALS

In this section, some general considerations concerning the effect of pressure on chemical reactions will be presented. Since at the high pressures considered here materials are commonly in condensed phases (liquids or solids), the attention will be mainly confined to such systems. First, the thermodynamic relations between the equilibrium constant, the reaction rate, and pressure will be discussed. Since, as already mentioned, the electronic structure of the reactants can change considerably with increasing pressure, the application of general thermodynamic equations can trun out to be nontrivial because the type and number of feasible processes can be very different at high pressures. In connection with this, some generalities about the changes of the electronic structure induced by the pressure in molecular liquids and solids will be discussed. This point, together with the discussion of the effect of pressure on reaction rates, will bring into evidence the importance of environmental and intermolecular interactions on the dynamics of high-pressure reactions. The importance of the environment is of particular relevance in crystalline solids, where the spatial arrangements of the molecules imposes geometrical constraints that can play a basic role in determining the reaction pathway. Therefore, the connection of the topology of the system with the reaction mechanism will also be discussed. The viscosity and the mobility of the molecules are greatly affected at high pressure; consequently the potential barriers for molecular rearrangements needed in the various reaction steps are considerably enhanced, and in many instances it may be necessary to consider the pressure and temperature parameters simultaneously.

A. Effect of Pressure on Chemical Equilibria

The equilibrium condition of a chemical reaction

$$aA + bB + \cdots \rightleftarrows lL + mM + \cdots$$

is determined by the minimum of the Gibbs free energy, and the basic thermodynamic relation for the equilibrium is

$$\Delta G^\circ = -RT \ln K \tag{5}$$

where ΔG° is the free energy difference between the products and the reactants in their standard state:

$$\Delta G^\circ = l\mu_L^0 + m\mu_M^0 + \cdots - a\mu_A^0 - b\mu_B^0 - \cdots \tag{6}$$

where μ_i^0 is the chemical potential of i in its standard state (to be defined as the pure state of the component at the T and P of the reaction; for components in solution, the standard state of the solute i is the state of the single component i at infinite dilution in the chosen solvent, at the T and P of the reaction) and K is the equilibrium constant given in terms of the activities a_i of the components in the system at the equilibrium:

$$K = \frac{(a_L)^l (a_M)^m \cdots}{(a_A)^a (a_B)^b \cdots} \tag{7}$$

The chemical potentials μ_i^0 depend on pressure, temperature, and chemical composition, and in particular

$$\left(\frac{\partial \mu_i^0}{\partial P}\right)_T = V_i^0 \tag{8}$$

where V_i^0 is the molar volume in the standard state. Therefore the dependence of the equilibrium constant on pressure is given by

$$\left(\frac{\partial RT \ln K}{\partial P}\right)_T = -\Delta V^0 \tag{9}$$

where ΔV^0 is the difference of the molar volumes in the standard state between the products and the reactants. Expressing the equilibrium constant in terms of molecular concentrations, the relevant quantity for its pressure dependence can be shown to be rather the variation of the partial molar volumes in the reaction mixture. Equation (9) shows that for reactions involving a volume contraction, the equilibrium constant increases exponentially with pressure at constant temperature. Increasing the pressure from ambient conditions to 1 GPa, along with a volume contraction of 10 cm^3/mol, the equilibrium constant increases by almost two orders of magnitude. This extremely large variation shows the importance of pressure as a regulatory tool for chemical equilibrium in the case where the reaction is associated with a volume variation and is not controlled by dynamic factors. The above estimate of the effect of pressure on the equilibrium constant was obtained according to Eq. (9), therefore neglecting that the molar volumes can change with pressure (i.e., assuming that ΔV^0 is constant). Application of Eq. (9) in the more general case requires the knowledge of compressibility data over the pressure range of interest. This information is not available in most cases.

In the simplest approximation, one could calculate the volume contraction of a reaction as the difference of the van der Waals molar volumes of the products and reactants. As described by Hamann [207], this allows a microscopic interpretation in terms of types of reactions (like, for instance, dimerizations, cyclizations, condensations, isomerizations)—each having, on average, a characteristic volume change—and in terms of formation of an additional number of chemical bonds in the products compared to the reactants. As an example, we report in Table II the volume changes for different classes of reaction involving pure liquid hydrocarbons, which are nearly nonpolar molecules. Such a procedure is qualitatively very useful but can be applied with a considerable degree of confidence only to reactions of nonpolar molecules giving rise to nonpolar products. In this case, in fact, the intermolecular interactions are comparable in reactants and products and environmental effects are negligible.

A quite different situation is encountered when the reaction involves polar or ionized species where the intermolecular interactions can involve rather significant volume effects. In essence, polar or ionic species produce a closer packing of the surrounding molecules and thus a volume contraction. The electrostriction of a dielectric medium in the presence of ions or polar molecules has been discussed at various levels of approximation. For the purposes of the present chapter, it can be noted that one of the general effects of pressurization can be the ionization of the starting molecules. Indeed ionization, or likewise the formation of polar species, leads to a closer packing of the material and thus to an increase of the equilibrium constant in agreement with Le Chatelier's principle. In the following, examples will be reported demonstrating that indeed

TABLE II
Volume Change ΔV^0 Accompanying the Formation of One Mole of Product in Reactions of Hydrocarbons in the Liquid State [271]

Reactions	Net Increase of Covalent Bonds	ΔV^0 (cm^3/mol)
Dimerizations		
2(1-Pentene) → 1-Decene	1	−30
2(1-Hexene) → 1-Dodecene	1	−28
2(1-Octene) → 1-Hexadecene	1	−27
Trimerizations		
3(1-Hexene) → 1-Octadecene	2	−36
Cyclizations		
1-Pentene → Cyclopentane	1	−15
1-Hexene → Cyclohexane	1	−17
1-Heptene → Cycloheptane	1	−20
3(Acetaldehyde) → Paraldehyde	3	−36
Isomerizations		
n-Pentane → 2-Methyl butane	0	+1.2
n-Hexane → 2-Methyl pentane	0	+1.2
n-Hexane → 2:2-Dimethyl Butane	0	+2.0
cis-2-Butene → *trans*-2-Butene	0	+2.6

ab initio molecular dynamics simulations show that at high pressure even nonpolar molecules (like hydrocarbons) can acquire a significant molecular dipole moment as a result of cooperative intermolecular interactions. This is found to occur before the onset of any chemical reaction.

The general considerations discussed above apply also to equilibria in condensed phases. In liquids the increase of viscosity with pressure and the reduced diffusion of molecules may prevent the attainment of the equilibrium in reasonable times. In solids, because of the high-energy barriers and the hindrance of the environment on the reacting molecules, the system may remain in a metastable state for an indefinite time. For the cases where the equilibrium is reached, Drickamer has proposed a thermodynamic theory [11], originally developed to describe electronic transitions in iron compounds and then applicable to describe a wide variety of equilibria as a function of temperature and pressure. The Gibb's free energy of a mixture of two species at equilibrium, with Gibbs energies G_0 and G_1, is

$$G = N_0[(1 - c)G_0 + cG_1 + c(1 - c)\Gamma] - T\sigma_{\text{mix}} \qquad (10)$$

where N_0 is the total number of molecules, c is the fraction of component 0 converted to 1, and Γ is an interaction term indicating that G in condensed phases is not simply the difference between the free energies of formation, but also depends on the fraction converted. Γ therefore accounts for environmental effects. G_0, G_1, and Γ are a function of pressure and temperature. σ_{mix} is the mixing term

$$\sigma_{\text{mix}} = k_B[N_0 \ln N_0 - N_0 c \ln N_0 c - N_0(1-c) \ln N_0(1-c)]$$

Applying the equilibrium condition $\left(\frac{\delta G}{\delta c}\right)_{P,T} = 0$ and defining $K = \frac{c}{1-c}$ and $\Delta G = G_1 - G_0$, one finds

$$\ln K = -\frac{\Delta G + (1-2c)\Gamma}{k_B T} \tag{11}$$

To obtain a description of $\ln K$ as a function of pressure and temperature, Γ should be made explicit and various degrees of approximations may be introduced. For example, the interaction term Γ can be neglected and a bulk modulus $B = -V\left(\frac{\delta P}{\delta V}\right)_T$ is assumed to be constant with pressure. These are very strong approximations and lead to results that are only in qualitative agreement with experiments. Otherwise, Γ is expanded as

$$\Gamma(P, T) = \Gamma_0(T) + P\Gamma_1(T) + P^2\Gamma_2(T)$$

where

$$\Gamma_1 \propto \left(\frac{\Delta V}{V}\right)\left(\frac{\Delta B}{B}\right)$$

and

$$\Gamma_2 \propto \left(\frac{\Delta B}{B}\right)^2$$

where ΔV and ΔB are the differences in volume and bulk modulus between the components 0 and 1 [272]. In analogy with the theory of ferromagnetism, graphical solutions are found for c at the various levels of approximation. The possibility of having discontinuous jumps in c on raising or lowering temperature and pressure, and hysteresis when metastable states are maintained for some time, is predicted [272], as experimentally observed for the low-spin–high-spin transitions in ferrous phenanthroline and ferrous bipyridyl compounds [273]. Actually, the interaction term Γ describes the cooperative effects typical of the solid phase, which will be discussed in the following.

B. Effect of Pressure on Reaction Rates

The important effect of increasing pressure on the kinetics of chemical reactions has been noted since the first chemical experiments at high pressure. The simplest expectation derives from the observation that in liquids the viscosity rapidly increases with pressure. As a result, in strongly compressed liquids, and finally in glasses, diffusion-controlled processes can be retarded. In contrast, however, other reaction pathways can be substantially accelerated. In general, the evolution of a reaction at high pressure can be heavily controlled by kinetic aspects, and these deeply involve intermolecular effects.

A chemical reaction is in most cases the result of an overall balance of a number of steps, called elementary reactions, whose rate law can be deduced from the stoichiometry. The rate law of an elementary reaction has the form

$$v(t) = k \prod_{\text{reagents}} c_i^{m_i}(t) \tag{12}$$

where c_i and m_i are the concentration and stoichiometric coefficient, respectively, of component i. Understanding the mechanism of chemical reactions requires the determination of the sequence of its elementary steps, a difficult task to be realized in most cases on the basis of experimentally determined rate laws. For reversible elementary reactions the equilibrium constant (in terms of concentrations) equals the ratio of backward to forward reaction rates (principle of detailed balance). For simple reactions, and in many other cases as well, it turns out that the overall reaction rate coincides with that of the slowest elementary reaction, which is thus referred to as the *rate-determining step*. Interpretation of the pressure dependence of the reaction rate should primarily concentrate on the pressure effect on the rate-determining step. In turn, studies of the pressure dependence of the reaction rate can give very useful information on the reaction mechanism.

Besides its temperature dependence described by an Arrhenius-type equation

$$\frac{\partial \ln k}{\partial T} = -\frac{E_a}{RT} \tag{13}$$

where, according to the transition state theory, E_a is the activation energy (the energy difference between the transition state and the reactants), the reaction rate strongly depends on pressure according to the relation

$$\frac{\partial \ln k}{\partial P} = -\frac{\Delta V^{\neq}}{RT} \tag{14}$$

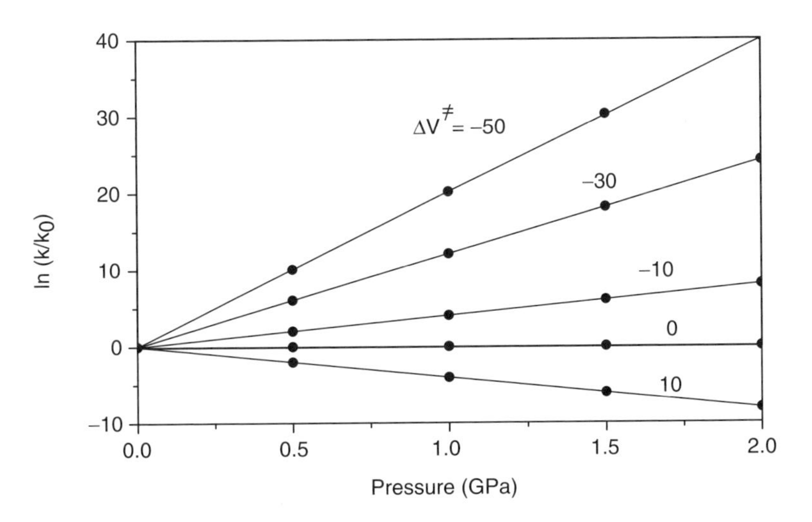

Figure 15. Room temperature pressure evolution of the rate constant for different activation volume values that have been assumed to be constant over the entire pressure range.

where $\Delta V^{\neq}$ is the activation volume—that is, the volume difference between the transition state complex and the reactants. In Fig. 15 the variation of the reaction rate at room temperature for typical values of the activation volume is shown. It can be seen that for activation volumes between -10 and $-50\ \mathrm{cm^3/mol}$, values actually encountered in many cases, the reaction rate increases quite steeply with pressure. The activation volume is, therefore, the basic quantity that can help us to understand the pressure dependence of the reaction rate. The effect of pressure on kinetics of solid-state reactions, up to 1 GPa, has been described in terms of $\Delta V^{\neq}$ and isothermal compressibility

$$\beta = -\frac{1}{V}\left(\frac{\partial V}{\partial P}\right)_T \tag{15}$$

(where V is the molar volume of the solid), since their values and sign and their variation with temperature and pressure give information on the reaction mechanisms.

A simplified approach to the description of $\Delta V^{\neq}$ starts from the statement made by Hamann [270] that the partial molar volume of any dissolved species in solution is the sum of the intrinsic volume of the species, corresponding to its van der Waals radius (intrinsic contribution), and of the contribution due to interaction with the solvent and with the other dissolved species (environmental contribution). Applying the same concept to the transition state, one may as well define two different contributions to the activation volume [274, 275]:

$$\Delta V^{\neq} = \Delta_s V^{\neq} + \Delta_m V^{\neq}$$

where $\Delta_s V^{\neq}$ is the intrinsic, or structural, contribution representing the volume difference arising from the different molecular structures of the transition state and of the reacting molecules, and $\Delta_m V^{\neq}$ is an environmental contribution arising from the different packing of the solvent or of the reacting molecules themselves around the transition state complex and around the reactants. In Table III we report, as an example, the activation volume range for different classes of reactions. As can be seen, reactions whose rate-determining step involves the formation of a covalent bond (associative process) have a negative structural contribution to $\Delta V^{\neq}$, while reactions whose rate-determining step involves the breaking of a covalent bond (dissociative process) have a positive structural contribution to $\Delta V^{\neq}$. Reactions where charged species are formed from uncharged reagents in the transition state of the rate-determining step have a negative electrostriction contribution to $\Delta V^{\neq}$, whose absolute value increases as the polarity of the solvent decreases, as will be discussed below. Analogously, reactions where the transition state of the rate-determining step involves a neutralization of charges have a positive environmental contribution to $\Delta V^{\neq}$. For multistep reactions, the measured $\Delta V^{\neq}$ is the sum of the activation volume of the rate-determining step and of the reaction volumes of all the pre-equilibrium steps eventually occurring prior to the rate-determining step. A nearly complete review of experimentally determined $\Delta V^{\neq}$ for organic reactions in solution are reported in Refs. 276–278.

TABLE III
Activation Volume Values (cm^3/mol) for Various Kinds of Reactions [279]

Reaction	$\Delta V^{\neq}$
Homolysis	5 to 20
Polymerization (radical propagation)	~ -20
Cycloadditions	
Diels–Alder	-25 to -40
Intramolecular	-25 to -30
Dipolar	-40 to -50
(2 + 2)	-40 to -55
Ester hydrolysis	-10 to -15 (basic)
	> -10 (acid)
Epoxide-ring opening	-15 to -20
Wittig reactions	-20 to -30

Jenner [275] has presented a thorough description of several possible contributions to both the intrinsic and the environmental parts of the activation volumes, based on accurate experimental observation of pressure effect on reactions in solutions. The intrinsic contribution to the activation volume essentially derives from the differences in structure between the transition state and the reacting species, so it is directly related to the partial cleavage and formation of chemical bonds in the transition state. In cases where the environmental contribution is negligible, the activation volume variation gives a direct insight in the molecular mechanism [275, 280]. In this case in fact, considering

$$\overline{\Delta V} = \overline{V}_{\text{prod}} - \overline{V}_{\text{reag}}$$

we may define the quantity

$$\theta = -\frac{\Delta V^{\neq}}{\overline{\Delta V}}$$

that directly indicates the position of the transition state along the reaction coordinate.

It has been observed that the kinetics of pericyclic reactions is often quite solvent-insensitive, and it can be assumed that $\Delta V^{\neq} = \Delta_s V^{\neq}$. Therefore, the evaluation of θ can allow to locate the transition state. For example, Diels–Alder reactions occurring via a concerted mechanism have $\theta \approx 1$. In this case the volume contraction of the overall reaction is the same as the activation volume, indicating that the structure of the transition state is close to that of the product molecule. For pericyclic reactions, both concerted or stepwise-diradical pathways are possible in principle and $\theta \approx 1$ when the mechanism is concerted, while it deviates from unity ($\theta > 1$) in stepwise mechanisms [281]. Moreover, in a concerted mechanism the transition state always has a smaller volume than the reactant (because its structure is similar to the cyclic product), so that pressure strongly accelerates pericyclic reactions occurring via a concerted path. On the other hand, a stepwise mechanism may have a transition state with a similar or even larger volume than the reactants, so that it is only slightly accelerated or even retarded by pressure. Thus pressure dependence of the reaction rates for pericyclic reactions contributes to our understanding of whether the reaction follows a concerted or a stepwise mechanism. Mechanisms involving a volume contraction to reach the transition state are accelerated by pressure. Among competing pericyclic reactions, those with the more negative $\Delta V^{\neq}$ will be promoted. Diastereoselectivity may therefore be induced by pressure, if the formation of an enantiomer has a smaller activation volume than the other [282]. For example, the enantiomeric excess increases from 10%

at 0.1 MPa to 98% at 1.4 GPa for Michael reactions of amines to chiral crotonates [283].

For concerted Diels–Alder reactions, as discussed above, both $\Delta V^{\neq}$ and $\overline{\Delta V}$ are negative and $\theta \approx 1$. In some unhindered Diels–Alder reactions, such as those involving maleic anhydride, it was observed [275] that $|\Delta V^{\neq}| > |\overline{\Delta V}|$. This means that the transition state has an additional volume contraction with respect to the products. Since Diels–Alder cycloadditions are essentially solvent-insensitive and thus have negligible or small environmental contribution to the activation volume, this contraction seems to be of intramolecular origin, and it was suggested [284] that it could be due to secondary orbital interactions in the transition state. This contribution to $\Delta_s V^{\neq}$ has been indicated as $\Delta_\mu V^{\neq}$.

Another contribution to $\Delta_s V^{\neq}$ was observed to arise in some reactions involving high steric hindrance (see references cited in Ref. 275); this is indicated as $\Delta_\sigma V^{\neq}$, the steric volume of activation. According to observation, high pressure promotes the most hindered process. This effect has been ascribed to a displacement of the most hindered transition states toward the more compact products in the reaction profiles [285].

When very high pressures (>1 GPa) are applied to liquid phases, glasses, or molecular crystals, mobility is reduced and steric effects become more important both in equilibrium and in kinetic aspects. Equations (9) and (14) are still valid, but equilibria and kinetics of chemical reactions must take into account the energetic, structural, and dynamic properties of the environment as well.

Then, for systems at very high pressure, the neglect of volume changes arising from different packing of the surrounding molecules (the solvent in the case of solutions) around the reactants and the transition state complex is not justified in general. The interactions to be taken into account include dispersion and repulsive terms, electrostatic contributions, and more specific interactions like hydrogen bonds. When the intermolecular forces are only (or at least mostly) of the dispersion type, it is appropriate to assume that $\Delta_s V^{\neq}$ is the only contribution to the activation volume. If, on the contrary, the reaction involves charged or polar species, and in addition the charge distribution changes significantly in the transition state, the environmental effects contribute to a volume variation, referred to as *electrostriction volume*, that can be larger and even of the opposite sign from that of the structural contribution. For a spherical charged species with charge q and radius r in a medium of dielectric constant ϵ, the electrostriction volume can be approximated by the Drude–Nernst equation [286]:

$$\Delta_e V^{\neq} = -\frac{q^2}{2r\epsilon}\frac{\delta \ln \epsilon}{\delta P}$$

In a liquid and at pressures of up to several kilobars, the dielectric constant can be represented by the Owen–Brinkley equation [287]

$$1 - \frac{\epsilon^0}{\epsilon(P)} = A\epsilon(P)\ln\left(\frac{B+P}{B+1}\right)$$

where A and B are pressure-independent constants. The electrostriction effect of course depends on the polarity of the environment. Addition reactions having a negative structural contribution to the activation volume and involving a buildup of charge in the transition state are faster in polar than in apolar solvents at ambient pressure. But in apolar solvents these reactions are more strongly accelerated by the pressure than in polar ones, indicating a more negative environmental contribution to the activation volume [275]. This result indicates that pressure forces the formation of polar transient species, accelerating the reaction in apolar solvents, while the rate of the charge buildup is almost pressure-independent in the most polar solvents.

The most important nonelectrostatic contributions to $\Delta_m V^{\neq}$ derive from the solvophobic interactions, which may lead to a volume shrinkage $\Delta_\phi V^{\neq}$ along the reaction coordinate in organic reactions involving neutral substrates in polar solvents.

The change of polarity should be associated with some kind of structural change, and therefore there is in general a complex interplay of structural and environmental effects on the activation volume.

The effect of the viscosity η of the medium can be incorporated into the expression of the activation volume [288]:

$$\Delta V^{\neq} = \Delta_s V^{\neq} + \Delta_m V^{\neq} + RT\left(\frac{\delta \ln \eta}{\delta P}\right)_T \tag{16}$$

Experimental determination of $\Delta V^{\neq}$ for a reaction requires the rate constant k to be determined at different pressures. k is obtained as a fit parameter by the reproduction of the experimental kinetic data with a suitable model. The data are the concentration of the reactants or of the products, or any other coordinate representing their concentration, as a function of time. The choice of a kinetic model for a solid-state chemical reaction is not trivial because many steps, having comparable rates, may be involved in making the kinetic law the superposition of the kinetics of all the different, and often unknown, processes. The evolution of the reaction should be analyzed considering all the fundamental aspects of condensed phase reactions and, in particular, beside the strictly chemical transformations, also the diffusion (transport of matter to and from the reaction center) and the nucleation processes.

In some instances, one of these processes is much slower than the others, and the kinetics of the overall reaction is governed by its rate law. In these cases the kinetic data will fit simpler models describing the rate-determining step. When breaking and reforming of chemical bonds (i.e., a strictly chemical elementary process) is the rate-controlling step, reaction kinetics may be strongly related to the molecular rearrangement in the transition state, from which it is possible to estimate $\Delta V^{\neq}$. An example, which also underlines the differences in kinetics that may arise between similar reactions occurring in liquid and solid phase, is given by hydrogen atom transfer reactions [289]. In the liquid phase the activation volume for these reactions is negative. For example, reactions in alcohols or hydrocarbons involving hydrogen transfer are accelerated by pressure. On the other hand, recombination of macroradicals in solid polymers, whose rate-determining step is an H atom migration, are strongly retarded by pressure. There are convincing arguments in favor of the hypothesis that the pressure dependence in these reactions is relative to the H transfer process rather than to diffusion-controlled encounters of macroradicals. Therefore, the rate-controlling factor seems to be the orientational reorganization of the reacting molecules in the transition state, through the motion of segments of a macromolecule that requires an extra volume $\Delta V^{\neq} > 0$. Moreover, the activation volume in the solid increases as the temperature increases, whereas in the similar liquid phase reactions it is temperature-independent. A similar example is the bimolecular recombination of polymer radicals in single crystals (generated by gamma radiolysis, while the decay was observed by ESR). The process follows a second-order kinetic law and has a positive activation volume, again meaning that the rate-determining step is the reorientation and disposition along the reaction coordinate of the reacting species.

If the transport process is rate-determining, the rate is controlled by the diffusion coefficient of the migrating species. There are several models that describe diffusion-controlled processes. A useful model has been proposed for a reaction occurring at the interface between two solid phases A and B [290]. This model can work for both solids and compressed liquids because it doesn't take into account the crystalline environment but only the diffusion coefficient. This model was initially developed for planar interface reactions, and then it was applied by Jander [291] to powdered compacts. The starting point is the so-called parabolic law, describing the bulk-diffusion-controlled growth of a product layer in a unidirectional process, occurring on a planar interface where the reaction surface remains constant:

$$\frac{\partial y}{\partial t} = \frac{kD}{y}$$

where y is the thickness of the product layer formed at the interface, D is the diffusion coefficient of the migrating species and k is a constant. If D is independent of time, and if $y = 0$ for $t = 0$, we obtain the parabolic law

$$y^2 = 2kDt$$

With the assumption that all the particles of phase A are spheres with uniform and constant radius r_0 that are immersed in a melt of B, and that the activities of A and B are constant with time, Jander derived the volume of unreacted material as

$$\frac{4\pi}{3}(r_0 - y)^3 = \frac{4\pi r_0^3}{3}(1 - x)$$

and by using the parabolic rate law for y:

$$k_J t = \frac{2kD}{r_0^2} t = [1 - (1 - x)^{1/2}]^2$$

where k_J is the rate constant and $x(t)$ is the fraction of reaction completed at time t. It is often found that the Jander's equation does not well represent solid-state reaction kinetics, indicating that a more complicate situation actually exists. In fact, D is not constant during the reaction, particularly in its earlier stages, and the activities of A and B should not be considered constant but proportional to the fraction of unreacted species $(1 - x)$. Most of all, the parabolic rate law assumes that the reaction surface remains constant with time, but since we assumed that the surface of phase A is spherical, this surface actually changes with time. The volume of unreacted A is assumed to decrease with time following Barrer's equation for the flux of heat diffusing through a spherical shell of radius $r_2(t) - r_1(t)$:

$$\frac{\partial V(t)}{\partial t} = -\frac{4\pi k D r_1 r_2}{r_2 - r_1}$$

where $r_1(t)$ is the radius of the particle of unreacted phase A at time t and $r_2(t)$ the total radius of the particle constituted by the formed product and unreacted A at time t. Introducing the parameter Z as the volume of the product per unit volume of A reacted, the Valensi–Carter equation was obtained [292], which, modified to take into account that D is inversely proportional to time, becomes [290]:

$$kt = \frac{Z - [1 + (Z - 1)x]^{2/3} - (Z - 1)(1 - x)^{2/3}}{Z - 1}$$

D is assumed to be inversely proportional to time because during the formation of the product, various imperfections are created, whose rate of removal is in turn assumed inversely proportional with time (Tamman theory). This equation was found appropriate to represent the formation of spinels but is not of easy general application due to the lack of density data for calculating Z.

Many high-pressure reactions consist of a diffusion-controlled growth where also the nucleation rate must be taken into account. Assuming a diffusion-controlled growth of the product phase from randomly distributed nuclei within reactant phase A, various mathematical models have been developed and the dependence of the nucleation rate I on time formulated. Usually a first-order kinetic law $I = fN_0e^{(-ft)}$ is assumed for the nucleation from an active site, where $N(t) = N_0e^{(-ft)}$ is the number of active sites at time t. Different shapes of the reactant A, which means different geometry of the formed nuclei and different forms for the diffusion law for the growth, can be assumed. In all cases, the rate law for nuclei growth takes the general form

$$\ln\frac{1}{1-x} = kt^n \tag{17}$$

where n is a parameter depending on the reaction mechanism, on the nucleation rate, and on the geometry of the nuclei. The form of this equation is the same as derived by Avrami for describing the growth of a crystal from a liquid phase [293].

As has already been noted, polymerization is a common output of high-pressure reactions. The kinetics of solid-state pressure-induced polymerizations have been treated within the nuclei growth [see eq. (17)] model. These reactions, as we will discuss in Section IV, are a typical example of how the crystal structure plays a fundamental role in solid-state chemistry. Kinetic data of polymerizations are usually analyzed according to Eq. (17) by inserting an additional parameter t_0 accounting for the nucleation step:

$$\ln\frac{1}{1-x} = k(t-t_0)^n \tag{18}$$

or, rearranging:

$$x = \frac{x(t)}{x(\infty)} = 1 - e^{\left[-\frac{(t-t_0)^n}{\tau}\right]} \tag{19}$$

where $\frac{x(t)}{x(\infty)}$ is the fraction of the product formed at time t relative to the total amount that is formed at equilibrium, $\frac{1}{\tau} = k$ is the reaction rate, and n is a parameter related to the spatial evolution of the reaction and increases with the dimensionality of the growth. Depending on the nucleation rate and the growth geometry, different n values (see Table IV) are obtained, making the determination of this parameter fundamental in understanding the microscopic evolution of the process.

TABLE IV
Summary of the n Values Found in the Diffusion-Controlled Nuclei Growth Model for Different Growth Geometries and Nucleation Rates.

3D growth (spheres)	
A	2.5
B	1.5
C	1.5–2.5
2D growth (plates)	
A	2.0
B	1.0
C	1.0–2.0
1D growth (rods)	
A	1.5
B	0.5
C	0.5–1.5

A, B, and C indicate a constant, a zero, and a decreasing nucleation rate, respectively.

Anticipating the discussion on acetylene polymerization [98], extensively reported in Section IV, a value of $n = 0.6$ has been found, which implies a linear diffusion-controlled growth where the molecular librational and translational oscillations control the approach of the monomers to the active sites (chain terminations).

Kinetic curves relative to polymerization reactions in the solid state commonly show a sigmoidal shape with a slow initiation step followed by a steep increase, even by two orders of magnitude, of the reaction rate. A reaction with this kind of kinetic curve is said to have an autocatalytic behavior.

A kinetic model for single-phase polymerizations—that is, reactions where because of the similarity of structure the polymer grows as a solid-state solution in the monomer crystal without phase separation—has been proposed by Baughman [294] to explain the experimental behavior observed in the temperature- or light-induced polymerization of substituted diacetylenes R–C≡C–C≡C–R. The basic feature of the model is that the rate constant for nucleation is assumed to depend on the fraction of converted monomer $x(t)$ and is not constant like it is assumed in the Avrami model discussed above. The rate of the solid-state polymerization is given by

$$\frac{\partial x}{\partial t} = By(x)(1 - x)\exp\left[-\frac{E_I(x)}{RT}\right]$$

where B is a constant essentially dependent on the structure of the monomer, y is the average length of the polymer at conversion x, and $E_I(x)$ is the activation energy for nucleation. The dependence of both y and E_I on the conversion degree arises mainly from the structural changes during the polymerization. On the other hand, the number of monomeric units z that can be added to the ends of the growing polymer, during its lifetime τ, also depends on conversion, and therefore the rate of the chain propagation is given by

$$k_P(x) = k_0 \exp\left[-\frac{E_P(x)}{RT}\right] = \frac{z(x)}{\tau}$$

where E_P is the activation energy for the polymer growth. The important feature exposed by the model is that the growing process of the polymer involves all the molecules in the sample, even though the model is worked out for a growth in one direction. The model has been applied to the polymerization of substituted diacetylenes and has been found sufficiently accurate in reproducing the experimental sigmoidal kinetics. For a practical application, an additional parameter is necessary, namely, the concentration of defects or impurities that will act as termination centers. For the calculation of the activation energies, one can rely on a crystal strain approximation, which is appropriate when major structural changes occur in the polymerization direction and can be obtained as the local strain energy deriving from interaction of the reacting monomer with the chain initiation species or with growing chain ends. The relaxation of the remainder of the crystal follows the overcome of the potential barrier. Application of the model within the crystal strain approximation shows that the autocatalytical sigmoidal behavior is observed only when both chain initiation rate and chain propagation length depend on the crystal strain. When one or both of these parameters are independent of the crystal strain, the kinetic law can lose the sigmoidal trend and eventually become rather flat. Changes of the autocatalytic character with changes of the experimental conditions have been experimentally observed and will be discussed in the next section. In conclusion, a microscopic important parameter is the difference in the unit cell parameters of the monomer and polymer in the direction of growth. Alternatively, and when no major changes in the dimensions occur during the polymerization, a nearest-neighbor approximation can be adopted, neglecting long-range monomer–polymer interactions. In this case, calculation of activation energies only includes the interaction of a monomer with two nearest neighbors, either along the closest packing direction or according to some strong and specific directional interactions. Also, within this approximation the kinetic curve can display a more or less autocatalytic behavior determined by the dependence of the nearest-neighbor interaction on the conversion. Extension of the model to high-pressure reactions is in principle straightforward but in practice its application requires a knowledge of the crystal structure parameters that are not easily available. In addition, the condition of single-phase polymerization without phase separation is not easily encountered. However, this model and other approaches mentioned above take into consideration many of the relevant aspects that should be taken into account in chemical reactions, and in particular in polymerizations, at high pressure in the crystalline phase.

C. Effects of High Pressure on the Electronic Structure

A general consequence of an increase in pressure is the observation of remarkable changes in the electronic structure of molecules. We already mentioned that the dielectric constant will increase with pressure, an effect that implies a redistribution of the charges in the system. Additional clear evidence is obtained for the behavior at high pressure of the electronic absorption spectra, of the infrared and Raman spectra, and of the electrical resistance. Analysis of the electronic spectra is particularly significant and has been discussed in detail by Drickamer [11, 295]. The main effects that are observed are a shift of the peak absorption (and of the absorption edge) with pressure, and a broadening of the absorption bands. The first effect, which can be particularly large in magnitude, is in general (but not necessarily) a shift to the red and can be appreciated as a change of colour (piezochromism). This is a sharp evidence of the effect of pressure in the orbital energies and therefore in the HOMO–LUMO separation corresponding to the first electronic transition. It is possible to distinguish two concurring contributions. The first is a vertical shift of the ground and excited state with respect to each other that can produce the important result of changing the configuration interaction to a considerable extent. As a consequence, the structure of the ground and excited states is altered. This can affect the frequency of the electronic transition, governed by the Franck–Condon principle and by the selection rules, but can also change substantially the probability of thermal population of the excited state even at room temperature, thus having important consequences for the chemical reactivity. The second contribution can be termed as a lateral shift corresponding to a change of the relative position of the ground and excited state along an appropriate configuration coordinate. This can also give rise to a frequency shift of the absorption band in the Franck–Condon approximation and can be viewed as arising from different compressibilities of the two electronic states. It has been found that the frequency shift can be relevant, and it is of interest to note that the shift increases more than linearly with density. This means that the effect of pressure on the electronic structure becomes much more pronounced at very high pressures. For its possible effects on chemical reactivity, the broadening of the absorption band at high pressures can be even more important.

In Fig. 16 the profile of the molecular orbital energies of propene at different densities is reported. The results have been obtained by an *ab initio* molecular

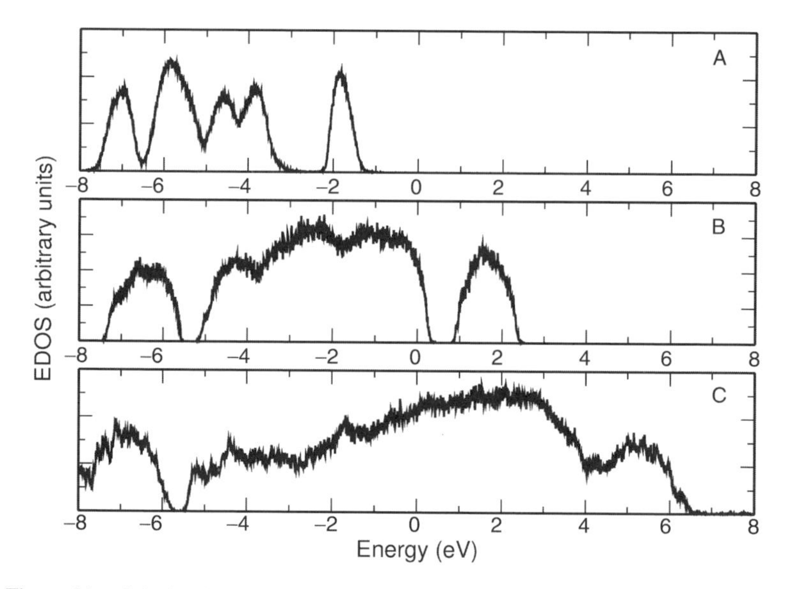

Figure 16. Calculated evolution of the electronic density of states of propene with increasing density (pressure): (A) 0.86 g/cm^3; (B) 1.45 g/cm^3; (C) 2.08 g/cm^3.

dynamics simulation in the density functional approach on a sample of 27 molecules treated with periodic boundary conditions [296]. From the figure the rapid broadening of the molecular orbitals energy as the pressure increases can be observed. At the same time the HOMO–LUMO gap decreases and new energy levels appear within the gap. It is evident that at still higher pressure the electronic density of state could degenerate into a continuum. The observed trend clearly explains the experimental observation of the red shift of the electronic transition bands at high pressure and, on the other hand, makes the possibility of thermally populating the excited states evident. In fact, the broadening of the energy levels is such that even at ambient temperature, thermal fluctuations could be sufficient to excite electrons in the lower tail of the unoccupied molecular orbitals. An accurate analysis of the electronic structure changes induced at high pressure in molecular liquids and solids reveals other important features. It has been found that even in assemblies of nonpolar (centrosymmetric) molecules, like ethylene and butadiene, the increase in pressure leads to a buildup of an electric dipole moment [297, 298].

Fig. 17 shows the distribution of the electric dipole moment in a sample of *trans*-butadiene, composed by 27 molecules, at a density of 1.87 g/cm^3 corresponding to an estimated pressure much higher than the pressure threshold where the reaction is experimentally observed (0.6 GPa). In Fig. 18 the

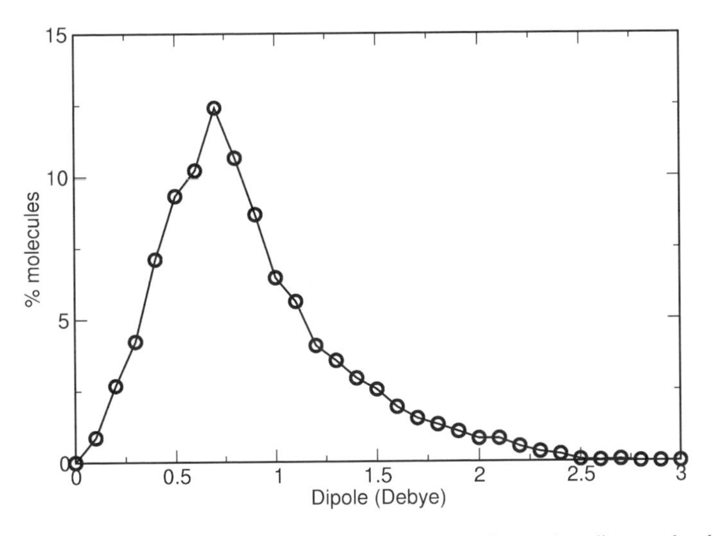

Figure 17. Distribution of the electric dipole moments of *trans*-butadiene molecules, at a density of 1.87 g/cm^3, calculated just before the occurrence of the polymerization reaction.

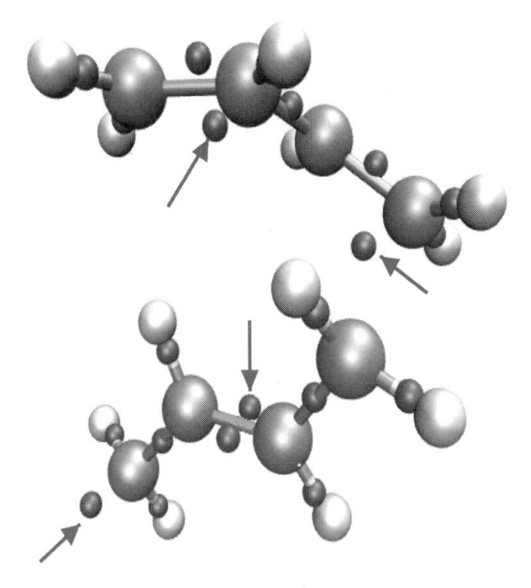

Figure 18. Calculated polymerization dynamics of *trans*-butadiene. Arrows indicate the electronic doublets that are moving from the 1–2 and 3–4 π bonds characteristic of the isolated molecule, to the new polymer position. In particular, the doublets that migrated to the molecule ends and that are ready to give the new C–C bond with the neighboring molecule should be noted.

arrangement of the electron doublets in the butadiene molecules, obtained from the analysis of the maximally localized Wannier functions, is shown, and it can be seen that at high pressure a charge separation occurs in the molecules of the sample, leading to the formation of zwitterionic species.

There are some important implications of this output of the *ab initio* molecular dynamics simulation. First it should be noted that the charge separation and rearrangement is observed well in advance of the occurrence of any chemical reaction (condensation and polymerization in the cases investigated). The onset of the high-pressure chemical reaction in the simulation is monitored by a sudden drop of the total energy of the system. A quite remarkable observation is that the rearrangement of the electronic cloud and the buildup of the electric dipole moment, as a response of the system to the highly repulsive potential experienced by the molecules at high pressure, is in general a strictly intermolecular effect. According to a Le Chatelier principle point of view, the formation at high pressure of highly polar species produces a favorable closer packing of the molecules (electrostriction) that has an obvious intermolecular character or is even a collective or cooperative response of the system. Actually, in the molecular dynamics simulation it has also been observed that the relative orientation of the induced molecular dipoles suggests that cooperative effects are at work in the prereactive region. However, the most important implication of these results, particularly if the sample polarization at high pressure can be assumed as a feature of general occurrence, refers to high-pressure reaction mechanism. In the few cases where the electronic structural changes have been analyzed in sufficient details by molecular dynamics simulation, it has been found that high-pressure condensation or polymerization reactions occur via an ionic mechanism. Furthermore, it is interesting to note that the evolution of the reaction also stresses the collective character of the process, in the sense that the polymerization does not occur as a simple step-by-step condensation of monomeric units but rather involves the participation of more molecules of the array in each elementary chemical event in a complex interplay of electronic doublets and hydrogen ion migrations. Concerning the effectiveness of ionic mechanisms in high-pressure chemical reactions, and in particular in polymerization reactions, it is worth mentioning the experimental observation reported by Zharov that in a sample of pressurized trioxane, treatment with ionizing radiation, able to produce ionic centers, readily produces the polymerization of the material [299].

The above discussion was based on the results of molecular dynamics simulations on unsaturated or conjugated hydrocarbons. Although the general features can be extended to molecular structures of more general types, in practice it is appropriate to consider the specific form of the electron orbitals involved. For instance, $d \rightarrow d$ transitions in transition metal ion complexes involve orbitals mainly localized on the metal ion that, in the crystal field theory, are identified with the 3d atomic orbitals whose degeneracy is lifted by the electrostatic field of the ligands. The effect of pressure on the electronic structure of these compounds has been discussed by Drickamer [11] with particular reference to high-spin octahedral complexes, but the conclusions can be easily generalized. The 3d orbitals in an octahedral field are split in two sets of t_{2g} (d_{xy}, d_{xz} and d_{yz}) and e_g ($d_{x^2-y^2}$ and d_{z^2}) symmetry, respectively, the former occurring at lower energy with a separation given by the crystal field splitting Δ. If Π is the mean spin pairing energy, the system is in the high-spin configuration when $\Delta < \Pi$ and in the low-spin configuration when $\Delta > \Pi$. Electrostatic theory predicts a linear dependence of Δ upon R^{-5}, where R is the ion–ligand distance. Thus the possibility arises that pressure may induce high-spin to low-spin transitions in transition metal complexes, as it has actually been observed experimentally. Δ values for some metallic ions in a host lattice have been experimentally observed to increase with pressure following a $(a_0/a)^5$ law in Al_2O_3, where a is the cubic cell parameter measured by x-ray diffraction. However, also a steeper dependence on pressure was measured in the case of MgO. Also, the mean spin pairing energy Π depends on pressure and because it can be expressed as a function of electronic repulsion terms (the Racah parameters), which can be obtained from the electronic spectra and are found to decrease with pressure. This behavior is also confirmed in the case of ^{57}Fe by Mössbauer spectroscopy. The observation of a decrease in the isomer shift values with increasing pressure indicates an electron density increase at the nucleus. This occurrence must be correlated to a less effective shielding of the nucleus charge on the 3s electrons due to a spread out of the 3d orbitals with a consequent reduction of the electron repulsion.

For highly oxidizing metals and reducing ligands, increase of pressure may induce an electronic transition resulting in the oxidation of the ligand and reduction of the metal ion. In all the other cases of electronic transitions (metal-to-ligand or ligand-to-metal charge transfers, transitions between the highest filled molecular orbital of the donor and the lowest unfilled orbital of the acceptor in electron donor–acceptor complexes, intraligand or intramolecular transitions, band-gap transitions) the main effect of pressure, by reducing the volume of the system, is to increase the overlap between the interacting orbitals, so that the transition energy generally decreases.

Among the various types of possible transitions ($\sigma \rightarrow \sigma^*$, $\sigma \rightarrow \pi^*$, $\pi \rightarrow \pi^*$, $n \rightarrow \sigma^*$, $n \rightarrow \pi^*$), the $\pi \rightarrow \pi^*$ transitions are generally the most sensitive to pressure and occur in unsaturated compounds, which are for this reason particularly reactive at high pressure. The lowest energy transition of an unsaturated system is usually a $\pi \rightarrow \pi^*$ transition. The energy shift with pressure of the $\pi \rightarrow \pi^*$ transitions has been reported for several conjugated systems up to 12 GPa [300, 301] and up to 40 GPa, well above the reaction pressure threshold, in the case of benzene and furan [302, 303]. For anthracene,

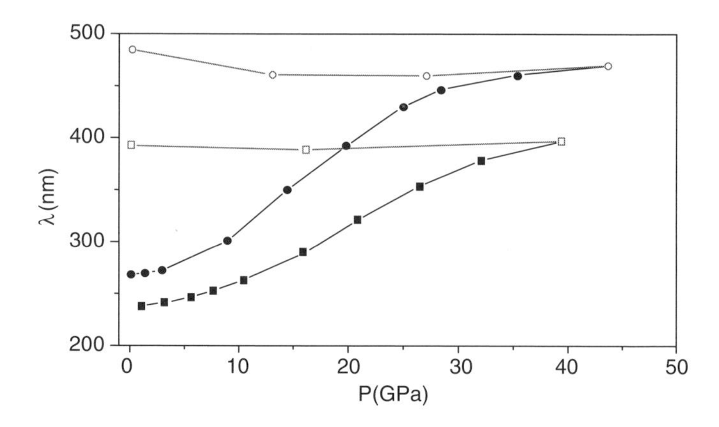

Figure 19. Pressure shift of the absorption edge of benzene (circles) and furan (squares) during room-temperature compression experiments. Empty symbols correspond to the value of the absorption edge measured on the reacted material on releasing pressure.

tetracene, and pentacene, the slope of the transition energy versus pressure or density (estimated by compressibility data) increases rapidly with density. In these systems the ground state is nonpolar while the excited state has a dipole moment and is thus stabilized by pressure. Anyway, the shift is steeper than $1/R^3$, and probably the pressure has the further effect of increasing the charge separation. For azulene, on the contrary, the transition shifts to higher energies with pressure up to a maximum at 5–6 GPa, and then it rapidly decreases with further increasing pressure. The ground state of azulene in fact has a dipole moment, which is higher than that of the first excited state, so the ground state is more stabilized by pressure, but at pressures higher than 6 GPa, probably the charge separation induced by pressure in the excited state becomes dominant and the transition energy rapidly decreases. In the case of benzene and furan the transition edge exhibits a consistent red shift with increasing pressure, but it flattens when the onset of the chemical reaction is reached (see Fig. 19).

In several cases—such as, for example, phthalocyanine and protoporphyrin IX derivatives—the energy red shift and broadening of the transition with pressure may be sufficient to ensure a thermal population of the excited state, by occupation of a π^* orbital. As discussed above, the thermal occupation of excited states may lead in principle to the formation of a reaction product.

D. Environmental and Steric Effects at High Pressure

The importance of the environment of the reacting centers in high-pressure reactions has already been dealt with in connection with the determination of the reaction volume and of the activation volume, and also in connection with the pressure induced changes of the electronic structure. There are, however, other relevant aspects of environmental and steric effects that deserve a discussion.

At the ultrahigh pressures that are of interest in the present chapter, the reactant will generally be in a glass or crystalline phase or, in some cases, in a liquid state with rather reduced diffusion coefficient. Therefore, reactions will eventually occur among reactants with fixed molecular orientations and intermolecular separation. In such circumstances, geometrical constraints and volume restrictions are imposed on the high-pressure reactants and can result in a twofold effect. On the one hand, reactions requiring a high mobility of the reacting molecules (for example, rotations, reorientations, diffusion, or isomerizations) can be completely prevented unless the high potential barriers are overcome by high temperatures. On the other hand, the relative orientational confinement of the molecules can selectively favor some particular reaction pathways. These concepts can be illustrated by some of the features reported for the high-pressure amorphization of benzene occurring through the opening of the aromatic ring. Omitting the details of this reaction, which will be discussed in the following, it has been reported that the reaction can be induced starting from different crystalline phases of benzene selecting appropriate points in the P–T phase diagram of the system and that the reaction product differs to some extent depending on the initial crystal phase [304]. For instance, the mechanical properties and the color of the reaction product differ when it is obtained from the III, III′, and IV crystal phases, being orange, white, and black, respectively. The different arrangement of the molecules in different crystal polymorphs is responsible for the differences in the reaction products. Other cases have been reported showing clearly that the same monomer crystallized into different phases gives rise to different solid-state reactivities [294]. For example, in hydrogen cyanide the molecules in the crystal are kept aligned by hydrogen bonds in linear chains with a head-to-tail alignment, and the pressure-induced reaction produces only one of the several possible polymers that are obtained in solution under different catalysis and solvent conditions [305]. Another example is the crystal of acetylene [98], where the formation of the *trans*-polyacetylene occurs along the diagonal of the *bc* plane of the orthorhombic cell, where the intermolecular distances are smaller and the relative orientation resembles the polymer structure thus allowing the reaction. Polymerization of the diacetylenes is another instructive example of how the crystalline structure can affect the reaction product and also in some cases the resulting crystalline phase [294]. Diacetylene monomers polymerize by thermal, UV, or x-ray activation in the solid state by a 1,4-addition reaction to give a linear polymer. On studying these reactions, it was found that in some cases the polymer crystal had similar dimensions and structural perfection as the starting crystal (this means that the polymer chains and the monomer form a solid solution during all the reaction), while in other cases a polymer-rich crystalline phase and a monomer-rich one were formed (a phase separation occurred during the reaction). These examples show that the crystal geometry may have effects in high-pressure reactions that have been described for general solid-state reactions as topochemical effects [306, 307]. Topochemical reactions are indeed defined as the solid-state reactions occurring with a minimum displacement of the reacting molecules. The reactions just mentioned in this discussion are examples of topochemistry. It is evident that reactions occurring at high pressure can be considered topochemical at various degree in the sense that concurring effects can oppose the regular growth of the products. For instance, cross-linking in polymerization processes may obscure the regular growth of polymer chains. In discussing these effects, many of the concepts of solid-state syntheses can be applied. In organic solid-state synthesis, crystal engineering procedures have been developed [308] intended to introduce in the reacting molecules functional groups not directly involved in the reaction but able to modify the crystal structure in order to topochemically convey the reaction toward some desired products. Several examples have been reported where it turns out that an ultrahigh pressure by itself produces the condition for a regular growth of the products.

For the high-pressure reaction of benzene, it has consistently been reported that at any given pressure the reaction proceeds up to a certain stage but not to completion. However, upon releasing the pressure the reaction accelerates and comes to completion rapidly [309]. Evidently, pressurization above a given threshold is sufficient to trigger the reaction; however, as the reaction proceeds, there is no volume available anymore for the reaction propagation unless some barrier is overcome. Alternatively, one could guess that as the formation of a polymeric network proceeds, the mobility of the unreacted molecules becomes insufficient as they are screened by the interaction with each other. Actually, some evidence of unreacted molecules screening has been found in the high-pressure polymerization of acetylene, as will be discussed in the next section.

In the same way as the environment may govern the reaction path, the molecular transformation and the connected molecular rearrangement can perturb the environment. The perturbation caused by the reacting molecules is of mechanical and electrical nature. Condensed phases have the important property that a perturbation arising at one site is transmitted to the other sites by strong intermolecular interaction. The importance of cooperative effects in a pressurized sample has already been stressed in the previous section. On the basis of Raman studies on reacting crystals, Dwarakanath and Prasad [310] proposed that reactions in molecular crystals could be activated by the softening of phonon modes involved in the reaction coordinate (phonon activation), suggesting that large-amplitude phonons are the analogous of collisions in the gas phase. The softening of a mode actually results in large-amplitude molecular translations or rotations that can bring two molecules close enough for a reaction to occur. Thus phonons play the role of diffusion in crystals and also account for the cooperativity of the processes. This is a critical point in high-pressure reactions. Even if the compressibility of molecular condensed systems is rather large, the static nearest-neighbor approach that can be realized at high pressure does not reach the values needed for a reaction to occur. For instance, in the polymerization of acetylene, extrapolation of the data available on the crystal structure at various pressures shows that at the pressure of the reaction ($\sim$3 GPa) the shortest intermolecular contact is 3.05 Å. *Ab initio* molecular dynamics simulations show that in this system a reaction occurs only when the nonbonded C$\cdots$C distance attains values smaller than 2.6 Å. At room temperature, this can occur considering the mean-square amplitude of the phonon modes. When we decrease the temperature, the reaction requires a higher pressure to occur that compensates the reduction of the mean-square amplitude of translational and librational motions. Therefore in high-pressure chemical experiments the temperature parameter can be critical.

The effect of the environment can be modeled starting by the definition of the reaction cavity as the region containing the reacting species and that experiences distortions due to the reaction. The exact definition of the reaction cavity is critical in modeling a reaction. It can be assumed to be passive if it only has a steric effect on the reactive process, via nonspecific and nondirectional van der Waals interactions, or active when it is involved in directional and specific interactions, such as charge transfer or hydrogen bonds, with the reacting molecules. The cavity may be assumed to be rigid but on the basis of the above consideration it should be considered flexible to account for the distortion produced by the reaction [311, 314].

The model of a reacting molecular crystal proposed by Luty and Eckhardt [315] is centered on the description of the collective response of the crystal to a local strain expressed by means of an elastic stress tensor. The local strain of mechanical origin is, for our purposes, produced by the pressure or by the chemical transformation of a molecule at site n. The mechanical perturbation field couples to the internal and external (translational and rotational) coordinates $Q(n)$ generating a non local response. The dynamical variable Q can include any set of coordinates of interest for the process under consideration. In the model the system Hamiltonian includes a single molecule term, the coupling between the molecular variables at different sites through a force constants matrix W, and a third term that takes into account the coupling to the dynamical variables of the operator of the local stress. In the linear approximation, the response of the system is expressed by a response function X to a local field that can be approximated by a mean field V:

$$\langle Q(n) \rangle = \sum_{n'} X(n, n') V^{Q}(n')$$

According to the model, a perturbation at one site is transmitted to all the other sites, but the key point is that the propagation occurs via all the other molecules as a collective process as if all the molecules were connected by a network of springs. It can be seen that the model stresses the concept, already discussed above, that chemical processes at high pressure cannot be simply considered mono- or bimolecular processes. The response function X representing the collective excitations of molecules in the lattice may be viewed as an effective mechanical susceptibility of a reaction cavity subjected to the mechanical perturbation produced by a chemical reaction. It can be related to measurable properties such as elastic constants, phonon frequencies, and Debye–Waller factors and therefore can in principle be obtained from the knowledge of the crystal structure of the system of interest. A perturbation of chemical nature introduced at one site in the crystal (product molecules of a reactive process, ionized or excited host molecules, etc.) acts on all the surrounding molecules with a distribution of forces in the reaction cavity that can be described as a chemical pressure.

The excess energy produced by the introduction of a perturbation at site n can be viewed as a work against the chemical pressure. The deformation energy can be separated into various components. Besides the interaction between perturbations at different sites, one should consider the self-energy of the perturbed site n—that is, the energy ΔE_0 needed to create separate perturbations and which is proportional to the concentration of perturbed sites and will always be negative. In addition, there is a contribution associated with the interaction of the perturbation with the molecular degrees of freedom. This term, which is of major interest in the present discussion, can be conveniently expressed in terms of an elastic multipole representation. Defining the force $\mathbf{v}(n\xi)$ acting on the perturbing molecule n and due to the atom ξ in the cavity at distance $\mathbf{r}(n\xi)$ from the molecular center of mass, we may define the net force on the molecule

$$\mathbf{V}(n) = \sum_{\xi} \mathrm{v}(n\xi)$$

and the elastic dipole moment with components

$$P_{ij}(n) = \sum_{\xi} r_i(n\xi)\mathrm{v}_j(n\xi)$$

Then the energy needed to introduce the perturbation will be given by

$$H_n = \sigma(n)\left[\Delta E_0 + \sum_i Q_i(n)V_i(n) + \sum_{ij} \epsilon_{ij}P_{ji}\right]$$

where ϵ is a local strain that can be assumed to be homogeneous within the cavity. The advantage of this formalism is that the stress multipoles explicitly contain the shape of the reaction cavity and the distribution of forces (chemical pressure) within it. In addition, since the force on each molecules is defined as $V(n) = \sum_{n'} V(nn')$, all collective effects are included in the model.

Some examples of direct measurement of the crystal distortion during the reaction by optical and electron microscopy, IR absorption, and x-ray diffraction have been reported (see references in Ref. 311). The observed (anisotropic) changes in lattice parameters have been used for calculating the kinetic parameters of the transformations. For some reactions, it was possible to follow by x-ray diffraction the change in lattice parameters and also in atomic coordinates, so the relaxation could be observed at atomic level.

IV. REACTIONS AT VERY HIGH PRESSURE

A. Molecular Systems

1. Diatomics and Triatomics

Diatomic and triatomic molecules are interesting cases to start a description of chemical reactions at very high pressures. In fact, in the condensed phases they exhibit an extremely high compressibility with observable volume changes by a factor of up to 2–3. In the case of solid nitrogen, compression above 100 GPa produces a volume change able to bring the ratio of inter- to intramolecular force constants to the value of 0.1 [316]. The extended range of variation of intermolecular distances offers unique opportunities of studying fundamental properties of condensed phases. In addition, the simple molecular structure of small molecules can reveal finer details of the electronic configuration changes induced at high pressure. As a whole the chemical transformation observed at high pressure in diatomic and triatomic molecules are in several cases rather surprising and are the most appropriate to illustrate the unique features that can be induced at ultrahigh pressures. In the following survey, no reference will be made to the extensive work that has been done on the properties of hydrogen and water at high pressures since these topics, in spite of their primary importance in fundamental physics [317, 318], still have a minor role in high pressure chemistry.

Nitrogen. Molecular nitrogen N_2 has a dissociation energy of ~950 kJ/mol, and the N–N triple bond is one of the strongest known chemical bond. Shock-wave experiments disclosed the possibility of N–N dissociation in condensed phases [224, 319–322]. From this an interest arose in the possible obtainment of arrays of N–N single bonds that could form in potentially energetic materials. *Ab initio* calculations of various kinds [323–327] showed that actually at high pressure single-bonded nitrogen arrays could be thermodynamically more stable than molecular nitrogen. In particular, Mailhiot et al. [324] from *ab initio* pseudopotential total energy calculations found that a polymeric form of single-bonded triply coordinated nitrogen, corresponding to a deformation of a simple cubic lattice, was the high-pressure stable phase. This structure resembles the tridimensional networks well known for other V group elements. The phase boundary between molecular and polymeric nitrogen was estimated from calculations to fall at room temperature in the range between 50 and 100 GPa. However, the polymeric phase of nitrogen escaped experimental observation for some time. Reichlin et al. [328] carried out room-temperature experiments at pressures of up to 130 GPa and found that molecular nitrogen is stable in this pressure range even if several phase transformations are observed [316, 328]. Later, however, a phase transformation, occurring above 150 GPa, was monitored by infrared and Raman spectroscopy and by optical absorption and electrical resistance measurements [316, 329, 330]. Visual observation shows a change of color at 130 GPa, and the material becomes completely opaque at 160 GPa. Correspondingly, the optical absorption band shifts to the red with pressure. The shape of the absorption cutoff and the electrical resistance measurement give a clear evidence that the new phase of nitrogen (named the η phase) is an amorphous semiconductor with an estimated optical gap in the range 0.6–0.7 eV [329]. The infrared and Raman spectra show the complete disappearance in the η phase of the vibron bands and of the phonon modes as well [316]. New bands appear in the infrared ($\sim$1700 cm^{-1}) and Raman (640 and 1750 cm^{-1}) spectra at much lower frequencies than in molecular nitrogen. The infrared and Raman bands are rather broad, and this is in agreement with the finding that the η phase is actually an amorphous polymeric form of nitrogen.

The pressure threshold for the transformation to the η phase has been found to decrease considerably with increasing temperature [330], a feature common to most of the high-pressure transformations of interest in this chapter. As theoretically predicted, a large hysteresis has been observed in the backward transformation upon decreasing the pressure. This stimulated studies intended to ascertain the range of pressure (and temperature) where polymeric nitrogen is stable or metastable. It has been found that the η polymorph is stable up to 240 GPa, and a threshold for an eventual transition to a metallic state has not been identified. The back transformation to molecular nitrogen on releasing pressure is somehow hindered, and the polymeric solid has been recovered at 10 K and at a pressure of 10–20 GPa. Once recovered in these conditions, the sample appears to be stable on successive heating up to 175 K [329].

Initially the η phase was assigned as the nonmolecular form of nitrogen, assuming the cubic *gauche* structure suggested by Mailhiot as the most likely [324]. However, the value of the band gap and the opaque appearance of the sample are not suited for an extended array of N–N single bonds. As a matter of fact, recent experiments by Eremets et al. [331] have shown that the η phase of nitrogen most likely consists of a mixture of small clusters of bonded nitrogen atoms. This is in keeping with both the disorder in the sample and its small band gap. Eremets et al. finally succeeded in actually obtaining the nonmolecular cubic *gauche* phase of nitrogen. In fact, they reported that using an appropriately designed laser-heated diamond anvil cell, the darkened sample obtained at 300 K and 140 GPa (the η phase) upon heating at temperatures above 2000 K becomes transparent and the pressure reduces to 115 GPa. At the same time, the vibron band and all the other broad bands disappear and a new sharp peak develops in the Raman spectrum at 840 cm^{-1}. The x-ray diffraction pattern of the transparent sample could only be fit to the cubic *gauche* structure with determination of the space group ($I2_13$) and of the unit cell parameter ($a = 3.4542$ Å). The nitrogen network in the cubic *gauche* structure is shown in Fig. 20. The N–N bond length at 115 GPa is 1.346 Å. The bulk modulus of nonmolecular nitrogen is found to be in the 300- to 340-GPa range, comparable with that of boron nitride and characteristic of covalent crystals. The cubic *gauche* form of nitrogen is metastable down to 42 GPa, and in this pressure range it reconverts to molecular nitrogen upon laser irradiation. Attempts to recover the nonmolecular form of nitrogen at low temperature were not successful.

Attempts have been made to obtain the nonmolecular form of nitrogen by pressurizing solid sodium azide—that is, a compound where nitrogen is present in another molecular form as a linear triatomic species N_3^-, with the expectation that the different starting molecular structure could facilitate the polymerization. On the other hand, one could also note that the nitride ion is isoelectronic with

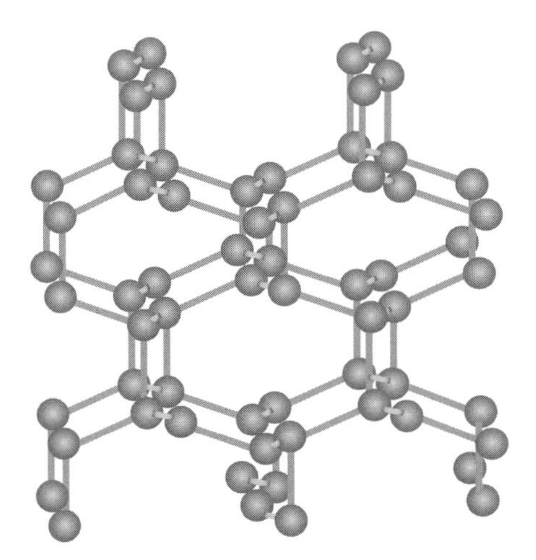

Figure 20. Atoms arrangement in the cubic *gauche* form of nonmolecular nitrogen.

carbon dioxide, and by comparison with the latter (see in the following) a transformation to a polymeric form should be possible. Experiments by Eremets et al. [332] on the high-pressure transformation of sodium azide show some remarkable similarities with findings obtained in nitrogen, but the details are not easily amenable to a clear-cut interpretation in the absence of direct structural determination. At room temperature starting from the rhombohedral phase, a transformation to the monoclinic α phase is first observed at ~1 GPa. In the 15- to 50-GPa range, NaN_3 transforms into a new phase, denoted as phase I, characterized by a reduced symmetry (loss of the inversion center) and by major changes of the Raman spectrum. While the modes of the N_3^- ion are still present in the Raman spectrum of phase I, bands that should be assigned to a new species also appear. Above 50 GPa the sample darkens and eventually becomes black at 120 GPa. In this range, NaN_3 enters in a new phase II where the nitride vibrons disappear. Phase II has a disordered structure, as it can be argued from the broadness of the Raman bands, and is characterized by photoconductivity. The transformation to phase II is greatly accelerated by shear deformations applied by rotating the diamonds of the DAC. The formation of phase II is also accelerated by heating the sample such that shear deformation and laser heating play very much the same role. The disappearance of the nitride vibrons and the increased photoconductivity are evidences that in phase II a non-molecular form of nitrogen must be present. This disordered nonmolecular form can be transformed in a crystalline form by laser heating at high temperatures (>2000 K), where the Raman bands become significantly sharper. As a whole, the appearance of the disordered black form and its crystallization at high-temperatures parallel observations in nitrogen. However, precise hypotheses on the structure of the darkened and crystalline forms are somehow speculative. Among other things, the fate of the charges present in the ionic starting material has not been discussed. The complexity of the chemical transformation occurring in sodium azide is best illustrated by the observed behavior on pressure release. From phase I the sample transforms back to β-azide. On the contrary, from the opaque phase on decompression a mixture of two new phases (phases II and IV) is obtained when the sample remains confined in the diamond cell. After separation of the anvils a further transformation is observed to two new phases V and VI and to β-azide.

Carbon Monoxide. There are close similarities between carbon monoxide and nitrogen. The molecules are isoelectronic, and the bond lengths and dissociation energies are quite comparable. The phase diagrams of the two compounds show the same trends in the moderate pressure range with a variety of phase transitions between essentially alike crystal structures [333], when allowance is made for the lack of the inversion center and the presence of a weak electric dipole moment in carbon monoxide. However, the behavior and stability at higher pressure are quite different. While nitrogen at room temperature is stable in its molecular form well above 1 Mbar, carbon monoxide has been found to react and polymerize at a much lower pressure giving a product recoverable at ambient conditions. Katz et al. [334] reported that in the 80–297 K temperature range, CO reacts above 4.6 GPa to give a colorless material that on aging, or at higher pressure, turns yellow. On irradiation with Ar or Kr visible laser lines, the yellow products photoreacts turning finally brownish. It has also been found that, upon laser irradiation, CO reacts at lower pressure [335] and that cooling at 15 K completely prevents the polymerization. The photopolymerization of carbon monoxide at high pressure has later been investigated by Lipp et al. [336] under more intense laser irradiation (several watts) characterizing the product with infrared and Raman spectroscopy. The reaction has also been studied by *ab initio* molecular dynamics [337]. There is no consensus on the structure of polymerized carbon monoxide. This can, at least in part, be due to the different conditions of the reported experiments and of the molecular simulation as well. Katz et al. argued, from the observed absorption edge and from the infrared transmission window, that the product could be poly-carbon suboxide formed after an initial disproportion of CO to give C_3O_2 [334]. By comparing the infrared spectra of their product with that of poly-C_3O_2, Lipp et al. ruled out this possibility, and from the analysis of the infrared spectrum they instead suggested that the polymer could be a poly-vinylester with the same stoichiometry as poly-carbon suboxide [336]. They also observed the formation of a significant amount of carbon dioxide. The similarity of the infrared spectrum of the recovered compound in the mid-infrared region 900–2000 cm^{-1} with that of the product of furan polymerization at high pressure [303, 338] (see the following discussion in this chapter) is rather striking. In this latter case the spectrum was assigned to an amorphous tridimensional polymer. Another remarkable observation is that also in the high-pressure polymerization of furan the laser irradiation results in the formation of carbon dioxide. It can also be noted that the broad feature at $\sim$3300 cm^{-1} present in the infrared spectrum of the recovered compound [336] can be taken as an overtone—as suggested by Lipp, with some difficulty due to its intensity—and could indicate some water contamination.

It is interesting to consider the results of the molecular dynamics simulation showing that the reaction occurs in two stages that could well correspond to the formation of the transparent yellow and to the brownish products. As a whole, the calculations indicate that the final product is a bent polycarbonyl chain with five-membered rings attached and with interconnection between the various chains in a quasi-two-dimensional array. Such a rather extended array is not much different from the suggested structure of the polymer obtained from furan at high pressure [303, 338]. In the molecular dynamics simulation the formation of carbon dioxide molecules is not observed in agreement with the experimental result that CO_2 formation only occurs upon intense laser irradiation.

There is no simple explanation for the much more pronounced instability to pressure of CO compared to N_2. Since the only structural difference arises from the heteroatomic character of CO, one could expect that the molecular dipole moment increases with pressure leading to a higher compressibility of CO. But no evidence for this is obtained from either the *ab initio* calculation or experimentally. In fact the equation of state of nitrogen and carbon monoxide are practically coincident in the pressure range of interest. One other point of interest is the head-to-tail disorder present in carbon monoxide because it has been observed in several high pressure experiments that defects and disorder can play an important role.

Carbon Dioxide. The high-pressure behavior of carbon dioxide is of considerable interest in many respects. In organic synthesis, supercritical CO_2 has proved to be a quite appealing solvent in comparison with the other frequently toxic organic solvents and considering also that the solvent properties can be finely tuned by pressure [339]. Carbon dioxide is also found in the temperature–pressure extreme conditions of some giant planets environments, and the high-pressure properties can be of interest in planetary science. Interest in carbon dioxide at high pressure also arises from the perspective of obtaining extended tridimensional arrays, very much like it has been described above for nitrogen, since superhard materials should mostly be associated with networks of bonds between light elements, as is the case for cubic boron nitride [340]. Carbon dioxide is therefore a possible candidate to produce a material with interesting hardness. From a more basic point of view, carbon dioxide is one of the simpler compounds with unsaturated chemical bonds, and its stability and behavior at high pressure can be a source of fundamental information on chemical bonding.

Shock wave [341] and laser heating [342] experiments have shown that carbon dioxide dissociates into atomic carbon and oxygen. Below the threshold for dissociation, solid carbon dioxide exhibits a variety of phase transformations among structures that have not all been completely resolved sofar. Above 10 GPa and at room temperature, carbon dioxide enters in the CO_2-III phase with an orthorhombic structure, where the molecules are arranged in planes to form typical herring bone arrays. Of interest in the present work are the transformations occurring above 30 GPa on increasing the temperature. In fact, in this pressure range, laser heating results in a transformation to a nonmolecular phase, CO_2-V, where the carbon atoms are tetrahedrally coordinated to the oxygen atoms by single bonds to form a tridimensional extended network of the quartz-like type. Therefore, as has been found in nitrogen, in an appropriate range of pressure and temperature, the structure of carbon dioxide resembles that of the heavier analog, SiO_2, of the same group of the periodic table. The formation of nonmolecular carbon dioxide was first reported by Iota et al. [343] by laser heating at an estimated temperature of 1800 K above 40 GPa. CO_2-V appears as a translucent solid. The main experimental findings include the following: (a) the observation of second harmonic generation that implies the lack of the inversion center and limits the range of silica polymorphs to be considered as analogs; (b) the disappearance of the molecular vibron bands and the appearance of new sharp Raman bands (with the most intense at 790 cm^{-1} assigned as O–C—O stretching mode) showing that CO_2-V is crystalline; (c) from the fit of the pressure–volume dependence, the value of the bulk modulus was obtained and found to be very close to the bulk modulus of cubic BN, supporting the idea that CO_2-V is actually a superhard material. It appears that CO_2-V can be formed from phase III independently from a specific selected pressure threshold above 40 GPa and of a specific temperature of heating above a threshold. The formation of nonmolecular carbon dioxide has been recently reinvestigated in some considerable detail using a specially deviced DAC apparatus that enabled an accurate determination of the phase boundary between phase III and phase V [188]. The reaction has been found to occur at considerably lower temperature than reported by Iota [343]. The phase boundary between CO_2-III and CO_2-V has a negative slope up to 50 GPa. This is in agreement with findings for polymerization reactions of simple hydrocarbons (to be discussed in the following) and suggests that the phase boundary has a kinetic rather than a thermodynamic character. At higher pressure the phase boundary has a positive slope as a result of the prevailing of the high barrier preventing the atomic rearrangement.

The structure of CO_2-V has been determined by x-ray diffraction [343], and the observed pattern could be reasonably fitted by using a tridymite-type structure (orthorhombic $P2_12_12_1$ lattice) shown in Fig. 21. The formation and structure of polymeric carbon dioxide has been studied by computational methods [344–348] in order to fully characterize this novel material; however,

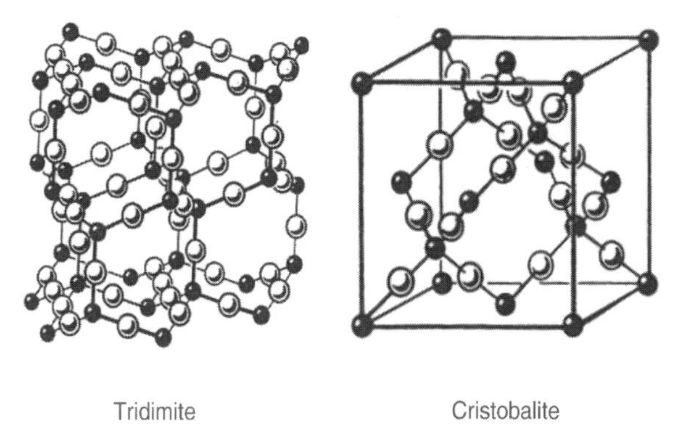

Figure 21. Experimentally (trydimite) and theoretically (α-cristobalite) structures proposed for the polymeric carbon dioxide.

there has been no consensus on the actual structure of CO_2-V, and a full agreement has not been obtained on the x-ray diffraction pattern or on the Raman spectrum. Yoo et al. [347] reported that the suggestion of a tridymite-like structure is also supported by an *ab initio* molecular dynamics simulation using density functional theory. However, within the same general computational approach, different conclusions have been reported by other authors. Serra et al. [344] compared the energetics of several silica-type polymorphs, coming to the conclusion that the most stable polymorph of nonmolecular CO_2-V is of the quartz or m-chalcopyrite type. However, their simulation produced an amorphous polymer. Dong et al. [346] found instead that the α-cristobalite-type structure (see Fig. 21) is energetically favored in the pressure range of interest. However, they also report that in any case it is very difficult to find a complete agreement with the measured x-ray diffraction pattern. Also in favor of the α-cristobalite structure is the agreement between the calculated and observed Raman spectra as well as the calculated dependence on pressure of the Raman frequencies. However, the model was found to reproduce a bulk modulus that is only one-half or one-third of the experimental value. Among the reasons for this variance are the possible differences in the conditions of the experiment and of the various calculations and the overall similarities of the several SiO_2 polymorphs assumed as possible candidates for the structure of polymeric CO_2. Actually, the idea that polymeric CO_2 is necessarily an analog of one of the known structures of silica has been questioned [349–351], considering the different length and polarity of the C—O bond compared to Si–O and, overall, the greater stiffness of the C—O–C bond angle. Of particular interest are the results of Holm et al. [348] reporting the electron distribution in various candidate structures of CO_2-V and their changes from molecular to nonmolecular structures. However, the dynamics of the evolution of the electron distribution along the molecular to nonmolecular transformation has not been explored. As to the mechanism of the reaction, two new phases on carbon dioxide have been discovered recently, CO_2-II and CO_2-IV, that can be obtained by heating CO_2-III in the pressure ranges 12–30 GPa [352] and above 20 GPa [353], respectively. It has been found that these new phases are characterized by strong intermolecular interactions and by the onset of tetrahedral coordination of the carbon atoms [352–356]. In particular, the loss of the inversion center and the Raman activation of the ν_2 mode in phase IV were interpreted in terms of a nonlinear bent structure of carbon dioxide in phase IV. Since the formation of CO_2-V can occur starting from these new phases, it has been suggested that CO_2-II and CO_2-IV are intermediates of the reaction leading to non-molecular carbon dioxide. However, this conclusion has been questioned by some recent calculations [356] and experiments [357] showing that the Raman spectra in CO_2-IV are actually consistent with a linear rather than with a bent structure of the molecules.

Nitrogen Oxides. As discussed in a previous section, the usual response of a molecular system to the increase of pressure can be the formation of additional chemical bonds—as has been described above for nitrogen, carbon monoxide, and dioxide—or the ionization of the system. The latter in general is favored by electrostriction. In the case of carbon dioxide, it has been anticipated that a disproportion can occur at high pressure with the formation of an ionic product $CO^+CO_3^-$ even though this reaction has not been described in detail [358]. The formation at high pressure of ionic crystals by disproportion has, on the contrary, been observed in nitrogen oxides. The high-pressure behavior of nitrogen oxides can give important information on the early chemical reactions and on intermediates in detonation processes. In this respect, NO is particularly important since it is the simplest explosive, at least in the solid state.

Here, the reports on high-pressure reactions of NO_2 and N_2O will be discussed. At room temperature and at 2.3 kbar, NO_2 is in a fluid phase in the form of the dimer N_2O_4, which is planar with D_{2h} symmetry [359]. At 4.6 GPa the sample solidifies in a crystalline form denoted as α-N_2O_4. This phase seems to have the same structure ($Im3$) as the low-temperature solid at ambient pressure. α-N_2O_4 is found to be stable at pressures up to 7.6 GPa except that laser irradiation produces blue spots in the sample that are due to the reversible formation of N_2O_3. Extensive or intensive laser irradiation produces a phase transformation, revealed by the Raman spectra, into a colorless β-N_2O_4 form that is further insensitive to laser irradiation. In the 1.5- to 3-GPa pressure range, β-N_2O_4 experiences a reversible transformation into the ionic compound nitrosonium nitrate, $NO^+NO_3^-$, that is stable up to 7.6 GPa. This reaction is not surprising considering that a similar disproportion has been observed to occur spontaneously in solidified N_2O_5 at zero pressure ($N_2O_5 \rightarrow NO_2^+NO_3^-$). The formation of nitrosonium nitrate has also been found to be induced by temperature or light from N_2O_4 in the solid phase or dispersed in matrices [360–362]. Some interesting observations have been made as to the mechanism of this reaction. One suggestion has been that the reaction could proceed through the nitrite form $ONONO_2$ as an intermediate that further transforms by a simple electron transfer. However, there seems to be clear evidence that the process is intermolecular. It is particularly revealing that the reaction only occurs at high pressure from phase β where possibly the molecules are parallel arranged. It has even been suggested that this could be an example of a topochemical reaction.

Nitrous oxide, N_2O, a triatomic isoelectronic with carbon dioxide, has been reported to react irreversibly [363]. In the range $P < 10$ GPa and $P > 40$ GPa, when laser-heated at temperatures between 1000 and 3000 K, it dissociates into nitrogen and oxygen. The same dissociation reaction is observed at $T > 2000$ K in the pressure range 10–30 GPa. At temperatures $T < 2000$ K and in the pressure range of 10–30 GPa, a new phase is formed that has been characterized on the basis of the infrared and Raman spectra and from synchrotron x-ray diffraction. The infrared and Raman spectra show the disappearance of the N_2O modes and the presence of the N_2 vibron and of the vibrational modes of the NO_3^- and NO^+ ions, very much like it has been observed in the reaction of N_2O_4 discussed above. The results are interpreted considering that the reaction $4N_2O \rightarrow NONO_3 + 3N_2$ occurs, leading to the formation of the unconventional ionic compound nitrosonium nitrate, $NO^+NO_3^-$. The structure of the product of the reaction has been characterized by x-ray diffraction. It has been found that the diffraction pattern is consistent with a superposition of scattering from a mixture of N_2 crystal (in the ϵ form) and an ionic crystal with an aragonite-type structure. This new phase is actually disordered (orientational disorder of the NO^+ ions), and a phase transition to an ordered phase at higher densities is suggested from the pressure dependence of the infrared and Raman spectra. It is also found that the structure is noncentrosymmetric, and from the change in the infrared intensities it is argued that a strongly temperature-dependent charge transfer is present.

Hydrogen Cyanide and the Nitriles. Hydrogen cyanide, HCN, is isoelectronic with nitrogen; but at variance with nitrogen, it is characterized by a high value of the electric dipole moment. Therefore, the structure of the solid will be governed by strong electrostatic interactions, and the molecules will tend to align accordingly. This same basic feature is shared by other molecules containing the $-C\equiv N$ group. We therefore have a class of molecules that, because of the highly directional intermolecular forces, could display at high-pressure topochemical reactions. In addition, the high-pressure transformations of hydrogen cyanide could be of some relevance in connection with reactions in prebiotic environments.

At normal pressure, hydrogen cyanide crystallizes in a tetragonal structure [305, 364] with the molecules aligned along the unique axis held together in linear chains by strong hydrogen bonds. At lower temperature a slight distortion of the plane perpendicular to the unique axis occurs with a phase transition to an orthorhombic structure. The same phase transition is observed at room temperature as a function of pressure with crystallization at 0.2 GPa and the tetragonal-to-orthorhombic transformation at 0.8 GPa. The transition is accompanied by a change in slope of the pressure dependence of the vibrational frequencies and is most likely a second-order transition. The system is stable up to 1.3 GPa. Above this threshold a change of color is observed—to red and finally to black—associated with a polymerization reaction. In a density functional theory *ab initio* calculation [365], it has been found that at 0 K the orthorhombic structure is stable up to 50 GPa when it reverts to a tetragonal structure stable up to 100 GPa, an unexpected finding even considering the underestimate of the van der Waals forces in the model and the temperature effects. However, it has been reported that the polymerization reaction slows down decreasing the temperature, and at 90 K the polymerization reaction has been observed above 4 GPa [366, 367]. The polymer obtained from HCN at high pressure has not been characterized completely. However, the infrared spectrum of the compound exhibits both the stretching mode of the CN triple bond and bands due to modes of the NH_2 group. These same groups are present in the products of the HCN polymerization in solution and in the presence of catalysts. These polymerization products have been named azulmic acids. It is interesting to note that the infrared spectrum of the high-pressure polymer closely resembles the infrared spectrum of azulmic acid 5 reported by Volker [368] which has the structure of a linear array of condensed hexagonal rings with NH_2 and CN side groups. If not coincidental, it is worth recalling that in the theoretical study [337] of CO condensation at high pressure, a somehow analogous type of reaction product has been found. In any case, from the infrared spectrum it is clearly seen that a hydrogen migration occurs during the polymerization, and therefore the reaction is to some extent a collective process. It is thus evident that substituting the hydrogen with a methyl group the reaction pathway should change substantially. Indeed, it has been reported [366] that the polymerization reaction of acetonitrile occurs only at 15 GPa at room temperature. The reaction product, however, has not been characterized in this case.

The high-pressure polymerization reaction of cyanogen $(CN)_2$ is of great interest because it has been suggested to occur topochemically and also because it has been possible to carry a kinetic study at very high pressure. At room temperature, $(CN)_2$ solidifies at 0.3 GPa [369] in a structure that is most likely identical with that of the zero-pressure low-temperature solid (*Pbca*) [370]. On increasing the pressure, two first-order phase transitions are observed at 0.5 and at 2.0 GPa [369]. The steep increase of the vibrational frequencies with pressure indicates that solid $(CN)_2$ is highly compressible. Above 3.5 GPa a further transformation is observed. The transformation occurs in two stages [369]. The first stage of the reaction is reversible: Above 3.5 GPa and up to 6.0 GPa the sample develops a yellow color that intensifies with increasing pressure, and new features develop in the infrared and Raman spectra while the C≡N triple bond stretching frequency is still observed. On the basis of the vibrational spectra, it has been suggested that this reversible reaction leads to the formation of a polymeric chain $(-N{=}C{-}C{=}N-)_n$ with side $-C{\equiv}N$ groups attached. The repeating unit of this poly (2,3-diaminosuccinnonitrile) polymer is quite reactive and it is not surprising that the polymer formed in the first stage of the reaction further reacts at pressures above 6.0 GPa. The color of the samples progressively darkens and becomes black at 10 GPa. The transformation is irreversible, and the product can be recovered at ambient pressure and is found to be thermally and chemically stable. It has been suggested that this second stage of the reaction leads to the formation of paracyanogen. From the comparison of the infrared spectra of the product with those of paracyanogen, it can be argued that the high-pressure polymer should instead be the so-called ladder paracyanogen.

The integrated intensities of the infrared bands of the black polymer forming during the reaction have been measured as a function of pressure (in the range 10–12 GPa) and temperature (in the range 290–350 K) [371]. Even if there is some scattering of the data, it has been possible to obtain both the activation enthalpy $\Delta H^{\ddagger} = 28$ kJ/mol and the activation volume $\Delta V^{\ddagger} = -3.3\,\text{cm}^3/\text{mol}$. The kinetic data have been analyzed with the Avrami model for nucleation and growth [293], obtaining a value of the exponential parameter n close to 0.5. This is the expected value for a diffusion/reorientational-controlled process. It has therefore been concluded that the reaction proceeds through a preliminary reorientation of the poly(2,3-diaminosuccinnonitrile), as a rate-determining step, followed by a condensation that is essentially a cycloaddition. On the other hand, the typical activation volume for cycloaddition reactions is negative but in the range of $-10\,\text{cm}^3/\text{mol}$. The much smaller value of the experimental activation volume can be interpreted as arising from the sum of a positive contribution for the diffusion process and of a negative contribution for the cycloaddition.

Carbon Disulfide. Carbon disulfide is an analog of carbon dioxide. However, the different chemical nature of the sulfur atom is such that chemical and structural properties of carbon disulfide are rather different. Because of the larger van der Waals radius of sulfur, carbon disulfide does not crystallize in the classical *Pa*3 structure of the linear molecules and at zero pressure is found in the orthorhombic *Cmca* structure that is similar to the structure of CO_2-III, which is stable only above 12 GPa. Thus, it can be expected that carbon disulfide should be more reactive than carbon dioxide. It is indeed well known that carbon disulfide can easily polymerize on exposure to light and in other mild conditions as well [372, 373]. Measurements of the Raman spectra in the temperature range 6–300 K have shown that carbon disulfide at room temperature crystallizes at 1.26 GPa in the same structure as at low temperature and zero pressure and that it is stable up to 8 GPa [374]. However, it has been found that the sample quality greatly depends on the method of preparation and that the crystal can easily turn irreversibly into an amorphous material. Bolduan et al. [374] reported a larger range of stability of CS_2 compared to the previous determination of a room-temperature transformation threshold of 4.5 GPa by Shimizu and Onishi [375]. A chemical transformation of carbon disulfide was first observed by Bridgman [376], who described the formation of a black polymeric material from the liquid at 4 GPa and at 150 K. In the Nobel lecture, Bridgman reports on his observations, saying "*I have found that ordinary liquid CS_2 may similarly be* *changed permanently into a black solid at temperatures in the neighborhood of 200°C and by pressures of the order of 40.000 kg/cm^2. This black substance is definitely not a mixture of sulfur and carbon, which one might at first expect, but is apparently a unitary substance, truly a black solid form of carbon disulfide. It has been suggested that the structure may be that of a single giant molecule like the known structure of SiO_2 which from the atomic point of view is very similar. It is fascinating to speculate that there may be many other common substances which may be pushed by sufficiently high pressures over a potential hill of some kind permanently into some hitherto unknown form.*" The reaction has later been reproduced, although in somehow different pressure–temperature conditions [377, 378]. The product of the reaction, known as Bridgman black carbon disulfide, has been characterized in some detail by Butcher et al. [379]. The polymer has a $(CS_2)_x$ composition and is amorphous, giving a diffuse x-ray diffraction pattern. It has a very low conductivity that increases with temperature, behaving as a semiconductor with a 1.15-eV energy gap. The infrared spectra show the presence of C=S groups and are identical with that of the $(CS_2)_x$ polymer prepared photochemically [372, 373], and the likely structure consists of highly cross-linked chains with a –S–C=S repeating unit. Experiments seem to indicate that the Bridgman black carbon disulfide can be obtained only in a limited temperature range.

2. *Hydrocarbons*

Hydrocarbons have attracted a considerable attention since the early days of high-pressure research. For instance, polymerization reactions of hydrocarbons at high pressures were already studied by Conant and Tongberg in 1930 [4]. Shock-wave experiments have shown that under extreme pressure and temperature conditions, hydrocarbons—and, in particular, unsaturated hydrocarbons—decompose in carbon and hydrogen [380–383]. These reactions can be important as a source of carbon aggregates in planetary and extraterrestrial environments [384, 385]. Ree discussed a unified model of hydrocarbons behavior at high pressures and temperatures in terms of the formation of different proportions of carbon and condensed hydrogen [386]. However, experiments at both dynamic and static high pressures [380–383, 387–390] and alternative modelings [391–395] have clearly shown that at high pressure/temperature, hydrocarbons display a more subtle and complicated chemical behavior as a result of the balance between the effect of temperature (which favors the dissociation into the constituent elements), and pressure (which favors instead condensation processes). This occurrence has been demonstrated, for instance, also in the case of methane, which, besides the dissociation in carbon and hydrogen, has been shown to produce, in appropriate regimes of pressure and temperature, higher hydrocarbons [396, 397]. Experiments in a laser-heated DAC (15–50 GPa and 2000–3000 K) led to the formation of two products: (1) an opaque product that x-ray diffraction and Raman spectroscopy revealed to be a diamond-like carbon, and (2) a transparent one, mainly formed in the cooler area around the hot spot, consisting of a mixture of doubly and triply bonded carbon atoms, with hydrogen probably bound as hydride to the rhenium gasket. These experiments confirmed the results of molecular dynamics simulation that showed a discontinuity in the equation of state associated with a reaction giving a mixture of various hydrocarbons with chain length increasing with the pressure of the simulation up to C_{12}.

Several aspects of the hydrocarbons behavior at high pressure and temperature are known, and in particular there is evidence that before complete amorphization or dissociation into the elements, several intermediates (dimers, oligomers, polymers) are formed. However, the microscopic mechanism of the reactions occurring in these conditions has not been elucidated. In the last two decades the improvements of the experimental methods in static high-pressure experiments with *in situ* structural determinations—including the simultaneous accurate control of the temperature and the use, as an additional activation tool, of laser irradiation—has allowed us to accumulate a wealth of data on hydrocarbons reaction at high pressure that will be helpful to start building a coherent picture of these materials in extreme conditions. In the following we shall consider various examples corresponding to the categories of unsaturated hydrocarbons including isolated and substituted double bonds (ethylene, propylene, styrene), conjugated polyenes (butadiene), triple bonds (acetylene, phenylacetylene, cyanoacetylene), aromatics (benzene), and heteroaromatics (furan, tiophene).

Acetylene. Acetylene, HC≡CH, is isoelectronic with nitrogen, but there is a fundamental structural difference between the two molecules. The ratio between the triple- and the single-bond energy is 5.9 for nitrogen atoms, while it is only 2.4 for carbon atoms. Therefore, acetylene is energetically unstable compared to compounds with single-bonded carbon atoms, and it can be expected that acetylene will polymerize at a low-pressure threshold. Pressure polymerization of acetylene can be of importance because the product, polyacetylene, has unique properties bound to the high conductivity that can be induced by doping and photoexcitation. Polyacetylene can be obtained by several routes and mainly by polymerization of aceytylene on Ziegler–Natta catalysts [398]. It could be of interest to explore if a pressure induced condensation of acetylene could exhibit some kind of selectivity. In the solid state, acetylene is found in two crystalline forms [399, 400]. At room temperature, C_2H_2 crystallizes at 0.7 GPa in the *Pa*3 cubic structure and has a transition at 1 GPa to an orthorhombic *Cmca* structure that is stable up to 3.6 GPa. Above this threshold a chemical reaction is observed. By Raman spectroscopy the polymerization threshold was located at 3.5 GPa, and the sample was observed to develop a deep red color. In these conditions the reaction was complete in 13 h [401, 402]. The reaction has also been monitored by infrared spectroscopy, and a slightly higher threshold (4.2 GPa) was observed with the reaction proceeding very slowly and being successively accelerated on further increasing the pressure up to 14 GPa [403]. The reaction has been further reconsidered both at room and lower temperature (200 K), with a more accurate control of the pressure and of the conditions of the experiments in a membrane DAC to allow kinetic studies [98]. All these experiments raised several problems that turned out to be of general interest. With the Raman monitoring of the reaction, it was concluded that the obtained polymer was *trans*-polyacetylene with amounts of the *cis* isomer. However, it is known that the Raman spectrum of polyacetylene is resonantly enhanced; and if, as is likely, the product of the reaction is a mixture of polyenes of different length and of different absorption band frequencies, the characterization with the Raman spectroscopy can be misleading. In addition, as has been shown by Ceppatelli et al. [98], the reaction kinetics and mechanism can be affected photochemically even at the low irradiance necessary for the measurement of the Raman spectrum or for the pressure monitoring by the ruby fluorescence method. These matters, which in any case do not change substantially the overall picture of the reaction, have been clarified by experiments in our laboratory [98]. The in situ pressure has been monitored with infrared sensors based on the ν_3 frequency shift of the nitrite ion, thus completely avoiding laser irradiation of the sample. It has been found that at each pressure above 3.5 GPa, the reaction proceeds up to a saturation and can be restarted by further increasing the pressure. The evolution of the reaction is illustrated in the lower section of Fig. 22 by the growth of the infrared intensity of a band of the product. The trend is the same when using either the infrared sensor or the ruby fluorescence. However, the nucleation step of the process is clearly observed with the former while it is completely smeared out with the latter. When the reaction is at the saturation and the sample is irradiated with a laser line of appropriate wavelength even at low power (1–5 mW), the reaction accelerates again. This effect is displayed in the upper part of Fig. 22, showing clearly that the reaction is activated photochemically.

A careful analysis of the infrared spectrum of the polymer, along with comparison with the spectra of linear conjugated polyenes and with the results of density functional theory calculations on polyenes [404], has shown that the reaction product is composed by *trans* polyenic chains containing fragments with more than 12 C=C double bonds. The reaction product also contains a certain amount of saturated sp^3 carbon atoms. This is evident from the presence in the spectrum of saturated CH strething and bending modes. An important finding is that the laser irradiation not only accelerates the reaction but changes the composition of the product, thereby increasing considerably the number of the saturated sp^3 carbons. This produces a cross-linking of the larger polyacetylenic chains at the expenses of the conjugation length.

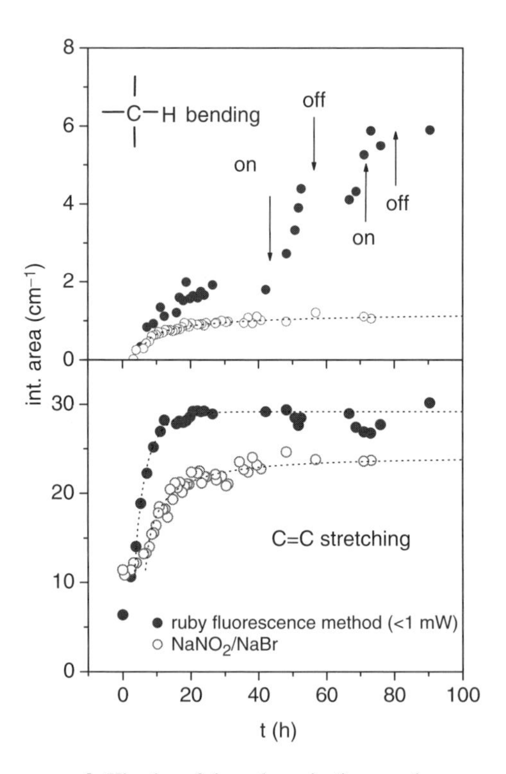

Figure 22. **Lower panel:** Kinetics of the polymerization reaction measured through the time evolution of the IR absorption relative to the C=C stretching modes. Full and empty symbols correspond to different experiments where the local pressure calibration was performed by using (ruby method) or not using laser light (IR calibrants), respectively. Differences in the early stages of the reaction reveal the effect of laser light in the activation of the reaction. **Upper panel:** Effect played by laser irradiation in the branching of the chains. The comparison between experiments where laser light is employed (full symbols) or is completely absent (empty symbols) shows that the absorption of the C—H stretching modes involving saturated C atoms sharply increases when the laser light is hitting the sample (irradiation cycles are delimited by the arrows).

These findings are in agreement with other experiments and with the results of calculations on acetylene polymerization. The formation of a prevalently *trans* polymer agrees with the high-pressure behavior of polyacetylene crystals reported by Balzaretti et al. [405], showing that in a high-pressure treatment of *cis–trans* mixtures the *cis* component completely disappears most likely by a transformation in the *trans* isomer. In *ab initio* molecular dynamics simulations [400, 406] it has been found that the reaction product is polyacetylene with some degree of cross-linking [406] and with a composition consisting of a mixture of *sp*, sp^2, and sp^3 carbons in the ratio 22:41:37. Branching of the chains, up to a transformation into an amorphous hydrogenated carbon, has also been reported in a molecular dynamics simulation of polyacetylene at high pressure [407]. The molecular dynamics calculation [406] also confirms that the system is photosensitive as experimentally observed, because injection of a molecule in an excited triplet state greatly reduces the pressure threshold of the reaction. The cross-linking of the polyenic chains can possibly produce screening of the unreacted monomers, thus leading to the observed saturation of the reaction. A very interesting observation reported by Sakashita [403] and confirmed in our laboratory is that the vibrational frequencies of acetylene harden with increasing pressure but abruptly revert to the liquid phase values at the onset of the reaction. This can be explained as a screening effect that disrupts the crystalline array and facilitates the formation of a disordered product.

Concerning the reaction mechanism, there is a consensus from experimental evidence [98, 402] and from molecular dynamics simulations [400, 406] that the formation of *trans*-polyacetylene occurs mainly along the diagonal of the *bc* crystal plane of the *Cmca* structure. In fact, in this direction the orientation of the molecules is the most favorable for a *trans* polymerization, and it has been estimated that at room temperature and at 4 GPa the nearest neighbor C···C distance is the shortest (3.05 Å). Unpublished results of *ab initio* molecular dynamics calculations in our laboratory [408] have shown that the reaction occurs only when the intermolecular C—C distance is shorter than 2.6 Å. This translates the reaction threshold in a microscopic language. The necessary close approach in the experiment can be ensured by the translational phonon modes whose mean-square amplitude at room temperature has been estimated to be 0.5 Å [98]. Indeed it has been found that at 200 K, where the translational mean-square amplitude is reduced to only 0.2 Å, the reaction considerably slows down and can be induced only at 9 GPa [98]. A kinetic analysis of the reaction has been carried measuring the integrated infrared intensities of the product bands as a function of pressure and analyzing the data with Avrami model [293]. A value of the n exponent of 0.5 has been obtained, implying that the reaction involves a diffusion step that is here correlated for the first time to the translational oscillations of the molecules.

The acetylene reaction has been studied also at 77 K [409]. It has been reported that at this temperature the reaction occurs at 12.5 GPa and that the reaction proceeds to saturation and then accelerates on pressure release. In this low-temperature experiment, it has been found that the product contains also *cis*-polyacetylene and that a transformation to the *trans* isomer occurs on heating.

Cyanoacetylene. As has been discussed above, the polymerization of acetylene in the solid at high pressure occurs along the diagonal of a definite crystal plane because in this direction the molecular reorientation required is minimum and the necessary close intermolecular approach is realized by the librational and translational thermal fluctuations. The reaction can therefore be considered to occur topochemically even though alternative pathways can lead to the branching of the chains. The increased anisotropy of the intermolecular interactions and the tighter binding of the molecules in the crystal makes the topochemical character of the cyanoacetylene polymerization definitely more pronounced. This reaction has been studied at high pressure by Aoki et al. [410] and is a quite exceptional case for molecular crystals because the reaction has been studied in a single crystal. At room temperature, cyanoacetylene crystallizes at 70 MPa in a monoclinic $P2_1/m$ structure identical to the zero-pressure low-temperature phase [411]. The molecular arrangement in the crystal is determined by strong dipole–dipole interactions and hydrogen bonding, making the molecules aligned in parallel linear chains along the *a* crystal axis as shown in Fig. 23(a). This arrangement determines a larger compressibility of the crystal in the directions perpendicular to the chain's axis.

A chemical reaction occurs above 1.5 GPa: The sample turns black, new peaks develop in the Raman spectrum, and the absorption edge moves below 11,000 cm^{-1}. The recovered material has an optical band gap of ~1.39 eV, smaller than the band gap of polyacetylene. From the analysis of the Raman spectrum, it is seen that the C≡C stretching mode completely disappears in the reaction product, while the C≡N stretching band is present but at a different frequency than in cyanocetylene. In addition, the Raman bands of polyacetylene are observed with their characteristic frequency dependence on the wavelength

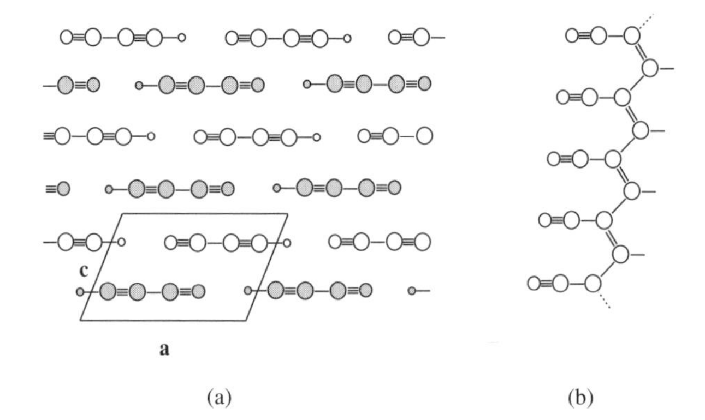

Figure 23. (*a*) Arrangement of the cyanoacetylene molecules in the monoclinic P_{21}/m crystal as observed along the *b* axis. Full and empty molecules lie on different, adjacent *ac* planes. (*b*) The proposed mechanism of the high-pressure reaction: The molecules lying on the same *ac* plane react through the opening of the triple bond along a direction perpendicular to the *a* axis and not involving the cyano groups.

of the exciting line. From these data the product has been identified as polycyanoacetylene. Observation of the crystal structure allows the identification of the reaction mechanism consisting of the opening of the C—C triple bond and a cross-linking of the hydrogen-bonded chains in a direction perpendicular to the a crystal axis, as is shown in Fig. 23b. As can be seen from the figure, the condensation requires a minimum dislocation of the molecules in the starting material once the compressibility of the material brings the molecules to a sufficiently short separation. The formation of a quite regular polymer, deduced from the sharpness of the Raman bands, is obviously favored by the single-crystal nature of the starting sample such that random initiation and termination at defects and grain boundaries is greatly reduced.

Phenylacetylene. From a simple consideration of average bond energies, the expectation that in cyanoacetylene the opening and polymerization of the C—C triple bond should be favored is proven, because it is actually observed in the experiments described above. It has been found that the substitution of a hydrogen with a —CN group reduces the stability of the C—C triple bond, bringing the reaction threshold from 3.5 GPa in acetylene to 1.5 GPa in cyanoacetylene. Also, in phenylacetylene the C—C triple bond reacts at high pressure well before the benzene ring. In fact, as will described in the following, the pressure threshold for the benzene reaction is higher than 20 GPa, much larger than the acetylene threshold. However, it has been found that in phenylacetylene the threshold for the opening of the triple bond is considerably higher than in acetylene. The high-pressure polymerization of phenylacetylene has been studied by Santoro et al. [412]. Crystallization of phenylacetylene at room temperature is observed at 1 GPa. The crystal is stable up to 8 GPa and the structure is most likely the same as the zero-pressure monoclinic $P\bar{1}$ structure [413] with one out of the five molecules in the unit cell orientationally disordered, a circumstance that is of importance for the reaction mechanism. Above 8 GPa a reaction occurs and the sample assumes a bright red color. The reaction is very slow and can be completed only by further pressurization at 12.2 GPa and waiting for 20 h. The reaction has been followed by infrared spectroscopy, and it is seen that the bands due to the C—C triple bond gradually disappear during the reaction. Analysis of the infrared spectra clearly show that the reaction product is polyphenylacetylene identical to the polymer obtained with more traditional chemical methods. The infrared spectra compare reasonably with those of polystyrene, confirming that the benzene ring is not affected by the reaction. From phenylacetylene, like from monosubstituted acetylenes, eight polymeric structures can be obtained, differing for the relative orientation of the substituent in the polymeric chain. For instance, the polycyanoacetylene polymer discussed above would be classified as a *trans–transoid* isomer. Comparison with the spectra of known polymers shows that the high-pressure polymer obtained from phenylacetylene has most likely a *cis–transoid* conformation. However, the sample is at least partly amorphous and shows also evidence of some degree of chain branching.

Ethylene. Ethylene is a highly interesting system for high-pressure polymerization. It is in fact the simplest hydrocarbon with an isolated double bond. Polyethylene is one of the most important polymeric materials; depending on the conditions of preparation from the monomer, its properties change considerably. From the structural point of view, one can distinguish between polymers in the form of extended linear chains and polymers with branching of the chains. The properties of the material are also affected by cross-linking of the chains. Consequently, polyethylene is classified as (a) high-density polyethylene (HDPE) corresponding mainly to linear chains that in turn can give rise to crystalline materials and (b) low-density polyethylene (LDPE) corresponding to branched polymeric chains. Considering the technological importance of this material, it can be of interest to explore whether high-pressure polymerization, in the absence of the traditional catalysts, could show some kind of peculiarity, as actually turned out to be the case. It has been known for a long time that at high pressure, ethylene can spontaneously polymerize; the reaction in the solid has been studied at pressure >3 GPa, obtaining a product identified as polyethylene but not characterized further [414, 415]. The high-pressure reaction of ethylene has been recently studied in our laboratory, both in the liquid and in the solid state at room temperature [416]. At 0.7 GPa, ethylene is liquid and no reaction occurs. When the liquid is irradiated with laser lines in the visible or near-UV region, a reaction occurs. The bands of ethylene gradually disappear, and new bands develop in the infrared spectrum. The reaction accelerates by increasing the pressure or by increasing the laser intensity or also moving the laser wavelength in the ultraviolet. The reaction of the irradiated sample proceeds despite the pressure drop occurring in the cell, and it has been found that the reaction can occur at a pressure as low as 0.2 GPa. Characterization of the recovered product has been made primarily by angle-resolved x-ray diffraction using monochromatic synchrotron radiation (see Fig. 24). The diffraction pattern closely agrees with that of polyethylene and shows that the sample is fully crystalline, and the absence of any background due to amorphous components indicates that the crystallinity of the sample is close to 1. The product has been further characterized by infrared and Raman spectroscopy. The sharpness of the bands, the well-resolved splitting of the vibrational modes, and their relative intensity clearly show the complete crystallinity of the high-pressure polymer. The result is remarkable in several respects. For the first time at high pressure a completely crystalline polymer has been obtained at room temperature from a liquid sample in the absence of branching and without the use of catalysts. The pressure threshold is sufficiently low that applications of the procedure are

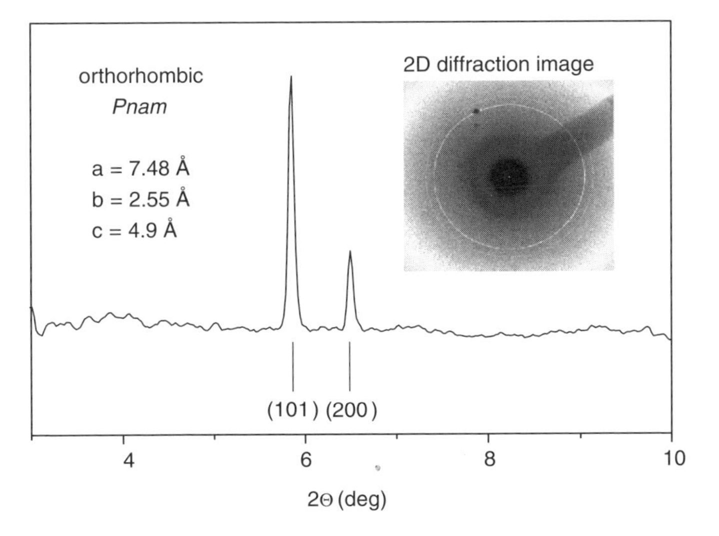

Figure 24. X-ray diffraction pattern (in the inset the 2D image) of the polyethylene sample recovered by the laser-assisted high-pressure reaction in the pure liquid monomer. The two measured sharp lines nicely fit the polymer diffraction pattern having a orthorhombic cell (*Pnam*) defined by the lattice parameters reported in the figure.

possible. In the solid the reaction has been studied at 3.3 and 5.4 GPa. The rate of transformation is very low in both cases. The reaction product from the solid is a low-density form of polyethylene with a considerable degree of branching.

Because ethylene is transparent in the visible region, the photochemical activation of the high-pressure reaction must occur through a two-photon absorption process. From the known structure of the excited states of ethylene and the shift of the absorption bands with pressure, it has been argued that the candidate excited state is the $^1B_{3u}$ antibonding state with a twisted geometry and a C—C bond distance in the 1.4-Å range. Because this state has a very short lifetime, the reaction is likely triggered by excited molecules after relaxation in the ground state and before rearrangement of the geometry in the planar structure. The high-pressure reaction mechanism has been studied in some details by *ab initio* molecular dynamics simulation with the Car–Parrinello method [297] on a sample of 52 molecules at a density of 1.91 g/cm^3. In the pressurized sample and before any reaction occurs, a considerable broadening of the electronic density of states is observed. At the same time, the response to high pressure produces a rearrangement of the electronic distribution within the molecule that is monitored by the molecular dipole moment that has a distribution ranging up to 1.25 D with a maximum at 4 D. The electronic rearrangement shows that the polymerization reaction occurs through an ionic mechanism that facilitates the regular head-to-tail growth. The product of the reaction in the simulation depends on the size of the sample. In the simulation with 52 molecules, an oligomer with 10 monomeric units is obtained, besides smaller oligomers. The important point is that a linear oligomer is obtained, without branching, as is found experimentally. It is likely that increasing the size of the simulated sample longer chains will be obtained. The molecular dynamics simulation has been extended to crystalline ethylene. It has been found that condensation can occur between ethylene molecules lying either on equivalent lattice sites or on nonequivalent sites. In the former case a linear chain is produced while the reaction between nonequivalent sites is responsible for branching. Because the ethylene crystal compressibility is highly anisotropic, the degree of branching will change by changing the pressure.

Tetracyanoethylene. The interest in the high-pressure reactivity of tetracyanoethylene (TCNE) is essentially due to the lack of chemical means capable of inducing the polymerization of TCNE at ambient pressure, while only an amorphous product has been obtained by plasma polymerization [417]. At ambient pressure, TCNE exists in a cubic phase ($Im3$) below 318 K and exists in a monoclinic phase ($P2_1/n$) above this temperature [418]. The molecular packing is significantly different in the two phases: The planes of the TCNE molecules are perpendicularly arranged in the cubic structure, but they are parallel in the monoclinic phase. The high energetic barrier associated with the rearrangement necessary for the phase transformation makes the cubic–monoclinic transition irreversible on cooling, allowing the compression of both crystalline phases at room temperature. An IR study revealed the formation of C=N stretching modes above 6 GPa in association with an intense darkening of the initially colorless sample [419]. These observations were correlated to the occurrence of a chemical reaction—in particular, the polymerization through the cyano groups. Room-temperature x-ray diffraction experiments have been performed compressing both the cubic and the metastable monoclinic phase [420]. In the first case the reaction is observed directly from this phase at 6 GPa, while after compressing the monoclinic crystal a reversible phase transition to a new phase is observed at 3.5 GPa, but also in this case the reaction threshold pressure was located near 6 GPa by the IR study [419]. The x-ray diffraction data provide an important insight about the reaction mechanism and confirm the reaction to involve the cyano groups of neighboring TCNE molecules. Assuming the molecules to be rigid and incompressible, the static distance between the carbon and nitrogen atoms of nearest-neighbor molecules is calculated in the cubic phase at the reaction threshold pressure of 6 GPa to be the shortest contact (2.83 Å) [420]. Raman spectra show the progressive disappearance of the lattice modes above 4.4 GPa; the TCNE internal modes are no longer observable, and therefore the reaction is concluded, near 10 GPa [421]. The Raman analysis of the product does not provide any evidence of the polymer formation, but a broad Raman peak at 1580 cm^{-1} (characteristic of microcrystalline sp^2 carbon) is the only signature of the product. Diffraction patterns showing evidence of a microcrystalline graphitic phase have been measured in black reflective samples produced by compressing TCNE in a multianvil apparatus to 16 GPa [422]. Microprobe analysis reveals that the nitrogen incorporation leads to the chemical composition C_6N_3. Following this result, TCNE has been employed for high-pressure–high-temperature synthesis of sp^3-bonded carbon nitrides [423].

Propylene. The high-pressure reaction of propene has been studied by *ab initio* molecular dynamics [296]. The possibility to induce also in the case of propylene a polymerization at high pressure is very important because polypropylene is one of the most versatile polymers because it serves both as a resistant plastic (melting above 170°C) and as a fiber. It is generally prepared from the propylene monomer by Ziegler–Natta polymerization and by metallocene catalysis polymerization. One of the relevant problems in the polypropylene synthesis is the possibility to produce samples with different tacticities depending on the position of the methyl groups with respect to the polymer backbone. In the molecular dynamics simulation, it has been found that with increasing pressure the electronic density of state changes completely and at sufficiently high density the HOMO–LUMO gap is absent. This implies a substantial increase of the chemical potential of the system and suggests the possibility of a chemical transformation at high pressure. Actually a reaction is observed in the simulation leading to the formation of a saturated and an unsaturated dimer and of a pentamer. As always in molecular dynamics simulations, the exact nature of the final products can depend on the size of the simulation box. It is therefore of more interest to analyze the information on the reaction mechanism that shows two basic features. The reaction mechanism is ionic, and a significant increase of the molecular dipole moment before the onset of the reaction is observed. Before the reaction occurs, there is a migration of a π-electron doublet that precedes the formation of a dimer that appears in dicationic form able to further propagate the reaction, while the termination occurs by hydrogen ion dislocations. The reaction mechanism has in any case a collective character involving more than two monomers in each step, and the formation of the three products cannot be described independently. It is interesting to note that while in cationic reactions generally a cationic initiator is introduced in the reaction mixture, at high pressure a charge separation occurs as a response to the repulsive intermolecular interaction. The transformation of propylene under high external compression has been studied in our laboratory [424]. At room temperature a chemical reaction is observed at 3.1 GPa when the sample is still liquid as argued by extrapolation of the low-temperature phase diagram [425]. The formation of polymeric chains is revealed by the infrared spectrum showing a strong broad

C–H stretching band involving sp^3 carbon atoms that masks the structured absorption of the monomer, along with the growth of two bands at 1385 and 1470 cm^{-1}. These are the most intense bands in the transparent soft material recovered at the end of the reaction. The reaction reaches a saturation (28 h) when a large amount of the unreacted monomer is still present, and a further pressure increase does not produce a substantial acceleration of the reaction. A full characterization of the viscous recovered product has not been possible, although some indications are found that only oligomers with up to 18–20 C atoms are formed. The amount of transformed material significantly increases when the reaction is laser-assisted (458 nm). Also in this case the reaction threshold pressure lowers under irradiation, and the transformation is observed already at 1.5 GPa. Therefore, also in this case the catalytic role of the laser light plays a twofold role: it increases the reaction yield and allows the reaction to occur under less drastic pressure conditions.

Butadiene. The reaction of *trans*-butadiene at high pressure has been recently reported and shows some unique features [426–429]. It has been found that two distinct reaction pathways are possible, leading to completely different products. By adjusting the conditions of the reaction (phase of the sample, pressure, laser irradiation), it is possible to choose between the two pathways in an extremely selective way. On the other hand, butadiene is an interesting system by itself: it is the prototype and the simplest of the conjugated hydrocarbons. The product of butadiene polymerization, polybutadiene, is a polymer of primary technological importance. Butadiene is a rather unstable species. At ambient conditions it can easily react to give a mixture of different dimers obtained through different reaction mechanisms, with the Diels–Alder reaction giving 4-vinylcyclohexene as the most abundant dimer. The polymerization of butadiene is generally carried in the liquid and in solution in the presence of appropriate catalysts and radical initiators. In general, polybutadiene is obtained as a mixture of the *trans* and *cis* isomers.

Experiments at high pressure have shown that the P–T phase diagram of butadiene is comparatively simple. The crystal phase I is separated from the liquid phase by an orientationally disordered phase II stable in a narrow range of pressure and temperature. The structure of phase I is not known, but the analyses of the infrared and Raman spectra have suggested a monoclinic C_{2h}^5 structure with two molecules per unit cell as the most likely [428]. At room temperature, butadiene is stable in the liquid phase at pressures up to 0.7 GPa. At this pressure a reaction starts as revealed by the growth of new infrared bands (see the upper panel of Fig. 25). After several days a product is recovered, and the infrared spectrum identifies it as 4-vinylcyclohexene. No traces of the other dimers can be detected, and only traces of a polymer are present. If we increase the pressure to 1 GPa, the dimerization rate increases but the amount of polymer

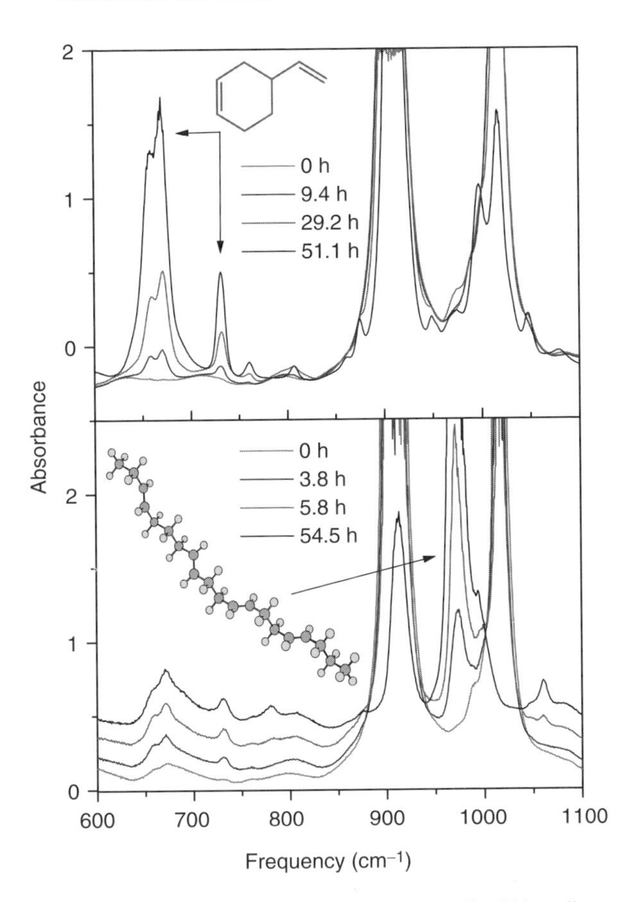

Figure 25. Evolution with time of the chemical reaction in liquid butadiene at 0.6 GPa and 300 K. **Upper panel:** Purely pressure-induced reaction, the formation of vinylcyclohexene is revealed by the growth of the bands of the dimer in the 650- to 750-cm^{-1} frequency range. **Lower panel:** In this case the reaction is assisted by the irradiation with few milliwatts of the 488-nm line of an Ar^+ laser. The fast increase of the characteristic polymer band at 980 cm^{-1} indicates the selective formation of polybutadiene.

remains negligible. At temperatures lower than 280 K, pressurization solidifies the material without any reaction. When the liquid is in the pressure regime close to 0.6 GPa, irradiation of the sample with the 488-nm Ar^+ laser line completely inhibits the formation of the dimer; in a few hours, butadiene completely transforms into a transparent solid that is identified as polybutadiene by the infrared spectra (see Fig. 25). The remarkable finding is that the polymer formed at high pressure is fully in the *trans* form, and no traces of the *cis* isomer can be detected in the infrared spectra. The response to laser irradiation is quite selective and ineffective at 514.5 nm. The growth of a linear polymer is confirmed by a kinetic analysis of the reaction carried following the growth of the infrared intensities of the polymer bands as a function of pressure. A very good fit of the data to the Avrami law is obtained also in the first stages of the reaction with a value of the n exponent of 1.86, which implies a linear propagation of the reaction.

Since butadiene is transparent in the visible the photochemical activation of the polymerization reaction involves again a two-photon excitation of the S_1 (2^1Ag) state that is symmetry-allowed with a high cross section and a long lifetime. In the excited state the outer C—C bonds considerably increase their lengths and become longer than the inner bond, thus lowering the rotational barrier of the terminal CH_2 groups and facilitating the propagation of the reaction along the molecular backbone. At the same time, the structural change is unfavorable to cyclization that occurs through an overlap of the π-electron densities.

The mechanism of the polymerization reaction of butadiene has been studied by a first principles molecular dynamics simulation [298]. In the simulation the reaction is induced solely by the pressure, and it is found that at high pressure the elongation of the outer bonds occurs with a distribution of bond lengths reaching the value characteristic of polybutadiene chains (1.52 Å). As already observed in the molecular dynamics simulation of propene [296], the reaction follows an ionic mechanism. The system responds to the high pressure with a dislocation of the electron doublet and the formation of a very reactive zwitterion that greatly facilitates the 1–4 reaction scheme. The characteristic features of the reaction (as seen in the simulation) also in this case are the ionic character and the collective interaction of the molecules leading to thermal fluctuations of the charge distributions induced by the neighboring molecules. The charge redistribution produces, before the reaction, a distribution of the molecular dipole moments peaked at 0.7 D but extending up to 3.0 D. The reaction product is, as observed experimentally, *trans*-polybutadiene; at the end of the simulation, 20 out of the 27 molecules in the simulation box have reacted. The reaction pressure in the simulation is much higher than the experimental one, but this difference should be associated to the short time duration of the simulation. Once again it is worth mentioning that in this and other simulations, and in a number of experiments as well, it seems that the photochemical effect of laser irradiation is an alternative to a large increase of the pressure.

The butadiene reaction at high pressure has also been studied in the solid [429]. The reaction has been followed at several pressures ranging from 2.1 and 6.6 GPa and has been monitored by infrared spectroscopy. It has been found that below 4.0 GPa, only vinylcyclohexene is formed with trace amounts of the polymer. Above this threshold pressure, the amount of polymer formed is not

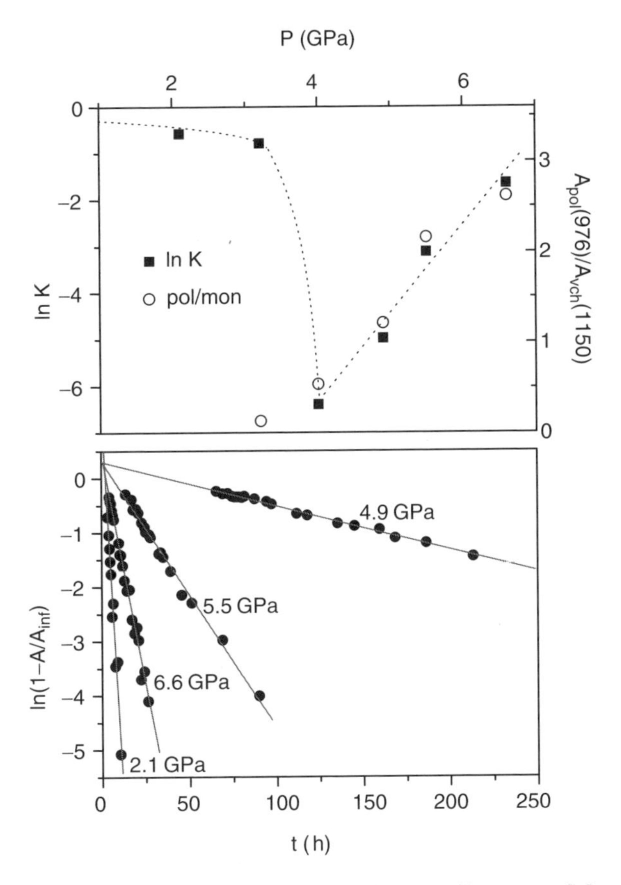

Figure 26. Dimerization of butadiene in the crystalline phase. **Lower panel:** Logarithmic plots of the room-temperature evolution of the integrated absorption of characteristic vinylcyclohexene absorption bands at different pressures. The linear evolution unambiguously demonstrates the first-order kinetics of the reaction. **Upper panel:** Evolution of the natural logarithm of the dimerization rate constant as a function of pressure (full squares, left axis; the dotted line is intended as a guide for the eye) and evolution of the intensity ratio between selected polymer and dimer (vinylcyclohexene) bands (empty dots, right axis).

negligible and rapidly increases with pressure. The reaction evolution in the crystal at higher pressures is therefore more complicated. The rate constants for the dimerization have been determined as a function of pressure, and the results are shown in Fig. 26. Up to 3.5 GPa the dimerization rate constant decreases smoothly, corresponding to a positive activation volume. Between 3.5 and 4.0 GPa, there is a large drop in the rate constant. Above 4.0 GPa the rate constant increases and recovers the low pressure value at 7.0 GPa. The reaction is also found to follow a first-order law all over the pressure range investigated. Assuming that the two inequivalent molecules in the hypothesized unit cell are almost perpendicular to each other, the following three-step reaction mechanism has been suggested. In the first step, head-to-tail condensation of the two inequivalent molecules occurs. This step is governed by the orientational and translational displacement of the molecules and gives rise to the observed temperature dependence of the rate constant. A second step of the reaction requires an isomerization and rearrangement of the intermediate dimer. The third step leads to the fast closure of the hexagonal ring. The second step needs the overcome of a considerable energy barrier in the crystal and is therefore the rate-limiting step. This mechanism explains the first-order reaction rate and the negative value of the activation volume in the lower-pressure regime. Within this scheme the change in slope of the rate constant above 4.0 GPa is not interpreted as a change of the reaction mechanism; however, with the onset of the polymerization reaction, vacancies are created in the crystal, thus facilitating the intramolecular rearrangement. The geometrical constraints of the crystal lattice suggest that while dimerization involves nonequivalent molecules in the unit cell, the polymerization occurs among equivalent molecules of different unit cells. Laser irradiation of the crystal produces an acceleration of both reactions, but at present it is difficult to establish whether the irradiation affects the dimerization directly or indirectly through the formation of vacancies due to the polymerization reaction as discussed above.

Aromatics

BENZENE. Benzene is the aromatic molecule par excellence. Crystalline benzene, among molecular crystals, is the more thoroughly studied at high pressure, and its P–T phase diagram has been extensively discussed. At large intermolecular interactions, the packing of aromatic rings is governed by electrostatic (quadrupole–quadrupole in the case of benzene) interactions and by dispersive forces with the π–π interactions playing a minor role [430]. In fact, structural and dynamical properties of crystalline benzene at low temperature and pressure are nicely reproduced by empirical intermolecular potentials [431]. At high pressure the interactions between the π electronic distributions of neighboring molecules will gain increasing importance, and the interplay between the various terms of the intermolecular potential may be the source of the complexity of the benzene phase diagram. In addition, the transition between the various crystal phases of benzene is sluggish and the persistence of crystal phases beyond their stability field has been reported, thus making the determination of phase boundaries a nontrivial problem. According to the phase diagram of Cansell et al. [304] at 300 K, increasing the pressure the phase boundaries between the liquid and form I (0.8 GPa), form I and form II (1.2 GPa), form II and form III (4.5 GPa), form III and form III′ (12 GPa), and finally form III′ and a polymeric reaction product (22.5 GPa) are crossed. At higher temperature an additional crystalline form IV is expected. At present, only the structure of phase I ($Pbca - D_{2h}^{5}$) [432, 433] and of phase II ($P21c - C_{2h}^{5}$) [434] have been determined with certainty by x-ray diffraction, with the whole of the phase diagram having been determined on the basis of infrared and Raman spectroscopy measurements. Recent experiments in our laboratory [435] carried by accurately annealing the samples, as indeed suggested by the work of Thiery and Legér [436], seem to indicate a simpler phase diagram not unlike the one originally proposed by Block, Weir, and Piermarini [388]. These preliminary considerations are important for the purposes of the present work because the reaction occurring for benzene at high pressure has been reported to give different products depending on the reaction conditions [304]. Even though the products have not always been characterized microscopically in detail, it has been found that at low temperature from phase III′ a hard, insoluble compound, orange in transmission and white in reflection, is obtained. From phase III at high temperature a friable and colorless compound is recovered. Low-pressure chemical reaction leads to a black friable compound. These differences are believed to be related to the different structure of the starting material.

The occurrence of a chemical reaction of benzene at high pressure and temperature has been known for long time and has been investigated both in shock waves [380, 384, 385, 389] and under static high pressures [388, 437–439]. Reports on modeling of benzene decomposition, particularly under shock loading, has already been mentioned [386, 390–394]. In more recent times a finer control of the experimental conditions (pressure, temperature, laser irradiation) allowed a more precise microscopic characterization of the reaction product, and some information on the mechanism leading to the aromatic ring opening have been obtained. Here we shall concentrate on this more recent results. On increasing the pressure above 23 GPa at room temperature, the chemical reaction proceeds only partly and the amount of the product formed increases with the pressure [309, 440–442]. It has been reported that the benzene bands in the infrared spectrum are still detected when the sample is pressurized at 50 GPa [309]. The reaction accelerates on downloading and benzene reacts completely [309, 440–442], as is clearly shown in Fig. 27.

The product obtained from the high-pressure reaction of benzene has been identified as amorphous [309]. The amorphous character of the sample prevents the obtainment of the Raman spectra. Other physical–chemical properties of the reaction product are the following: refractive index $n = 1.75$; density $\rho = 1.39\ \mathrm{g/cm^3}$; elastic constant $B_0 = 80$ GPa; optical gap 2.5 eV. These values must to be considered only as typical values of the properties because, as described above, the reaction product is reported to change according to the

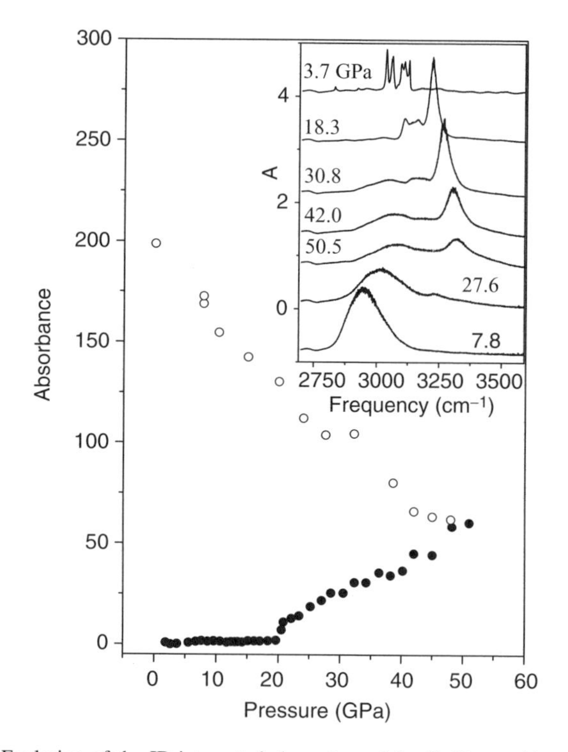

Figure 27. Evolution of the IR integrated absorption of the C—H stretching modes involving sp^3 carbon atoms. The corresponding absorption band is indicated by the star in the IR spectra reported in the inset during a compression (full dots)–decompression (empty dots) cycle.

reaction conditions. However, the reported entries clearly show that the product no longer has a molecular character. The pressure threshold for the reaction decreases on increasing the temperature, and the following values are reported by Cansell et al [304]: 4 GPa at 590°C, 7 GPa at 515°C, 12 GPa at 410°C, and 25 GPa at 25°C. There is evidence that the H/C ratio in the product obtained at room temperature is close to 1, and in fact in all recent experiments the development of hydrogen has not been reported. From the analysis of the infrared spectra in the C—H stretching region, a ratio of sp^3 to sp^2 carbon atoms of 75% has been deduced. This should be compared with the ratio of 60% for amorphous hydrogenated a:C-H carbons prepared by chemical vapor deposition (CVD). Indeed the infrared spectrum of the recovered product nicely compares with that of carbons obtained by CVD, as can be seen from Fig. 28. Some of the differences between the two spectra arise from the residual stress present in the product of the high-pressure experiment.

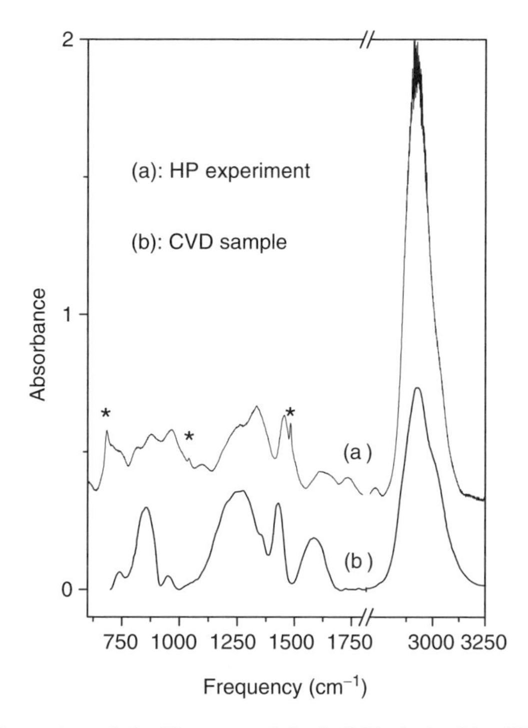

Figure 28. Comparison of the IR spectra of the (a-C:H) obtained by CVD preparation and from the high-pressure transformation of benzene. Asterisks indicate the absorption bands of unreacted benzene.

The reaction mechanism of the benzene ring opening is not completely clarified, but some experimental data on the process are now available. From the analysis of the vibrational frequencies as a function of pressure, it has been found that the totally symmetric vibrations continuously harden with increasing pressure [436, 442, 443]. On the contrary, the asymmetric E_{2g} vibrations show a different behavior, and their frequency remains constant or softens above a certain pressure threshold; furthermore, it has been found that it is impossible to fit their pressure dependence with the same intermolecular potentials that fit the symmetric vibrations. From this the authors deduced that the instability leading to the ring opening is not mechanical (vibrational) but rather electronic and suggested that the mixing between the ground and the first triplet excited state led to the ring instability. Later it was noted [309] that the benzene vibrations that show an "anomalous" behavior with pressure are those that significantly soften in the first singlet excited state. This suggested that a mixing between the ground and first singlet excited states is important. Experimentally, it was then found [302] that when benzene is pressurized at room temperature at 15 GPa (a value well below the reaction threshold of 25 GPa) and is irradiated with laser lines falling on the edge of the first electronic excited singlet state, the reaction rapidly proceeds and the same product that is formed in the pure pressure-induced reaction is obtained. At this pressure the reaction proceeds only as far as the sample is laser-irradiated. This clearly shows that a mixing with the first singlet excited state, induced by the thermal fluctuations and favored by the usual lowering and broadening of the energy levels at high pressure, plays a key role in the activation of the reaction. In the modeling of benzene condensation mentioned above, it has been found that in the hydrocarbon reactions at high pressure some intermediates, like dimers or other small aggregates, can be formed. Although there is no clear evidence of this in the available static experiments, it has been reported that in the initial stage of the reaction, some bands appear in the infrared spectrum that later completely vanish. These bands should probably be associated with some not yet identified intermediates of the complete disruption of the aromatic system. The laser-induced experiments show that laser irradiation, or photochemical effects, seem to play the same role as an increase of pressure. However, while in the case of benzene the reaction is activated only while the sample is irradiated, in the case of butadiene, discussed above, the irradiation only triggers the reaction that then proceeds independently. A difference of the two cases could be bound to the different aggregation state of the monomer in the two reactions (liquid for butadiene and solid for benzene).

STYRENE. Styrene represents a very interesting system for studying pressure-induced reactivity because, similarly to the already discussed phenylacetylene, two different reactive centers, the vinyl and phenyl groups, are present in the molecule. Monostyrene was compressed up to 32 GPa, and the effects of the applied pressure were monitored through the infrared and Raman spectra [444]. A product that can be recovered at ambient conditions is obtained when the pressure is raised above 10 GPa. The amount of transformed material increases with rising pressure up to 25 GPa even though the onset of the reaction can be placed at 15 GPa where a sharp increase of the fractional conversion to the final product is observed. This compound is identified as polystyrene by the direct comparison with the polymer spectra even at high pressure. Also in the present case the phenyl group seems not to be involved in the chemical reaction, and this result is particularly relevant if one considers that the experiment shows the aromatic ring to be stable up to 32 GPa, therefore well above the threshold pressure of the transformation of benzene that occurs at 23 GPa. This greater stability is explained by Gourdain et al. on the basis of the amorphous character of the polymer under pressure that prevents the parallelism of the double bonds belonging to nearest-neighbor rings. Also in this case, photochemical effects were noticed both during the pressure calibration by the ruby fluorescence technique and during the acquisition of Raman spectra. These effects were indicated to be probably responsible for the chemical transformation below 15 GPa.

ETEROAROMATICS: FURAN AND THIOPHENE. The chemical transformation of thiophene at high pressure has not been studied in detail. However, an infrared [441, 445] study has placed the onset of the reaction at 16 GPa when the sample becomes yellow-orange and the C—H stretching modes involving sp^3 carbon atoms are observed. This reaction threshold is lower than in benzene, as expected for the lower stability of thiophene. The infrared spectrum of the recovered sample differs from that of polythiophene, and the spectral characteristics indicate that it is probably amorphous. Also, the thiophene reaction is extremely sensitive to photochemical effects as reported by Shimizu and Matsunami [446]. Thiophene was observed to transform into a dark red material above 8 GPa when irradiated with 50 mW of the 514.5-nm Ar^+ laser line. The reaction was not observed without irradiation. This material was hypothesized to be polythiophene because the same coloration is reported for polymeric films prepared by electrochemical methods, but no further characterization was carried out.

More detailed information is available on furan, which is reported to react at room temperature near 10 GPa [338]. On increasing the pressure two stages of the reaction are identified. First, new bands develop in the infrared and Raman spectra that grow in intensity with pressure and then disappear. These bands, resembling those already mentioned in benzene, should be assigned to intermediates that can be tentatively be associated with some sort of bifuran or polyfuran. In a second stage, and at higher pressures, bands of a recoverable product are observed. Furan completely transforms at 25 GPa. The sample is highly anisotropic, as indicated by the broadening and asymmetry of the ruby fluorescence spectrum. In furan, as well as in benzene, unloading accelerates the reaction and produces considerable changes in the infrared spectrum, particularly in the high-frequency region. The spectrum is similar to that of the product recovered from the benzene reaction, as can be seen from Fig. 29. However, the spectrum of the compound produced from furan shows additional features due to the presence of the oxygen atom and to be assigned to the O–H ($\sim$3450 cm^{-1}), CO ($\sim$1720 cm^{-1}), and C—O–C and O–C—O ($\sim$1080 cm^{-1}) stretching vibrations. Most likely the high-pressure decomposition of furan produces an amorphous carbon containing alkylpolyether-type segments. The estimated abundance of sp^3 carbon atoms (67%) is lower than for benzene, as expected for the presence of functional groups involving the oxygen. The presence of OH groups is particularly significant because their formation implies hydrogen migration during the reaction.

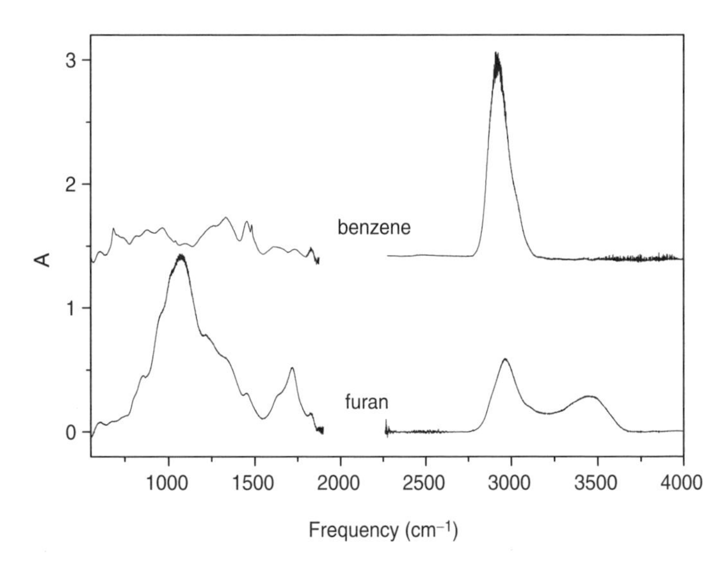

Figure 29. Comparison of the IR spectra of the recovered materials from the high-pressure transformation of benzene and furan.

Irradiation of the sample with the 458-nm or shorter wavelength laser lines considerably reduces the pressure threshold for the chemical transformation [303]. Since furan is transparent in the visible spectrum, the photochemical effect is produced also in this case by a two-photon absorption process, as also indicated by the extreme selectivity to the irradiation wavelength. The gross features of the laser irradiation are similar to those in benzene. However, in furan the reaction product shows remarkable differences with respect to the purely pressure-induced reaction, as can be seen from Fig. 30. In fact, the laser-assisted reaction produces carbon dioxide, whose amount depends appreciably on the pressure of the experiments. At the same time, the abundance of the sp^3 carbons in the product obtained in laser-assisted experiments is much higher (90%). This means that the hydrogen atoms migrating from the carbon involved in the formation of carbon dioxide saturate the carbons of the polymeric chains increasing the number of sp^3 carbons. Since the purely pressure-induced and the laser-assisted reactions occur at quite different pressures, the reaction in the two cases is induced from different crystalline phases of furan. The different molecular arrangements could actually favor different reaction mechanisms. Unfortunately, the unknown structure of the high-pressure phases of furan does not allow us to substantiate this assumption.

In conclusion, it has been found that the high-pressure reaction of heteroaromatics has some considerable similarities with that of benzene, but

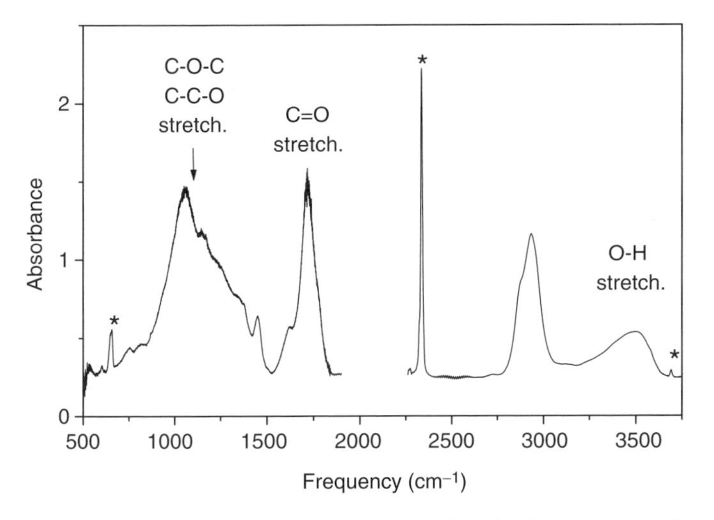

Figure 30. IR spectrum of the recovered product after furan has been compressed at 3.5 GPa and irradiated with 50 mW of the 458-nm line of an Ar^+ laser. Asterisks indicate the CO_2 absorption bands.

the presence of the heteroatom and the asymmetry introduced in the ring produces some distinct features.

3. *Fullerenes*

Fullerenes are the third natural form of carbon. These have been found to exist in interstellar dust and in geological formations on Earth, but only in 1985 did Smalley, Kroto and co-workers discovered this class of carbon solids and their unusual properties [447, 448]. It has been shown that C_{60}, the most common fullerene, could be transformed under high pressure into the other forms of carbon, diamond, and graphite [449] or, at moderately high pressures and temperatures, into new various metastable forms [450–453]. C_{60} crystals, fullerites, have *fcc* structure with weak van der Waals interactions. This structure is stable at ambient temperature up to 20 GPa and at ambient pressure up to 1800 K [454, 455].

Polymerization of fullerite was first obtained at ambient pressure by means of sample irradiation [456]. Visible or UV light leads molecules to link together in a covalently bonded *fcc* structure. From x-ray diffraction data a 2 + 2 cycloaddition mechanism was proposed in which the van der Waals interactions are replaced by single covalent bonds between adjacent C_{60} molecules [456]. More recent results indicate that linear chains with two square rings per molecule or branched chains with three square rings per molecule are obtained from photoinduced reaction [457]. However, the sample polymerization is not complete and only oligomers are synthesized. The $2+2$ cycloaddition mechanism has also been observed in polymerization of C_{60} doped with alkali metals where the metal catalyzes the reaction [458].

Using high-pressure–high-temperature (HP–HT) treatment, different metastable polymeric structures have been synthesized from fullerite C_{60} in large-volume presses [452, 453]; this polymerization reaction is mainly studied by powder x-ray diffraction and vibrational spectroscopy (IR and Raman). For these compounds the same $2+2$ cycloaddition mechanism of the photoinduced transformation has been proposed. The polymerization reaction was observed to be reversible when the sample is heated at 500–600 K because the square ring connecting the C_{60} units in the polymeric phases can be broken and the monomeric form reobtained at ambient conditions [452, 459]. Fullerite C_{60} reacts at low pressure, below 9 GPa, in the temperature range 370–600 K. The powder diffraction pattern of the recovered sample indicates an orthorhombic structure. The spectroscopy data indicate that the local environment of the C_{60} molecule is different, with a substantial reduction of symmetry. At higher temperatures, above 600 K, two different polymers have been synthesized depending on the pressure range. X-ray diffraction data indicate a tetragonal network at $P < 4$ GPa and a rhombohedral structure at higher pressures [451, 452]. *Núñez-Regueiro* et al. suggested a one-dimensional (1D) chain for the orthorhombic phase and two types of two-dimensional polymers for the tetragonal and the rhombohedral phases (see Fig. 31) [451]. Comparative analysis of the IR and Raman spectra [460] and calculations of the structural and energy properties [461] are in good agreement with these models.

When compressed above 12–13 GPa at temperatures exceeding 800 K or at ambient temperature and higher pressures, above 20–22 GPa, C_{60} fullerites give rise to three-dimensional (3D) networks with extremely high hardness, with the calculated bulk modulus of 288 GPa [462], and low density [463–468]. These superhard phases contain both amorphous and polycrystalline carbon but they are not well characterized, and different data are reported from different groups. XRD patterns exhibit no signals or few lines that have not been assigned. Raman spectrum is also typically featureless [469]. A recent XRD investigation has been performed using synchrotron radiation. Debye–Sherrer ellipses were observed instead of the typical diffraction rings and attributed to very large cell distortions. These distortions are irreversible because of the formation of covalent bonds [462]. Diffraction pattern of the heated sample at 12 GPa shows that the C_{60} cage remains up to 900–1000 K in a very high covalently bonded polymer. At more extreme conditions the C_{60} molecules collapse to graphite or amorphous carbon.

Reactivity of fullerite powder under shock compression has been investigated in the 10- to 110-GPa pressure range [470], and recovered samples of multiple

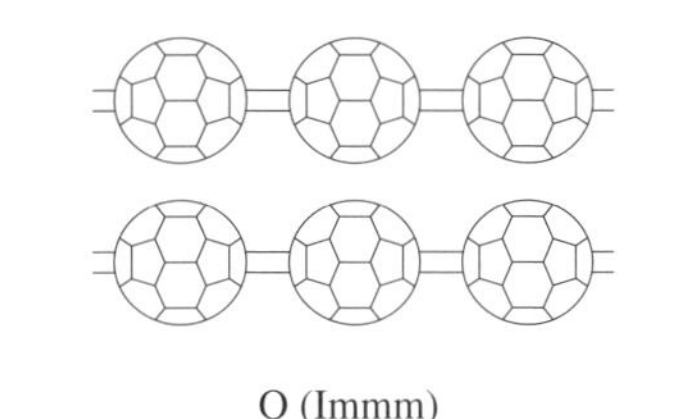

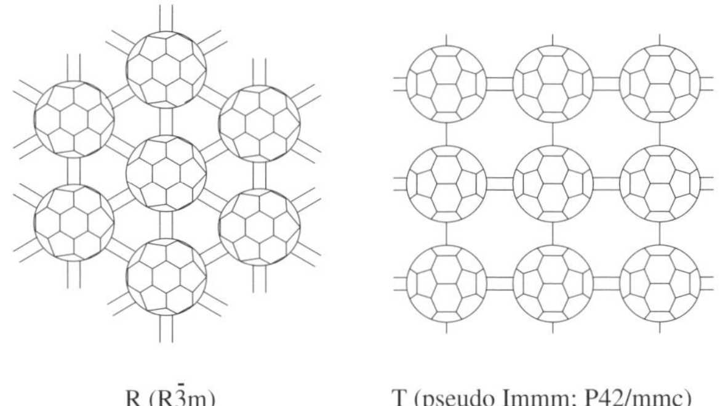

Figure 31. Proposed structures for polymeric C_{60}: The orthorhombic (O) phase is formed by one-dimensional chains; the rhombohedral (R) and tetragonal (T) structures are two-dimensional networks of C_{60} units.

shock experiments have been characterized by Raman spectroscopy. Raman spectra of shocked samples at 10 GPa and at 13 GPa are reported to be very similar to those measured at ambient conditions. At 17 GPa the Raman spectrum indicates the partial conversion to graphite, while this transformation is completed at 27 GPa, as indicated by the typical Raman peaks at 1350 cm^{-1} and at 2710 cm^{-1}. The synthesis of graphite at 17 GPa is consistent with the static pressure results when the interfullerene C–C distance becomes close to C=C bond length in C_{60} and in graphite [470]. At higher pressures the relative intensity of the 2710-cm^{-1} band decreases, indicating the presence of mixtures of graphite and disordered carbon, but the nature of the disorder is not known. The content of disordered carbon in the mixture increases with pressure, and from the 110 GPa experiments, only the disordered phase is recovered.

4. *Hydrocarbons in Geosciences*

The contribution of petroleum of abiotic origin to the natural deposits is a topic of enormous importance and actuality. The geologic conditions of the Earth's upper mantle can be easily reproduced by using the diamond anvil cell in combination with different heating methods. The first experimental study of an abiotic hydrocarbons synthesis was performed by Kenney et al. [471]. The formation of a hydrocarbon mixture was observed when FeO, marble, and water were compressed above 3 GPa at temperatures exceeding 700 K. These conditions correspond to those encountered at depths greater than 100 km well into the Earth mantle. The mixture of hydrocarbons resulting from the reaction overlaps in distribution with that characteristic of natural petroleum, with methane being about one order of magnitude more abundant than any other component. In this study it was also found that the amount of all the different hydrocarbons produced, including methane, steadily increase up to 1200°C. The increase with temperature of the methane produced in the reaction was recently disputed on the basis of an analogous study where resistive and laser heating methods were employed to rise temperature up to 1500°C in a compressed mixture of natural $CaCO_3$–calcite, FeO–wüstite, and H_2O [472]. As a matter of fact, a greater amount of methane was found to form when the reaction was induced at 600°C with respect to the situation in which the sample was laser heated at 1500°C. The bubbles formed in the sample during the reaction, due to gaseous compounds, were analyzed by *in situ* optical microscopy and Raman spectroscopy, and they were revealed to be entirely due to methane. X-ray diffraction measurements were employed to analyze the other reaction products; in particular, the presence of Fe_3O_4–magnetite and calcium–iron oxides was observed. The presence of this last compound suggested a complex reaction path. Thermochemical calculations indicate at 500°C the H_2 formation (which is not experimentally detected, perhaps because of the low concentration) and its consequent reaction with $CaCO_3$ to give methane and calcium oxide. On the contrary, as the temperature rises, a high CO_2 concentration is anticipated from the calculations, according to the calcite decomposition, and the methane reforming reaction

$$4H_2 + CO_2 \rightleftarrows CH_4 + 2H_2O$$

is expected to take place. The direction of this reaction strictly depends on temperature, and it is found that at 1 GPa methane formation is favored below 1200°C while H_2 and carbon dioxide form above 1200°C, therefore confirming the larger experimental formation of methane at temperatures below 600°C.

5. *Chemical Reactions of Explosives*

High-pressure techniques have been of considerable importance for the comprehension of the properties of energetic materials and explosives, as already mentioned in the case of the high-pressure reactions of nitrogen oxides. In this field, experiments have been carried out mostly in shock waves studying the mechanics of the process that lead to the rapid release of energy in the detonation. However, the chemistry of explosives is also of particular importance for the identification and characterization of the products forming during and after the explosion and for the mechanism of the initiation. Unfortunately, measurements in the extreme conditions of pressure and temperature during the detonation of the explosive are particularly challenging. Here we will only mention the results of some static experiments on nitromethane because this is the prototype of a mono-propellent, and the results reported nicely illustrate the general concept that has been discussed in this review. A comprehensive introduction to the experimental work on the chemical decomposition of nitromethane under both shock and static compression is reported in Ref. 473.

The high-pressure thermal decomposition of nitromethane has been studied by Piermarini et al. [474], characterizing the kinetics of the reaction by infrared spectroscopy. It is found that the kinetics of the decomposition process follows a sigmoidal trend, characteristic of an autocatalytic process. The sigmoidal character decreases at higher temperatures. The same behavior has been found for explosives of the nitroamine type [475]. A careful kinetic study reveals a change in the reaction mechanism in the 4 to 5-GPa pressure range. However, a positive pressure dependence of the reaction rate (i.e., a negative activation volume) is found in all the kinetic studies, and a bimolecular reaction mechanism is proposed [474]. It has been suggested that a strong interaction between the nitro group of a molecule and the methyl group of a companion molecule first leads to a transfer of an oxygen atom and to the formation of nitrosomethane (CH_3–NO) and nitromethanol. The latter rapidly decomposes into carbon and gaseous products while the former reacts to give HCN which later hydrolyzes to produce ammonia and formic acid. This mechanism has been confirmed by theoretical modeling [476]. The intermolecular character of the reaction and the variable reaction mechanism has given rise to an interest on the structure of the solid nitromethane, and various crystalline phases have been proposed [477–479]. As a matter of fact, the molecular arrangement and the structure of the crystal could be of relevance for the initiation process. The chemical reaction of nitromethane has been studied at high static pressure by Courtecuisse et al. [478], who have identified a slow decomposition process below 200°C leading to the formation of a transparent amorphous compound that on heating converts to a new dark product. The infrared spectrum of the yellow-brown product obtained at 50 GPa has been characterized by Pruzan et al. [480].

The important result of the studies on nitromethane at high pressure is the bimolecular character of the reaction and the relevance of intermolecular interactions. The importance of the anisotropy and strains in the sample has been stressed in recent electronic structure calculations in the solid [481]. The leading role of intermolecular interactions has been most clearly demonstrated in studies of the chemical transformations of nitramines [475, 482, 483].

6. *Piezochromism*

The following examples will illustrate the case where the pressure-induced chemical reaction is a change of molecular conformation. Many organic and organometallic substances exhibit a dramatic color change increasing temperature due to the conversion between two different conformations [484]. This thermochromic behavior has been extensively studied with physical and spectroscopic methods. The process is completely reversible. The same color change, with the same mechanism, can be induced under a pressure change [485]. In this case the process is described as piezochromism, and it is exploited for crack detection systems: When a crack occurs, there is a pressure change, and thus a color change, of the piezochromic material.

All the systems reported in the following examples are solid, and in this case every conformational change implies the rearrangement of the whole crystal. This means that the change of conformation is a first-order transition where the degree of cooperativity changes depending on the strength of the interactions with the chemical environment.

The main diagnostic tools employed in studying molecular isomerizations are electronic absorption and infrared spectroscopy. Changes of the molecular conformation can be indicated by variations in energy, intensity, or number of electronic peaks, as well as by shifts, splitting, and appearance–disappearance of vibrational bands.

Salicylidene Anilines. The optical properties of salicylidene aniline derivatives have been widely studied [486–496]. These systems illustrate the principle that, contrary to the common thermodynamic behavior, an increase in temperature or pressure produces the same effect; in other words, temperature and pressure do not seem to work as conjugate variables. Moreover, the crystalline anils are both thermochromic and photochromic; that is, a new absorption peak in the visible spectrum appears either by changing temperature or by irradiating in the main absorption band in the UV. At ambient conditions the salicylidine anils have enolic structure and they absorb light in the near UV range, at about 380 nm. When heated at 150–180°C a new absorption peak appears in the visible spectrum with 6–12% of the intensity of the UV peak. NMR and IR investigations indicate that the observed color change is due to the hydrogen transfer to the nitrogen—that is, due to the formation of the *cis–keto* structure of the molecule (see Fig. 32).

The effect of pressure on the optical spectra of these salicylidene anilines has been investigated in the range 0–10 GPa, showing a common behavior in all the compounds. Drickamer describes the 5-Br-salicylidene aniline as a model system [485, 497]. Compression at ambient temperature induces the same conformation change induced by temperature or irradiation, as indicated by the

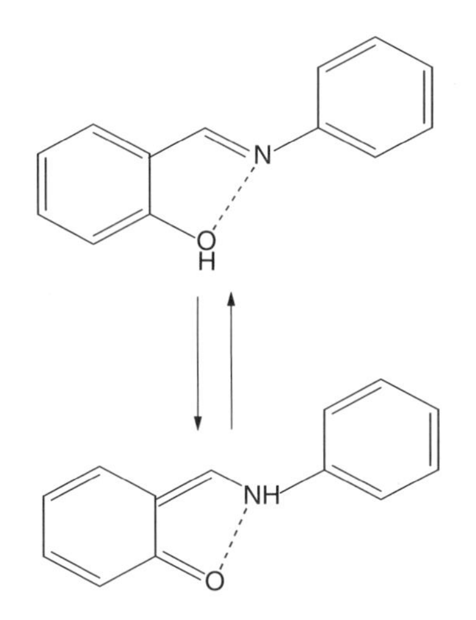

Figure 32. Enolic and *cis*-ketonic forms of salicylidine anils.

visible peak at 490 nm. The absorption band appears at low pressure, below 0.4 GPa, and the intensity increases with compression. At the same time, the UV band area decreases and both the UV and visible peaks shift to lower energy. The shift of the visible peak with increasing pressure is typical for $\pi \rightarrow \pi^*$ transitions. The small measured volume change suggests that the product is geometrically similar to the initial molecule, and the high-pressure IR spectrum confirms the identification of the piezochromic species as *cis–keto* form. The comparison of the relative intensity of the two bands indicates that at 6 GPa the *cis–keto* form is already dominant, and at 10 GPa the conversion is complete.

Cu(II) Complexes with Diethylethylenediamine. The complexes $[Cu(dieten)_2]X_2$, where X is an anion and dieten is the bivalent ligand N,N-diethylethylendiamine, well illustrate the effect of the medium on the extent of cooperativity of the transformation. These systems present interesting properties, and they have been investigated extensively [498–503]. When X is BF_4^-, ClO_4^-, or NO_3^- the complexes have thermochromic behavior and the color changes from red, at low temperature, to blue in the high-temperature form [499, 504].

The symmetry is square-planar D_{4h} with respect to the ligands, but in practice the Cu(II) is subjected to an essentially octahedral field. When the ligand is ethylene diamine (not substituted), the four nitrogen atoms are about 2 Å from the Cu(II) ion while the counterions bind 2.2–2.3 Å above and below. In this description the electronic spectra have three characteristic bands relative to the $d_{xy} \rightarrow d_{x^2-y^2}, d_{x^2y^2} \rightarrow d_{x^2-y^2}$, and $d_{z^2} \rightarrow d_{x^2-y^2}$ excitations, with the last

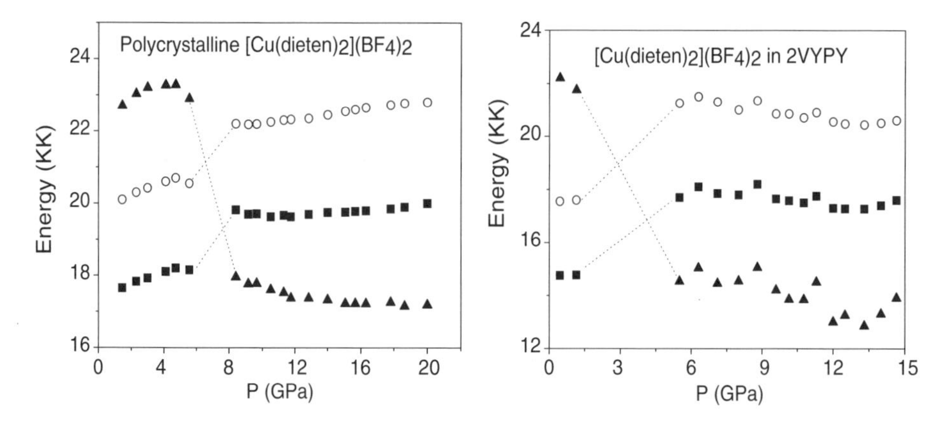

Figure 33. Pressure dependence of the electronic transition energies of $[Cu(dieten)_2](BF_4)_2$ polycrystalline (**left panel**) and in polyvinyl pyridine matrix (**right panel**).

absorption being much less intense because of the weak overlap and lying at lower energy. With the diethyldiethylene diamine ligand, which is bigger, the symmetry around the Cu(II) ion is effectively square planar. We can observe the same excitation lines; but in this geometry, at normal conditions of pressure and temperature, the weak $d_{z^2} \rightarrow d_{x^2-y^2}$ peak is at higher energy.

In numerous experiments, Cu(II)–diethyldiethylene diamine complexes with different counterions have been subjected to increasing pressures revealing a similar behavior. The case of $Cu(dieten)_2(BF_4)_2$ is discussed as representative [485, 505]. The peak shift as a function of pressure is reported in Fig. 33 for pure systems and for complexes dissolved in a liquid medium. In the former case we can observe that between 6.5 and 8.0 GPa the weak peak shifts from the highest to the lowest energy, while the other two electronic absorptions have a significant blue shift.

At the same time, the color changes from red to orange and then to yellow. The infrared spectra measured up to 12 GPa indicate a progressive reduction of symmetry as evidenced by the splitting of many degenerate modes [505]. All the experimental data are consistent with an increasing anion–copper cation interaction, and the dramatic red shift of the weak electronic band is interpreted on the basis of a molecular rearrangement in which the symmetry becomes octahedral. The same conformational change is reported for the $Cu(dieten)_2(BF_4)_2$ complex in polyvinyl pyridine [485, 505], but, as we can appreciate from the pressure evolution of the peak position (Fig. 33), the discontinuity occurs in this case below 2 GPa and extends over a wider pressure range (4–5 GPa), indicating that the extent of cooperativity increases with the density.

Tetrachlorocuprates X_2CuCl_4. Many complex ions exhibit a rich stereochemistry at ambient pressure. In this section the effect of high pressure on the molecular rearrangement and on the range of configurations for the X_2CuCl_4 molecules will be discussed. The $CuCl_4^-$ ion can assume different configurations from regular tetrahedral to square-planar geometry, depending on the nature of the counterion. The stereochemical structure can be described by the dihedral angle, which correspondingly varies from 90° to 0°, and there is a strict relationship between geometry and electronic spectrum [506, 507]. Pressure effect on electronic spectra has been studied for five different tetrachlorocuprates having regular or intermediate geometry [485, 505, 508]. For distorted tetrahedral geometry, two peaks are expected from energy state analysis, while the square-planar configuration gives three peaks centered at relatively high energy. The three large dihedral angle complexes (see Fig. 34) are reported to have a nearly discontinuous distortion occurring between 5 and 6.5 GPa for Cs and tmba cations, or 4–6 GPa for the $NphpipzH_2$ cation. At higher pressure, three peaks are necessary to fit the electronic spectrum and the dihedral angle is estimated to be 35–40° for the Cs^+ ion and 30–35° for the $(tmba)^+$. The $(nmpH)^+$ counterion gives strong hydrogen bonds that stabilize a square-planar geometry, and the low-pressure electronic spectrum exhibits three adsorption bands. This system presents a discontinuity in the range 7–8 GPa, where three peaks are still observed and the spectrum is significantly shifted at lower energy. The spectrum of $(NbzpipzH_2Cl)_2CuCl_4$ (see Fig. 34) exhibits only intensity changes of the peaks and the behavior of this complex under compression is more difficult to describe. However, it is clear that there is no considerable conformational difference between the low-pressure tetrahedrally distorted geometry (dihedral angle of 19°) and the high-pressure form.

These examples clearly demonstrate that all tetrachlorocuprate complexes evolve to a similar pressure-stabilized conformation, with dihedral angle of about 25–45°. The range of conformations is significantly smaller compared to 0–90° at ambient pressure, and this is because the high pressure diminishes the weak attractive forces, and the configuration is essentially determined by packing and repulsive forces.

B. Superhard Materials

Geophysics and geochemistry, in the purpose to understand the composition and the dynamic behavior of the earth's interior, have given a great impulse to the very-high-pressure research. In particular, many ultrahard materials have been synthesized at extreme high-pressure and high-temperature conditions in large-volume and multianvil cells and have become available for industrial applications. Among these materials, completely new substances unknown in nature and having surprising optical and mechanical properties have been synthesized. Some of them might be harder than diamond, and they are currently under further

Cation		Dihedral angle	Symmetry
Regular tetrahedron		90°	T_d (A)
Cs^{2+}		67.9°	D_{2d} (B)
$[C_6H_5-CH_2-N^{+}-(CH_3)_3]_2$	(tmba)	66.6°	D_{2d} (B)
$[C_6H_5-NH^{+}(CH_2-CH_2)_2NH_2]$	($NphpipzH_2$)	51.6°	D_{2d} (B?)
$\{[C_6H_5-CH_2-NH^{+}(CH_2-CH_2)_2NH_2^{+}]Cl^{-}\}_2$	($NbzpipzH_2$)	19°	D_{2d} (C)
$[C_6H_5-CH_2-CH_2-N^{+}H_2(CH_3)]_2$	(nmpH)	0°	D_{4h} (D)

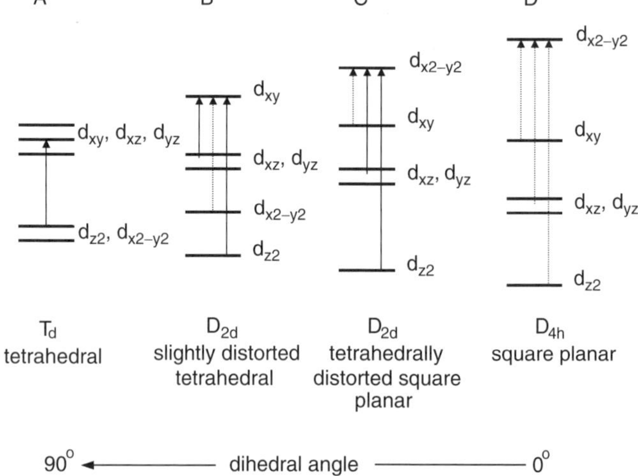

Figure 34. Cations, dihedral angles, and symmetry of the tetrachlorocuprates complexes at ambient pressure. The *d* orbital calculated structures (A–D) reported in the lower section correspond to the different geometries of the $CuCl_4^{2-}$ complex.

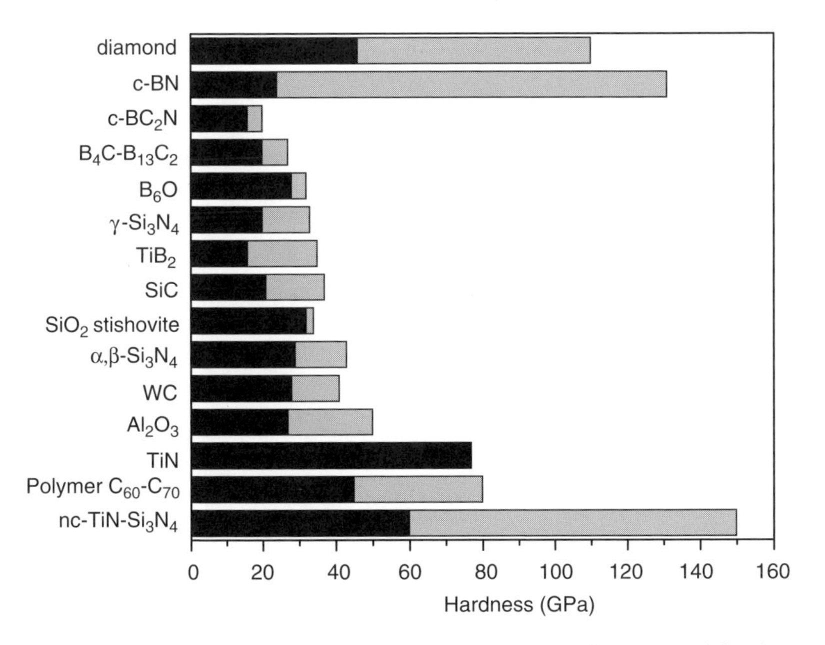

Figure 35. Low hardness (black) and high hardness (gray) values reported for the most important superhard materials. (Adapted from Ref. 7.)

investigations. An idea of the present knowledge in this field is provided by the diagram in Fig. 35.

1. Synthesis of Diamond

The synthesis of diamond is the most famous high-pressure and high-temperature industrial process, and vast quantities of this material are produced using modern industrial technology. The small synthetic crystals obtained are principally used for cutting tools and abrasives.

Diamond is the hardest and the most incompressible known material with a bulk modulus, B_0, of 443 GPa. Diamond is metastable at room temperature, and the diamond formation from pure graphite is theoretically possible. It is well known that natural graphite consists of parallel layers of carbon atoms forming hexagonal rings. The structure has ...*ababab*... sequence, and atoms in alternate layers occupy equivalent positions. The structure of the natural diamond can be described as a network of equidistant carbon atoms arranged in puckered, hexagonal rings lying approximately in the 111 crystallographic plane. The sequence is ...*abcabc*... where every fourth plane duplicates the position of the atoms in the first one. Because of the similarities in crystal

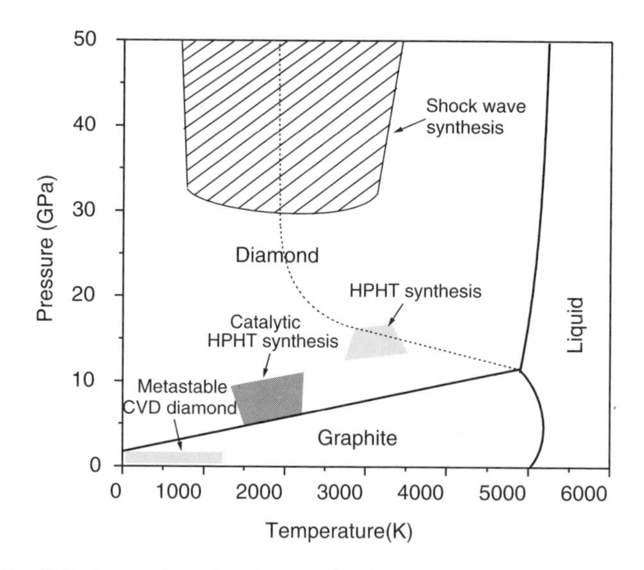

Figure 36. P–T phase and reaction diagram of carbon as results from Refs. 509 and 510. Solid lines represent equilibrium phase boundaries. The dashed line is the threshold for conversion of hexagonal diamond and both hexagonal and rhombohedral graphite into cubic diamond.

structure, we can conclude that graphite might be converted to diamond by applying a suitable pressure. The (P, T) phase diagram of carbon has been widely investigated. The most recent version of the one-component diagram is that of Bundy [509, 510]. In these works, two crystallographic structures are reported for both graphite and diamond: In addition to the previously described forms, graphite has a rhombohedral high-pressure structure with layering sequence ...*abcabc*..., while diamond has a hexagonal form with ...*ababab*... sequence (see phase diagram in Fig. 36). The diamond/graphite equilibrium line has been accurately studied by thermodynamic calculations [511–513] and by experiments on graphitization of diamond [514, 515].

From a thermodynamic point of view, the transformation of graphite is accessible with the available experimental apparatuses, but it is kinetically impossible. Geological times, hundreds of years, are required for spontaneous formation of diamond in appropriate conditions, and kinetic factors prevent the observation of the reaction in any practical time scale. H. T. Hall has demonstrated that for graphite → diamond conversion, carbon–carbon bonds must be broken in a solvent and on December 1954 realized the first synthesis of diamond, at approximately 2000 K and 10 GPa, in molten troilite (FeS) solvent, using a belt-type high-pressure–high-temperature apparatus [516–519]. Since then, many substances, minerals, and transition metals, in particular, have been successfully used in diamond production [520]. Transition metals such as nickel, cobalt, and their alloys are the most common catalysts in modern industrial synthesis of diamonds, but it has been reported that inorganic compounds such as carbonates and hydroxides are effective catalysts for forming diamonds [521, 522]. According to the generally accepted model for catalytic diamond formation under high pressure and high temperature, the catalyst first must melt. This molten material then dissolves the graphite until the graphite solution is saturated. There is a slight solubility difference between graphite and diamond; thus this situation is supersaturated for diamond. As a result, small crystals begin to grow. If the temperature and pressure are maintained in the region where the diamond is in the stable phase, the growing crystals are diamonds [523].

It has been reported that when C_{60} is rapidly and nonhydrostatically compressed above 20 GPa at room temperature, it transforms into polycrystalline diamond [524]. Although C_{60} can be considered as a folded graphite sheet, we must take into account that in the pentagons there is an important tetrahedral distortion making the transformation of C_{60} into diamond likely easier than the HP–HT conversion from graphite, and it is possible to use this reaction for industrial production of diamonds.

2. *Ultrahard Materials of the B–C–N–O Group*

Cubic Phase of Boron Nitride c-BN. The cubic phase of boron nitride (c-BN) is one of the hardest materials, second only to diamond and with similar crystal structure. It is the first example of a new material theoretically predicted and then synthesized in laboratory. From automated synthesis a microcrystalline phase of cubic boron nitride is recovered at ambient conditions in a metastable state, providing the basic material for a wide range of cutting and grinding applications. Synthetic polycrystalline diamonds and nitrides are principally used as abrasives; but in spite of the greater hardness of diamond, its employment as a superabrasive is limited by a relatively low chemical and thermal stability. Cubic boron nitride, on the contrary, has only half the hardness of diamond but an extremely high thermal stability and inertness.

The c-BN phase was first obtained in 1957 [525] by exposing hexagonal boron nitride phase (h-BN) to high pressures and low temperatures. A pressure of more than 11 GPa is necessary to induce the hexagonal to cubic transformation, and these experimental conditions prevent any practical application for industrial purposes. Subsequently, it has been found that the transition pressure can be reduced to approximately 5 GPa at very high temperature (1300–1800°C) by using catalysts such as alkali metals, alkali metal nitrides, and Fe–Al or Ag–Cd alloys [526–528]. In addition, water, urea, and boric acid have been successfully used for synthesis of cubic boron nitride from hexagonal phase at 5–6 GPa and temperature above 800–1000°C [529]. It has been demonstrated that controlling the grain size or crystallinity of the starting material (h-BN) can significantly facilitate the high-pressure and high-temperature-induced phase transition [530–532].

Yoo et al. [533] reported the observation of the direct reaction of molecular nitrogen and pure boron induced at high pressure by laser heating. Boron and nitrogen are reported to not react at ambient temperature even at 50 GPa, which implies a large kinetic barrier. The laser-heating techniques allow the overcoming of the activation barrier, and a highly exothermic reaction—typical of many elementary reactions leading to oxides, nitrides, and carbides—is observed. In this experiment, boron powders were loaded into a DAC cell, and liquid nitrogen was added by an immersion technique. The samples were heated by a cw Nd:YAG laser at a constant pressure of 2 GPa, and the chemical reaction has been characterized by means of *in situ* x-ray diffraction measurements. The threshold temperature for the reaction resulted 1800 K. Before heating, and at lower temperatures, the x-ray diffraction pattern exhibits diffuse features corresponding to amorphous boron and/or fluid nitrogen. During heating, the diffuse rings disappear and several sharp diffraction rings appear. The persistence after heating of all the new rings, along with the transparency of the product, is clearly interpreted as the signatures of an irreversible chemical transformation of the reactants in a crystalline material [533]. This is in contrast with catalytic synthetic methods that lead to highly colored and low crystalline products. The analysis of the integrated diffraction pattern indicates that the final product after laser heating at 2 GPa is the hexagonal boron nitride phase (h-BN). From other similar experiments, it has been evidenced that the BN products change with pressure, depending on its thermodynamic stability. Only pure h-BN is obtained below 7.5 GPa, while the c-BN phase seems to be the unique product above 10 GPa [533].

Cubic BC_2N. Hetero-diamond B—C—N compounds have recently received a great interest because of their possible applications as mechanical and optical devices. The similar properties and structures of carbon and boron nitrides (graphite and hexagonal BN, diamond, and cubic BN) suggested the possible synthesis of dense compounds with all the three elements. Such new materials are expected to combine the best properties of diamond (hardness) and of c-BN (thermal stability and chemical inertness). Several low-density hexagonal phases of B,C, and N have been synthesized [534] while with respect to the high-density phases, different authors report contradictory data [535–538], but the final products are probably solid mixtures of c-BN and dispersed diamonds [539].

Very recently, Solozhenko [540] reported the high-pressure–high-temperature synthesis of cubic BC_2N with *in situ* control of the reaction by x-ray diffraction measurement. The first high-density product has been obtained in a laser-heated diamond anvil cell (DAC). The starting material was g-BC_xN, prepared according to a standard method [541] from g-BC_2N (graphite-like phase) and BC_4N. Laser-heating experiments at different pressures and temperatures have indicated the formation of c-BC_2N only at pressures above 18 GPa and temperatures above 2100 K. A recovered sample of c-BC_2N synthesized at 25.8 GPa and 3000 K was compressed up to 30 GPa at ambient temperature forcompressibility measurements. The extrapolated bulk modulus (B_0) is 282 ± 15 GPa, which is smaller than the calculated value of 420–438 GPa for ideal mixing between diamond and c-BN [542–544]. For hardness measurements, another synthesis has been performed in a large-volume multianvil cell starting from graphite-like BC_4N in a MgO capsule [540]. The recovered sample from compression up to 25 GPa and 2100 K has been characterized and identified as c-BC_2N. The observed reduction in the carbon content is attributed to a chemical reaction between g-BC_4N and oxygen [540]. Zhao et al. [545] also reported the high-pressure and high-temperature synthesis of millimeter-sized bulks of superhard BC_2N and BC_4N compounds. The experimental hardness of the cubic phase, measured by different methods, is lower than that of diamond but higher than that of c-BN single crystal [540, 545]. This result, also supported by recent semiempirical calculations of Vickers hardness [546], indicates that the high-density cubic phase of the ternary B—C—N system is almost comparable with diamond in hardness, and it could then replace c-BN as the second hardest material. This conclusion is confuted by other first-principles calculations of the ideal strength of the optimal BC_2N superhard phase [547]. The lower hardness values found with respect to c-BN are explained on the basis of an overestimate of the measured BC_2N hardness due to (a) the nanocrystalline size of the synthesized material and (b) the effect of the matrix in which the product is contained.

Oxides and Nitrides of the B–C–N–O System. Boron-rich compounds form a large family of refractory compounds with very interesting physical and chemical properties. The common characteristics of these materials are the crystal structure (based on that of α-rhombohedral boron with B_{12} icosahedra [548, 549]), short interatomic bond lengths, and the strongly covalent character, which are responsible of their high hardness. In particular, boron carbides (B_xC) and boron suboxides (B_6O_{1-x}) combine high hardness with low density and chemical inertness so they are largely used as abrasives and for high-wear applications [7, 548–552]. Because of their mechanical properties, boron carbides with composition ranging from $B_{10}C$ to near B_4C are used as armor components and grinding materials and as neutron absorbers in nuclear reactors [548].

Experimental HP–HT synthesis of solid solutions in the B_6O-B_4C system and the first conclusive synthesis of a new boron subnitride B_6N_{1-x} are reported by Hubert et al. [548, 549]. All the experiments have been performed in a large-volume multianvil cell. The recovered samples have been investigated by means of powder x-ray diffraction (XRD) to identify the resulting phases, as well as by parallel electron energy loss spectroscopy (PEELS) to determine their composition [548].

B_6O. The microhardness of B_6O is 38 GPa, greater than TiB_2 (34 GPa) and W_4C (30 GPa) [7, 548]. The formation of B_6O at room pressure was studied by several authors [553–555], but the pure stoichiometric B_6O phase was difficult to obtain; more correctly, these products are indicated as B_6O_{1-x}. Nevertheless, the HP–HT syntheses give rise to stoichiometries very close to match the nominal formula. Maintaining mixtures of amorphous B and B_2O_3 for 30 min at 1700° and at pressures between 4 and 5.5 GPa, the reduction of B_2O_3 by B leads to the formation of B_6O according to the reaction $16B + B_2O_3 \rightarrow 3B_6O$. Orange-red crystalline grains form in a soft, water-soluble matrix of B_2O_3, and the pure product can be recovered as the only insoluble material. Scanning electron microscopy (SEM) and transmission electron microscopy (TEM) images show almost perfect icosahedral particles up to 30 μm of diameter [556]. Oxigen atoms are inserted between the B_{12} units [549, 556]. A part of B_6O structure can be described by $B_{12}(B_{12})_{12}$ units with O atoms in the close-packed layers, forming a "super-icosahedron." The color is consistent with a bandgap of ~2 eV that makes boron suboxide a wide-bandgap semiconductor [7, 548, 557].

$B_xC_yO_z$. Mixed phases of the B–C—O system have been prepared from amorphous B, B_2O_3, and C in various ratios at 1700°C and up to 7.5 GPa. A wide range of compositions from $B_{6.5}C$ and $B_6C_{0.45}O_{0.77}$ was observed with intermediate structures between B_4C and B_6O [548].

B–N SYSTEM. The new hexa-boron nitride B_6N_{1-x} has been prepared at high pressure and high temperature, starting from amorphous B and hexagonal BN in ethanol according to the proposed reaction $5B + BN \rightarrow B_6N$ [548]. The recovered sample from compression at 7.5 GPa and 1700°C has a powder diffraction pattern very similar to that of B_6O along with other reflections indicating the presence of B, h-BN and c-BN as byproducts. For the high-pressure sample, B_6N_{1-x} the PEELS analysis confirms a chemical composition close to the B_6N stoichiometry. Recent results indicate that this boron subnitride is metallic and perhaps a high-T_c superconductor [7, 552].

Technological Nitrides. The nitrides form another class of materials synthesized at high pressure with interesting technological properties. They possess a very large variety of intriguing properties, and many efforts have been made to characterize the natural existing compounds and synthesize new technological materials based on the N^{3-} units. Group III and group IV elements form nitrides that are widely used in optoelectronics and as semiconductors. Transition metal nitrides frequently exhibit superconduction properties [558]. Most of them—for example, ZrN, VN, NbN, and MoN—are binary compounds obtained at extreme conditions of pressure and temperature. Among these compounds, new forms of crystalline C_3N_4 have been theoretically predicted that could be harder than diamond [559, 560]. The following examples are only a little selection of the theoretical and experimental studies of nitrides.

$(Si, Ge, Sn)_3N_4$ SPINELS. Recent HP–HT experiments have resulted in the synthesis of new spinel structure compounds where the nitride anion, N^{3-}, is in the place of oxygen ion [13, 561–563]. Three new nitrides—Si_3N_4, Ge_3N_4, and Sn_3N_4—are now well-characterized, and their potential technological applications are under evaluation. Each of the three materials, indeed, possesses high density, elevated hardness, and a very wide bandgap (3–3.5 eV). Preliminary studies indicate that they could be used as abrasives, semiconductors, and optoelectronic materials tunable over a wide range of visible and ultraviolet wavelengths [564, 565]. The spinel structure is common among AB_2O_4 oxides. In "normal" spinels the A^{2+} cation is in tetrahedral coordination and B^{3+} is in octahedral; in "inverse" spinels, A^{2+} and B^{3+} are in octahedral sites and one B^{3+} is in tetrahedral. Only recently, spinel structure based upon the anion N^{3-} have been discovered. In these compounds, termed γ-phases, group IV elements are both in tetrahedral and octahedral coordination. The α and β phases were already known as low-pressure structures with tetrahedral geometry, while the spinel γ-phases are specific new products from very-high-pressure treatment.

In 1999 Zerr et al. [561] reported the synthesis of the cubic phase (γ) of Si_3N_4 from the elements or from low-pressure phases. Pure silicon crystals or amorphous Si_3N_4 with α- and β-Si_3N_4 were placed in a diamond anvil cell with molecular nitrogen as pressure medium and reactant. The sample was compressed and laser-heated at constant pressure. The *in situ* high-pressure and high-temperature transformation was characterized by Raman spectroscopy, and the recovered samples were investigated by transmission electron microscope (TEM) and energy-dispersive x-ray (EDX) analysis [561]. Compression of pure silicon at 5.2 GPa and at 2100 K gives a pale gray product that resulted in a mixture of α and β phases in addition to the starting silicon. From silicon at 15 GPa, heated at 2200 K, the entire recovered light-yellow product was identified as a new cubic [561] phase with spinel structure. Laser-heating high-pressure experiments with low-pressure phases of Si_3N_4 were also performed up to 2800 K, revealing that the γ-Si_3N_4 is stable at high temperatures for pressures between 15 and 30 GPa [561]. The estimated bulk modulus is of 300 GPa, practically the same value calculated for B_6O.

In the same way as for the γ-Si_3N_4 phase, the new spinel structure of Ge_3N_4 (and of the analogous Sn_3N_4) has been synthesized with the HP–HT technique from elemental germanium (or Sn) and nitrogen in DAC or multianvil cells [562, 563]. Under compression at ambient temperature, the germanium sample undergoes a transition to the metallic β-tin structure at 10 GPa. The black metallic phase was pressurized up to 19 GPa and heated to temperatures exceeding 2000 K. The recovered sample at ambient conditions appears grayish-white, and TEM and EDX analysis revealed a new cubic germanium nitride [562, 563]. The structural x-ray diffraction data are in agreement with a spinel structure. Subsequent Raman experiments have evidenced that the cubic Ge_3N_4 phase can be recovered after the reaction at pressures exceeding 14 GPa [564]. The *in situ* EDX experiment clearly shows that the γ-Ge_3N_4 spinel form can be obtained from α- or β- phase at pressures greater than 12 GPa and temperatures above 1000–1200°C.

Soignard et al. [565] have studied the stability of the two spinels in the system Si_3N_4–Ge_3N_4, at pressures up to 20 GPa and temperatures >2000°C. The x-ray patterns analysis shows that the two new spinel nitrides remain essentially stable over the range of conditions examined.

CARBON NITRIDES. The synthesis of new materials based on carbon and nitrogen atoms have received particular attention in the last two decades. Liu and Cohen [566, 567] first predicted that carbon nitrides formed by tetrahedral sp^3 carbon and with the structure of β-Si_3N_4 could have a bulk modulus of 414 GPa, which is comparable to diamond. More recently, theoretical calculations by Teter and Hemley [568] indicate that this β-C_3N_4 structure, as well as other low-pressure phases (α- and graphite-C_3N_4), could have a high-pressure transition to a new potentially superhard cubic phase of C_3N_4. This structure consists of fourfold coordinated carbon and threefold coordinated nitrogen atoms, and the calculated bulk modulus, 496 GPa, is greater than the experimental value for diamond. These results have stimulated a great number of experiments to synthesize crystalline C_3N_4 structures with carbon atoms in sp^3 coordination. Different chemical routes have been attempted, such as thermal decomposition of molecular precursors [569–572], beam deposition and laser-ablation [573, 574], and shock-wave and static high-pressure synthesis [575–578]. In spite of these numerous experiments, the "superhard" materials have not been achieved to date. However, research on C—N-based solids has yielded new low-density, layered structures that are structurally similar to graphite. These compounds, named "graphenes" (C_xN_y), do not have the extraordinary optical and mechanical characteristics predicted for sp^3-bonded carbon nitrides; nevertheless, they may have equally useful properties and potential applications as electrodes, hard coatings, and catalyst supports.

HP–HT techniques have been used to synthesize and recover metastable forms of carbon nitrides. A mixture of C_{60} and N_2, laser-heated up to 2000–2500 K at about 30 GPa, gives rise to a chemical reaction, visually indicated by texture and color change [579]. X-ray diffraction analysis on the recovered sample indicates a crystalline nature for the new carbon nitride, but the composition is not determined and the observed pattern cannot be indexed on the basis of a unique structure. Well-characterized planar graphenes have been crystallized at high pressure and high temperature from molecular precursors—in particular, tetracyanoethylene (C_6N_4) [575, 578]. Pyrolysis of C_6N_4 for 30 min at 5 GPa and about 1700 K in a belt-type apparatus has yielded to a carbon nitride with pure sp^2 character and composition close to C_5N [575]. Pyrolysis of triazine ($C_3H_3N_3$) in the same experimental conditions leads to a nitrogen-free compound [575]. More recently, Nesting and Badding [578] reported the high-pressure synthesis from tetracyanoethylene of a much more nitrogen-rich carbon nitride. The C_6N_4 powder has been compressed in a diamond anvil cell, ranging from 18 to 42 GPa, and has been laser-heated up to 2300 K. Immediately after quenching to ambient conditions, the recovered samples appear transparent, indicating the presence of a certain amount of sp^3 bonds, but slowly transform in sp^2-bonded materials and become opaque. Electron microscopy, x-ray diffraction, and spectroscopic analysis indicate a two-phase product containing a crystalline phase of carbon and an amorphous carbon nitride in sp^2 coordination. The estimated content of nitrogen increases with the reaction pressure, from 24% (C_3N) at 18 GPa to 38% ($C_3N_{1.9}$) at 42 GPa [578].

PtN: THE FIRST BINARY NOBLE METAL NITRIDE. Very recently, Gregoryanz et al. [580] reported the synthesis and characterization of platinum nitride (PtN)—that is, the first binary nitride of the noble metals group. This new compound has been obtained in a laser-heated diamond anvil cell (DAC) from metallic Pt and molecular N_2. Compression at pressures up to 50 GPa and temperatures exceeding 2000 K induce a complete chemical reaction, and the product can be recovered at ambient conditions. Electron microprobe analysis, Raman spectra, and x-ray diffraction patterns are consistent with an almost stoichiometric and homogeneous PtN compound with cubic *fcc* structure [580]. PtN has very intense pressure-dependent Raman peaks and hardness higher than pure Pt (bulk modulus, $B_0 = 372 \pm 5$ GPa), and it could be used as a pressure calibrant in both optical and structural measurements.

V. SUMMARY AND CONCLUSIONS

The equilibrium and kinetic behavior of chemical reactions are governed by the chemical potential, the temperature, and the pressure. The dependence on the former two thermodynamic variables has been studied quite extensively, both theoretically and experimentally, and it can be safely assumed that it is well understood in all details. The same cannot be said so far for the pressure variable for a number of reasons. As it has been discussed in previous sections, technical difficulties have delayed the availability of the necessary experimental devices to explore systematically the extremely extended range of pressure that matter experiences in diverse natural environments. Even if this full range is not yet attainable in high-pressure laboratories, the technological progresses make feasible the investigation of all kind of materials in a pressure range spanning almost seven orders of magnitude, with an accurate simultaneous control of the other thermodynamic variables. It would not be justified to assert that running high-pressure equipment is experimentally trivial. However, with the advent of the diamond anvil cells, high-pressure equipment has been greatly simplified, reduced in size down to almost miniaturized dimensions and made to operate in full safety. Therefore, it is possible that high-pressure equipment will spread among physics and chemistry laboratories for fundamental research. This is greatly facilitated by the fact that diamond anvil cells have been efficiently coupled with instrumentations for investigation of all kinds of physical and chemical properties on samples of very reduced dimensions. At the same time, large-volume high-pressure types of equipment are facilitating technological applications.

For chemical reactions at high pressure, a general statement resulting from the large variety of available experiments is that at sufficiently high pressures all materials decompose into the atomic components. Below the atomization threshold, a more sophisticated chemical reactivity can be displayed if the appropriate general reactions conditions are met. All of the experiments on chemical reactions at very high pressure reviewed in this chapter show that pressure cannot be considered only as a regulatory tool for the chemical equilibrium and the chemical kinetics, eventually favoring some of the alternative possible pathways. It has been shown that at very high pressure, novel reaction pathways become possible. At very high pressure the changes of the electronic structure of the reactants induced by intermolecular interactions can be substantial. These effects, jointly to the geometrical constraints (molecule orientations, limited molecular mobility, volume constraints, and high potential energy barriers), turn the reaction pathway along novel directions. In a number of experiments described in this chapter, the macroscopic threshold conditions (of pressure and temperature, but also of laser irradiation) to be overcome to induce chemical reactions have been identified. There is presently no simple correlation of the observed thresholds with microscopic parameters (like nearest-neighbor distances, molecular orientations). Identification of the microscopic conditions to be met for the occurrence of chemical reactions at high pressure is an important issue. It appears that *ab initio* molecular dynamics simulations in conjunction with experiments can be of great help in this respect. It is worth reconsidering the results emerging from the high-pressure reactions of ethylene, butadiene, cyanogen, and others showing that, by appropriate selection of the overall reaction conditions, high-pressure reactivity can be finely tuned to obtain products that are intrinsically of the highest interest (a highly crystalline polymer from liquid ethylene, a pure dimer, or, alternatively, a completely *trans* polybutadiene from butadiene, a strictly topochemical polymerization from crystalline cyanogen). In order to fully exploit the potentialities of the high-pressure chemistry, it would be very useful to translate the reaction threshold conditions found in the different experiments into a microscopic language. Recent unpublished results from our laboratory on the benzene reaction at high pressure show that this is indeed possible [435]. It has in fact been found that the different pressure and temperature conditions where the chemical transformation occurs, and described in a previous section, can actually be interpreted to give a unified picture of the intermolecular coupling necessary to induce a reaction. It appears therefore that there is still much experimental and theoretical work to be done to fully understand the dynamics of high-pressure reactions. The positive perspectives in this direction arise from the refinement of the experimental methods. In this respect the recent results on the polymerization of carbon dioxide at high pressure and high temperature [188] are very interesting because they show that a more accurate control of the thermodynamic variables is possible and this is fundamental for a better comprehension of high pressure transformation. As a whole, the high-pressure science and technology appear as a mature field of research, and it can reasonably be expected that chemistry at very high pressure will in the near future be better understood and will disclose new interesting phenomena.

References

1. P. W. Bridgman, *The Physics of High Pressure*, G. Bell and Sons, Ltd., London, 1931.
2. E. J. *Physico-Chemical Metamorphosis and Problems in Piezochemistry*, McGraw-Hill, New York, 1926.
3. A. L. T. Moesveld and W. A. T. DeMeester, *Z. Phys. Chem.* **138**, 169 (1928).
4. J. P. Conant and C. O. Tongberg, *J. Am. Chem. Soc.*, **52**, 1659 (1930).
5. A. J. Jayaraman, *Sci. Am.* **250**, 54 (1984).
6. R. J. Hemley and N. W. Ashcroft, *Phys. Today* **51**, 26 (1998).
7. P. F. McMillan, *Nat. Mater.* **1**, 19 (2002).
8. R. M. Hazen, *The Diamond Makers*, Cambridge University Press, Cambridge, 1999.
9. R. Winter and J. Jonas, eds., *High Pressure Molecular Science*, NATO ASI Ser. E, Vol. 338, Kluwer, Doordrecht, 1999.
10. K. Eremans, ed., *High Pressure Research in the Biosciences and Biotechnology*, Proceedings, XXXIVth Meeting EHPRG, Leuven University Press, Leuven, 1997.
11. H. G. Drickamer and C. W. Frank, *Electronic Transitions and the High Pressure Chemistry and Physics of Solids*, Chapman and Hall, London, 1973.
12. R. J. Hemley and H. K Mao, Overview of static high pressure science, in *High Pressure Phenomena*, Proceedings of the International School of Physics Enrico Fermi, Course CXLVII, R. J. Hemley, G. L. Chiarotti, M. Bernasconi, and L. Ulivi, eds. IOS Press, Amsterdam, 2002, p. 3.

13. P. F. McMillan, Solid state chemistry at high pressure and high temperatures, in *High Pressure Phenomena*, Proceedings of the International School of Physics Enrico Fermi, Course CXLVII, R. J. Hemley, G. L. Chiarotti, M. Bernasconi, and L. Ulivi, eds., IOS Press, Amsterdam, 2002, p. 477.
14. N. S. Isaacs, *Liquid Phase High Pressure Chemistry*, Wiley, Chichester, 1981.
15. G. Jenner, High pressure organic synthesis: overview of recent applications, in *High Pressure Phenomena*, Proceedings of the International School of Physics Enrico Fermi, Course CXLVII, R. J. Hemley, G. L. Chiarotti, M. Bernasconi, and L. Ulivi, eds., IOS Press, Amsterdam, 2002, p. 373.
16. W. F. Sherman and A. A. Stadmuller, *Experimental Techniques in High Pressure Research*, Wiley, Chichester, 1987.
17. A. Jayaraman, A. R. Hutson, J. H. McFee, A. S. Coriell, and R. G. Maines, *Rev. Sci. Instrum.* **38**, 44 (1967).
18. J. R. Ferraro and L. J. Basile, *Appl. Spectrosc.* **28**, 505 (1974).
19. R. Eckel, M. Buback, and G. R. Strobl, *Colloid. Polym. Sci.*, **259**, 326 (1981).
20. T. C. Poulter, *Phys. Rev.* **40**, 860 (1932).
21. M. I. Eremets, *High Pressure Experimental Methods*, Oxford University Press, New York, 1996.
22. J. M. Besson, J. P. Pinceaux, and R. Piottrzkowski, *High Temp. High Press.* **6**, 101 (1974).
23. J. C. Chervin, G. Syfosse, and J. M. Besson, *Rev. Sci. Instrum.* **65**, 2719 (1994).
24. R. A. Fitch, T. F. Slykhouse, and H. G. Drickamer, *J. Opt. Soc. Am.* **47**, 1015 (1957).
25. H. G. Drickamer and A. S. Balchan, Optical and electrical studies at high pressures, in *Modern Very High Pressure Techiques*, R. N. Wentorf, ed., Butterworths Scientific Publications Ltd., London, 1962.
26. W. B. Holzapfel and N. S. Isaacs, *High-Pressure Techniques in Chemsitry and Physics, A Practical Approach*, Oxford University Press, New York, 1997.
27. W. J. Le Noble and R. Schlott, *Rev. Sci. Instrum.* **47**, 770 (1977).
28. M. Buback, *Z. Naturforsch. A: Phys. Sci.* **32**, 1295 (1977).
29. M. Buback, M. Busch, K. Lovis, and F. O. Mähling, *Chem. Ing. Tech.* **66**, 510 (1994).
30. E. U. Franck and K. Roth, *Discuss. Faraday Soc.* **47**, 770 (1976).
31. M. Buback, M. Busch, K. Panten, and H. P. Vöegele, *Chem. Ing. Tech.* **64**, 352 (1992).
32. M. Buback, Kinetics and selectivity of chemical processes in fluid phases, in *Supercritical Fluids: Fundamentals for Application*, E. Kiran and J. M. H. Levelt-Sengers, eds., Kluwer, Dordrecht, 1994.
33. M. Buback, *Z. Naturforsch. A: Phys. Sci.* **39**, 399 (1984).
34. M. Buback and B. Degener, *Makromol. Chem.* **194**, 2875 (1993).
35. H. Brackermann and M. Buback, *Makromol. Chem.* **190**, 2471 (1989).
36. M. Buback and K. Panten, *Makromol. Chem.* **194**, 2471 (1993).
37. J. Lees, in *Advances in High Pressure Research*, Vol. 1., R. S. Bradley, ed., Academic Press, London, 1966.
38. A. Yoneda, *High Temp. High Press.* **19**, 531 (1987).
39. G. Fasol and J. C. Schilling, *Rev. Sci. Instrum.* **49**, 1722 (1978).
40. M. Reghu, R. S. Vaidyanathan, V. Prasad, and S. V. Subramanyam, *Rev. Sci. Instrum.* **61**, 1336 (1990).

41. A. S. Balchan and H. G. Drickamer, *Rev. Sci. Instrum.* **32**, 308 (1961).
42. F. P. Bundy and K. J. Dunn, *Rev. Sci. Instrum.* **46**, 1318 (1975).
43. T. Yagi, W. Utsumi, W. Yamakata, and O. Shimomura, *Phys. Rev. B* **46**, 6031 (1991).
44. W. F. Sherman, *J. Sci. Instrum.* **43**, 462 (1966).
45. L. G. Khvostantsev, L. F. Vereshchagin, and A. P. Novikov, *High Temp. High Press.* **9**, 637 (1977).
46. L. G. Khvostantsev, *High Temp. High Press.* **16**, 171 (1983).
47. L. G. Khvostantsev, *High Temp. High Press.* **16**, 165 (1983).
48. A. A. Semerchan, N. N. Kuzin, T. N. Davydova, and K. Kh. Bibaev, *Sverkhtverdye Mater.* **4**, 8 (1983).
49. J. M. Besson, R. J. Nelmes, G. Hamel, J. S. Loveday, G. Weill, and S. Hall, *Physica B* **180 & 181** 907–910 (1992).
50. S. Klotz, J. M. Besson, G. Hamel, R. J. Nelmes, J. S. Loveday, W. G. Marshall, and R. M. Wilson, *Appl. Phys. Lett.* **66**, 1735 (1994).
51. P. Grima, A. Polian, M. Gauthier, J. P. Iti, M. Mezouar, G. Weill, J. M. Besson, D. Haserman, and H. Hanfland, *J. Phys. Chem. Solids* **56**, 311 (1995).
52. S. Klotz, J. M. Besson, G. Hamel, R. J. Nelmes, J. S. Loveday, W. G. Marshall, and R. M. Wilson, Crystal structure studies to 10 GPa with the Paris-Edinburgh cell: high pressure aspects in *High-Pressure Science and Technology*, S. C. Schmidt, J. W. Shaner, G. A. Samara, and M. Ross, eds. AIP Press, New York, 1993, pp. 1577–1580.
53. M. Wakutsuki and K. Ichinose, in *High Pressure Research in Geophysics—Advances in Earth and Planetary Sciences*, Vol. 12, S. Akimoto, M. M. Manghnani, eds., D. Riedel, Dordrecht, 1982, p. 5.
54. T. Yagi, 'Experimental overview of large volume techniques', in *High Pressure Phenomena*, Proceedings of the International School of Physics Enrico Fermi, Course CXLVII, R. J. Hemley, G. L. Chiarotti, M. Bernasconi, and L. Ulivi, eds., IOS Press, Amsterdam, 2002, p. 41.
55. K. Inoue and T. Asada, *Jpn. J. Appl. Phys.* **12**, 1786 (1973).
56. N. Kawahara and T. Imuna, *Jpn. J. Appl. Phys.* **28**, L615 (1989).
57. N. Kawai and S. Endo, *Rev. Sci. Instrum.* **41**, 1178 (1970).
58. N. Kawai, M. Togaya, and A. Onodera, *Proc. Jpn. Acad.* **49**, 623 (1973).
59. D. Walker, M. A. Carpenter, and C. M. Hitch, *Am. Mineral.* **75**, 1020 (1990).
60. A. W. Lawson and T. Y. Tang, *Rev. Sci. Instrum.* **21**, 815 (1950).
61. J. C. Jamieson, A. W. Lawson, and N. D. Nachtrieb, *Rev. Sci. Instrum.* **30**, 1016 (1959).
62. C. E. Weir, E. R. Lippincott, A. Van Valkenburg, and E. N. Bunting, *J. Res. Natl. Bur. Stand. Sec A* **63**, 55 (1959).
63. G. J. Piermarini and C. E. Weir, *J. Res. Natl. Bur. Stand. Sec A* **66**, 325 (1962).
64. C. E. Weir, S. Block, and G. J. Piermarini *J. Res. Natl. Bur. Stand. Sec C*, **69**, 275 (1965).
65. H. K. Mao and P. M. Bell, *Science* **191**, 851 (1976).
66. Y. K. Vohra, S. J. Duclos, K. E. Brister, and A. L. Ruoff, *Phys. Rev. Lett.* **61**, 574 (1988).
67. H. K. Mao, Y. Wu, L. C. Chen, J. F. Shu, and A. P. Jephcoat, *J. Geophys. Res.* **95**, 21 (1990).
68. A. L. Ruoff, H. Xia, and Y. Vohra, *Rev. Sci. Instrum.* **61**, 3830 (1990).
69. A. Van Valkenburg, *Conference Internationale Sur Les Hautes Pressions*, LeCreusot, Saone-et-Loire, France, 1965.
70. R. A. Forman, G. J. Piermarini, J. D. Barnett, and S. Block, *Science* **176**, 284 (1972).

71. J. D. Barnett, S. Block, and G. J. Piermarini, *Rev. Sci. Instrum.* **44**, 1 (1973).
72. G. J. Piermarini, S. Block, J. D. Barnett, and R. A. Forman, *J. Appl. Phys.* **46**, 2774 (1975).
73. G. J. Piermarini, S. Block, and J. D. Barnett, *J. Appl. Phys.* **44**, 5377 (1973).
74. H. K. Mao and P. M. Bell, *Carnegie Inst. Yearb* **78**, 659 (1979).
75. J. M. Besson and J. P. Pinceaux, *Science* **206**, 1073 (1979).
76. M. G. Pravica and I. F. Silvera, *Phys. Rev. Lett.* **81**, 4180 (1998).
77. A. Jayaraman, *Rev. Mod. Phys.* **55**, 65 (1983).
78. A. Jayaraman, *Rev. Sci. Instrum.* **57**, 1013 (1986).
79. D. J. Dunstan and I. L. Spain, *J. Phys. E* **22**, 913 (1989).
80. K. Shimizu, K. Suhara, M. Ikumo, M. I. Eremets and K. Amaya, *Nature* **393**, 767 (1998).
81. V. V. Struzhkin, E. Gregoryanz, H. K. Mao, R. J. Hemley, and Y. A. Timofeev, New methods for investigating superconductivity", in *High Pressure Phenomena*, Proceedings of the International School of Physics Enrico Fermi, Course CXLVII, R. J. Hemley, G. L. Chiarotti, M. Bernasconi, and L. Ulivi, eds. IOS Press, Amsterdam, 2002, p. 275.
82. A. L. Ruoff, H. Xia, and Q. Xia, *Rev. Sci. Instrum.* **63**, 4342 (1992).
83. F. Moshary, N. H. Chen, and I. F. Silvera, *Phys. Rev. Lett.* **71**, 3814 (1993).
84. A. F. Goncharov, R. J. Hemley, H. K. Mao, and J. Shu, *Phys. Rev. Lett.* **80**, 101 (1998).
85. A. W. Webb, D. U. Gubser, and L. C. Towle, *Rev. Sci. Instrum.* **47**, 59 (1976).
86. R. W. Shaw and M. Nicol, *Rev. Sci. Instrum.*, **52**, 1013 (1981).
87. M. I. Eremets, E. Gregoryanz, H. K. Mao, R. J. Hemley, N. Mulders, and N. M. Zimmerman, *Phys. Rev. Lett.* **85**, 2797 (2000).
88. R. K. W. Haselwimmer, A. W. Tyer, and E. Pugh, in *Review of High Pressure Science and Technology*, M. Nakahara, ed., Japan Society of High Pressure Science and Technology, Kyoto, 1998, Vol. 7, p. 481.
89. R. Boehler, *Nature* **363**, 534 (1993).
90. L. C. Ming and W. A. Bassett, *Rev. Sci. Instrum.* **45**, 1115 (1974).
91. S. Block and G. J. Piermarini, *Physics Today* **29**, 44 (1976).
92. S. Block and G. J. Piermarini, *Vibrational spectroscopy at high external pressure*, Academic Press, Orlando, 1984.
93. Q. Williams and R. Jeanloz, Ultra-high-pressure experimental technique, in *Molten Salt Techniques*, Vol. 4, R. G. Gale and D. G. Lovering, eds., Plenum Press, New York, 1991, p. 193.
94. A. P. Jephcoat, H. K. Mao, and P. M. Bell, Operation of the megabar diamond anvil cell, in *Hydrothermal Experimental Techniques*, G. C. Ulmer and H. E. Barnes, eds., Wiley-Interscience, New York, 1978, p. 469.
95. I. L. Spain and D. J. Dunstan, *J. Phys. E: Sci. Instrum.* **22**, 903 (1989).
96. H. K. Mao, P. M. Bell, J. W. Shaner, and D. J. Steinberg, *J. Appl. Phys.* **49**, 3276–3283 (1978).
97. D. Klug and E. Whalley, *Rev. Sci. Instrum.* **54**, 1205 (1983).
98. M. Ceppatelli, M. Santoro, R. Bini, and V. Schettino, *J. Chem. Phys.* **113**, 5991 (2000).
99. H. K. Mao and P. M. Bell, *Carnegie Inst. Yearb.* **77**, 904 (1978).
100. W. A. Bassett, T. Takahashi, and P. W. Stock, *Rev. Sci. Instrum.* **38**, 37 (1967).
101. R. Le Toullec, J. P. Pinceaux, and P. Loubeyre, *High Press. Res.* **1**, 77 (1988).
102. H. K. Mao, P. M. Bell, K. J. Dunn, R. M. Chrenko, and R. C. de Vries, *Rev. Sci. Instrum.* **50**, 1002 (1979).

103. G. J. Piermarini and S. Block, *Rev. Sci. Instrum.* **46**, 973 (1975).
104. A. Jayaraman, *Phys. Rev. Lett.* **29**, 167 (1972).
105. W. A. Bassett and T. Takahashi, *Advanced High Pressure Research*, Vol. 4, R. H. Wentorf, ed., Academic Press, New York, 1974, p. 165.
106. H. K. Mao and P. M. Bell, *Science* **200**, 1145 (2000).
107. H. K. Mao and P. M. Bell, *Science* **203**, 1004 (1979).
108. H. K. Mao and P. M. Bell, *J. Geophys. Res.* **84**, 4533 (1979).
109. G. Huber, K. Syassen and W. B. Holzapfel, *Phys. Rev.* **B15**, 5123 (1977).
110. K. R. Hirsch and W. B. Hozapfel, *Rev. Sci. Instrum.* **52**, 52 (1981).
111. R. Keller and W. B. Hozapfel, *Rev. Sci. Instrum.* **48**, 517 (1977).
112. L. Merrill and W. A. Bassett, *Rev. Sci. Instrum.* **45**, 290 (1974).
113. T. Tanino, *Rev. Sci. Instrum.* **57**, 2993 (1986).
114. M. I. Eremets and Y. A. Timofev, *Rev. Sci. Instrum.* **63**, 3123 (1992).
115. R. Le Toullec, P. Loubeyre, J. P. Pinceaux, and H. K. Mao, J. Hu, *High Press. Res.* **8**, 691 (1993).
116. J. C. Chervin, B. Canny, J. M. Besson, and P. Pruzan, *Rev. Sci. Instrum.* **66**, 2595 (1995).
117. R. Bini, R. Ballerini, G. Pratesi, and H. J. Jodl, *Rev. Sci. Instrum.* **68**, 3154 (1980).
118. F. A. Gorelli, L. Ulivi, M. Santoro, and R. Bini, *Phys. Rev. Lett.* **83**, 4093 (1999).
119. G. Davies, *Chemistry and Physics of Carbon*, P. W. Phillips and P. A. Turner, eds., Marcel Dekker, New York, 1977.
120. J. E. Field, *Properties of Diamond*, Academic Press, London, 1979.
121. J. Walker, *Rep. Prog. Phys.* **42**, 1607 (1979).
122. M. I. Eremets, *Semicond. Sci. Technol.* **6**, 439 (1991).
123. J. Wilks and E. Wilks, *Properties and Applications of Diamond*, Butterworth–Heinemann, Oxford, 1991.
124. P. K. Briddon and R. Jones, *Phys. B* **185**, 179 (1993).
125. A. T. Collins, *Phys. B* **185**, 284 (1993).
126. P. T. T. Wong and D. D. Klug, *Appl. Spectrosc.* **37**, 284 (1983).
127. R. J. Wijngaarden and I. F. Silvera, in *High Pressure Science and Technology*, Vol. 1, B. Vodar and P. H. Martean, eds., Pergamon Press, Oxford, 1980, p. 157.
128. D. M. Adams, S. J. Payne, and K. M. Martin, *Appl. Spectrosc.*, **27**, 377 (1973).
129. D. M. Adams and K. M. Martin, *J. Phys.* **10**, 10 (1977).
130. D. M. Adams and S. K. Sharma, *J. Phys. E: Sci. Instrum.* **10**, 680 (1977).
131. Y. K. Vohra and S. S. Vagarali, *Appl. Phys. Lett.* **61**, 2860 (1992).
132. Y. K. Vohra, T. S. McCauley, G. Gu, and S. S. Vagarali, in *High Pressure Science and Technology*, S. C. Schmidt, J. W. Shaner, and M. Ross, eds., AIP Press, New York, 1993, p. 515.
133. A. F. Goncharov, V. V. Struzhkin, H. K. Mao, and R. J. Hemley, *Phys. Rev. Lett.* **83**, 1998 (1999).
134. S. Merkel, A. F. Goncharov, H. K. Mao, P. Gillet, and R. J. Hemley, *Science* **288**, 1626 (2000).
135. M. Born and E. Wolf, *Principles of Optics*, Pergamon Press, London, 1993.
136. J. Xu, S. Yeh, J. Yen, and E. Huang, *J. Raman Spectrosc.* **27**, 823 (1996).
137. J. Xu, J. Yen, Y. Wang, and E. Huang, *High Press. Res.* **15**, 127 (1996).
138. A. Onodera, M. Hasegawa, K. Furuno, M. Kobayashi, Y. Nisida, H. Sumija, and S. Yazu, *Phys. Rev. B* **44**, 12176 (1991).

139. A. Onodera, A. Nakatani, M. Kobayashi, Y. Nisida, and O. Mshima, *Phys. Rev. B* **48**, 2777 (1993).

140. I. A. Trojan, M. I. Eremets, M. Yu Korolik, V. V. Struzhkin, and A. N. Utjuzh, *Jpn. J. Appl. Phys.*, Suppl. 32-1, 282 (1993).

141. K. Furuno, A. Onodera, S. Kume, V. V. Struzhkin, and A. N. Utjuzh, *Jpn. J. Appl. Phys., Part 2* **25**, L646-7 (1986).

142. K. J. Takano and M. Wakatsuki, *Rev. Sci. Instrum.* **62**, 1576 (1991).

143. J. Xu and H. K. Mao, *Science* **290**, 783 (2000).

144. J. Xu, H. K. Mao, R. J. Hemley, and E. Hines *J. Phys. Condens. Matter* **14**, 11543 (2002).

145. H. K. Mao and R. J. Hemley, *High Press. Res.* **14**, 257 (1996).

146. R. Boehler, M. Ross, and D. B. Boercker *Phys. Rev. Lett.* **78**, 4589 (1997).

147. G. Zou, Y. Ma, H. K. Mao, R. J. Hemley, and S. A. Gramsh, *Rev. Sci. Instrum.* **72**, 1298 (2001).

148. H. K. Mao and P. M. Bell, *Science* **200**, 1145 (1978).

149. R. J. Hemley, H. K. Mao, G. Shen, J. Badro, Ph. Gillet, M. Hanfland, and D. Hausermann, *Science* **276**, 1242 (1997).

150. G. J. Piermarini, R. A. Forman, and S. Block, *Rev. Sci. Instrum.* **49**, 1061 (1978).

151. I. Fujishiro, G. J. Piermarini, S. Block, and R. G. Munro, in *High Pressure in Research and Industry*, Proceedings of the 8th AIRAPT Conference Uppsala, Vol. II, C. M. Backman, T. Johannisson, and L. Tegner, eds., ISBN Sweden, 1981, p. 608.

152. J. M. Besson and J. P. Pinceaux, *Rev. Sci. Instrum.* **50**, 541 (1979).

153. R. L. Mills, D. M. Liedenberg, J. C. Bronson, and L. C. Schmidt, *Rev. Sci. Instrum.* **51**, 891 (1980).

154. J. A. Schouten, N. J. Trappeniers, and L. C. van den Berg, *Rev. Sci. Instrum.* **54**, 1209 (1983).

155. H. K. Mao, Static compression of simple molecular system in the megabar range, in *Simple Molecular System at Very High Density*, A. Polian, P. Loubeyre, and N. Boccara, eds., Plenum Press, New York, 1989, pp. 221–236.

156. P. Loubeyre, R. Le Toullec, D. Hausermann, M. Hanfland, R. J. Hemley, H. K. Mao, and L. W. Finger, *Nature* **383**, 702 (1996).

157. H. K. Mao, J. Xu, and P. M. Bell, *J. Geophys. Res. B* **91**, 4673 (1986).

158. P. H. Bell and H. K. Mao, *Carnegie Inst. Yearb.* **80**, 404 (1981).

159. J. Haynes, private communication.

160. K. Asaumi and A. L. Ruoff, *Phys. Rev. B* **33**, 5633 (1986).

161. D. H. Liedenberg, *Phys. Lett.* **73A**, 74 (1979).

162. J. C. Chervin, B. Canny, M. Gauthier, and P. Pruzan, *Rev. Sci. Instrum.* **64**, 203 (1993).

163. L. Liu and W. A. Bassett, *J. Geophys. Res. B* **28**, 3777 (1975).

164. R. Boehler, *Geophys. Res. Lett.* **13**, 1153 (1986).

165. H. K. Mao, P. M. Bell, and C. Hadidiacos, Experimental phase relations of iron to 360 kbar, 1400°, determined in an internally heated diamond anvil apparatus, in *High-Pressure Research in Mineral Physics*, M. H. Manghnani and Y. Syono, eds., Terra Scientific Publishing Co., Tokyo, and American Geophysical Union, Washington DC, 1987, pp. 135–138.

166. C. S. Zha, W. A. Bassett, *Rev. Sci. Instrum.* **74**, 1255 (2003).

167. D. Schiferl, *Rev. Sci. Instrum.* **58**, 1316 (1987).

168. D. Schiferl, S. K. Sharma, T. F. Cooney, S. Y. Wang, and K. Mohanan, *Rev. Sci. Instrum.* **64**, 2821 (1993).

169. A. S. Zinn, D. Schiferl, and M. F. Nicol, *J. Chem. Phys.* **87**, 1267 (1987).

170. H. Arashi, Raman spectroscopic studies at high temperatures and high pressure: application to deternmination of P–T diagram of Zr_2O, in *High-Pressure Research in Mineral Physics*, M. H. Manghnani and Y. Syono, eds., American Geophysical Union, Washington, DC, 1987, pp. 335–339.

171. W. A. Bassett, A. H. Shen, M. Bucknum, and I. M. Chou, *Rev. Sci. Instrum.* **64**, 2340 (1993).

172. L. C. Ming, M. H. Manghnani, and J. Balogh, Resistive heating in the diamond anvil cell under vacuum conditions, in *High-Pressure Research in Mineral Physics*, M. H. Manghnani and Y. Syono, eds., American Geophysical Union, Washington, DC, 1987, pp. 69–74.

173. E. Gregoryanz, A. F. Goncharov, K. Matsuishi, H. K. Mao, and R. J. Hemley, *Phys. Rev. Lett.* **90**, 175701 (2003).

174. T. Kikegawa, X-ray diamond anvil press for structural studies at high pressures and high tenperatures, in *High-Pressure Research in Mineral Physics*, M. H. Manghnani and Y. Syono, eds., American Geophysical Union, Washington, DC, 1987, pp. 61–68.

175. L. C. Ming, M. H. Manghnani, J. Balogh, S. B. Qadri, E. F. Skelton, and J. C. Jamieson, *J. Appl. Phys.* **54**, 4390 (1983).

176. R. Boehler, D. Errandonea, and M. Ross, The laser-heated diamond cell: High P–T phase diagrams, in *High Pressure Phenomena* Proceedings of the International School of Physics Enrico Fermi, Course CXLVII, R. J. Hemley, G. L. Chiarotti, M. Bernasconi, and L. Ulivi, eds., IOS Press, Amsterdam, 2002, p. 55.

177. Q. Williams, R. Jeanloz, R. Bass, B. Svendsen, and T. J. Ahrens, *Science* **236**, 181 (1987).

178. R. J. Hemley, P. M. Bell, and H. K. Mao, *Science* **237**, 605 (1987).

179. H. K. Mao, G. Shen, R. J. Hemley, and T. S. Duffy, in *Properties of Earth and Planetary Materials at High Pressure and Temperature*, M. H. Manghnani and T. Yagi, eds., AGU, Washington, DC, 1998, pp. 27–34.

180. W. A. Bassett and M. S. Weathers, Temperature measurement in a laser heated diamond anvil cell, in *High-Pressure Research in Mineral Physics*, M. H. Manghnani and Y. Syono, eds., American Geophysical Union, Washington, DC, 1987, pp. 129–134.

181. R. Boehler and A. Chopelas, *Geophys. Res. Lett.* **18**, 1147 (1991).

182. D. L. Heinz and J. S. Sweeney, *Rev. Sci. Instrum.* **62**, 1568 (1991).

183. R. Jeanloz and D. L. Heinz, *J. Phys. Paris* **45**, 83 (1984).

184. Q. Williams, R. Jeanloz, J. Bass, B. Svendsen, and T. J. Ahrens, *J. Geophys. Res. B*, **96**, 2171 (1991).

185. D. L. Heinz and R. Jeanloz, Temperature measurements in the laser-heated diamond anvil cell, in *High-pressure research in mineral Physics*, M. H. Manghnani and Y. Syono, eds., American Geophysical Union, Washington, DC, 1987, pp. 113–128.

186. R. Boehler, Advances in high temperature research in a diamond anvil cell, in *Recent Trends in High Pressure Research.* A. K. Singh, ed., Oxford & IBH Publishing Co., New Dehli, 1992, pp. 591–600.

187. R. Boehler, N. von Bargen, and A. Chopelas, *J. Geophys. Res. B* **95**, 731 (1990).

188. M. Santoro, J. Lin, H. K. Mao, and R. J. Hemley, *J. Chem. Phys.* **121**, 2780 (2004).

189. J. Lin, M. Santoro, V. V. Struzhkin, H. K. Mao, and R. J. Hemley, *Rev. Sci. Instrum.* **75**, 3302 (2004).

190. W. J. Nellis, Planetary interiors, in *High Pressure Phenomena*, Proceedings of the International School of Physics Enrico Fermi, Course CXLVII, R. J. Hemley, G. L. Chiarotti, M. Bernasconi, and L. Ulivi, eds., IOS Press, Amsterdam, 2002, p. 607.

191. W. J. Nellis, N. C. Holmes, A. C. Mitchell, and D. C. Hamilton, *J. Chem. Phys.* **107**, 9096 (1997).
192. P. S. Fiske, W. J. Nellis, M. Lipp, H. Lorenzana, M. Kikuchi, and Y. Syono, *Science* **270**, 281 (1995).
193. R. Koch, W. J. Nellis, J. Hunter, H. Davidson, and T. H. Geballe, *Pract. Met.* **27**, 391 (1990).
194. W. J. Nellis, S. T. Weir, N. A. Hinsey, U. Balachandran, M. J. Kramer, and R. Raman, *High Pressure Science and Technology—1993, C. Scmidt, J. W. Shaner, G. A. Samara, and M. Ross, eds., American Institute of Physics Press, New York, 1994, pp. 695–697.*
195. R. Chau, M. B. Maple, and W. J. Nellis, *J. Appl. Phys.* **79**, 9236 (1996).
196. C. S. Yoo, W. J. Nellis, M. L. Sattler, and R. G. Musket, *Appl. Phys. Lett.* **61**, 273 (1992).
197. T. Sekine, *Proc. Jpn. Acad. Ser. B* **68**, 95 (1992).
198. G. R. Cowan, B. W. Dunnington, and H. A. Holtzman, U.S. Patent 3,401,019 (1968).
199. A. Benuzzi-Mounaix, M. Koenig, G. Huser, B. Faral, D. Batani, E. Henry, M. Tomasini, B. Marchet, T. A. Hall, M. Boustie, T. De Ressguier, M. Hallouin, F. Guyot, D. Andrault, and T. Charpin, *Phys. Plasmas* **9**, 2466 (2002).
200. G. W. Collins, L. B. Da Silva, P. Celliers, D. M. Gold, M. E. Foord, R. J. Wallaca, A. Ng, S. V. Weber, K. S. Budil, and R. Cauble, *Science* **281**, 1178–1181 (1998).
201. J. S. Wark, R. R. Whitlock, A. A. Hauer, J. E. Swain, and P. J. Solone, *Phys. Rev. B* **40**, 5705 (1989).
202. Ya. B. Zeldovich and Yu. P. Raizer, *Physics of Shock Waves and High Temperature Hydrodynamic Phenomena*, Vols. 1 and 2, Academic Press, New York, 1967.
203. R. A. Graham, *Solids Under High-Pressure Shock Compression: Mechanics, Physics and Chemistry*, Springer-Verlag, New York, 1993.
204. M. H. Rice, R. G. McQueen, and J. M. Walsh, *Solid State Phys.* **6**, 1–63 (1958).
205. R. G. McQueen, S. P. Marsh, J. W. Taylor, J. N. Fritz, and W. J. Carter, The equation of state of solids from shock waves studies, in *High Velocity Impact Phenomena*, R. Kinslow, ed., Academic Press, New York, 1970, pp. 293–417.
206. W. J. Murri, D. R. Curren, C. F. Peterson, and R. C. Crewdson, in *Advances in High Pressure Research*, R. H. Wentorf, Jr., ed., Academic Press, New York, 1974, Vol. 4, pp. 1–163.
207. W. J. Nellis, in *High Pressure Measurements Techniques*, C. N. Peggs, ed., Applied Science Publishers, London, pp. 68–89.
208. Y. M. Gupta, in *Encyclopedia of Physics*, 3rd ed., R. M. Bescanon, ed., Van Nostrand Reinhold, New York, 1985.
209. T. J. Ahrens, Shock waves techniques, in *Methods of Experimental Physics*, Vol. 24, part A, C. G. Sammis and T. L. Henvey, eds., 1987, pp. 185–235.
210. M. Ross and H. B. Radousky, Shock wave compression and metallization of simple molecules'', in *Simple Molecular Systems at Very High Density*, A. Polian, P. Loubeyre, and N. Boccara, eds., Plenum Press, New York, 1989, pp. 47–56.
211. E. N. Avrorin, B. K. Vodolago, V. A. Simonenko, and V. E. Fortov, *Usp. Fiz. Nauk* **163**, 1 (1993).
212. R. F. Trunin, *Usp. Fiz. Nauk* **164**, 1215 (1994).
213. W. J. Nellis, Dynamic experiments: An overview, in *High Pressure Phenomena* Proceedings of the International School of Physics Enrico Fermi, Course CXLVII, R. J. Hemley, G. L. Chiarotti, and M. Bernasconi, L. Ulivi, eds., IOS Press, Amsterdam, 2002, p. 109.
214. V. A. Simonenko, *High Press. Res.* **5**, 816 (1991).

215. W. J. Nellis, J. A. Moriarty, A. C. Mitchell, M. Ross, R. G. Dandrea, N. W. Ashcroft, N. C. Holmes, and G. R. Gathers, *Phys. Rev. Lett.* **60**, 1414 (1988).

216. C. E. Ragan, M. G. Silbert, and B. C. Diven, *J. Appl. Phys.* **48**, 2860 (1977).

217. V. E. Fortov and V. B. Minstev, Strongly coupled plasma physics at megabar pressures, in *High Pressure Phenomena* Proceedings of the International School of Physics Enrico Fermi, Course CXLVII, R. J. Hemley, G. L. Chiarotti, M. Bernasconi, and L. Ulivi, eds., IOS Press, Amsterdam, 2002, p. 127.

218. Y. Kato, K. Mima, N. Miyanaga, S. Arinaga, Y. Kitagawa, and M. Nakatsuka, *Phys. Rev. Lett.* **53**, 1057 (1984).

219. R. S. Hawke, D. E. Duerre, J. G. Huebel, H. Klapper, D. J. Steinberg, and R. N. Keeler, *J. Appl. Phys.* **43**, 2734 (1971).

220. A. I. Pavlovskii, M. I. Dolotenko, A. I. Bykov, N. I. Egorov, A. A. Karpikov, and N. P. Kolokolchikov, *High Press. Res.* **7**, 381 (1991).

221. C. A. Hall, *Phys. Plasmas* **7**, 2069 (2000).

222. C. A. Hall, J. R. Asay, M. D. Knudson, W. A. Stygar, R. B. Soielman, T. D. Pointon, D. B. Reisman, A. Toor, and R. C. Cauble, *Rev. Sci. Instrum.*, **72**, 3587 (2001).

223. M. D. Knudson, D. L. Hanson, J. E. Bailey, C. A. Hall, and J. R. Asay, *Phys. Rev. Lett.* **87**, 225501 (2001).

224. W. J. Nellis and A. C. Mitchell, *J. Chem. Phys.* **73**, 6137 (1980).

225. M. Ross, W. J. Nellis, and A. C. Mitchell, *Chem. Phys. Lett.* **68**, 532 (1979).

226. M. Ross, H. K. Mao, P. M. Bell, and J. A. Xu, *J. Chem. Phys.* **85**, 1028 (1986).

227. V. E. Bean et al., Toward an international practical presure scale: An AIRAPT task group report, in *Proceedings of the Xth AIRAPT International High Pressure Conference on Research in High Pressure Science and Technology*, Amsterdam, 1985, pp. 144–151.

228. O. L. Anderson, D. G. Isaak, and S. Yamamoto, *J. Appl. Phys.* **65**, 1534 (1989).

229. D. L. Heinz and R. Jeanloz, *J. Appl. Phys.* **55**, 885 (1994).

230. G. Chen, R. C. Liebermann, and D. J. Weidner, *Science* **280**, 1913 (1998).

231. C. S. Zha, H. K. Mao, and R. J. Hemley, *Proc. Natl. Acad. Sci.* **97**, 13494 (2000).

232. R. W. G. Wyckoff, *Crystal Structures*, Vol. 2, Interscience, New York, 1948.

233. D. S. McClure, *J. Chem. Phys.* **36**, 2757 (1962).

234. S. Sugano and Y. Tanabe, *Multiplets of Transition Metal Ions in Crystals*, Academic Press, New York, 1970.

235. J. H. Eggert, K. A. Goettel, and I. F. Silvera, *Phys. Rev. B* **40**, 5724 (1989).

236. J. C. Chervin, B. Canny, and M. Mancinelli, *High Press. Res.* **21**, 305 (2001).

237. D. L. Decker et al., *J. Phys. Chem. Ref. Data* **1**, 773 (1972).

238. P. M. Bell and H. K. Mao, Static compression of gold and copper and calibration of the ruby pressure scale to pressures to 1.8 megabars by x-ray diffraction, in *Shock Waves in Condensed Matter, Y. Gupta, ed., Plenum, New York, 1986, pp. 125–130.*

239. P. M. Bell, J. Xu, and H. K. Mao, in *Topical Conference on Shock Waves in Condensed Matter*, Y. Gupta, ed., Plenum Publishing Corp., New York, 1986, p. 125.

240. H. K. Mao and R. J. Hemley, in *Ultrahigh-Pressure Mineralogy, Reviews in Mineralogy*, R. Hemley and D. Mao, Eds., Vol. **37**, Mineralogical Society of America, Washington, DC, 1998, p. 1.

241. K. A. Goettel, J. H Eggert, I. F. Silvera, and W. C. Moss, *Phys. Rev. Lett.* **62**, 665 (1989).

242. R. J. Hemley and H. K. Mao, *Phys. Rev. Lett.* **61**, 857 (1988).

243. W. C. Moss, J. O. Hallquist, R. Reichlin, K. A. Goettel, and S. Martin, *Appl. Phys. Lett.* **48**, 1258 (1986).

244. J. A. Xu, H. K. Mao, and P. M. Bell, *Science* **232**, 1404 (1986).

245. W. L. Vos and J. A. Schouten, *J. Appl. Phys.* **69**, 6744 (1991).

246. S. Rekhi, S. Dubrovinski, and S. Saxena, *High Temp. High Press.* **31**, 299 (1999).

247. W. B. Holzapfel, *J. Appl. Phys.* **93**, 1813 (2003).

248. J. H. Eggert, K. A. Goettel, and I. F. Silvera, *Phys. Rev. B* **40**, 5733 (1989).

249. J. H. Eggert, F. Moshry, W. J. Evans, K. A. Goettel, and I. F. Silvera, *Phys. Rev. B* **44**, 7202 (1991).

250. Z. Liu, Q. Cui, and G. Zou, *Phys. Lett. A* **143**, 79 (1990).

251. N. H. Chen, and I. F. Silvera, *Rev. Sci. Instrum.* **67**, 4275 (1996).

252. J. Lin, O. Degtyareva, C. Prewitt, P. Dera, N. Sata, E. Gregoryanz, H. K. Mao, and R. J. Hemley, *Nat. Mater.* **3**, 389 (2004).

253. Y. K. Vohra, Spectroscopic studies on diamond anvil under extremes static stress, in *Recent Trends in High Pressure Research*, A. K. Singh, ed., Oxford and IBH Publishing Co., New Dehli, 1992.

254. H. Arashi and M. Ishigame, *Jpn. J. Appl. Phys.* **21**, 1647 (1982).

255. A. Lacam and C. Chateau, *J. Appl. Phys.* **66**, 366 (1989).

256. F. Datchi, R. LeToullec, and P. Loubeyre, *J. Appl. Phys.* **81**, 3333 (1997).

257. B. Lorenz, Y. R. Shen, and W. B. Holzapfel, *High Press. Res.* **12**, 91 (1994).

258. J. Liu and Y. K. Vohra, *J. Appl. Phys.* **79**, 7978 (1996).

259. N. J. Hess and D. Schiferl, *J. Appl. Phys.* **71**, 2082 (1991).

260. A. Grzechnik, P. Simon, P. Gillet, and P. McMillan, *Physica B* **262**, 67 (1999).

261. S. C. Schmidt, D. Schiferl, A. S. Zinn, D. D. Ragan, and D. S. Moore, *J. Appl. Phys.* **69**, 2793 (1991).

262. D. Schiferl, M. Nicol, J. M. Zaug, S. K. Sharma, T. F. Cooney, S. Y. Wang, T. R. Anthony, and J. F. Fleischer, *J. Appl. Phys.* **82**, 3256 (1997).

263. M. Popov, *J. Chem. Phys.* **95**, 5509 (2004).

264. T. Kawamoto, K. N. Matsukage, T. Nagai, K. Nishimura, T. Mataki, S. Ochiai, and T. Taniguchi, *Rev. Sci. Instrum.* **75**, 2451 (2004).

265. C. Schmidt and M. Ziemann, *Am. Mineral.* **85**, 1725 (2000).

266. J. C. Jamieson, J. N. Fritz, and M. H. Manghnani, in *High Pressure Research in Geophysics*, S. Akimoto and M. H. Manghnani, eds., Center for Academic Publications Japan, Tokyo, 1982, pp. 27–84.

267. D. L. Decker, *J. Appl. Phys.* **36**, 157 (1965).

268. C. S. Zha, W. A. Bassett, and S. H. Shim, *Rev. Sci. Instrum.* **75**, 2409 (2004).

269. N. C. Holmes, J. A. Moriarty, G. R. Gathers, and W. J. Nellis, *J. Appl. Phys.* **65**, 1534 (1989).

270. S. D. Hamann, *Rev. Phys. Chem. Jpn.* **50**, 147 (1980).

271. S. D. Hamann, Chemical equilibria in condensed systems, in *High Pressure Physics and Chemistry*, R. S. Badley, ed., Academic Press, London, 1963, p. 139.

272. C. P. Slichter and H. G. Drickamer, *J. Chem. Phys.*, **56**, 2142 (1972).

273. E. König and K. J. Madeja, *J. Am. Chem. Soc.* **90**, 1146 (1968).

274. W. J. Le Noble, and H. Kelm, *Angew. Chem. Int. Ed. Engl.* **11**, 841 (1980).

275. G. Jenner, in *High Pressure Molecular Science*, R. Winter and J. Jonas, eds., NATO Science Series, Series E: Applied Science, Vol. 358, 1998, p. 291 and references therein.
276. T. Asano and W. J. Le Noble, *Chem. Rev.* **78**, 407 (1978).
277. R. Van Eldik, T. Asano, and W. J. Le Noble, *Chem. Rev.* **89**, 549 (1989).
278. A. Drljaca, C. D. Hubbard, R. Van Eldik, T. Asano, M. V. Basilevsky, and W. J. Le Noble, *Chem. Rev.* **98**, 2167 (1998).
279. N. S. Isaac, Chemical transformations, in *High Pressure Techniques in Chemistry and Physics*, W. B. Holzapfel and N. S. Isaac, eds., Oxford University Press, New York, 1997, pp. 307–351
280. G. Jenner, *Organic High Pressure Chemistry*, W. J. Le Noble, ed., Elsevier, Amsterdam, 1988, pp. 143–203.
281. F. G. Klärner, B. Krawczyk, V. Ruster, and U. K. Deiters *J. Am. Chem. Soc.* **116**, 7646 (1994).
282. G. Jenner, *Tetrahedron* **53**, 2669 (1997).
283. F. Dumas, B. Mezrhab, J. D'Angelo, C. Riche, and A. Chiaroni, *J. Org. Chem.* **61**, 2293 (1996).
284. G. Jenner, *New J. Chem.* **15**, 897 (1991).
285. W. J. Le Noble and T. Asano, *J. Am. Chem. Soc.* **97**, 1778 (1975).
286. P. Drude and W. Nernst, *Z. Phys. Chem.* **15**, 79 (1894).
287. B. B. Owen and S. R. Brinkley, *Phys. Rev.* **64**, 32 (1943).
288. L. N. Kowa, D. Schwarzer, J. Troe, and J. Schroeder *J. Chem. Phys.* **97**, 4827 (1992).
289. A. L. Buchachenko, M. V. Motyakin, and I. I. Aliev, *Reactivity of Molecular Solids*, E. V. Boldyreva and V. V. Boldyrev, eds., John Wiley & Sons, Chichester, 1999, pp. 221–239.
290. S. F. Hulbert, *J. Br. Ceram. Soc.*, **6**, 11 (1969).
291. W. Jander, *Z. Anorg. Chem.* **163**, 1 (1927).
292. R. E. Carter, *J. Chem. Phys.* **34**, 2010 (1961).
293. M. Avrami, *J. Chem. Phys.* **7**, 1103 (1939), **8**, 212 (1940), **9**, 177 (1940).
294. R. H. Baughman, *J. Chem. Phys.* **68**, 3110 (1977).
295. H. G. Drickamer, C. W. Frank, and C. P. Slichter, *Proc. Natl. Acad. Sci. 69*, 933 (1972).
296. M. Mugnai, G. Cardini, and V. Schettino, *J. Chem. Phys.* **120**, 5327 (2004).
297. M. Mugnai, M. Pagliai, G. Cardini, V. Schettino, to be published.
298. M. Mugnai, G. Cardini, and V. Schettino, *Phys. Rev. B* **70**, 020101 (2004).
299. A. A. Zharov, *Rus. Chem. Rev. 53*, 140 (1984).
300. S. Wiederhorn and H. G. Drickamer, *J. Phys. Chem. Solids* **9**, 330 (1959).
301. G. Samara and H. G. Drickamer, *J. Chem. Phys.* **37**, 474 (1962).
302. L. Ciabini, M. Santoro, R. Bini, and V. Schettino, *Phys. Rev. Lett.* **88**, 085505 (2002).
303. M. Santoro, M. Ceppatelli, R. Bini, and V. Schettino, *J. Chem. Phys.* **118**, 8321 (2003).
304. F. Cansell, D. Fabre, and J. P. Petitet, *J. Chem. Phys.* **99**, 7300 (1993).
305. K. Aoki, B. J. Baer, H. C. Cynn, and M. Nicol, *Phys. Rev. B* **42**, 4298 (1990).
306. A. Gavezzotti and M. Simonetta, *Chem. Rev.* **82**, 1 (1982).
307. M. D. Cohen, *Angew. Chem., Int. Ed. Engl.* **14**, 386 (1975).
308. G. Desiraju, *Crystal Engineering. The Design of Organic Solids, Material Science Monographs*, Vol. 54, Elsevier, Amsterdam, 1989.
309. L. Ciabini, M. Santoro, R. Bini, and V. Schettino, *J. Chem. Phys.* **116**, 2928 (2002).
310. K. Dwarakanath and P. N. Prasad, *J. Am. Chem. Soc.* **102**, 4254 (1980).

311. E. V. Boldyreva, in *Reactivity of Molecular Solids*, E. V. Boldyreva and V. V. Boldyrev, eds., John Wiley & Sons., Chichester, 1999, pp. 1–50.

312. N. Nagaosa and T. Ogawa, *Phys. Rev. B* **39**, 4472 (1989).

313. M. D. Hollingsworth and J. M. McBride, *Mol. Cryst. Liq. Cryst.* **161**, 25 (1988).

314. T. Luty and R. Fouret, *J. Chem. Phys.* **90**, 5696 (1989).

315. T. Luty and C. J. Eckhardt, *J. Am. Chem. Soc.* **117**, 2441 (1995).

316. A. F. Goncharov, E. Gregoryanz, H. Mao, Z. Liu, and R. J. Hemley, *Phys. Rev. Lett.* **85** 1262, (2000).

317. P. Loubeyre, F. Occelli, and R. LeToullec, *Nature* **416** 613 (2002) and references therein.

318. R. J. Hemley, *Annu. Rev. Phys. Chem.* **51**, 763 (2000).

319. W. J. Nellis, N. C. Holmes, A. C. Mitchell, and M. Van Thiel, *Phys Rev. Lett.* **53**, 1661 (1984).

320. H. B. Radousky, W. J. Nellis, M. Ross, D. H. Hamilton, and A. C. Mitchell, *Phys Rev. Lett.* **57**, 2419 (1986).

321. M. Ross, *J. Chem. Phys.* **86**, 7110 (1987).

322. D. C. Hamilton and F. H. Free, *J. Chem. Phys.* **90**, 4972 (1989).

323. A. K. McMahan and R. Le Sar, *Phys Rev. Lett.* **54**, 1929 (1985).

324. C. Mailhiot, L. H. Yang, and A. K. McMahan, *Phys. Rev. B* **46**, 14419 (1992).

325. L. Mitas and R. M. Martin, *Phys Rev. Lett.* **72**, 2438 (1994).

326. A. K. McMahan, *Physica B + C* **138 & 139**, 31 (1986).

327. R. M. Martin and R. J. Needs, *Phys. Rev. B* **34**, 5082 (1986).

328. R. Reichlin, D. Schiferl, S. Martin, C. Vanderborgh, and R. L. Mills, *Phys Rev. Lett.* **55**, 1464 (1985).

329. M. I. Eremets, R. J. Hemley, H. K. Mao, and E. Gregoryanz, *Nature*, **411**, 170 (2001).

330. E. Gregoryanz, A. F. Goncharov, R. J. Hemley, and H. K. Mao, *Phys. Rev. B* **64**, 52103 (2001).

331. M. I. Eremets, A. G. Gavriliuk, I. A. Trojan, D. A. Dzivenko, and R. Boehler, *Nat. Mater.* **3**, 558 (2004).

332. M. I. Eremets, M. Y. Popov, I. A. Trojan, V. N. Denisov, R. Boehler, and R. J. Hemley, *J. Chem. Phys.* **120**, 10618 (2004).

333. R. L. Mills, B. Olinger, and D. T. Cromer, *J. Chem. Phys.* **84**, 2837 (1986).

334. A. I. Katz, D. Schiferl, and R. L. Mills, *J. Chem. Phys.* **88**, 3176 (1984).

335. R. L. Mills, D. Schiferl, A. I. Katz, and B. Olinger, *J. Phys. (Paris) Colloq.* **C-8**, 187 (1984).

336. M. Lipp, W. J. Evans, V. Garcia-Baonza, and H. E. Lorenzana, *J. Low Temp. Phys.* **111**, 247 (1998).

337. S. Bernard, G. L. Chiarotti, S. Scandolo, and E. Tosatti, *Phys Rev. Lett.* **81**, 2092 (1998).

338. M. Ceppatelli, M. Santoro, R. Bini, and V. Schettino, *J. Chem. Phys.* **118**, 1499 (2003).

339. M. W. A. Kuijpers, D. van Eck, M. F. Kemmere, and J. T. F. Keurentjes, *Science* **298**, 1969 (2002).

340. J. Haines, J. M. Lèeger, and G. Bocquillon, *Annu. Rev. Mater. Res.* **31**, 1 (2001).

341. W. J. Nellis, A. C. Mitchell, F. A. Ree, M. Ross, N. C. Holmes, R. J. Trainor, and J. Erskine, *J. Chem. Phys.* **95**, 5268 (1991).

342. O. Tschauner, H. K. Mao, and R. J. Hemley, *Phys Rev. Lett.* **87**, 075701 (2001).

343. V. Iota, C. S. Yoo, and H. Cynn, *Science* **283** 1510 (1999).

344. S. Serra, C. Cavazzoni, G. L. Chiarotti, S. Scandolo, and E. Tosatti, *Science*, **284**, 788 (1999).

345. J. Dong, J. K. Tomfohr, O. F. Sankey, K. Leinenweber, M. Somayazulu, and P. F. McMillan, *Phys. Rev. B* **62**, 14685 (2000).

346. J. Dong, J. K. Tomfohr, and O. F. Sankey, *Phys. Rev. B*, **61**, 5967 (2000).

347. C. S. Yoo, H. Cynn, F. Gygi, G. Galli, V. Iota, S. Carlson, D. Hausermann, and C. Mailhiot , *Phys Rev. Lett.* **83**, 5527 (1999).

348. B. Holm, R. Ahuja, A. Belonoshko, and B. Johansson, *Phys Rev. Lett.* **85**, 1258 (2000).

349. J. Dong, J. K. Tomfohr, and O. F. Sankey, *Science* **287**, 11a (2000).

350. C. Cavazzoni, G. L. Chiarotti, S. Scandolo, S. Serra, and E. Tosatti, *Science* **287**, 11a (2000).

351. C. S. Yoo, *Science* **287**, 11a (2000).

352. V. Iota and C. Yoo, *Phys Rev. Lett.* **86**, 5922 (2001).

353. C. S. Yoo, V. Iota, and H. Cynn, *Phys Rev. Lett.* **86**, 444 (2001).

354. C. S. Yoo, H. Kohlmann, H. Cynn, M. F. Nicol, V. Iota, and T. LeBihan, *Phys. Rev. B* **65**, 104103 (2002).

355. J. H. Park, C. S. Yoo, V. Iota, H. Cynn, M. F. Nicol, and T. Le Bihan, *Phys. Rev. B* **68**, 014107 (2003).

356. S. Bonev, F. Gygi, T. Ogitsu, and G. Galli, *Phys Rev. Lett.* **91**, 065501 (2003).

357. F. A. Gorelli, V. M. Giordano, P. R. Salvi, and R. Bini, *Phys Rev. Lett.* **93**, 205503 (2004).

358. C. S. Yoo, in *Science and Technology of High Pressure*, M. H. Manghnani, V. J. Nellis, and M. Nicol, Universities Press, Hyderabad, 2000.

359. S. F. Agnew, B. I. Swanson, L. H. Jones, R. L. Mills, and D. Schiferl, *J. Phys. Chem.* **87**, 5065 (1983).

360. S. F. Agnew, B. I. Swanson, L. H. Jones, and R. L. Mills, *J. Phys. Chem.* **89**, 1678 (1985).

361. F. Bolduan, H. J. Jodl, and A. Loewenschuss, *J. Chem. Phys.* **80**, 1739 (1984).

362. A. Givan and A. Loewenschuss, *J. Chem. Phys.*, **93**, 7592 (1990).

363. M. Somayazulu, A. Madduri, A. F. Goncharov, O. Tschauner, D. F. McMillan, H. Mao, and R. J. Hemley, *Phys Rev. Lett.* **87**, 135504 (2001).

364. W. J. Bulmage and W. N. Lipscomb, *Acta Cryst.* **4**, 330 (1951).

365. M. Chall, B. Winkler, and V. Milman, *J. Phys. Condens. Matter* **8**, 9049 (1996).

366. K. Aoki, and H. Yamawaki, *Rev. High Pressure Sci. Technol.* **5**, 137 (1996).

367. H. Yamawaki, M. Sakashita, and K. Aoki, *J. NIMC* **6**, 169 (1998).

368. T. Volker, *Angew. Chem.* **72**, 379 (1960).

369. C. Yoo and M. Nicol, *J. Phys. Chem.* **90**, 6726 (1986).

370. A. S. Parkes and R. E. Hughes, *Acta Cryst.* **16**, 734 (1963).

371. C. Yoo and M. Nicol, *J. Phys. Chem.* **90**, 6732 (1986).

372. P. B. Zmolek, H. Sohn, P. K. Gantzel, and W. C. Trogler, *J. Am. Chem. Soc.* **123**, 1199 (2000).

373. J. J. Colman and W. C. Trogler, *J. Am. Chem. Soc.*, **117**, 11270 (1995).

374. F. Bolduan, H. D. Hocheimer, and H. J. Jodl, *J. Chem. Phys.* **84**, 6997 (1986).

375. H. Shimizu and T. Onishi, *Chem. Phys. Lett.* **99**, 507 (1983).

376. P. W. Bridgman, *J. Appl. Phys.* **12**, 461 (1941).

377. W. S. Chan and A. K. Jonscher, *Phys. Stat. Sol.* **32**, 749 (1968).

378. E. Whalley, *Can. J. Chem.*, **38**, 2105 (1960).

379. E. G. Butcher, M. Alsop, J. A. Weston, and H. A. Gebbie, *Nature* **199**, 756 (1963).

380. R. D. Dick, *J. Chem. Phys.* **52**, 6021(1970).

381. R. D. Dick, *J. Chem. Phys.* **71**, 3203 (1979).
382. R. H. Wentorf Jr., *J. Phys. Chem.* **69**, 3063 (1965).
383. R. H. Warnes, *J. Chem. Phys.* **53**, 1088 (1970).
384. K. Mimura, M. Ohashi, and R. Sugisaki, *Earth. Planet. Sci.* **133**, 265 (1995).
385. K. Mimura, *Geochim. Cosmochim. Acta* **59**, 579 (1995).
386. F. H. Ree, *J. Chem. Phys.* **70**, 974 (1979).
387. H. G. Drickamer, *Science* **156**, 3779 (1967).
388. S. Block, C. E.Weir, and G. J. Piermarini, *Science* **169**, 586 (1970).
389. M. Nicol, M. L. Johnson, and N. C, Holmes, *Physica* **139 & 140B**, 582 (1986).
390. R. Engelke *J. Am. Chem. Soc.* **108**, 579 (1986).
391. R. Pucci and N. H. March, *J. Chem. Phys.* **74**, 1373 (1981).
392. R. Engelke, P. J. Hay, D. A. Kleier, and W. R. Wadt, *J. Chem. Phys.* **79**, 4367 (1983).
393. R. Engelke, P. J. Hay, D. A. Kleier, and W. R. Wadt, *J. Am. Chem. Soc.* **106**, 5439 (1984).
394. R. Engelke and N. C. Blais, *J. Chem. Phys.* **101**, 10961 (1994).
395. S. R. Bickham, J. D. Kress, and L. A. Collins, *J. Chem. Phys.* **112**, 9695 (2000).
396. L. R. Benedetti, J. H. Nguyen, W. A. Caldwell, H. Liu, M. Kruger, and R. Jeanloz, *Science* **285**, 100 (1999).
397. F. Ancilotto, G. L. Chiarotti, S. Scandolo, and E. Tosatti, *Science* **275**, 1288 (1997).
398. H. Shirakawa, *Synth. Met.* **69**, 3 (1995).
399. K. Aoki, Y. Kakudate, S. Usuba, M. Yoshida, K. Tanaka, and S. Fujiwara, *J. Chem. Phys.* **88**, 4565 (1988).
400. R. LeSar, *J. Chem. Phys.* **86**, 1485 (1987).
401. K. Aoki, Y. Kakudate, S. Usuba, M. Yoshida, K. Tanaka and S. Fujiwara, *Synth. Met.* **28**, D91 (1989).
402. K. Aoki, Y. Kakudate, S. Usuba, M. Yoshida, K. Tanaka, and S. Fujiwara, *J. Chem. Phys.* **89**, 529 (1988).
403. M. Sakashita, H. Yamawaki, and K. Aoki, *J. Chem. Phys.* **100**, 9943 (1996).
404. V. Schettino, F. L. Gervasio, G. Cardini, and P. R. Salvi, *J. Chem. Phys.* **110**, 1 (1999).
405. N. M. Balzaretti, C. A. Perottoni, and J. A. Herz da Jornada, *J. Raman Spectrosc.* **34**, 259 (2003).
406. M. Bernasconi, G. L. Chiarotti, P. Focher, M. Parrinello, and E. Tosatti, *Phys Rev. Lett.* **78**, 2008 (1997).
407. M. Bernasconi, M. Parrinello, G. L. Chiarotti, P. Focher, and E. Tosatti, *Phys Rev. Lett.* **76**, 2081 (1996).
408. G. Cardini, unpublished observations.
409. C. C. Trout and J. V. Badding, *J. Phys. Chem.* **104**, 8142 (2000).
410. K. Aoki, Y. Kakudate, S. Usuba, M.Yoshida, K. Tanaka, and S. Fujiwara, *J. Chem. Phys.* **91**, 778 (1989).
411. K. Aoki, Y. Kakudate, S. Usuba, M. Yoshida, K. Tanaka, and S. Fujiwara, *J. Chem. Phys.* **91**, 2814 (1989).
412. M. Santoro, L. Ciabini, R. Bini, and V. Schettino, *J. Raman Spectrosc.* **34**, 557 (2003).
413. H. C. Weiss, D. Blser, R. Boese , B. M. Doughm, and M. M. Hley, *Chem. Commun.*, 1703 (1997).

414. H. Wieldraaijer, J. A. Schouten, and N. J. Trappenier, *High Temp. High Press.* **15**, 87 (1983).
415. L. Van der Putten, J. A. Schouten, and N. J. Trappenier, *High Temp. High Press.* **18**, 255 (1986).
416. D. Chelazzi, M. Ceppatelli, M. Santoro, R. Bini, and V. Schettino, *Nat. Mater.* **3**, 470 (2004).
417. Y. Osada, Q. S. Yu, H. Yasunaga, and Y. Kagami, *J. Polym. Sci. Polym. Chem.* **27**, 3799 (1989).
418. W. L. Chaplot and R. Mukhopadhyay, *Phys. Rev. B* **33**, 5099 (1986).
419. H. Yamawaki, K. Aoki, Y. Kakudate, M. Yoshida, S. Usuba, and S. Fujiwara, *Chem. Phys. Lett.* **198**, 183 (1992).
420. H. Yamawaki, M. Sakashita, K. Aoki, and K. Takemura, *Phys. Rev. B* **53**, 11403 (1996).
421. J. V. Badding, L. J. Parker, and D. C. Nesting, *J. Solid State Chem.* **117**, 229 (1995).
422. E. Ohtani, T. Irifune, W. O. Hibberson, and A. E. Ringwood, *High Temp. High Press.* **19**, 523 (1987).
423. D. C. Nesting and J. V. Badding, *Chem. Mater.* **8**, 1535 (1996).
424. M Citroni, M. Ceppatelli, R. Bini, and V. Schettino, to be published.
425. L. E. Reeves, G. J. Scott, and S. E. Babb, *J. Chem. Phys.* **40**, 3662 (1964).
426. M. Citroni, M. Ceppatelli, R. Bini, and V. Schettino, *Science* **295**, 2058 (2002).
427. M. Citroni, M. Ceppatelli, R. Bini, and V. Schettino, *High Press. Res.* **22**, 507 (2002).
428. M. Citroni, M. Ceppatelli, R. Bini, and V. Schettino, *Chem. Phys. Lett.* **367**, 186 (2003).
429. M. Citroni, M. Ceppatelli, R. Bini, and V. Schettino, *J. Chem. Phys.* **118**, 1815 (2003).
430. F. L. Gervasio, R. Chelli, P. Procacci, and V. Schettino, *J. Phys. Chem. A* **106**, 2945 (2002) and references therein.
431. R. Torre, R. Righini, L. Angeloni, and S. Califano, *J. Chem. Phys.* **93**, 2967 (1990).
432. E. G. Cox, D. W. J. Cruickshank, J. A. Smith, *Proc. R. Soc. A* **24**, 1 (1958).
433. G. E. Bacon, N. A. Curry, and S. A. Wilson, *Proc. R. Soc. A* **279**, 98 (1964).
434. G. J. Piermarini, A. D. Mighell, C. E. Weir, and S. Block, *Science* **165**, 1250 (1969).
435. L. Ciabini, F. A. Gorelli, M. Santoro, R. Bini, and V. Schettino, to be published.
436. M. M. Thiéry and J. M. Léger, *J. Chem. Phys.* **89** 4255 (1988).
437. P. W. Bridgman, *Phys. Rev.* **3**, 153 (1914).
438. P. W. Bridgman, *J. Chem. Phys.* **9**, 794 (1941).
439. J. Akella and G. C. Kennedy, *J. Chem. Phys.* **55**, 793 (1971)
440. P. Pruzan, J. C. Chervin, M. M. Thiéry, J. P. Itié, and J. M. Besson, *J. Chem. Phys.* **92**, 6910 (1990).
441. M. Gauthier, J. C. Chervin, and P. Pruzan, in *Frontiers of High Pressure Research*, H. D. Hochheimer and R. D. Etters, eds., Plenum, New York, 1991, p. 87.
442. J. M. Besson, M. M. Thiéry, and P. Pruzan, in *Molecular Systems Under High Pressure*, R. Pucci and G. Piccitto, eds., Elsevier, Amsterdam, 1991, p. 341.
443. M. M. Thiéry, J. M. Besson, and J. L. Bribes, *J. Chem. Phys.* **96**, 2633 (1992).
444. D. Gourdain, J. C. Chervin, and P. Pruzan, *J. Chem. Phys.* **105**, 9040 (1996).
445. P. Pruzan, J. C. Chervin, and J. P. Forgerit, *J. Chem. Phys.*, **96**, 761 (1992).
446. H. Shimizu and N. Matsunami, *J. Raman Spectrosc.* **19**, 199 (1988).
447. H. W. Kroto, J. R. Heath, S. C. O'Brien, R. F. Curl, and R. E. Smalley, *Nature* **318**, 162 (1985).
448. H. W. Kroto, *Science* **242**, 1139 (1988).
449. J. L. Hodeau, J. M. Tonnerre, B. Bouchet-Favre, M. Núñez-Regueiro, J. J. Capponi, and M. Perroux, *Phys. Rev. B* **50**, 10311 (1994).

450. O. Béthoux, M. Núñez-Regueiro, L. Marques, J. L. Hodeau, and M. Perroux, in *Proceedings of the Materials Research Society*, Materials Research Society, Pittsburgh, 1993, p. 202.

451. M. Núñez-Regueiro, L. Marques, J. L. Hodeau, O. Béthoux, and M. Perroux, *Phys. Rev. Lett.* **74**, 278 (1995).

452. Y. Iwasa, T. Arima, R. M. Fleming, T. Siegrist, O. Zhou, R. C. Haddon, L. J. Rothberg, K. B. Lyons, H. L. Carter, A. F. Hebard, R. Tycko, G. Dabbagh, J. J. Krajewski, G. A. Thomas, and T. Yagi, *Science* **264**, 1570 (1994).

453. L. Marques, J. L. Hodeau, M. Núñez-Regueiro, and M. Perroux, *Phys. Rev. B* **54**, R12633 (1996).

454. S. J. Duclos, K. Brister, R. C. Haddon, A. R. Kortan, and F. A. Thiel, *Nature* **351**, 380 (1991).

455. Q. Z. Zhang, J. Y. Yi, and J. Bernholc, *Phys. Rev. Lett.* **66**, 2633 (1991).

456. A. M. Rao, P. Zhou, K. A. Wang, G. T. Hager, J. M. Holden, Y. Wang, W. T. Lee, X. X. Bi, P. C. Eklund, D. S. Cornett, M. A. Duncan, and I. J. Amster, *Science* **259**, 955 (1993).

457. T. Wågberg, P. Jacobsson, and B. Sundqvist, *Phys. Rev. B* **60**, 4535 (1999).

458. S. Pekker, A. Janossy, L. Mihaly, P. W. Stephens, O. Chauvet, M. Carrard, and L. Forro, *Science* **265**, 1077 (1994).

459. P. Nagel, V. Pasler, S. Lebedkin, A. Soldatov, C. Meingast, B. Sundqvist, P. A. Persson, T. Tanaka, K. Komatsu, S. Buga, and A. Ihaba, *Phys. Rev. B*, **60**, 16920 (1999).

460. V. A. Davydov, L. S. Kashevarova, A. V. Rakhmanina, V. M. Senyavin, R. Céolin, H. Szwarc, H. Allouchi, and V. Agafonov, *Phys. Rev. B* **61**, 11936 (2000).

461. C. H. Xu and G. E. Scuseria, *Phys. Rev. Lett.* **74**, 274 (1995).

462. M. Mezouar, L. Marques, J. L. Hodeau, V. Pischedda, and M. Núñez-Regueiro *Phys. Rev. B* **68**, 193414 (2003).

463. V. D. Blank, S. G. Buga, G. A. Dubitsky, N. R. Serebryanaya, M. Y. Popov, and B. Sundqvist, *Carbon* **36**, 319 (1998).

464. V. D. Blank, M. Y. Popov, S. G. Buga, V. Davydov, V. N. Denisov, A. N. Iblev, B. N. Mavrin, A. V. Agafonov, R. Ceolin, H. Szwarc, and A. Rasat, *Phys. Lett. A* **188**, 281 (1994).

465. V. D. Blank, S. G. Buga, N. R. Serebryanaya, V. N. Denisov, G. A. Dubitsky, A. N. Ivlev, B. N. Martin, and M. Y. Popov, *Phys. Lett. A* **205**, 208 (1995).

466. V. D. Blank, S. G. Buga, N. R. Serebryanaya, G. A. Dubitsky, S. N. Sulyanov. M. Y. Popov, V. N. Denisov, A. N. Ivlev, and B. N. Martin, *Phys. Lett. A*, **220**, 149 (1996).

467. V. D. Blank, S. G. Buga, N. R. Serebryanaya, G. A. Dubitsky, R. H. Bagramov, M. Yu. Popov, V. M. Pokhorov, and S. A. Sulyanov, *Appl. Phys. A* **64**, 247 (1997).

468. V. V. Brazkin, A. G. Lyapin, S. V. Popova, Y. A. Klyuev, and A. M. Naletov *J. Appl. Phys.* **84**, 219 (1998).

469. C. Z. Zhang, K. M. Wang, A. Ho, and C. T. Chan, *Europhys. Lett.* **28**, 219 (1994).

470. C. S. Yoo and W. J. Nellis, *Science* **254**, 1489 (1991).

471. J. F. Kenney, V. A. Kutcherov, N. A. Bendeliani, and V. A. Alekseev, *Proc. Natl. Acad. Sci.* **99**, 10976 (2002).

472. H. P. Scott, R. J. Hemley, H. K. Mao, D. R. Herschbach, L. E. Fried, W. M. Howard, and S. Bastea, *Proc. Natl. Acad. Sci.* **101**, 14023 (2004).

473. N. C. Blais, R. Engelke, and S. A. Sheffield, *J. Phys. Chem. A* **101**, 8285 (1997).

474. G. J. Piermarini, S. Block, and P. J. Miller, *J. Phys. Chem.* **93**, 457 (1989).

475. G. J. Piermarini, S. Block, P. J. Miller, *J. Phys. Chem.* **91**, 3872 (1987).

476. M. R. Manaa, E. J. Reed, L. E. Fried, G. Galli, and F. Gygi, *J. Chem. Phys.* **120**, 10146 (2004).

477. S. Courtecuisse, F. Cansell, D. Fabre, and J. P. Petitet, *J. Chem. Phys.* **108**, 7350 (1998).

478. S. Courtecuisse, F. Cansell, D. Fabre, and J. P. Petitet, *J. Chem. Phys.* **102**, 968 (1995).

479. J. P. Pinan-Lucarré, R. Ouillon, B. Canny, P. Pruzan, and P. Ranson, *J. Raman Spectrosc.* **10**, 819 (2003).

480. P. Pruzan, B. Canny, C. Power, and J. C. Chervin, Infrared Spectroscopy of Nitromethane up to 50 GPa, in *Proceedings of the Seventeenth International Conference on Raman Spectroscopy (ICORS 2000)*, S.-L. Zhang and B.-f. Zhu, eds., Peking University, Beijing, China, 2000, p. 142.

481. D. Margetis, E. Kaxiras, M. Elstner, T. Fraunheim, and M. R. Manaa, *J. Chem. Phys.* **117**, 788 (2002).

482. S. Bulusu, D. I. Weinstein, J. P. Autera, and R. W. Velicky, *J. Phys. Chem.* **90**, 4121 (1986).

483. T. B. Brill and R. J. Karpowicz, *J. Phys. Chem.* **86**, 4260 (1982).

484. H. Bouas-Laurent and H. Dürr, *Pure Appl. Chem.* **73**(4), 639 (2001).

485. H. G. Drickamer, in *High Pressure Chemistry, Biochemistry and Materials Science*, R. Winter and J. Jonas, eds., Kluwer Academic Publishers, Dordrecht, 1993, pp. 67–77.

486. M. D. Cohen and G. M. J. Schmidt, *J. Phys. Chem.* **66**, 2442 (1962).

487. T. Rosenfeld, M. Ottolenghi, and A. Y. Meyer, *Mol. Photochem.* **5**, 39 (1973).

488. R. Potashnik and M. Ottolenghi, *J. Chem. Phys.* **51**, 3671 (1969).

489. R. S. Becker and W. F. Richey, *J. Am. Chem. Soc.* **89**, 1298 (1967).

490. M. D. Cohen, G. M. J. Schmidt, and S. Flavian, *J. Chem. Soc.* 2041 (1964).

491. J. W. Ledbetter, Jr., *J. Phys. Chem.* **72**, 4111 (1968).

492. J. W. Ledbetter, Jr., *J. Phys. Chem.* **71**, 2351 (1967).

493. G. O. Dudek and E. P. Dudek, *J. Am. Chem. Soc.* **88**, 2407 (1966).

494. W. F. Richey and R. S. Becker, *J. Chem. Phys.* **49**, 2092 (1968).

495. M. D. Cohen and S. Flavian, *J. Chem. Soc. B* 317 (1967).

496. M. D. Cohen, Y. Hirshberg, and G. M. J. Schmidt, *J. Chem. Soc.* 2051 (1964).

497. E. M. Hockert and H. G. Drickamer, *J. Chem. Phys.* **67**, 5178 (1977).

498. L. Fabbrizzi, M. Micheloni, and P. Paoletti, *Inorg. Chem.* **13**, 3019 (1974).

499. A. B. P. Lever, E. Mantovani, and J. C. Donini, *Inorg. Chem.* **10**, 2424 (1971).

500. J. R. Ferraro, L. J. Basile, L. R. Garcia-Ineguez, P. Paoletti, and L. Fabbrizi, *Inorg. Chem.* **15**, 2342 (1976).

501. B. P. Kennedy and A. B. P. Lever, *J. Am. Chem. Soc.* **95**, 6907 (1973).

502. I. Grenthe, P. Paoletti, M. Sandstrom, and S. Glikberg, *Inorg. Chem.* **18**, 2687 (1979).

503. R. J. Pilkki, R. D. Willett, and H. W. Dodgen, *Inorg. Chem.* **23**, 594 (1984).

504. P. Pfeiffer and H. Glaser, *J. Prakt. Chem.* **151**, 134 (1938).

505. K. L. Bray, H. G. Drickamer, E. A. Schmitt, and D. N. Hendrickson, *J. Am. Chem. Soc.* **111**, 2849 (1989).

506. L. P. Battaglia, A. Bonamartini Corradi, G. Marcotrigiano, L. Menabue, and G. C. Pellicani, *Inorg. Chem.* **18**, 148 (1979).

507. M. C. Molla, J. Garcia, J. Borras, C. Foces-Foces, F. H. Cano, and M. M. Ripoll, *Transition Met. Chem.* **10**, 460 (1985).

508. H. G. Drickamer and K. L. Bray, *Acc. Chem. Res.* **23**, 55 (1990).

509. F. P. Bundy, *J. Geophys. Res.* **85**, 6930 (1980).

510. F. P. Bundy, W. A. Bassett, M. S. Weathers, R. J. Hemley, H. K. Mao, and A. F. Goncharov, *Carbon* **34**, 141 (1996).

511. F. D. Rossini and R. S. Jessup, *J. Nat. Bur. Stds.* **C21**, 491 (1938).

512. O. I. Leipunskii, *Usp. Khim.* **8**, 1519 (1939).

513. R. Berman and F. Simon, *Z. Elektrochem.* **59**, 333 (1955).

514. F. P. Bundy, H. P. Bovenkerk, H. M. Strong, and R. H. Wentorf, Jr., *J. Chem. Phys.* **35**, 383 (1961).

515. C. S. Kennedy and G. C. Kennedy, *J. Geophys. Res.* **81**, 2467 (1976).

516. H. T. Hall, *Rev. Sci. Instr.* **31**, 125 (1960).

517. H. T. Hall, *Proceedings of the Third Conference on Carbon*, Pergamon Press, London, 1957, pp. 75–84.

518. H. T. Hall, U.S. Patent No. 2,941,248.

519. H. T. Hall, *Progress in Very High Pressure Research*, F. P. Bunday, W. R. Hibbard, Jr., and H. M. Strong, eds., John Wiley & Sons, New York, 1961, pp. 1–9.

520. H. T. Hall, *J. Chem. Educ.* **38**, 484 (1961).

521. M. Ajaishi, *J. Cryst. Growth.* **104**, 578 (1990).

522. M. Ajaishi, *Japan. J. Appl. Phys.* **29**, L1172 (1990).

523. W. Utsumi, T. Okada, T. Taniguchi, K. Funakoshi, T. Kikegawa, N. Hamaya, and O. Shimomura, *J. Phys. Condens. Matter* **169**, S1017 (2004).

524. M. N. Regueiro, P. Monceau, and J. L. Hodeau, *Nature* **355**, 237 (1992).

525. R. H. Wentorf, Jr., *J. Chem. Phys.* **26**, 956 (1957).

526. K. Kuroda, H. Konno, and T. Matoba, *Kogyo Kagaru Zasshi* **69**, 365 (1966).

527. R. H. Wentorf, Jr., *J. Chem. Phys.* **34**, 809 (1961).

528. H. Saito, M. Ushio, and S. Nagano, *Yogyo Kyokai Shi* **78**, 7 (1970).

529. T. Kobayashi, *J. Chem. Phys.* **70**, 5898 (1979).

530. M. Wakatsuki, K. Ichinose, and T. Aoki, *Mater. Res. Bull.* **7**, 999 (1972).

531. H. Sumiya, T. Iseki, and A. Onodera, *Mater. Res. Bull.* **18**, 1203 (1983).

532. J. Y. Huang and Y. T. Zhu, *Chem. Mater.* **14**, 1873 (2002).

533. C. S. Yoo and J. Akella, H. Cynn, *Phys. Rev. B* **56**, 140 (1997).

534. M. Kawagichi, *Adv. Mater.* **9**, 615 (1997).

535. A. R. Badzian, *Mater. Res. Bull.* **16**, 1385 (1981).

536. E. Knittle, R. B. Kaner, R. Jeanloz, and M. L. Cohen, *Phys. Rev. B* **51**, 12149 (1995).

537. S. Nakano, *Proceedings of the 3rd NIRIM International Symposium on Advanced Materials*, NIRIM, Tsukuba, Japan, 1996, p. 287.

538. T. Komatsu, M. Nomura, Y. Kakudate, and S. Fujiwara, *J. Mater. Chem.* **6**, 1799 (1996).

539. A. V. Kurdyumov and V. L. Solozhenko, *J. Superhard Mater.* **21**, 1 (1999).

540. V. L. Solozhenko, *Appl. Phys. Lett.* **78**, 1385 (2001).

541. M. Hubaček, and T. Sato, *J. Solid State Chem.* **114**, 258 (1995).

542. Y. Tateyama, T. Ogitsu, K. Kusakabe, S. Tsuneyuki, and S. Itoh, *Phys. Rev. B* **55**, 10161 (1997).

543. P. Gillet, G. Fiquet, I. Daniel, B. Reynard, and M. Hanfland, *Phys. Rev. B* **60**, 14660 (1999).

544. V. L. Solozhenko, D. Häusermann, M. Mezouar, and M. Kunz, *Appl. Phys. Lett.* **72**, 1691 (1998).

545. Y. Zhao, D. W. He, L. L. Daemen, J. Huang, T. D. Shen, R. B. Schwarz, Y. Zhu, D. L. Bish, J. Zhang, G. Shen, J. Qian, and T. W. Zerda, *J. Mater. Res.* **17**, 3139 (2002).

546. F. Gao, J. He, E. Wu, S. Liu, D. Yu, D. Li, S. Zhang, and Y. Tian, *Phys. Rev. Lett.* **91**, 015502 (2003).

547. Y. Zhang, H. Sun, and C. Chen, *Phys. Rev. Lett.* **93**, 195504 (2004).

548. H. Hubert, L. A. J. Garvie, P. R. Buseck, W. T. Petuskey, and P. F. McMillan, *J. Solid State Chem.* **133**, 356 (1997).

549. H. Hubert, L. A. J. Garvie, B. Devouard, P. R. Buseck, W. T. Petuskey, and P. F. McMillan, *Chem. Mater.,* **10**, 1530 (1998).

550. P. F. McMillan, *High Press. Res.* **23**, 7 (2003).

551. P. F. McMillan, *High Press. Res.* **24**, 67 (2004).

552. P. F. McMillan, *Chem. Commun.*, 919 (2003).

553. I. Higashi, M. Kobayashi, J. Bernhard, C. Brodhag, and F. Thévenot, in *Boron-Rich Solids, A. I. P. Conference and Proceedings*, Vol. 231, D. Emin, T. Aselage, C. L. Beckel, I. A. Howard, and C. Wood, eds., American Institute of Physics, New York, 1991, p. 201.

554. D. R. Petrack, R. Ruh, and B. F. Goosey, in *5th Materials Research Symposium*, Vol. 364, NBS Special Publication, 1972, p.605.

555. H. Bolmgren, T. Lundström, and S. Okada, in *Boron-Rich Solids, A.I.P. Conference and Proceedings*, Vol. 231, D. Emin, T. Aselage, C. L. Beckel, I. A. Howard, and C. Wood, eds., American Institute of Physics, New York, 1991, p. 197.

556. H. Hubert, B. Devouard, L. A. J. Garvie, M. O'Keeffe, P. R. Buseck, W. T. Petuskey, and P. F. McMillan, *Nature* **391**, 376 (1998).

557. S. Lee, S. W. Kim, D. M. Bylander, and L. Kleinman, *Phys. Rev. B* **44**, 3550 (1991).

558. E. Soignard, P. F. McMillan, T. D. Chaplin, S. M. Farag, C. L. Bull, M. S. Somayazulu, and K. Leinenweber, *Phys. Rev. B* **68**, 132101 (2003).

559. J. Ortega and O. F. Sankey, *Phys. Rev. B* **51**, 2624 (1995).

560. Y. Miyamoto, M. L. Cohen, and S. G. Louie, *Solid State Commun.* **102**, 605 (1997).

561. A. Zerr, G. Miehe, G. Serghiou, M. Schwarz, E. Kroke, R. Riedel, H. Fuess, P. Kroll, and R. Boehler, *Nature* **400**, 340 (1999).

562. G. Serghiou, G. Miehe, O. Tschauner, A. Zerr, and R. Boehler, *J. Chem. Phys.* **111**, 4659 (1999).

563. K. Leinenweber, M. O'Keeffe, M. Somayazulu, H. Hubert, P. F. McMillan, and G. H. Wolf, *Chem. Eur. J.* **5**, 3076 (1999).

564. J. Dong, O. F. Sankey, S. K. Deb, G. Wolf, and P. F. McMillan, *Phys. Rev. B* **61**, 11979 (2000).

565. E. Soignard, M. S. Somayazulu, H. K. Mao, J. Dong, O. F. Sankey, and P. F. McMillan, *Solid State Commun.* **120**, 237 (2001).

566. A. Y. Liu and M. L. Cohen, *Science* **245**, 841 (1989).

567. A. Y. Liu and M. L. Cohen, *Phys. Rev. B* **41**, 10727 (1990).

568. D. M. Teter and R. J. Hemley, *Science* **271**, 53 (1996).

569. J. Kouvetakis, A. Bandari, M. Todd, and B. Wilkens, *Chem. Mater.* **6**, 811 (1994).

570. Z. J. Zhang, S. Fan, and C. M. Lieber, *Appl. Phys. Lett.* **66**, 3582 (1995).

571. D. Marton, K. J. Boyd, A. H. Al-Bayati, S. S. Todorov, and J. W. Rabalais, *Phys. Rev. Lett.* **73**, 118 (1994).

572. E. G. Gillan, *Chem. Mater.* **12**, 3906 (2000).

573. C. Niu, Y. Z. Lu, and C. M. Lieber, *Science* **261**, 334 (1993).

574. K. M. Yu, M. L. Cohen, E. E. Haller, W. L. Hansen, A. Y. Liu, and I. C. Wu, *Phys. Rev. B* **49**, 5034 (1994).

575. T. Sekine, H. Kanda, Y. Bando, M. Yokoyama, and K. Hojou, *J. Mater. Sci. Lett.* **9**, 1376 (1990).

576. J. V. Badding and D. C. Nesting, *Chem. Mater.* **8**, 535 (1996).

577. J. V. Badding and *Annu. Rev. Mater. Sci.* **28**, 631 (1998).

578. D. C. Nesting and J. V. Badding, *Chem. Mater.* **8**, 1535 (1996).

579. J. H. Nguyen and R. Jeanloz, *Mater. Sci. Eng. A* **209**, 23 (1996).

580. E. Gregoryanz, C. Sanloup, M. Somayazulu, J. Badro, G. Fiquet, H. K. Mao, and R. J. Hemley, *Nature* **3**, 294 (2004).

CLASSICAL DESCRIPTION OF NONADIABATIC QUANTUM DYNAMICS

GERHARD STOCK

Institute of Physical and Theoretical Chemistry
J. W. Goethe University, Frankfurt, Germany

MICHAEL THOSS

Department of Chemistry, Technical University
of Munich Garching, Germany

CONTENTS

Advances in Chemical Physics, Volume 131, edited by Stuart A. Rice

I. INTRODUCTION

Since the birth of quantum theory, there has been considerable interest in the transition from quantum to classical mechanics. Because the two formulations are given in a different theoretical framework (nonlinear classical trajectories versus expectation values of linear operators), this transition is far more involved than the naive limit $\hbar \to 0$ suggests. By exploring the classical limit of quantum mechanics, new theoretical concepts have been developed, including path integrals [1], various phase-space representations of quantum mechanics [2], the semiclassical propagator and the trace formula [3], and the notion of quantum chaos [4]. Moreover, there are practical reasons to describe quantum phenomena by a classical approach. While the numerical effort of a quantum-mechanical basis-set calculation increases exponentially with the number of nonseparable degrees of freedom (DoF), classical mechanics scales linearly with the number of DoF.

In cases where both the system under consideration and the observable to be calculated have an obvious classical analog (e.g., the translational-energy distribution after a scattering event), a classical description is a rather straightforward matter. It is less clear, however, how to incorporate discrete quantum-mechanical DoF that do not possess an obvious classical counterpart into a classical theory. For example, consider the well-known spin–boson problem—that is, an electronic two-state system (the spin) coupled to one or many vibrational DoF (the bosons) [5]. Exhibiting nonadiabatic transitions between discrete quantum states, the problem apparently defies a straightforward classical treatment.

In this chapter, we are concerned with various theoretical formulations that allow us to treat nonadiabatic quantum dynamics in a classical description. To introduce the main concepts, we first give a brief overview of the existing methods and then discuss their application to ultrafast molecular photoprocesses.

A. Classification of Methods

In order to incorporate quantum DoF into a classical formulation, a number of *mixed quantum-classical models* have been proposed. The first mixed quantum-classical formulation was proposed by Mott [6], who considered the excitation of atoms in collision reactions. Inspired by the work of Born and Oppenheimer [7], he described the dynamics of the electrons employing an expansion of the total wave function in adiabatic electronic wave functions, while the motion of the heavy nuclei was considered in a classical manner. As the electronic dynamics is evaluated along the classical path of the nuclei, this ansatz is often referred to as *classical-path* approximation. Although the classical-path approximation describing the reaction of the quantum DoF to the dynamics of the classical DoF is common to most mixed quantum-classical formulations, there are several ways to describe the "back-reaction" of the classical DoF to the dynamics of the quantum DoF. One way is to employ Ehrenfest's theorem [8] and calculate the effective force on the classical trajectory through a mean potential that is averaged over the quantum DoF [9–22]. As with most mixed quantum-classical formulations, the resulting mean-field trajectory method employs a *quasiclassical* approximation to the heavy-particle DoF; that is, the quantum nature of the initial state of the classically treated subsystem is simulated through quasiclassical sampling of the corresponding probability distribution [23–26]. In contrast to a rigorous *semiclassical* description (i.e., in the sense of the Van Vleck–Gutzwiller formulation [3]), possible quantum-mechanical interferences between individual classical paths are therefore not included.

In general, a mixed quantum-classical description may be derived by starting with a quantum-mechanically exact formulation for the complete system and performing a partial classical limit for the heavy-particle DoF. This procedure is not unique, however, since it depends on the particular quantum formulation chosen as well as on the specific way to achieve the classical limit. In the *mean-field trajectory* method explained above, for example, the wave-function formulation of quantum mechanics is adopted and the Ehrenfest classical limit is performed for the heavy-particle DoF. Alternatively, one may consider the Liouville equation of the density operator and perform a classical Wigner limit for the heavy-particle DoF. This leads to the *quantum-classical Liouville* description, which has recently received considerable attention [27–44]. Furthermore, one may start with a path-integral formulation and treat the heavy-particle DoF by the stationary-phase approximation, thus yielding Pechukas' theory [45]. Quite recently, also the hydrodynamic or Bohmian formulation of quantum mechanics has been used as a starting point for a mixed quantum-classical description [46–51]. It should be made clear at the outset that the mixed quantum-classical formulations differ greatly, depending on whether the problem is approached via the wave-function, density-operator, path-integral, or hydrodynamic formulation of quantum mechanics.

A different way to combine classical and quantum mechanics is the "connection approach," which was proposed independently by Landau [52], Zener [53], and Stückelberg [54] and has later been adopted and generalized by many authors [55–65]. In this formulation, nonadiabatic transitions of classical trajectories are described in terms of a connection formula of the semiclassical WKB wave functions associated with the two coupled electronic states. While the true semiclassical evaluation of these formulations has mostly been concerned with one-dimensional problems, the intuitively appealing picture of trajectories hopping between coupled potential-energy surfaces gave rise to a number of quasiclassical implementations of this idea [66–84]. In the popular *surface-hopping* scheme of Tully and co-workers [66–68], classical trajectories are propagated on a single adiabatic potential-energy surface until, according to some "hopping criterion," a transition probability $p_{1\to 2}$ to another potential-energy surface is calculated and, depending on the comparison of $p_{1\to 2}$ with a random number, the trajectory "hops" to the other adiabatic surface. The many existing variants of the method mainly differ in choice and degree of sophistication of the hopping criteria. In recent years, the term "surface hopping" and its underlying ideas have also been used in the stochastic modeling of a given deterministic differential equation—for example, the quantum-classical Liouville equation [35, 38, 40].

As a conceptionally different approach to incorporate quantum DoF in a classical formulation, one may exploit *continuous analogs* of discrete quantum systems. The simplest example of such an analogy was discovered 1927 by Dirac [85], who noted that the time-dependent Schrödinger equation of an N-level system (i.e., N equations for N complex coefficients $c_i(t) = [X_i(t) + P_i(t)]/\sqrt{2}$) is completely equivalent to Hamilton's equation for a system of N coupled harmonic oscillators with positions $X_i(t)$ and conjugated momenta $P_i(t)$ [86]. In the late 1970s, McCurdy, Meyer, and Miller used this and other quantum-classical analogies to construct classical models for the description of nonadiabatic dynamics [87–91]. The most popular version, the so-called classical electron analog model of Meyer and Miller [89], is formally similar to an Ehrenfest mean-field formulation, which, however, was then treated in a semiclassical manner in the theoretical framework of classical S-matrix theory [92]. The model has been found to describe a variety of processes at least semiquantitatively, including the evaluation of nonadiabatic collision cross sections [89], the calculation of non-Born–Oppenheimer dynamics of conical intersections [93, 94], and photodissociation of nonadiabatic coupled states both in the gas phase [95] and in solution [96].

Although the idea of a classical analog of quantum DoF is conceptionally appealing, the approach is not completely satisfying from a theoretical point of view. Starting out with an approximate classical (rather than an exact quantum-mechanical) formulation, there are two interrelated problems: The nature of the approximations involved is difficult to specify, and the formulations are not unique; that is, various analogies result in different classical models. To avoid these problems, the equivalence of discrete and continuous DoF should be established on the *quantum-mechanical* (rather than on a classical) level. This can be achieved, for example, by employing quantum-mechanical bosonization techniques such as the Holstein–Primakoff transformation [97] or Schwinger's theory of angular momentum [98]. Representing spin operators by boson operators, the discrete quantum DoF are hereby mapped onto continuous variables. Since the latter possess a well-defined classical limit, the problem of a classical treatment of discrete quantum DoF is bypassed. Exploiting this idea, we recently proposed a "mapping approach" to the semiclassical description of nonadiabatic dynamics [99, 100]. The approach consists of two steps: a quantum-mechanical exact transformation of discrete onto continuous DoF (the "mapping") and a standard classical or semiclassical treatment of the resulting dynamical problem.

On a purely classical level, it has been shown that the mapping formalism recovers the classical electron analog model of Meyer and Miller [89]. The "Langer-like modifications" [101] that were empirically introduced in this model could be identified as a zero-point energy term that accounts for quantum fluctuations in the electronic DoF [102, 103]. This in practice quite important feature was found to be the main difference between the classical mapping formulation and standard quantum-classical mean-field models. On a semiclassical level, the mapping goes beyond the original classical electron analog model, because it is exact for an N-level system and unambiguously defines the Hamiltonian as well as the boundary (or initial) conditions of the semiclassical propagator [99, 100]. Employing an initial-value representation for the semiclassical propagator [104–111], the formulation has recently been applied to a variety of systems with nonadiabatic dynamics [99–103, 112–122]. The approach also allows us to study the classical phase space of a nonadiabatically coupled system [123, 124] and to introduce classical periodic orbits of such a system [125–127]. Furthermore, the mapping formalism has been used to facilitate the treatment of nonadiabatic dynamics in various theoretical approaches, including centroid molecular dynamics [128], the coupled coherent state method [129], the coherent-state path integral [130], and the evaluation of time correlation functions [131, 132]. Finally, there exists a close connection between the semiclassical mapping approach and the spin-coherent state representation [100, 133–135]. Spin-coherent states [136] provide an alternative way to obtain a continuous representation of discrete quantum systems. In particular, the spin-coherent state path integral has been used to investigate the semiclassical description of spin systems [137–147].

In summary, we have made an attempt to classify existing quantum-classical methods in formulations resulting from (i) a partial classical limit, (ii) a connection ansatz, and (iii) a continuous analog of discrete quantum DoF. In this chapter, we focus on essentially classical formulations that may be relatively easily applied to multidimensional surface-crossing problems. On the other hand, it should be noted that there also exist a number of essentially quantum-mechanical formulations which at some point use classical ideas. A well-known example are formulations that combine quantum-mechanical time-dependent perturbation theory with a classical evaluation of the resulting correlation functions—for example, Golden Rule-type formulations [148, 149]. Furthermore, several workers have suggested time-dependent quantum propagation schemes that employ Gaussian wave packets as a time-dependent basis set [150–154]. For example, Martinez and co-workers [151–153] have developed such a method ("multiple spawning") that borrows ideas from surface-hopping and is designed to be effectively interfaced with quantum-chemical methods, thus providing an *ab initio* molecular dynamics description of nonadiabatic photochemical processes.

B. Application to Ultrafast Molecular Photoprocesses

Reflecting personal preferences, we focus in this review on the modeling of ultrafast bound-state processes following photoexcitation such as electron transfer, internal-conversion via conical intersections, and nonadiabatic photoisomerization. Keeping in mind that the validity of an approximate description depends to a large extent on the specific physical application under consideration, one may ask whether a classical or mixed quantum-classical approach appears promising for this purpose. In particular, this holds when one considers the electronic and vibrational relaxation dynamics associated with conical intersections. These systems exhibit several characteristic features, which represent a hard challenge for an approximate theoretical description: (i) The dynamics is caused by strong intramolecular interactions that cannot be accounted for in a perturbative manner. (ii) Due to the large anharmonicity of the adiabatic potential-energy surfaces, the vibrational motion is highly correlated, thus hampering the application of simple self-consistent-field schemes. Within the limits of the underlying classical approximation, these requirements are fulfilled by a mixed quantum-classical formulation that is a nonperturbative description and also fully includes the correlation between the individual DoF. In fact, a mixed quantum-classical description appears to be one of the few approximations that may be expected to work. Furthermore, it should be stressed that a classical (and therefore local) description is readily combined with an "on-the-fly" *ab initio* evaluation of the potential energy [152, 153, 155–160]. Nonadiabatic *ab initio* molecular dynamics methods are one of the few promising strategies to explore excited-state potential-energy surfaces of multidimensional system in an unbiased manner.

Let us briefly review what methods have so far been applied to ultrafast nonadiabatic photodynamics. Here the surface-hopping method has been the most popular approach [69, 76–78], in particular in combination with an on-the-fly *ab initio* evaluation of the potential energy [155–160]. Furthermore, various self-consistent-field methods have been employed, including the mean-field trajectory method, the classical electron analog model, and the quasiclassical mapping formulation [19, 93, 102, 103]. Since there is no sampling problem due to oscillating phases, all these methods are readily implemented and typically converge for a moderate number of trajectories ($\sim 10^2$–10^4). Because of rapidly oscillating phases and the representation of nonlocal operators, however, the numerical propagation of the quantum-classical Liouville equation represents a significantly more tedious problem [38, 39, 41, 42].

Due to the development of efficient initial-value representations of the semiclassical propagator, recently there has been considerable progress in the semiclassical description of multidimensional quantum processes [104–111, 161]. Considering the semiclassical description of nonadiabatic dynamics, only the mapping approach [99, 100] and the equivalent formulation that is obtained by requantizing the classical electron analog model of Meyer and Miller [112] appear to be amenable to a numerical treatment via an initial-value representation [114, 116, 117, 121, 122]. Other semiclassical formulations such as Pechukas' path-integral formulation [45] and the various connection theories [55–59, 61, 62, 64, 65] have been conceptionally illuminating, but so far of limited computational value.

C. Outline of the Chapter

The goal of this chapter is twofold. First we wish to critically compare—from both a conceptional and a practical point of view—various classical and mixed quantum-classical strategies to describe non-Born–Oppenheimer dynamics. To this end, Section II introduces five multidimensional model problems, each representing a specific challenge for a classical description. Allowing for exact quantum-mechanical reference calculations, all models have been used as benchmark problems to study approximate descriptions. In what follows, Section III describes in some detail the mean-field trajectory method and also discusses its connection to time-dependent self-consistent-field schemes. The surface-hopping method is considered in Section IV, which discusses various motivations of the ansatz as well as several variants of the implementation. Section V gives a brief account on the quantum-classical Liouville description and considers the possibility of an exact stochastic realization of its equation of motion.

Second, the mapping approach to nonadiabatic quantum dynamics is reviewed in Sections VI–VII. Based on an exact quantum-mechanical formulation, this approach allows us in several aspects to go beyond the scope of standard mixed quantum-classical methods. In particular, we study the classical phase space of a nonadiabatic system (including the discussion of vibronic periodic orbits) and the semiclassical description of nonadiabatic quantum mechanics via initial-value representations of the semiclassical propagator. The semiclassical spin-coherent state method and its close relation to the mapping approach is discussed in Section IX. Section X summarizes our results and concludes with some general remarks.

II. MOLECULAR SYSTEMS

A. Model Hamiltonian

In this section, we introduce the model Hamiltonian pertaining to the molecular systems under consideration. As is well known, a curve-crossing problem can be formulated in the adiabatic as well as in a diabatic electronic representation. Depending on the system under consideration and on the specific method used, both representations have been employed in mixed quantum-classical approaches. While the diabatic representation is advantageous to model potential-energy surfaces in the vicinity of an intersection and has been used in mean-field type approaches, other mixed quantum-classical approaches such as the surface-hopping method usually employ the adiabatic representation.

In what follows we introduce the model Hamiltonian using both diabatic and adiabatic representations. Adopting diabatic electronic basis states $|\psi_k\rangle$, the molecular model Hamiltonian can be written as [162, 163]

$$H = \sum_{k,k'} |\psi_k\rangle h_{kk'} \langle\psi_{k'}| = T + \sum_{k,k'} |\psi_k\rangle V_{kk'} \langle\psi_{k'}| \tag{1}$$

Here the $V_{kk'}$ represent the electronic matrix elements of the diabatic potential matrix and T denotes the kinetic-energy operator of the nuclei.

We wish to consider molecular models describing (i) photophysical processes in which the molecular system undergoes electronic relaxation (e.g., through internal conversion) and (ii) photochemical processes in which the molecular system additionally changes its chemical identity (e.g., through isomerization). To account for the first kind of photoprocess, we introduce the normal modes of the electronic ground state x_j and approximate the diabatic potential matrix elements $V_{kk'}$ by a Taylor expansion with respect to the electronic ground-state equilibrium geometry [162, 163]. In lowest order we thus obtain (if not noted otherwise, we use atomic units where $\hbar = 1$)

$$T = \tfrac{1}{2}\textstyle\sum_j \omega_j p_j^2 \tag{2a}$$

$$V_{kk} = E_k + \tfrac{1}{2}\sum_j \omega_j x_j^2 + \sum_j \kappa_j^{(k)} x_j, \tag{2b}$$

where ω_j is the vibrational frequency and x_j and p_j are the dimensionless position and momentum of the jth vibrational mode. E_k denotes the vertical transition energy of the diabatic state $|\psi_k\rangle$, and $\kappa_j^{(k)}$ represents the gradient of the corresponding excited-state diabatic potential-energy surface (PES) at the ground-state equilibrium geometry.

To account for photochemical processes, we adopt a simple model that was proposed by Seidner and Domcke for the description of *cis–trans* isomerization processes [164]. In addition to the normal-mode expansion above, they introduced a Hamiltonian exhibiting torsional motion. The diabatic matrix elements of the Hamiltonian are given as

$$h_{kk}^{\mathrm{R}} = -\frac{1}{2I}\frac{\partial^2}{\partial\varphi^2} + \tfrac{1}{2}V_k^{\mathrm{R}}[1 - \cos(n\varphi)] \tag{3}$$

Here φ represents the torsional coordinate, I denotes the moment of inertia, and V_k^{R} is the first coefficient of the Fourier series expansion of the isomerization potential of periodicity $n\pi$.

To specify the off-diagonal elements $V_{kk'}$ of the diabatic potential matrix, we consider two cases of the diabatic coupling $V_{kk'}$ which are of particular interest in molecular physics:

1. The coupling is constant,

$$V_{kk'} = g_{kk'} \tag{4}$$

   leading to an *avoided crossing* of the adiabatic PES. Such models are often used to describe electron-transfer processes, where the different electronic states describe the donor, bridge, and acceptor states of an electron-transfer reaction [165–167], respectively. Considering specifically two electronic states, the resulting model is often referred to as a "spin–boson problem" [5, 168].
2. The coupling is linear in one or several coordinates x_c (the so-called coupling modes),

$$V_{kk'} = \lambda_{kk'} x_c \tag{5}$$

   leading to a *conical intersection* of the adiabatic PES. This case represents the standard situation of internal conversion through a photochemical funnel [169, 170].

To specify the model Hamiltonian in the adiabatic representation, we introduce adiabatic electronic states

$$|\psi_k^{\mathrm{ad}}\rangle = \sum_{k'} S_{kk'} |\psi_{k'}\rangle \tag{6}$$

where the $S_{kk'}$ are matrix elements of the unitary transformation S between the diabatic and adiabatic wave functions. The Hamiltonian in the adiabatic representation can then be written as [162, 163]

$$H = \tfrac{1}{2}\sum_j \omega_j p_j^2 + \sum_{k,k'} |\psi_k^{\mathrm{ad}}\rangle (W_k \delta_{k,k'} + \Lambda_{kk'}) \langle \psi_{k'}^{\mathrm{ad}}| \tag{7}$$

where W_k denotes the adiabatic Born–Oppenheimer PESs that are obtained by diagonalizing the diabatic potential matrix defined above. The non-Born–Oppenheimer operator

$$\Lambda_{kk'} = -\sum_j \omega_j \left(i T_{kk'}^{(j)} p_j + \tfrac{1}{2} G_{kk'}^{(j)} \right) \tag{8}$$

is given in terms of the nonadiabatic coupling matrices of first and second order:

$$T_{kk'}^{(j)} = \left\langle \psi_k^{\text{ad}} \middle| \frac{\partial}{\partial x_j} \middle| \psi_{k'}^{\text{ad}} \right\rangle \tag{9a}$$

$$G_{kk'}^{(j)} = \left\langle \psi_k^{\text{ad}} \middle| \frac{\partial^2}{\partial x_j^2} \middle| \psi_{k'}^{\text{ad}} \right\rangle \tag{9b}$$

For further information on general concepts of non-Born–Oppenheimer dynamics, we refer to recent reviews [162, 163] and textbooks [169, 170].

B. Observables of Interest

The various aspects of photoinduced nonadiabatic dynamics are reflected by different time-dependent observables. Following a brief introduction of the observables of interest, we discuss how these quantities are evaluated in a mixed quantum-classical simulation.

In general, the time evolution of the molecular system is governed by the Liouville–von Neumann equation

$$i\dot{\rho}(t) = [H, \rho(t)] \tag{10}$$

where H represents the molecular Hamiltonian defined in Eqs. (1) or (7) and $\rho(t)$ is the density operator. The time-dependent expectation value of the observable A is then given by

$$A(t) = \text{tr}\{\rho(t)A\} = \text{tr}\{\rho(0)A(t)\} \tag{11}$$

where $\text{tr}\{\ldots\}$ represents the quantum-mechanical trace over all degrees of freedom of the system and $A(t)$ denotes the operator A in the Heisenberg representation. Assuming that at time $t = 0$ the molecular system is photoexcited from the electronic ground state (i.e., $|\psi_0\rangle$) to an upper state (e.g., $|\psi_2\rangle$) by an ultrashort laser pulse, the initial state is given as

$$\rho(0) = |\psi_2\rangle\rho_{\text{vib}}\langle\psi_2| \tag{12}$$

Here, the initial state of the vibrational DoF, ρ_{vib}, is described by

$$\rho_{\text{vib}} = Z^{-1}e^{-\beta h_0}, \qquad Z = \text{tr}(e^{-\beta h_0}) \tag{13}$$

where β denotes the inverse temperature, $\beta = 1/k_B T$, and h_0 is given by

$$h_0 = \tfrac{1}{2}\sum_j \omega_j(p_j^2 + x_j^2) \tag{14}$$

Most of the models considered in this chapter involve vibrational modes with frequencies that are large compared to typical thermal energies. In such situations, thermal effects can be neglected and the initial state of the vibrational DoF is given by the vibrational ground state $|\mathbf{0}\rangle$ in the electronic ground state $|\psi_0\rangle$ [163], that is,

$$\rho(0) = |\psi_2\rangle|\mathbf{0}\rangle\langle\mathbf{0}|\langle\psi_2| \tag{15}$$

In a mixed quantum-classical calculation the trace operation in the Heisenberg representation is replaced by a quantum-mechanical trace (tr_q) over the quantum degrees of freedom and a classical trace (i.e., a phase-space integral over the initial positions $\mathbf{x}_0$ and momenta $\mathbf{p}_0$) over the classical degrees of freedom. This yields

$$A(t) = \int d\mathbf{x}_0 \int d\mathbf{p}_0 \, \text{tr}_\text{q}\{\rho(\mathbf{x}_0, \mathbf{p}_0) A[\mathbf{x}(t), \mathbf{p}(t)]\}, \tag{16}$$

where $\rho(\mathbf{x}_0, \mathbf{p}_0)$ denotes a phase-space representation of the vibrational initial state [23–26]. For example, to represent the ground state $|0\rangle\langle 0|$ of a one-dimensional harmonic oscillator $h_0 = \frac{1}{2}\omega(p^2 + x^2)$, one may employ the Wigner distribution [25, 171]

$$\rho(x_0, p_0) = \frac{1}{\pi} e^{-p_0^2 - x_0^2} \tag{17}$$

Alternatively, one may change to classical action-angle variables n, q (using the transformation $x = \sqrt{2n+1}\sin q$, $p = \sqrt{2n+1}\cos q$, $q \in [0, 2\pi]$ [23–25]) and consider initial conditions with fixed action $n_0 = \frac{1}{2}$:

$$\rho(n_0) = \delta\left(n_0 - \tfrac{1}{2}\right) \tag{18}$$

reflecting the zero-point energy of the harmonic oscillator.

As a first example, let us consider the time-dependent mean position of a normal mode x_j of the system. In a mixed quantum-classical calculation, this observable is directly given by the quasiclassical average over the nuclear trajectories $x_j^{(r)}(t)$, that is,

$$\langle x_j(t)\rangle_C = \frac{1}{N_\text{traj}} \sum_{r=1}^{N_\text{traj}} x_j^{(r)}(t) \tag{19}$$

where the phase-space integral over the classical initial conditions in Eq. (16) has been approximated by a sum over N_traj trajectories with equal weights. Similarly, the corresponding mean momenta and other vibrational quantities such as normal-mode energies are readily evaluated.

The time evolution of a photoisomerization process can be visualized by considering the time-dependent probability of finding the system in either the *cis* or the *trans* conformation [163, 172]. Assuming a 2π-periodic torsional potential as used in Model III below, we assign a torsional angle φ between $-\pi/2$ and $\pi/2$ to the *cis* conformation and assign a torsional angle between $\pi/2$ and $3\pi/2$ to the *trans* conformation, resulting in the projectors

$$\begin{aligned} P_{trans} &= \Theta(|\varphi| - \pi/2) \\ P_{cis} &= \mathbf{1} - P_{trans} \end{aligned} \tag{20}$$

where Θ denotes the Heaviside step function and φ is restricted to $-\pi/2 \leq \varphi \leq 3\pi/2$.

To describe the electronic relaxation dynamics of a photoexcited molecular system, it is instructive to consider the time-dependent population of an electronic state, which can be defined in a diabatic or the adiabatic representation [163]. The population probability of the diabatic electronic state $|\psi_k\rangle$ is defined as the expectation value of the diabatic projector

$$\begin{aligned} P_k^{\text{dia}}(t) &= \text{tr}\{\hat{P}_k^{\text{dia}} \rho(t)\} \\ \hat{P}_k^{\text{dia}} &= |\psi_k\rangle\langle\psi_k| \end{aligned} \tag{21}$$

In a mixed quantum-classical simulation such as a mean-field-trajectory or a surface-hopping calculation, the population probability of the diabatic state $|\psi_k\rangle$ is given as the quasiclassical average over the squared modulus of the diabatic electronic coefficients $d_k(t)$ defined in Eq. (27). This yields

$$\begin{aligned} P_k^{\text{dia}}(t) &= \langle d_k^*(t) d_k(t) \rangle_C \\ &= \frac{1}{N_{\text{traj}}} \sum_{r=1}^{N_{\text{traj}}} |d_k^{(r)}(t)|^2 \end{aligned} \tag{22}$$

In complete analogy, the adiabatic population probability is defined as the expectation value of the adiabatic projector $\hat{P}_k^{\text{ad}} = |\psi_k^{\text{ad}}\rangle\langle\psi_k^{\text{ad}}|$ and is quasi-classically given as average over the squared modulus of the adiabatic electronic coefficients $a_k(t)$ defined in Eq. (29).

A measure for the electronic coherence of the wave function are the nondiagonal elements of the electronic density matrix, which, for example, in the diabatic representation are given by ($k \neq k'$)

$$C_{kk'}^{\text{dia}} = \text{tr}\{\rho(t)(|\psi_k\rangle\langle\psi_{k'}| + |\psi_{k'}\rangle\langle\psi_k|)\} \tag{23}$$

Finally, in the semiclassical formulation introduced in Section VII, it is of interest to calculate the autocorrelation function of the initially prepared state

$$J(t) = \langle \mathbf{0} | \langle \psi_2 | e^{-iHt} | \psi_2 \rangle | \mathbf{0} \rangle \tag{24}$$

the Fourier transform of which yields the electronic absorption spectrum.

C. Model Systems

In order to discuss various aspects of a mixed quantum-classical treatment of photoinduced nonadiabatic dynamics, we consider five different kinds of molecular models, each representing a specific challenge for a mixed quantum-classical modeling. Here, we introduce the specifics of these models and discuss the characteristics of their nonadiabatic dynamics. The molecular parameters of the few-mode models (Model I–IV) describing intramolecular nonadiabatic dynamics are collected in Tables I–V. The parameters of Model V describing various aspects of nonadiabatic dynamics in the condensed phase will be given in the text.

TABLE I

Parameters of Model I, Which Represents a Three-Mode Model of the $S_1 - S_2$ Conical Intersection in Pyrazine [173][a]

	E_k	ω_1	$\kappa_1^{(k)}$	ω_{6a}	$\kappa_{6a}^{(k)}$	ω_{10a}	λ
$\lvert\psi_1\rangle$	3.94	0.126	0.037	0.074	−0.105	0.118	0.262
$\lvert\psi_2\rangle$	4.84	0.126	−0.254	0.074	0.149	0.118	

[a]All quantities are given in electron-volts.

TABLE II

Parameters of Model II, Which Represents a Three-State Five-Mode Model of the Ultrafast $\widetilde{C} \rightarrow \widetilde{B} \rightarrow \widetilde{X}$ Internal-Conversion Process in the Benzene Cation [179, 180][a]

	E_k	ω_2	$\kappa_2^{(k)}$	ω_{16}	$\kappa_{16}^{(k)}$	ω_{18}	$\kappa_{18}^{(k)}$	ω_8	$\lambda_8^{(12)}$	ω_{19}	$\lambda_{19}^{(23)}$
$\lvert\psi_1\rangle$	9.65	0.123	−0.042	0.198	0.246	0.075	0.125	0.088	0.164	0.12	
$\lvert\psi_2\rangle$	11.84	0.123	−0.042	0.198	0.246	0.075	0.1	0.088		0.12	0.154
$\lvert\psi_3\rangle$	12.44	0.123	−0.301	0.198	0	0.075	0	0.088		0.12	

[a]All quantities are given in electron-volts.

TABLE III
Parameters of Model III, Which Represents a Three-Mode Model Exhibiting Nonadiabatic Photoisomerization [164][a]

	E_k	I^{-1}	V_k	ω_t	$\kappa^{(k)}$	ω_c	λ
$\|\psi_1\rangle$	0.0	1.11×10^{-3}	4.5	0.2	0.0	0.17	0.34
$\|\psi_2\rangle$	5.0	1.11×10^{-3}	−4.5	0.2	0.3	0.17	

[a]The torsional potentials are characterized by the reciprocal moment of inertia I^{-1} and the potential parameters E_k and V_k. All quantities are given in electron-volts.

TABLE IV
Parameters of Model IVa, Which Represents a One-Mode Model of Intramolecular Electron Transfer [125][a]

	E_k	ω_1	$\kappa_1^{(k)}$	g_{12}
$\|\psi_1\rangle$	0	0.05	−0.05	0.1
$\|\psi_2\rangle$	0	0.05	0.05	

[a]All quantities are given in electron-volts.

1. *Model I: $S_2 \to S_1$ Internal Conversion in Pyrazine*

Model I represents a three-mode model of the $S_1(n\pi^*)$ and $S_2(\pi\pi^*)$ electronic states of pyrazine [173], which has been adopted by several authors as a standard example of ultrafast electronic relaxation [19, 174–176]. Taking into account a single coupling mode (ν_{10_a}) and two totally symmetric modes (ν_1, ν_{6a}), Domcke and co-workers [174] have identified a low-lying conical intersection of the two lowest excited singlet states of pyrazine, which has been shown to trigger internal conversion and a dephasing of the vibrational motion on a femtosecond timescale. A variety of theoretical investigations, including quantum wave-packet studies [174, 175], absorption and resonance Raman spectra [177], as well as time-resolved pump-probe spectra [163, 178], have been reported for this system. Exhibiting complex electronic and vibrational relaxation dynamics, the model provides a stringent test for an approximate description.

Let us briefly discuss the characteristics of the nonadiabatic dynamics exhibited by this model. Assuming an initial preparation of the S_2 state by an ideally short laser pulse, Fig. 1 displays in thick lines the first 500 fs of the quantum-mechanical time evolution of the system. The population probability of the diabatic S_2 state shown in Fig. 1b exhibits an initial decay on a timescale of ~20 fs, followed by quasi-periodic recurrences of the population, which are

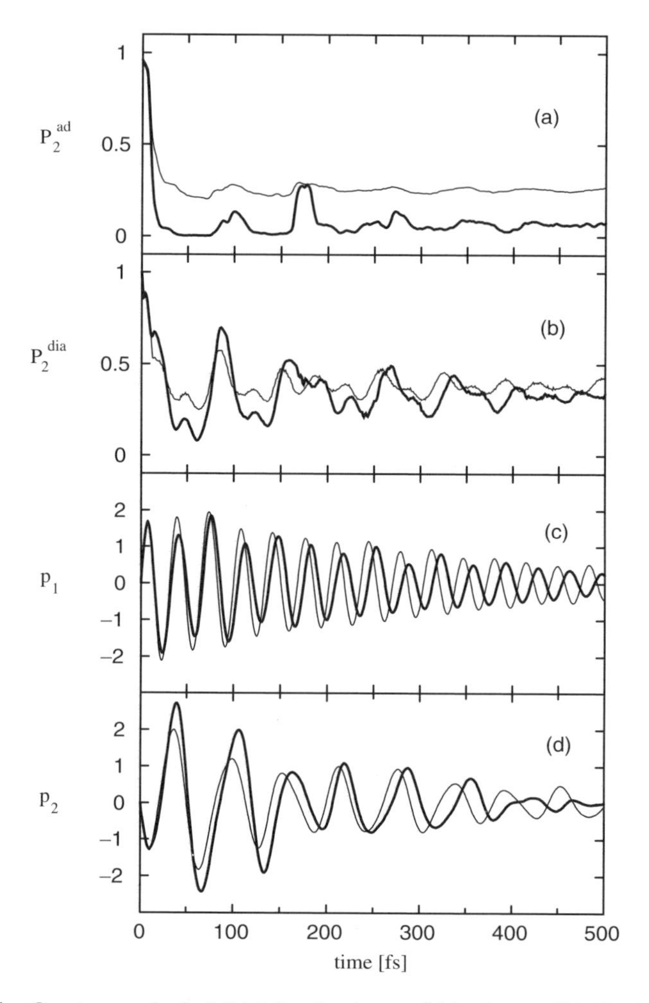

Figure 1. Quantum-mechanical (thick lines) and mean-field-trajectory (thin lines) calculations obtained for Model I describing the $S_2 \to S_1$ internal-conversion process in pyrazine. Shown are the time-dependent population probabilities $P_2^{\mathrm{ad}}(t)$ and $P_2^{\mathrm{dia}}(t)$ of the initially prepared adiabatic and diabatic electronic state, respectively, as well as the mean momenta $p_1(t)$ and $p_2(t)$ of the two totally symmetric modes ν_1 and ν_{6a} of the model.

damped on a timescale of a few hundred femtoseconds. Beyond ~500 fs (data not shown) the S_2 population probability becomes quasi-stationary, fluctuating statistically around its asymptotic value of ~ 0.3. The time-dependent population of the adiabatic S_2 state, displayed in Fig. 1a, is seen to decay even faster than the diabatic population—essentially within a single vibrational period—and to attain an asymptotic value of ~ 0.05. The finite asymptotic value of $P_2^{\rm ad}$ is a consequence of the restricted phase space of the three-mode model. The population $P_2^{\rm ad}$ is expected to decay to zero for systems with many degrees of freedom (see Section VI.D).

The vibrational dynamics of the model is illustrated in Figs. 1c and 1d, showing the expectation values of the momenta of the tuning modes ν_1 and ν_{6a}, respectively. It is seen that the oscillations of the two modes are damped on a timescale of a few hundred femtoseconds. As has been discussed in detail in Ref. 174, this damping reflects the strong nonadiabatic coupling of the system, which causes a large anharmonicity of the lower adiabatic potential-energy surface. It should be pointed out that the decay of the amplitude of $\langle p \rangle_t$ is not related to the dissipation of vibrational energy but rather reflects a vibrational pure dephasing process.

2. *Model II:* $\widetilde{C} \rightarrow \widetilde{B} \rightarrow \widetilde{X}$ *Internal-Conversion Process in the Benzene Cation*

The second system pertains to an *ab initio*-based model of the ultrafast $\widetilde{C} \rightarrow \widetilde{B} \rightarrow \widetilde{X}$ internal-conversion process in the benzene cation (Bz^+) which has been proposed by Köppel, Domcke, and Cederbaum [179, 180]. The diabatic potential matrix elements of the Hamiltonian have been approximated by a quadratic Taylor expansion in normal coordinates of the electronic ground state of neutral benzene. The resulting model system has been shown to give rise to multidimensional conical intersections of the corresponding adiabatic potential-energy surfaces. Model II is obtained by taking into account the five most important vibrational modes ($\nu_2, \nu_8, \nu_{16}, \nu_{18}$, and ν_{19} in Herzberg numbering) and ignoring the degeneracy of the $\widetilde{X}$ and $\widetilde{B}$ electronic states as well as of the modes ν_{16}, ν_{18}, and ν_{19}.

Köppel [180] has performed exact time-dependent quantum wave-packet propagations for this model, the results of which are depicted in Fig. 2A. He showed that the initially excited $\widetilde{C}$ state decays irreversibly into the $\widetilde{X}$ state within $\sim$250 fs. The decay is nonexponential and exhibits a pronounced beating of the $\widetilde{C}$ and $\widetilde{B}$ state populations. This model will allow us to test mixed quantum-classical approaches for multistate systems with several conical intersections.

3. *Model III: Nonadiabatic Photoisomerization*

As a last example of a molecular system exhibiting nonadiabatic dynamics caused by a conical intersection, we consider a model that recently has been proposed by Seidner and Domcke to describe ultrafast *cis–trans* isomerization processes in unsaturated hydrocarbons [172]. Photochemical reactions of this type are known to involve large-amplitude motion on coupled potential-energy surfaces [169], thus representing another stringent test for a mixed quantum-classical description that is complementary to Models I and II. A number of theoretical investigations, including quantum wave-packet studies [163, 164, 172], time-resolved pump-probe spectra [164, 181], and various mixed

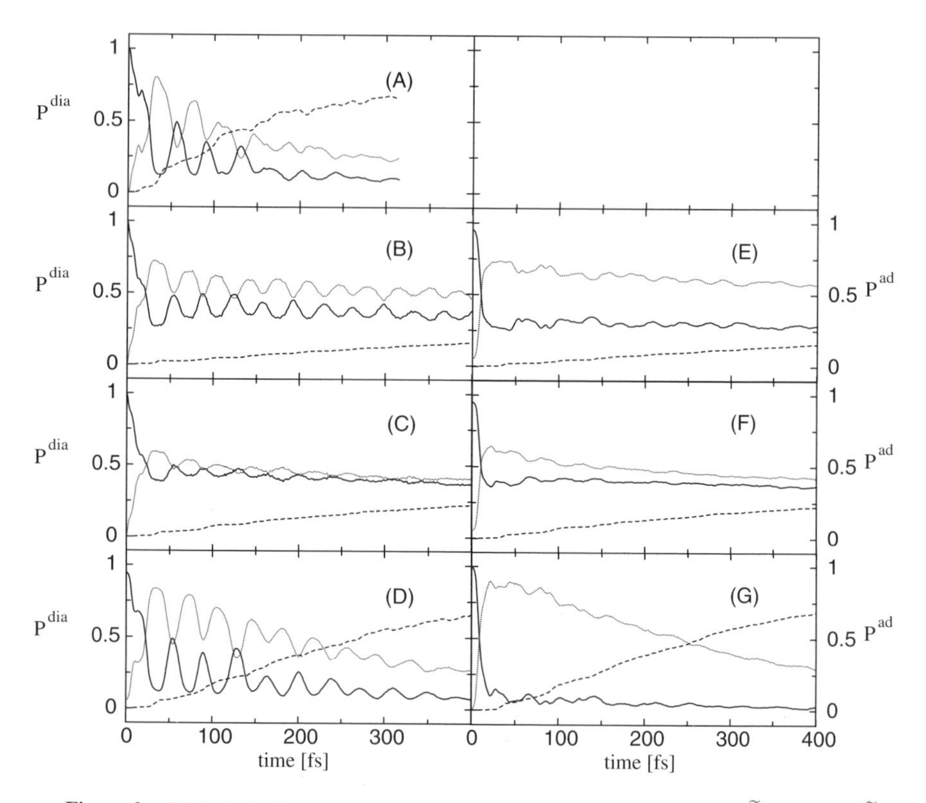

Figure 2. Diabatic (left) and adiabatic (right) population probabilities of the $\widetilde{C}$ (full line), $\widetilde{B}$ (dotted line), and $\widetilde{X}$ (dashed line) electronic states as obtained for Model II, which represents a three-state five-mode model of the benzene cation. Shown are (A) exact quantum calculations of Ref. 180, as well as mean-field-trajectory results [(B), (E)] and surface-hopping results [(C),(D),(F),(G)]. The latter are obtained either directly from the electronic coefficients [(C),(F)] or from binned coefficients [(D),(G)].

quantum-classical simulations [78, 127, 182, 183], have been reported for this system. Furthermore, the ansatz has been employed to model the isomerization of retinal in rhodopsin [184, 185] and to study the femtosecond dynamics and spectroscopy of a molecular photoswitch [186–188]. The model considered here includes three nuclear degrees of freedom consisting of a large-amplitude torsional motion, a vibronically active mode that couples the electronic ground state and the excited state, and a totally symmetric mode that modulates the energy gap of the interacting states.

Figure 3 shows the quantum results (thick full lines) for time-dependent population probabilities $P_2^{\rm ad}(t), P_2^{\rm dia}(t)$ of the initially prepared (a) adiabatic and (b) diabatic electronic state, respectively, as well as (c) the probability $P_{cis}(t)$

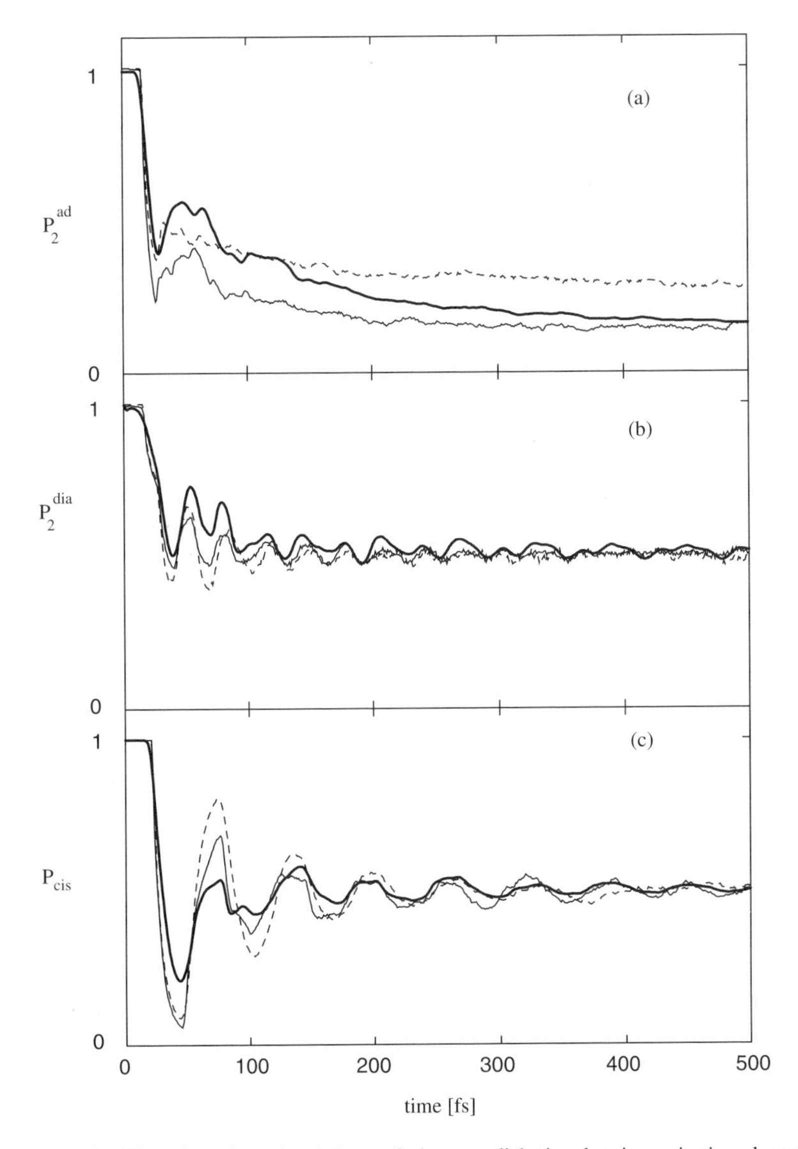

Figure 3. Time-dependent simulations of the nonadiabatic photoisomerization dynamics exhibited by Model III, comparing results of the mean-field-trajectory method (dashed lines), the surface-hopping approach (thin lines), and exact quantum calculations (full lines). Shown are the population probabilities of the initially prepared (a) adiabatic and (b) diabatic electronic state, respectively, as well as (c) the probability $P_{cis}(t)$ that the sytem remains in the initially prepared *cis* conformation.

that the system remains in the initially prepared *cis* conformation. The results illustrate the interplay of isomerization dynamics and internal-conversion dynamics of Model III: The ultrafast photoisomerization process as monitored by the probability $P_{cis}(t)$ is seen to directly result in a highly effective internal-conversion process as monitored by the adiabatic population $P_2^{\mathrm{ad}}(t)$. It is noted that the isomerization process and the decay of the diabatic population is virtually completed after 100 fs. The long-time values of $\approx \frac{1}{2}$ for $P_{cis}(\infty)$ and $P_2^{\mathrm{dia}}(\infty)$ are a consequence of the symmetric torsional potential chosen in Model III (cf. Table III). The adiabatic population probability, on the other hand, decays to its long-time value of 0.25 within $\sim$400 fs, thus reflecting the redistribution of vibrational energy occurring on this timescale [172].

4. *Model IV: Ultrafast Intramolecular Electron Transfer*

While in the first three models the nonadiabatic dynamics is induced by conical intersections of the respective PESs, the internal conversion processes described by Models IV and V are caused by avoided crossings of the PESs. Such models are widely used to describe electron-transfer reactions [165–167]. Specifically, Model IV represents the class of ultrafast intramolecular electron-transfer processes. As representative examples for such reactions, we consider models that include two electronic states (describing the donor and acceptor state of an electron-transfer reaction) and up to three vibrational modes. As a first example, we consider a simple two-state one-mode model (in the following referred to as Model IVa). The parameters of this model (cf. Table IV) correspond to a symmetric (e.g., self-exchange) electron-transfer reaction with a relatively small reorganization energy, given by $\lambda = 2(\kappa^{(1)})^2/\omega = 0.05$ eV. The diabatic and adiabatic potential energy curves of the model are depicted in Fig. 4. At $x = 0$, the diabatic potentials are seen to intersect, while the adiabatic potentials exhibit an avoided curve crossing. Due to the relatively large electronic coupling g_{12}, the adiabatic potentials are, furthermore, characterized by a single minimum.

Figure 5 shows quantum-mechanical results for the adiabatic and the diabatic population, as well as the diabatic electronic coherence, assuming that at time $t = 0$ the systems is prepared in the state $|\phi_0\rangle|\psi_2\rangle$, where $|\phi_0\rangle$ describes a nuclear Gaussian wave packet centered at $x = 3$. With this initial preparation, Model IVa may thus represent the situation of a photoinduced electron transfer promoted by a high-frequency vibrational mode. The initially prepared vibrational wave packet starts to slide down the upper adiabatic PES. Reaching the crossing region at $\sim$20 fs, the wave packet bifurcates into two components evolving on the upper and lower electronic surface, respectively. Performing quasi-periodic oscillations on coupled PESs, in the course of time the molecule undergoes nonadiabatic transitions whenever the wave packet approaches the curve crossing. The adiabatic population probability $P^{\mathrm{ad}}(t)$ clearly reflects this behavior: Whenever the wave packet reaches the curve crossing, the adiabatic

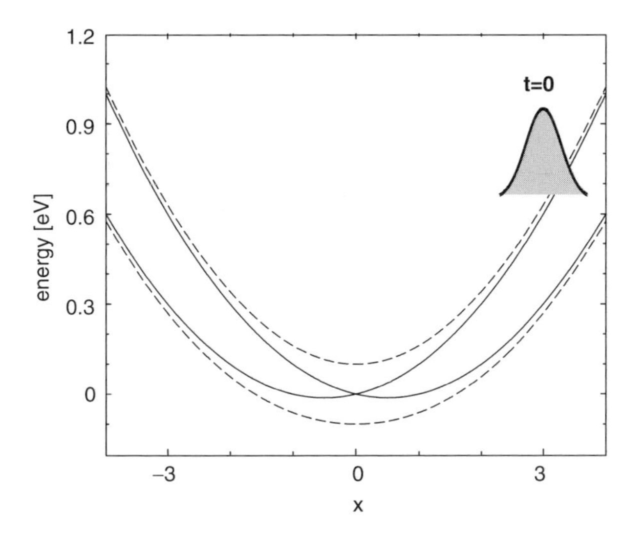

Figure 4. Diabatic (solid lines) and adiabatic (dashed lines) potential-energy curves of Model IVa. The Gaussian wave packet indicates the initial preparation of the system at time $t = 0$.

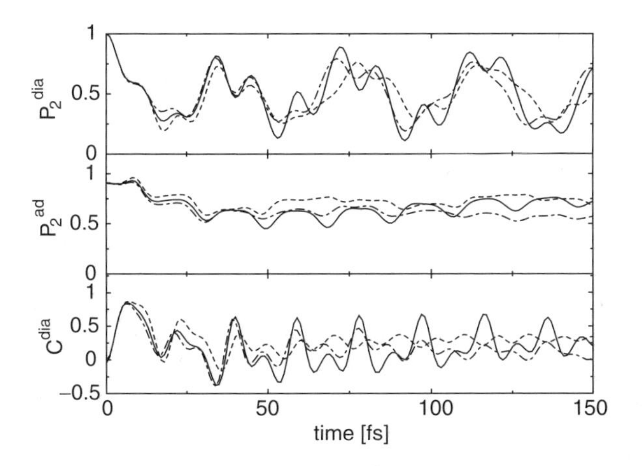

Figure 5. Quantum-mechanical (full lines), mean-field-trajectory (dashed lines), and mapping (dashed-dotted lines) calculations as obtained for Model IVa. Shown are the time-dependent population probabilities $P_2^{\text{dia}}(t)$ and $P_2^{\text{ad}}(t)$ of the initially prepared adiabatic and dibatic electronic state, respectively, as well as the diabatic electronic coherence C^{dia}.

TABLE V
Parameters of Model IVb, Which Represents a Three-Mode Model of Intramolecular Electron Transfer [189][a]

	E_k	ω_1	$\kappa_1^{(k)}$	ω_2	$\kappa_2^{(k)}$	ω_3	$\kappa_3^{(k)}$	g_{12}
$\vert\psi_1\rangle$	0	0.07	0	0.18	0	0.26	0	0.06
$\vert\psi_2\rangle$	0.91	0.07	0.09	0.18	0.22	0.26	0.34	

[a]All quantities are given in electron-volts.

population is seen to change rapidly, while it is approximately constant when the wave packet is away from the crossing region. In addition to this oscillation, the diabatic population $P^{\text{dia}}(t)$ and the diabatic electronic coherence $C^{\text{dia}}(t)$ exhibit a beating with a shorter period [125].

Due to the limited number of modes, the electronic population of the simple model studied above does not exhibit relaxation. To obtain electronic relaxation (and hence irreversible electron transfer), the coupling to more intramolecular modes and/or the interaction with an (solvent) environment needs to be taken into account. Wolfseder and Domcke proposed a three-mode model for the description of ultrafast intramolecular electron transfer [189] which exhibits electronic relaxation. It is noted that the parameters of this model (in the following referred to as Model IVb), collected in Table V, are quite similar to the parameters of Model I; however, the diabatic electronic coupling has been chosen constant, whereas it depends on a nontotally symmetric coupling mode in the case of Model I. Figure 6 presents in thick lines quantum-mechanical results for the first 200 fs of the time evolution of Model IVb, showing the time-dependent adiabatic (a) and diabatic (b) population probabilities as well as the mean momentum of a representative vibrational mode (c). While the adiabatic population probability $P_2^{\text{ad}}(t)$ is seen to exhibit a pronounced quasi-periodic beating, the diabatic population probability $P_2^{\text{dia}}(t)$ decays within 100 fs and fluctuates for larger times around its long-time value of ~ 0.25. It is interesting to note that the electronic relaxation dynamics of this model is thus complementary to the dynamics shown by Model I, which exhibits a fast decaying adiabatic population and an oscillatory diabatic population. Contrary to the behavior of Model I, the vibrational motion of the mode ν_2 of Model IVb is hardly damped on the timescale of the electronic relaxation process.

5. *Model V: Photoinduced Electron Transfer in Solution*

In the examples studied so far, the photoinduced short-time dynamics of a molecular system has been governed by a few high-frequency intramolecular vibrational modes that strongly couple to the electronic transition, a situation that

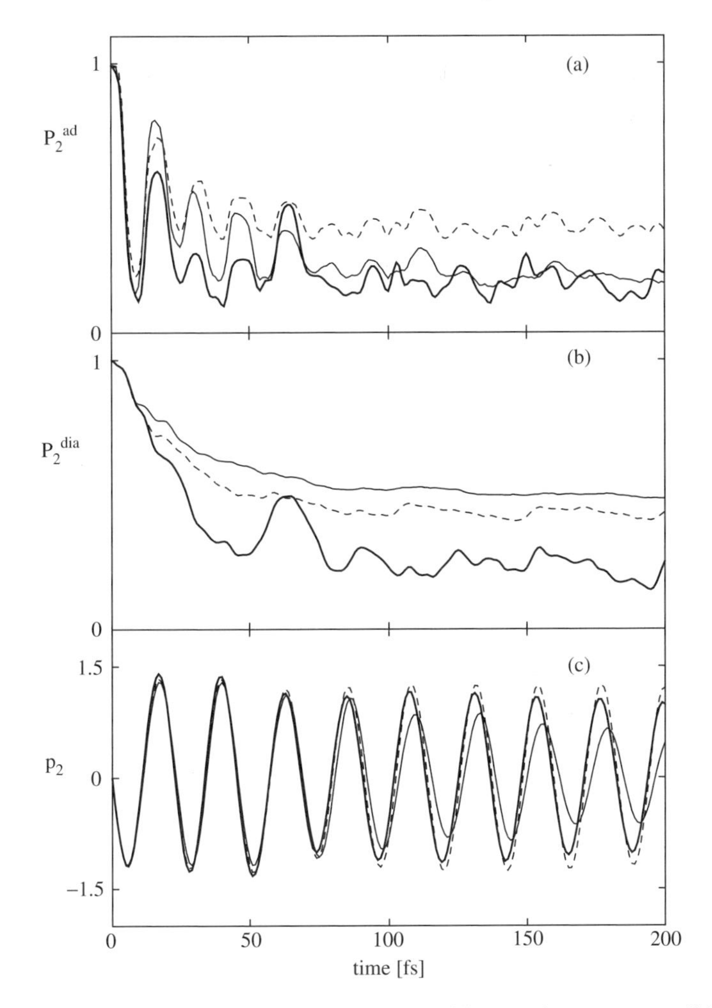

Figure 6. Comparison of SH (thin lines), MFT (dashed lines), and exact quantum (thick lines) calculations obtained for Model IVb describing ultrafast intramolecular electron transfer. Shown are the time-dependent population probabilities $P_2^{\mathrm{ad}}(t)$ and $P_2^{\mathrm{dia}}(t)$ of the initially prepared adiabatic (*a*) and diabatic (*b*) electronic state, respectively, as well as the mean momentum of a representative vibrational mode (*c*).

presumably is generic for the optical response of a polyatomic molecule on a femtosecond timescale [163]. In the last type of model, we wish to consider the case where the dynamics of a molecular system is mainly determined by the interaction with many weakly coupled low-frequency vibrational modes of the environment. Typical examples for this case are photoinduced electron-transfer reactions in the condensed phase such as, for example, in a mixed-valence system embedded in polar solvents [166, 190–192] or in photosynthetic reaction centers of bacteria and plants [193–195].

The strategy, usually adopted to achieve a theoretical description of this complex dynamics, is to describe the influence of the solvent environment on the electron-transfer reaction within linear response theory [5, 26, 196, 197] as linear coupling to a bath of harmonic oscillators. Within this model, all properties of the bath enter through a single function called the spectral density [5, 168]

$$J(\omega) = \frac{\pi}{2} \sum_j \kappa_j^2 \delta(\omega - \omega_j) \tag{25}$$

where ω_j and $\kappa_j = \kappa_j^{(2)} = -\kappa_j^{(1)}$ represent the frequency and the linear coupling of the jth vibrational mode, respectively. Here, we consider specifically an Ohmic bath described by the spectral density

$$J(\omega) = \frac{\pi}{2} \alpha \omega e^{-\omega/\omega_c} \tag{26}$$

where ω_c denotes the characteristic frequency of the solvent environment. The strength of the electronic–vibrational coupling is specified by the dimensionless Kondo parameter α, which is related to the reorganization energy λ of the electron-transfer reaction via $\lambda = 2\alpha\omega_c$.

To investigate the applicability of mixed quantum-classical methods to electron-transfer reactions in solution, we consider two examples that differ in their respective model parameters. The first example (referred to as Model Va) is described by the parameters (given in terms of the electronic-coupling matrix element g_{12}) $k_B T = 2.5 g_{12}$, $\omega_c = 2.5 g_{12}$, and $E_1 = E_2 = 0$, corresponding to a symmetric electron-transfer reaction in the nonadiabatic regime with a relatively high temperature. Figure 7 shows results of quantum Monte-Carlo path-integral calculations by Mak and Chandler [198] for this model for a electronic–vibrational coupling strength of $\alpha = 0.25$. It is seen that for this strength of the coupling the diabatic population exhibits an almost incoherent decay to its equilibrium value of 0.5.

The second, more challenging example (Model Vb) is a model for a nonsymmetric electron-transfer reaction with energy bias $E_1 - E_2 = 2g_{12}$, in the limit of weak electronic–vibrational coupling ($\alpha = 0.1$), low temperature ($k_B T = 0.2 g_{12}$), and a rather high characteristic frequency of the bath ($\omega_c = 7.5 g_{12}$) corresponding again to nonadiabatic electron transfer. The quantum-mechanical results depicted in Fig. 8 have been obtained by Makarov and Makri [199] employing numerical path-integral methods. In contrast to the example

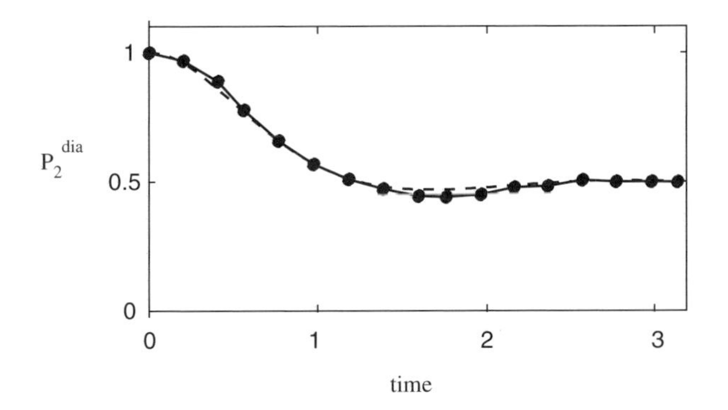

Figure 7. Comparison of SH (thin solid line), MFT (dashed line), and quantum path-integral (solid line with dots) calculations (Ref. 198) obtained for Model Va describing electron transfer in solution. Shown is the time-dependent population probability $P_2^{\rm dia}(t)$ of the initially prepared diabatic electronic state.

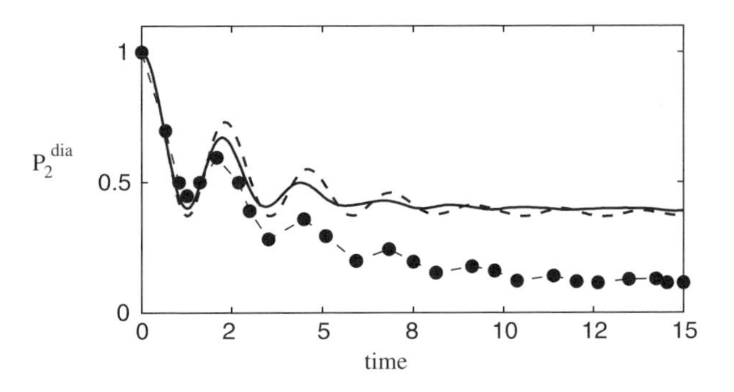

Figure 8. Comparison of SH (solid line), MFT (dashed line), and quantum path-integral (dots) calculations (Ref. 198) obtained for Model Vb describing electron transfer in solution. Shown is the time-dependent population probability $P_2^{\rm dia}(t)$ of the initially prepared diabatic electronic state.

considered above, the population of the initially prepared diabatic electronic state exhibits pronounced coherent oscillation before it relaxes to its equilibrium value.

III. MEAN-FIELD TRAJECTORY METHOD

A. Classical-Path Approximation

As explained in the Introduction, most mixed quantum-classical (MQC) methods are based on the classical-path approximation, which describes the reaction of the quantum degrees of freedom (DoF) to the dynamics of the classical DoF [9–22]. To discuss the classical-path approximation, let us first consider a diabatic electronic representation and expand the time-dependent molecular wave function $|\Psi(t)\rangle$ in terms of diabatic basis states $|\psi_k\rangle$

$$|\Psi(t)\rangle = \sum_k d_k(\mathbf{x}, t)|\psi_k\rangle \tag{27}$$

where $d_k(\mathbf{x}, t)$ denotes the vibrational wave function pertaining to the diabatic state $|\psi_k\rangle$. Insertion of Eq. (27) into the time-dependent Schrödinger equation yields the equations of motion in the diabatic representation

$$i\dot{d}_k = \sum_{k'}(T\delta_{k,k'} + V_{kk'})d_{k'} \tag{28}$$

Equation (28) is still exact. To introduce the classical-path approximation, we assume that the nuclear dynamics of the system can be described by classical trajectories; that is, the position operator $\widehat{\mathbf{x}}$ is approximated by its mean value, namely, the trajectory $\mathbf{x}(t)$. As a consequence, the quantum-mechanical *operators* of the nuclear dynamics (e.g., $V_{kk'}(\mathbf{x})$) become classical *functions* that depend parametrically on $\mathbf{x}(t)$. In the same way, the nuclear *wave functions* $d_k(\mathbf{x}, t)$ become complex-valued *coefficients* $d_k(\mathbf{x}(t), t)$. As the electronic dynamics is evaluated along the classical path of the nuclei, the approximation thus accounts for the reaction of the quantum DoF to the dynamics of the classical DoF.

In order to introduce the classical-path approximation in the adiabatic electronic representation, we expand the total wave function in terms of adiabatic basis states

$$|\Psi(t)\rangle = \sum_k a_k(\mathbf{x}, t)|\psi_k^{\mathrm{ad}}\rangle \tag{29}$$

In complete analogy to the diabatic case, the equations of motion in the adiabatic representation are then obtained by inserting the ansatz (29) into the time-dependent Schrödinger equation for the adiabatic Hamiltonian (7)

$$\begin{aligned} \dot{a}_k &= \sum_{k'} A_{kk'}a_{k'} \\ iA_{kk'} &= (T + W_k)\delta_{k,k'} + \Lambda_{kk'} \end{aligned} \tag{30}$$

As long as no approximation is introduced, it is clear that the equations of motion are equivalent in the diabatic and adiabatic representations. This is no longer true, however, once the classical-path approximation is employed; the resulting classical-path equations of motion in the adiabatic representation are *not equivalent* to the diabatic equations of motion. Depending on whether the approximation is employed in the diabatic or in the adiabatic representation, the resulting classical-path Hamiltonian contains identical first-order nonadiabatic couplings but different second-order nonadiabatic couplings [78]. The existence of several possibilities always raises the question on the "best" approximation. The numerical studies reported below suggest that the mean-field trajectory simulations for the model Hamiltonian (1) are best performed in a diabatic representation; employing the classical-path approximation in the adiabatic representation was found to lead to numerical instabilities in the vicinity of an intersection of the electronic potential-energy surfaces. The surface-hopping simulations, on the other hand, are best performed in the adiabatic representation. This appears to be the natural choice, since the surface-hopping model is defined in the adiabatic representation.

B. Mean-Field Trajectory Scheme

The classical-path approximation introduced above is common to most MQC formulations and describes the reaction of the quantum DoF to the dynamics of the classical DoF. The back-reaction of the quantum DoF onto the dynamics of the classical DoF, on the other hand, may be described in different ways. In the mean-field trajectory (MFT) method (which is sometimes also called Ehrenfest model, self-consistent classical-path method, or semiclassical time-dependent self-consistent-field method) considered in this section, the classical force $F = \dot{p}_j$ acting on the nuclear DoF x_j is given as an average over the quantum DoF

$$\dot{p}_j = -\left\langle \Psi(t) \left| \frac{\partial H}{\partial x_j} \right| \Psi(t) \right\rangle \tag{31}$$

where the wave function $|\Psi(t)\rangle$ and the molecular Hamiltonian H may be defined in a diabatic as well as in the adiabatic representation. If we adopt, for example, a diabatic electronic representation, the nuclear MFT equations of motion will read

$$\dot{p}_j = -\sum_{k,k'} d_k^* d_{k'} \frac{\partial V_{kk'}(\mathbf{x})}{\partial x_j} \tag{32}$$

That is, the classical DoF propagate according to a mean-field potential, the value of which is weighted by the instantaneous populations of the different quantum states. A MFT calculation thus consists of the self-consistent solution of the time-dependent Schrödinger equation (28) for the quantum DoF and Newton's equation (32) for the classical DoF. To represent the initial state (15) of the molecular system, the electronic DoF $d_k(0)$ as well as the nuclear DoF $x_j(0)$ and $p_j(0)$ are sampled from a quasi-classical phase-space distribution [23, 24, 26].

For example, the vibrational initial state may be represented by a Wigner distribution (17), while the initial electronic coefficients may be determined by $d_k(0) = \delta_{k2}e^{iq}$, q being an arbitrary phase [200].

In the MQC mean-field trajectory scheme introduced above, all nuclear DoF are treated classically while a quantum mechanical description is retained only for the electronic DoF. This separation is used in most implementations of the mean-field trajectory method for electronically nonadiabatic dynamics. Another possibility to separate classical and quantum DoF is to include (in addition to the electronic DoF) some of the nuclear degrees of freedom (e.g., high frequency modes) into the quantum part of the calculation. This way, typically, an improved approximation of the overall dynamics can be obtained—albeit at a higher numerical cost. This idea is the basis of the recently proposed self-consistent hybrid method [201, 202], where the separation between classical and quantum DoF is systematically varied to improve the result for the overall quantum dynamics. For systems in the condensed phase with many nuclear DoF and a relatively smooth distribution of the electronic–vibrational coupling strength (e.g., Model V), the separation between classical and quantum can, in fact, be optimized to obtain numerically converged results for the overall quantum dynamics [202, 203].

C. Results

To illustrate the quality of the MFT method, let us first consider Model I describing the $S_2 - S_1$ conical intersection in pyrazine. Figure 1 compares quantum-mechanical and MFT results obtained for the adiabatic and diabatic electronic population probabilities $P_2^{\text{ad}}(t)$ and $P_2^{\text{dia}}(t)$ as well as for the mean momenta of the two totally symmetric modes of the model. Generally speaking, the results are in qualitative agreement; in particular, the ultrafast initial decay of the adiabatic population and the oscillations of the diabatic population are quite well described by the MFT method. The correct electronic relaxation dynamics for longer time, however, is not reproduced. Moreover, the oscillations of the diabatic population are damped too strongly and the final value of the adiabatic population is too large. The comparison of the MFT and quantum results obtained for the mean momenta of the two vibrational modes v_1 and v_{6a} demonstrates that the MFT method is able to describe the dephasing of the vibrational motion caused by the internal conversion process. For longer times, however, the quality of the MFT result deteriorates. In particular, there is a mismatch of the phase of the oscillations. As discussed in detail in Section VI.C, the main deficiency of the method (i.e., insufficient electronic relaxation and phase mismatch in the nuclear dynamics) are related to the fact that the MFT approach treats electronic and nuclear DoF on an unequal dynamical footing.

Let us next turn to Model II, representing the $\widetilde{C} \rightarrow \widetilde{B} \rightarrow \widetilde{X}$ internal-conversion process in the benzene cation. Figure 2 demonstrates that this (compared to the electronic two-state model, Model I) more complicated process is difficult to describe with a MFT ansatz. Although the method is seen to catch the initial fast $\widetilde{C} \rightarrow \widetilde{B}$ decay quite accurately and can also qualitatively reproduce the oscillations of the diabatic populations of the $\widetilde{C}$- and $\widetilde{B}$-state, it essentially fails to reproduce the subsequent internal conversion to the electronic $\widetilde{X}$-state. In particular, the MFT method predicts a too-slow population transfer from the $\widetilde{C}$- and $\widetilde{B}$-state to the electronic ground state.

Figure 3 displays the results for Model III, describing ultrafast photoisomerization triggered by a conical intersection. Similarly as found for Model I, the MFT method is seen to reproduce the initial fast decay of the adiabatic population as well as the coherent oscillations of the diabatic population, but has problems to describe the long-time decay of the adiabatic population correctly. The probability P_{cis} that the system remains in the initially prepared *cis* conformation is also seen to be in quite good agreement with the quantum result. This finding demonstrates the ability of the MFT method to qualitatively describe the nonadiabatic dynamics in the presence of highly anharmonic potential-energy surfaces.

Finally, we consider the performance of the MFT method for nonadiabatic dynamics induced by avoided crossings of the respective potential energy surfaces. We start with the discussion of the one-mode model, Model IVa, describing ultrafast intramolecular electron transfer. The comparison of the MFT method (dashed line) with the quantum-mechanical results (full line) shown in Fig. 5 demonstrates that the MFT method gives a rather good description of the short-time dynamics (up to $\sim$50 fs) for this model. For longer times, however, the dynamics is reproduced only qualitatively. Also shown is the time evolution of the diabatic electronic coherence $C^{\mathrm{dia}}(t)$, which, too, is reproduced by the MFT method for short times.

Figure 6 shows the results for the more challenging model, Model IVb, comprising three strongly coupled vibrational modes. Overall, the MFT method is seen to give only a qualitatively correct picture of the electronic dynamics. While the oscillations of the adiabatic population are reproduced quite well for short time, the MFT method predicts an incorrect long-time limit for both electronic populations and fails to reproduce the pronounced recurrence in the diabatic population. In contrast to the results for the electronic dynamics, the MFT is capable of describing the almost undamped coherent vibrational motion of the vibrational modes.

In all results of the MFT method discussed so far, all vibrational modes have been treated classically. As has been mentioned above, an improved approximation of the overall dynamics can be obtained if some of the vibrational modes are included in the quantum-mechanical treatment. Figure 9

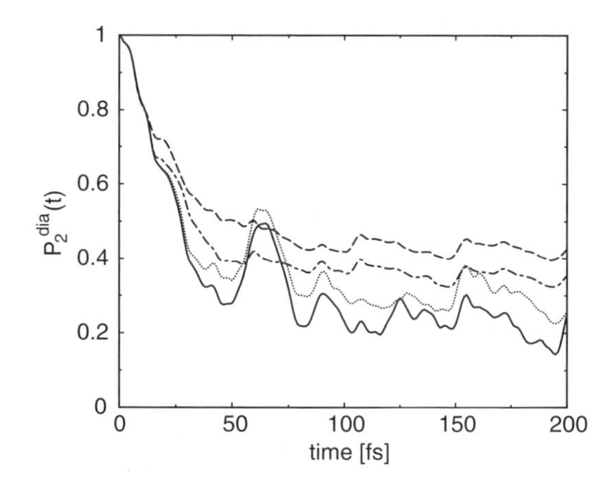

Figure 9. Population of the initially prepared diabatic state for Model IVb. Shown are results of a generalized MFT method where all modes (dashed line), the two lower frequency modes (dashed-dotted line), and the mode with the lowest frequency (dotted line) have been treated classically, respectively. The full line shows the quantum-mechanical result.

demonstrates this strategy for the diabatic electronic population of the electron-transfer Model IVb. It is seen that both high-frequency modes have to be treated quantum mechanically to obtain a semiquantitative description of the dynamics.

We next consider two examples of Model V describing outer-sphere electron transfer in solution. Figures 7 and 8 display results for the diabatic electronic population for Model Va and Model Vb, respectively. While for Model Va the MFT results are in excellent agreement with the quantum calculations, the MFT method can only describe the short-time dynamics for Model Vb. Similar as for the three-mode ET model discussed above, the MFT method in particular predicts an incorrect long-time limit for the diabatic population. The success of the MFT method for Model Va may be caused by the fact that the simulations are performed for relatively high temperature ($k_B T = 2.5g$), thus facilitating a classical description. Furthermore, the special case of a vanishing bias ($E_1 = E_2$) represents a simplifying assumption, because for nonzero temperature it results in a long-time limit of 0.5 for the diabatic population. As will be discussed in the next section, the failure of the MFT method to describe the long-time dynamics of Model Vb correctly may be a consequence of the fact that the *diabatic* electronic populations as calculated by the MFT methods do not obey the principle of microreversibility.

D. Discussion

The MFT equation of motion, Eq. (31), can be derived in many ways, including the WKB approximation [9], the eikonal method [13], a (semi)classical time-dependent self-consistent field ansatz [11, 15, 18, 19], density-matrix approaches [14, 16, 20], and the classical limit of algebraic quantization [29]. Depending on the specific approach used, slightly different MFT schemes may result. For example, the classical force can be described either by the average of the quantum force as in Eq. (31) or by the derivative of the average quantum potential [17, 21], that is,

$$\dot{p}_j = -\frac{\partial}{\partial x_j}\langle\Psi(t)|H|\Psi(t)\rangle \tag{33}$$

While most derivations focus on the equation of motion, an equally important aspect of the MFT method is the correct representation of the quantum-mechanical initial state. It is well known that the classical limit of quantum dynamics in general is represented by an ensemble of classical orbits [23, 24, 26, 204]. Hence it is not appropriate to use a single classical trajectory, but it is necessary to average over many trajectories, the initial conditions of which are chosen to mimic the quantum nature of the initial state of the classically treated subsystem. Interestingly, it turns out that several misconceptions concerning the theory and performance of the MFT method are rooted in the assumption of a single classical trajectory.

First, it is noted that it is this sampling that makes the MFT method quite different from the quantum-mechanical time-dependent self-consistent-field (TDSCF) approach (which is sometimes also called the time-dependent Hartree method). It is well known that the MFT equation of motion for a single trajectory may be derived as an classical approximation to the single-configuration TDSCF ansatz [11, 15, 18, 19]. To derive the MFT method including the correct quasi-classical averaging, however, the derivation needs to start from a multiconfiguration-type TDSCF ansatz and, furthermore, to assume a rapid randomization of the nuclear phases [18, 19]. Because of the ensemble average, the quasi-classical MFT method contains "static" correlation and actually may perform qualitatively better than a quantum-mechanical single-configuration TDSCF ansatz. This effect was explicitly shown in the case of Model I [176], where the quantum TDSCF approximation was found to completely fail to account for the long-time behavior of the electronic dynamics, which is at least qualitatively reproduced by the MFT results reported above. The results depicted in Fig. 10 obtained for Models Va and Vb, respectively, confirm this finding. The results of the quantum-mechanical TDSCF method exhibit artificial oscillation and cannot describe the incoherent character of the dynamics which is well-captured by the MFT method (cf. Figs. 7 and 8). This is because the standard TDSCF ansatz neglects all correlations between the dynamical DoF, whereas the MFT treatment may account for some correlation due to the sampling of the initial state.

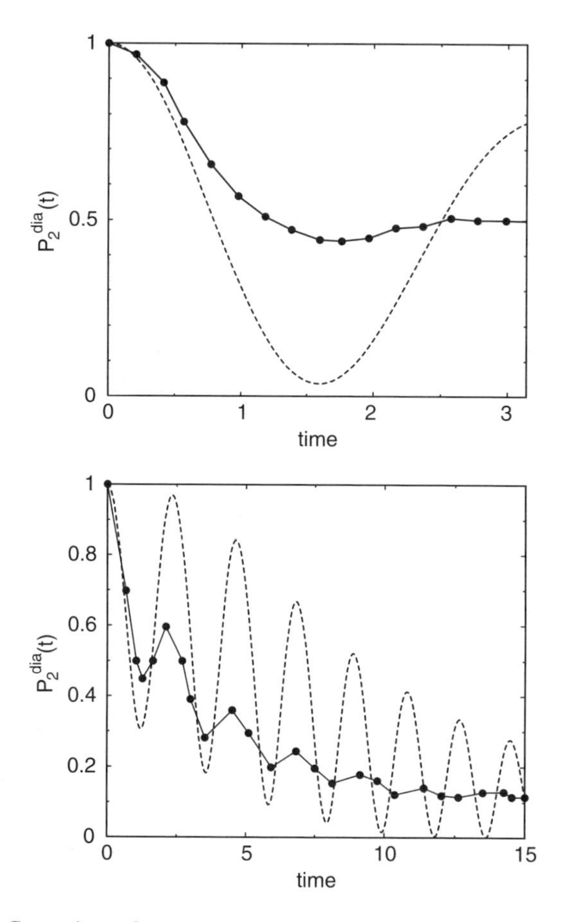

Figure 10. Comparison of quantum-mechanical time-dependent self-consistent field (time-dependent Hartree) (dashed line) and quantum path-integral (dots) calculations obtained for Model Va (upper panel) and Model Vb (lower panel), respectively. Shown is the time-dependent population probability $P_2^{\rm dia}(t)$ of the initially prepared diabatic electronic state.

Furthermore, a related and common criticism of the MFT method is that a mean-field approach cannot correctly describe the branching of wave packets at crossings of electronic states [67, 70, 82]. This is true for a *single* mean-field trajectory, but is not true for an ensemble of trajectories. In this context it may be stressed that an *individual* trajectory of an ensemble does not even possess a physical meaning—only the average does.

Employing a correct quasi-classical average, the numerical results obtained above for the Models I and III–Va have clearly demonstrated that the MFT method qualitatively reproduces the electronic population dynamics and therefore—at least in principle—is able to account for the branching of trajectories in a nonadiabatic calculation. The failure of the MFT method to describe the long-time dynamics of the three-state Model II and the condensed phase model Vb has a quite different origin. It is related to the fact that the MFT approach treats electronic and nuclear DoF on an unequal dynamical footing. This becomes clear within the mapping approach introduced in Section VI, where it is shown that the MFT method can be derived from the classical mapping formulation if the initial quantum fluctuations of the electronic and nuclear DoF are treated in a different way.

Let us finally discuss to what extent the MFT method is able to (i) obey the principle of microreversibility, (ii) account for the electronic phase coherence, and (iii) correctly describe the vibrational motion on coupled potential-energy surfaces. It is a well-known flaw of the MFT method to violate quantum microreversibility. This basic problem is most easily rationalized in the case of a scattering reaction occurring in a two-state curve-crossing system, where the initial and final state of the scattered particle may be characterized by the momenta p_i and p_f, respectively. We wish to calculate the probability $P_{1\to 2}$ that the system starts at $t = -\infty$ in the electronic state $|\psi_1\rangle$ with an initial momentum p_i and ends up in the electronic state $|\psi_2\rangle$ with a final momentum p_f. The principle of quantum microreversibility then states that the probability $P_{1\to 2}$ is equal to $P_{2\to 1}$, which is the probability that the system starts in the state $|\psi_2\rangle$ with an initial momentum p_f and ends up in the state $|\psi_1\rangle$ with a final momentum p_i. In a MFT calculation, the electronic transition probability $P_{1\to 2}$ is obtained by starting the propagation with the initial conditions $|d_1(-\infty)|^2 = 1, |d_2(-\infty)|^2 = 0, p(-\infty) = p_i$ and an arbitrary $x(-\infty)$ outside the interaction region and calculating the final values for the electronic coefficients, which directly yield $P_{1\to 2} = |d_2(\infty)|^2$. In complete analogy to this procedure, the probability $P_{2\to 1}$ is obtained as $P_{2\to 1} = |d_1(\infty)|^2$, whereby the initial conditions $|d_2(-\infty)|^2 = 1, |d_1(-\infty)|^2 = 0, p(-\infty) = p_f$ are used. The problem with the MFT propagation is that the initial electronic coefficients used in the backward propagation $2 \to 1$ are not equivalent to the final electronic coefficients as obtained from the forward propagation $1 \to 2$. Assuming for illustrative purposes that the electronic transition probability is small, the forward trajectory mainly propagates on the electronic $|\psi_1\rangle$ surface, while the back trajectory mainly propagates on the electronic $|\psi_2\rangle$ surface, thus leading to different dynamics and transition probabilities. Although the MFT equations of motion and their solutions are invariant with respect to the symmetry operation of time reversal, the MFT propagation therefore violates quantum microreversibility. It should be noted that this property of the MFT scheme is not caused by the classical-path approximation, but is common to any approximate time-dependent self-consistent-field description.

A further important property of a MQC description is the ability to correctly describe the time evolution of the electronic coefficients. A proper description of the electronic phase coherence is expected to be particularly important in the case of multiple curve-crossings that are frequently encountered in bound-state relaxation dynamics [163]. Within the limits of the classical-path approximation, the MFT method naturally accounts for the coherent time evolution of the electronic coefficients (see Fig. 5). This conclusion is also supported by the numerical results for the transient oscillations of the electronic population, which were reproduced quite well by the MFT method. Similarly, it has been shown that the MFT method in general does a good job in reproducing coherent nuclear motion on coupled potential-energy surfaces.

To summarize, the results presented for five representative examples of nonadiabatic dynamics demonstrate the ability of the MFT method to account for a qualitative description of the dynamics in case of processes involving two electronic states. The origin of the problems to describe the correct long-time relaxation dynamics as well as multi-state processes will be discussed in more detail in Section VI. Despite these problems, it is surprising how this simplest MQC method can describe complex nonadiabatic dynamics. Other related approximate methods such as the quantum-mechanical TDSCF approximation have been found to completely fail to account for the long-time behavior of the electronic dynamics (see Fig. 10). This is because the standard Hartree ansatz in the TDSCF approach neglects all correlations between the dynamical DoF, whereas the ensemble average performed in the MFT treatment accounts for the "static" correlation of the problem.

IV. SURFACE-HOPPING METHOD

As explained in the Introduction, one needs to distinguish the following kinds of "surface hopping" (SH) methods: (i) Semiclassical theories based on a connection ansatz of the WKB wave function [52–59, 61, 62, 64, 65], (ii) stochastic implementations of a given deterministic multistate differential equation (e.g., the quantum-classical Liouville equation [35, 38, 40]), and (iii) quasi-classical models such as the well-known SH schemes of Tully and others [68–84]. In this chapter, we focus on the latter type of SH method, which has turned out to be the most popular approach to describe nonadiabatic photodynamics.

A. General Idea

The quasi-classical SH model employs the simple and physically appealing picture in which a molecular system always evolves on a *single adiabatic* potential-energy surface (PES). When the trajectory reaches an intersection of the electronic PESs, the transition probability $p_{k \to k'}$ to the other PES is calculated according to some hopping criterion and, depending on the comparison of $p_{k\to k'}$ with a random number, the trajectory "hops" to the other adiabatic surface. Contrary to the mean-field trajectory scheme, where the trajectories evolve on an *averaged* electronic PES [cf. Eq. (31)], the SH equations of motion for the nuclear degrees of freedom therefore assume the simple form

$$\dot{x}_j = \omega_j p_j$$
$$\dot{p}_j = -\frac{\partial W_k}{\partial x_j} \tag{34}$$

where W_k denotes the adiabatic PES on which the trajectory currently evolves. In order to conserve the energy of the classical system during an electronic transition, the momenta of the nuclei have to be adjusted after every hop. We note in passing that SH simulations are performed in the adiabatic representation, because (i) the picture of instantaneous hops only appears plausible for highly localized interactions between the PESs such as the nonadiabatic kinetic-energy couplings, and (ii) a single adiabatic PES is expected to represent the coupled-surfaces problem better than a single diabatic one.

Since the hopping procedure accounts for the coupling of electronic and nuclear degrees of freedom, the key problem of the SH approach is to establish a dynamically consistent hopping criterion and momentum adjustment. As already mentioned, any rigorous way to derive such a formulation leads to complicated equations of motion that in general are quite cumbersome to implement (see, e.g., Section V.B). Many workers have therefore developed simple but practical models of SH [66–84]. Here the most popular formulation is the "fewest switches" algorithm proposed by Tully [67]. Since a detailed description can be found in many articles (see, e.g., Refs. 67, 68, 82, and 83), in the following only the main idea of this method is outlined.

B. Theory and Numerical Implementation

As a starting point, we consider the Schrödinger equation (30) in the adiabatic classical-path approximation. This equation can be recast in a density-matrix notation by

$$\dot{\rho}_{kk} = 2\mathrm{Re}\sum_{k'} A_{kk'}\rho_{kk'} \tag{35}$$

where the electronic density-matrix elements are given as $\rho_{kk'}(t) = a_k^*(t)a_{k'}(t)$, and $k, k' = 1, 2$. Introducing a time step Δt which is small enough that the change $\Delta\rho_{kk}$ of the diagonal matrix elements can be approximated by

$$\Delta\rho_{kk} = \dot{\rho}_{kk}\Delta t \tag{36}$$

the basic assumption is that the populations ρ_{kk} change according to the master equation

$$\Delta\rho_{kk} = -\sum_{k'} p_{k\to k'}\rho_{kk} + \sum_{k'} p_{k'\to k}\rho_{k'k'} \tag{37}$$

By combining Eqs. (35)–(37), the electronic transition probability $p_{1\to 2}$ is readily derived as

$$p_{1\to 2} = 2\mathrm{Re}\ A_{21}\Delta t\ \rho_{21}/\rho_{11} \tag{38}$$

Similarly, the transition probability $p_{2\to 1}$ is obtained by exchanging the indices 1 and 2.

The formulation outlined above allows for a simple stochastic implementation of the deterministic differential equation (35). Starting with an ensemble of trajectories on a given adiabatic PES W_k, at each time step Δt we (i) compute the transition probability $p_{k\to k'}$, (ii) compare it to a random number $\zeta \in [0, 1]$, and (iii) perform a hop if $p_{k\to k'} > \zeta$. In the case of a pure N-level system (i.e., in the absence of nuclear dynamics), the assumption (37) holds in general, and the stochastic modeling of Eq. (35) is exact. Considering a vibronic problem with coordinate-dependent $A_{kk'}$, however, it can be shown that the electronic transition probability depends in a complicated and nonlocal way on the nuclear dynamics (cf. Section V.B). In this case, the stochastic model represents an approximation to the true dynamics.

As a consistency test of the stochastic model, one can check whether the percentage $N_k(t)$ of trajectories propagating on the adiabatic PES W_k is equal to the corresponding adiabatic population probability $P_k^{\mathrm{ad}}(t)$. In a SH calculation, the latter quantity may be evaluated by an ensemble average over the squared modulus of the adiabatic electronic coefficients [cf. Eq. (22)], that is,

$$P_k^{\mathrm{ad}}(t) = \langle a_k^*(t) a_k(t)\rangle_C \tag{39}$$

To obtain a similar expression for $N_k(t)$, we introduce "binned" adiabatic coefficients $\tilde{a}_k(t)$, that is,

$$\tilde{a}_k(t) = \begin{cases} 1 & \text{if trajectory evolves on } W_k \\ 0 & \text{otherwise} \end{cases} \tag{40}$$

Hence the percentage $N_k(t)$ of trajectories propagating on the adiabatic PES W_k can be written as

$$N_k(t) = \langle \tilde{a}_k^*(t)\tilde{a}_k(t)\rangle_C \tag{41}$$

Combining these equation, the consistency condition of the stochastic model reads

$$\langle a_k^*(t) a_k(t) \rangle_C = \langle \tilde{a}_k^*(t) \tilde{a}_k(t) \rangle_C \tag{42}$$

As is shown below, the simple SH model in general does not satisfy this relation.

Besides the hopping criterion, one furthermore needs to establish a rule that ensures the energy conservation of the classical system during an electronic transition. Assuming that the positions of the nuclei do not change during an instantaneous hopping process, we need to find a procedure that allows us to calculate the momenta of the trajectory on the new electronic PES. Semiclassical analyzes [55, 57, 61] suggest that the momenta are adjusted in the direction of the nonadiabatic coupling $\mathbf{F}_{kk'} = \langle \psi_k^{\mathrm{ad}} | \nabla | \psi_{k'}^{\mathrm{ad}} \rangle$, thus yielding the ansatz

$$\mathbf{p}' = \mathbf{p} + \sigma \mathbf{F}_{kk'} \tag{43}$$

where the parameter σ is assumed to be real and identical for all momenta. Employing the condition of energy conservation, we obtain a quadratic equation for σ:

$$c_{kk'} \sigma^2 + b_{kk'} \sigma + (W_{k'} - W_k) = 0 \tag{44}$$

where we have used the abbreviations $c_{kk'} = \sum_j \frac{1}{2} \omega_i |F_{kk'}^j|^2$ and $b_{kk'} = \sum_j \omega_j p_j F_{kk'}^j$.

To obtain a unique and consistent matching condition for the vibrational momenta in a hopping process, a single and real solution of Eq. (44) is required. If the equation possesses two real roots, one may choose the solution with the smaller modulus. Problems arise, however, if there is no real solution to Eq. (44). This case may occur for transitions from a lower to a higher electronic state whereby there is less kinetic energy (in direction of the nonadiabatic coupling) available than the potential-energy difference $W_{k'} - W_k$ requires. Hence the hopping process has to be rejected although the electronic equations of motion (35) suggest an electronic transition. Since the situation is quite similar to quantum-mechanical tunneling, the rejected hops have been referred to as *classically forbidden electronic transitions* [78].

The breakdown of the SH scheme in the case of classically forbidden electronic transitions should not come as a surprise, but is a consequence of the rather simplifying assumptions [i.e., Eqs. (37) and (43)] underlying the SH model. On a semiclassical level, classically forbidden transitions may approximately be described within an initial-value representation (see Section VIII) or by introducing complex-valued trajectories [55]. On the quasi-classical level considered here, however, there is no rigorous way to solve the problem. This has spurred a large number of ideas to fix the situation. The standard proposal is to "reflect" the trajectory when a surface hop is rejected—that is, to reverse the component of momentum in direction of the nonadiabatic coupling vector [61, 68]. Other proposals include the combination of mean-field and SH ideas [70–73], the resetting of electronic coefficients [77], and the introduction of a position adjustment [65] or a time uncertainty [74] during a hop. Based on detailed numerical studies of the dynamics at a conical intersection, Müller and Stock [78] suggested the minimal solution of simply ignoring forbidden hops (see below).

C. Results and Discussion

As a first example, we again consider Model I describing a two-state three-mode model of the $S_1(n\pi^*)$ and $S_2(\pi\pi^*)$ states of pyrazine. Figure 11a shows the quantum-mechanical (thick line) and the SH (thin lines) results for the adiabatic population probability $P_2^{\rm ad}(t)$ of the initially prepared electronic state $|\psi_2\rangle$. As explained above, there are two ways to evaluate the adiabatic population in a SH calculation: One may either (a) evaluate the adiabatic population from the electronic coefficients [Eq. (39)] or (b) employ the binned adiabatic coefficients defined in Eq. (40) and calculate the percentage of trajectories $N_k(t)$ propagating on the adiabatic surface W_k [Eq. (41)]. According to Eq. (42), the results of both calculations should coincide if the SH model provides a consistent description of the nonadiabatic dynamics. Figure 11a clearly demonstrates that this is not the case. Although the general appearance (i.e., ultrafast initial decay to an approximately constant value) is roughly the same for both quantities, they are seen to differ significantly after only $\sim$20 fs.

As discussed above, this discrepancy may be caused by classically forbidden electronic transitions—that is, cases in which a proposed hopping process is rejected due to a lack of nuclear kinetic energy. Figure 11c supports this idea by showing the absolute numbers of successful (thick line) and rejected (thin line) surface hops. In accordance with the initial decay of the adiabatic population, the number of successful surface hops is largest during the first 20 fs. For larger times, the number of rejected hops exceeds the number of successful surface hops. This behavior clearly coincides with the onset of the deviations between the two classically evaluated curves $N_k(t)$ and $P_k^{\rm ad}(t)$. We therefore conclude that the observed breakdown of the consistency relation (42) is indeed caused by classically forbidden electronic transitions.

In direct analogy to the adiabatic case, the classical diabatic population probability is given by

$$P_k^{\rm di}(t) = \langle d_k^*(t) d_k(t) \rangle_C \tag{45}$$

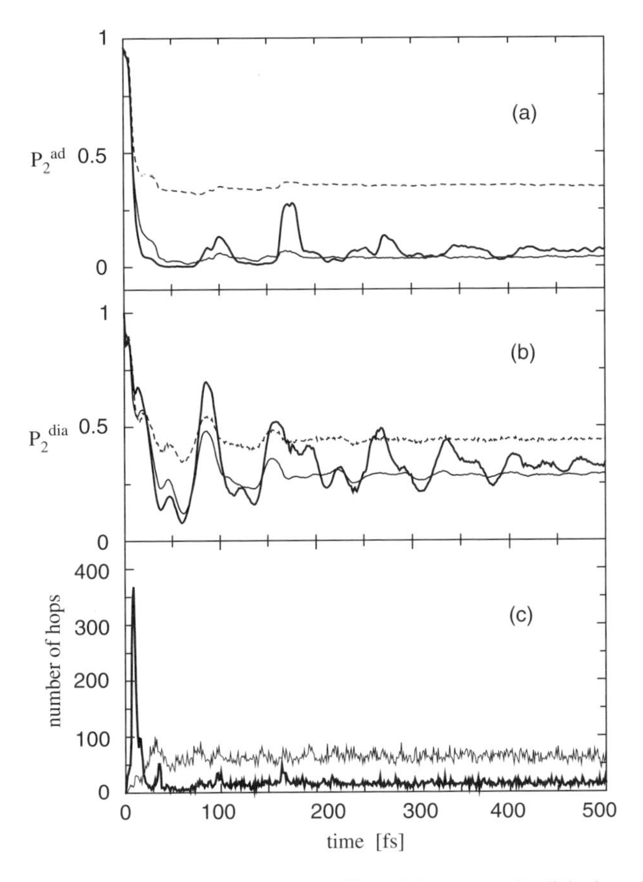

Figure 11. Time-dependent population probability of the upper (*a*) adiabatic and (*b*) diabatic electronic state of Model I. The quantum-mechanical results (thick lines) are compared to SH results obtained directly from the electronic coefficients (dashed lines) and to SH results obtained from binned coefficients (thin solid lines), reflecting the percentage $N_2(t)$ of trajectories propagating on the upper adiabatic surface. Panel (*c*) shows the absolute number of successful (thick line) and rejected (thin line) surface hops occurring in the SH calculation.

where the diabatic electronic coefficients $d_k(t)$ may be calculated from the adiabatic electronic coefficients $a_k(t)$ employing the transformation between the two representations. Figure 11b compares the SH results obtained for the diabatic population (dashed line) to the quantum-mechanical reference data (thick line). Similar to what occurs in the adiabatic case, the diabatic population calculated from the electronic coefficients can only roughly match the quantum results. Inspired by the finding that the quantity $N_k(t)$ provides a better approximation of the adiabatic population probability, we have also evaluated the "binned" diabatic population

$$\widetilde{P}_k^{\mathrm{di}}(t) = \langle \tilde{d}_k^*(t)\tilde{d}_k(t)\rangle_C \tag{46}$$

where the diabatic coefficients $\tilde{d}_k(t)$ have been calculated from the binned adiabatic coefficients $\tilde{a}_k(t)$. Interestingly, it is seen that the quantum results are reproduced much better by this approximation. In fact, in all simulations performed for conical intersections, it has been found that the binned quantities $N_k(t)$ and $\widetilde{P}_k^{\mathrm{di}}(t)$ provide the better approximation to the corresponding quantum-mechanical population probabilities. If not noted otherwise, henceforth we will always consider binned electronic population probabilities.

To obtain a more comprehensive picture of the performance of the SH method for Model I, Fig. 12 compares quantum (thick lines) and SH (thin lines) results for the adiabatic and diabatic population probabilities as well as for the mean momenta of the two totally symmetric modes of the model. Let us first consider the results obtained for the standard SH algorithm [67] shown on the left side and then focus on the vibrational dynamics of the system. The SH calculations of the momenta are seen to qualitatively match the quantum-mechanical results, although the classical approximation exaggerates the damping of the coherent vibrational motion. Interestingly, the onset of the deviations between quantum and classical results again coincides with the onset of rejected hops displayed in Fig. 11b. This finding indicates that the adjustment of momenta performed in this case might not be appropriate. In fact, we have found that the SH results improve significantly, if one simply omits the momentum adjustment in the case of a rejected hop. A comparison of the results obtained by the thus modified SH algorithm (right panels) and the standard algorithm (left panels) reveals that all time-dependent observables are reproduced much better by the modified algorithm. In particular, the vibrational momenta of the quantum calculation and the modified SH calculation are in excellent agreement, thus suggesting that the vibrational dephasing associated with the internal-conversion process in pyrazine is mainly caused by the anharmonicity of the lower adiabatic PES. Because the modified algorithm has been found to yield better results for *all* model systems considered [78], it has been used in all remaining calculations. Apart from this simple modification suggested above, there have been numerous suggestions to improve Tully's fewest switches algorithm in the case of classically forbidden transitions [70–74]. In particular, Truhlar and co-workers have performed systematic studies of various methods and test problems, revealing that the "best" hopping criterion depends largely on the problem under consideration.

Let us turn to Model II describing the $\widetilde{C} \rightarrow \widetilde{B} \rightarrow \widetilde{X}$ internal-conversion of the benzene cation. Figure 2 shows the diabatic population probabilities pertaining

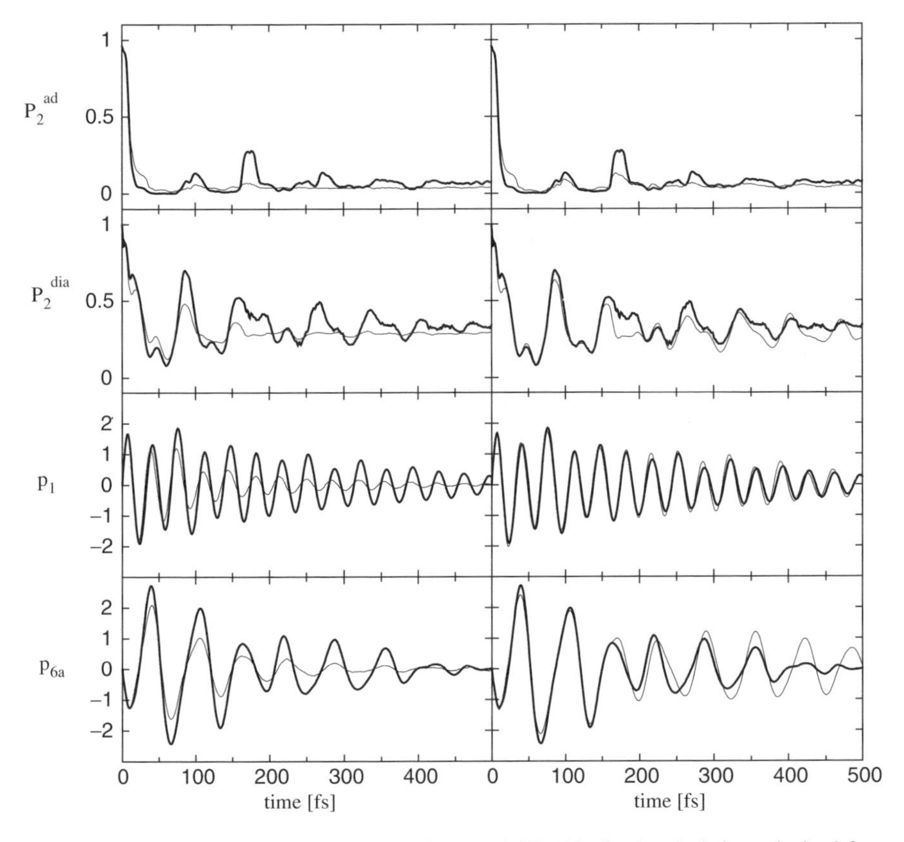

Figure 12. Quantum-mechanical (thick lines) and SH (thin lines) calculations obtained for Model I describing the $S_2 \rightarrow S_1$ internal-conversion process in pyrazine. Shown are the time-dependent population probabilities $P_2^{ad}(t)$ and $P_2^{dia}(t)$ of the initially prepared adiabatic and diabatic electronic state, respectively, as well as the mean momenta of the two totally symmetric modes ν_1 and ν_{6a}. The SH calculations used the standard algorithm of Tully [67] (left panels) as well as a modified algorithm described in Ref. 78 (right panels).

to the three electronic states as obtained from (A) the quantum-mechanical calculation of Köppel [180], (B) the mean-field trajectory method described in Section III, (C) the SH method using Eq. (4.12), and (D) the SH method using Eq. (46). Furthermore, the corresponding adiabatic population probabilities are shown on the right-hand side of Fig. 2, although a quantum-mechanical calculation of this observable is not available. Again, we observe that only the binned SH results match the quantum reference data. It is interesting to note that the standard SH results shown in panel (C) are quite similar to the mean-field trajectory results in that they essentially fail to reproduce the internal conversion to the electronic $\widetilde{X}$-state. Because both SH populations are obtained from the same simulation, this finding is somewhat disturbing, although the binned results in (D) look quite promising.

As a last example associated with a conical intersection, we consider model III describing ultrafast nonadiabatic *cis–trans* photoisomerization. Figure 3 compares quantum (thick line), mean-field (dashed line), and SH (thin line) results of (a) the adiabatic population, (b) the diabatic population, and (c) the probability $P_{cis}(t)$ that the system remains in the initially prepared *cis* conformation. Again, binned electronic population probabilities and no momentum adjustment for classically forbidden transitions are employed in the SH calculations. Similar to the mean-field calculations, the SH simulations are able to qualitatively reproduce the main features of the nonadiabatic photoisomerization, including initial decay, oscillatory transients, and long-time limits. It may be noted that, similar to Model I, the adiabatic relaxation dynamics of Model III is relatively simple, thus facilitating a modeling by the SH method. On the other hand, the simulations for Model III convincingly demonstrate that the SH technique is able to account for nonadiabatic relaxation dynamics in the presence of highly anharmonic PESs.

Although the systems discussed so far exhibit fairly complex vibrational and diabatic electronic relaxation dynamics, their *adiabatic* population dynamics is relatively simple and thus is well-suited for a SH description. To provide a challenge for the SH method, we next consider various spin–boson-type models, which may give rise to a quite complicated adiabatic population dynamics.

Let us begin with the one-mode electron-transfer system, Model IVa, which still exhibits relatively simple oscillatory population dynamics [205]. Similar to what is found in Fig. 5 for the mean-field description, the SH results shown in Fig. 13 are seen to qualitatively reproduce both diabatic and adiabatic populations, at least for short times. A closer inspection shows that the SH results underestimate the back transfer of the adiabatic population at $t \approx 50$ and 80 fs. This is because the back reaction would require energetically forbidden electronic transitions which are not possible in the SH algorithm. Figure 13 also shows the SH results for the electronic coherence $C^{\mathrm{ad}}(t)$, which are found to deviate drastically from the quantum reference. This failure of the SH description may be attributed to an approximate description of electronic coherences, which is not present in the mean-field trajectory method (see Fig. 5). As a remedy, it has been suggested to augment the SH method with a "decoherence" function that affects a rapid decay of the electronic coherence [75, 206, 207]. As is discussed in Section V.C, however, no such ansatz is necessary if the quantum coherences are treated in an appropriate way such as in the quantum-classical Liouville formulation.

Next we consider Model IVb, which describes ultrafast intramolecular electron transfer driven by three strongly coupled vibrational modes. Figure 6

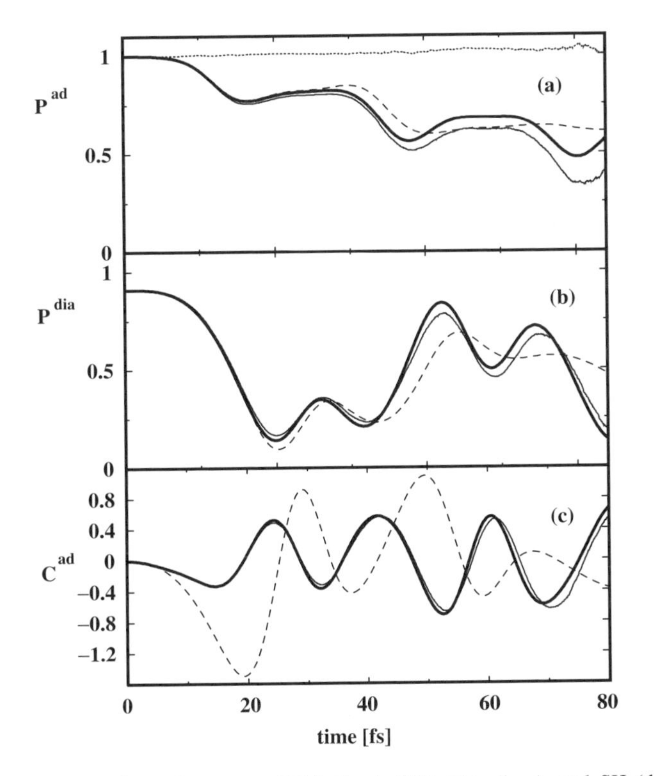

Figure 13. Comparison of quantum (thick lines), QCL (thin lines), and SH (dashed lines) results as obtained for the one-mode two-state model IVa [205]. Shown are (a) the adiabatic excited-state population $P^{\mathrm{ad}}(t)$, (b) the corresponding diabatic population probability $P^{\mathrm{dia}}(t)$, and (c) the coherence $C^{\mathrm{ad}}(t)$ of the adiabatic density matrix. The dotted line in panel (a) displays the norm of the QCL calculation.

presents the first 200 fs of the time evolution of Model IVb, showing the time-dependent adiabatic (a) and diabatic (b) population probabilities as well as the mean momentum of a representative vibrational mode (c). The SH calculation (thin lines) is seen to reproduce the quasi-periodic recurrences of the adiabatic population fairly well, thus stating that the SH method is capable of describing the back-crossing of trajectories to the upper PES. The diabatic population dynamics, however, is only qualitatively reproduced by the SH calculations. Interestingly, for all spin–boson-type models it has been found that binning of the diabatic electronic population probabilities [cf. Eq. (46)] hardly improves the agreement with quantum mechanics. Although the vibrational motion of the mode ν_2 of Model IVb hardly dephases on the timescale of the electronic relaxation process, the SH calculation predicts some damping of the coherent vibrational motion. This flaw of the SH calculation is presumably a consequence of the frequent hopping processes associated with the oscillatory adiabatic population dynamics of the system. Due to the approximations employed in the adjustment of momenta (e.g., the assumption that all momenta are adjusted in the same way), it cannot be expected that a SH modeling is capable of providing an accurate description of the vibrational dynamics for long times.

Finally, we consider Model V by describing two examples of outer-sphere electron-transfer in solution. Figures 7 and 8 display results for the diabatic electronic population for Models Va and Vb, respectively. Similar to the mean-field trajectory calculations, for Model Va the SH results are in excellent agreement with the quantum calculations, while for Model Vb the SH method only is able to describe the short-time dynamics. As for the three-mode Model IVb discussed above, the SH calculations in particular predict an incorrect long-time limit for the diabatic population. The origin of this problem will be discussed in more detail in Section VI in the context of the mapping formulation.

To summarize, it has been found that the SH method is able to at least qualitatively describe the complex photoinduced electronic and vibrational relaxation dynamics exhibited by the model problems under consideration. The overall quality of SH calculations is typically somewhat better than the quality of the mean-field trajectory results. In particular, this holds in the case of several curve crossings (see Fig. 2) as well as when the dynamics and the observables of interest are essentially of adiabatic nature—for example, for the calculation of the adiabatic population dynamics associated with a conical intersection (see Figs. 3 and 12). Furthermore, we have briefly discussed various consistency problems of a simple quasi-classical SH description. It has been shown that binned electronic population probabilities and no momentum adjustment for classically forbidden transitions help us to improve this matter. There have been numerous suggestions to further improve the hopping algorithm [70–74]; however, the performance of all these variants seems to depend largely on the problem under consideration.

V. QUANTUM-CLASSICAL LIOUVILLE DESCRIPTION

A. General Idea

The dynamics of classical as well as of quantum systems can be described by a Liouville equation for the time-dependent density. In quantum mechanics, the Liouville equation for the density operator $\hat{\rho}(t)$ reads

$$\frac{\partial}{\partial t}\hat{\rho}(t) = -i/\hbar[\hat{H}, \hat{\rho}(t)] \tag{47}$$

where $\hat{H} = \hat{H}(\hat{q}, \hat{p})$ is the quantum Hamiltonian depending on position and momentum operators $\hat{q}$ and $\hat{p}$, respectively, and $[\hat{A}, \hat{B}] = \hat{A}\hat{B} - \hat{B}\hat{A}$ denotes the usual quantum commutator. In order to obtain the classical limit of Eq. (47), one may introduce a quantum-mechanical phase-space representation of the operators such as the Wigner–Weyl formulation [2, 171]. If we retain only the classical ($\hbar = 0$) terms, the resulting equation of motion reads

$$\frac{\partial}{\partial t}\rho(t) = \{H, \rho(t)\} \tag{48}$$

where $\rho(t) = \rho(q, p, t)$ is the Wigner density function, $H = H(q, p)$ is the classical Hamiltonian, and $\{A, B\} = \partial_q A\, \partial_p B - \partial_p A\, \partial_q B$ denotes the classical Poisson bracket.

The similar appearance of the quantum and classical Liouville equations has motivated several workers to construct a mixed quantum-classical Liouville (QCL) description [27–44]. Hereby a partial classical limit is performed for the heavy-particle dynamics, while a quantum-mechanical formulation is retained for the light particles. The quantities $\rho(t)$ and H in the mixed QC formulation are then operators with respect to the electronic degrees of freedom, described by some basis states $|\phi_n\rangle$, and classical functions with respect to the nuclear degrees of freedom with coordinates $\mathbf{x} = \{x_j\}$ and momenta $\mathbf{p} = \{p_j\}$—for example,

$$\rho(t) = \sum_{nm} \rho_{nm}(\mathbf{x}, \mathbf{p}, t)|\phi_n\rangle\langle\phi_m| \tag{49}$$

The standard proposal for the Liouville equation for this QC density operator is

$$\frac{\partial}{\partial t}\rho(t) = [H, \rho(t)]_{QC} \tag{50a}$$

$$[A, B]_{QC} = -i/\hbar[A, B] + \tfrac{1}{2}\{A, B\} - \tfrac{1}{2}\{B, A\} \tag{50b}$$

where $[\cdot, \cdot]_{QC}$ denotes the QC bracket that contains the quantum commutator $[\cdot, \cdot]$ in (47) and the classical Poisson bracket $\{\cdot, \cdot\}$ in (48) as limiting cases. Equations (50) and (51) can be derived via a number of approaches—for example, by requiring certain axioms for the QC bracket to be satisfied [28], by replacing nuclear commutators by Poisson brackets and nuclear anticommutators by products of operators [31], by performing the partial classical limit in the Wigner phase-space representation [30, 33, 34], and by linearizing the forward–backward action in the influence functional of the path integral [44].

Although the QC bracket (50) is not unique and there has been some discussion on the overall consistency of the approach [29, 30, 33], the QCL formulation appears to be a quite promising formulation. As shown below, the QCL description naturally provides a correct treatment of electronic coherences and of the momentum change associated with an electronic transition, and it is therefore not plagued by the obvious inconsistencies arising in simpler MQC approaches. Furthermore, since the classical Wigner approximation is exact for quadratic Hamiltonians, it is clear that the QCL equation is actually *exact* in the case of the linear vibronic-coupling Hamiltonian of Models I and III, describing a multimode model of a conical intersection, as well as for the spin–boson-type Models IV and V.

The latter observation, however, indicates that the general solution of the QCL equation may require a similar effort as the solution of an exact quantum-mechanical description. In fact, the numerical implementation of the partial differential equation (50a) is only straightforward if the density is represented by some sort of a grid in phase space. The grid may be fixed [31] or moving in phase space, for example, via Gaussian wave packets [32, 38, 39]. Similar to a full quantum calculation, however, the numerical effort for these strategies appears to scale exponentially with the number of quantum *and* classical degrees of freedom. To obtain a numerical method that is directly applicable to truly multidimensional systems, it therefore is desirable to employ a stochastic scheme involving a Monte Carlo sampling of local classical trajectories [35, 40–43]. Such an approach has to cope with two major complications—that is, the representation of nonlocal phase-space operators and the convergence of the sampling procedure, which is cumbersome due to complex-valued trajectories with rapidly oscillating phases. The latter is related to the so-called dynamical sign problem, which is well known from real-time path-integral [208] and time-dependent semiclassical [11] calculations.

To give an impression of the virtues and shortcomings of the QCL approach and to study the performance of the method when applied to nonadiabatic dynamics, in the following we briefly introduce the QCL working equation in the adiabatic representation, describe a recently proposed stochastic trajectory implementation of the resulting QCL equation [42], and apply this numerical scheme to Model I and Model IVa.

B. Theory and Numerical Implementation

Performing a partial Wigner transformation with respect to the nuclear variables, the molecular Hamiltonian can be written as

$$H(\mathbf{x}, \mathbf{p}) = \sum_j \frac{p_j^2}{2m_j} + h(\mathbf{x}) \tag{51}$$

where the nuclear coordinates $\mathbf{x}$ and momenta $\mathbf{p}$ are now classical degrees of freedom and $h(\mathbf{x}) = T_q + V(\mathbf{q}, \mathbf{x})$ is the electronic Hamiltonian comprising the electronic coordinates $\mathbf{q}$. If we adopt the adiabatic electronic representation with basis states $\{|\psi_n^{\text{ad}}(\mathbf{x})\rangle\}$, the electronic Hamiltonian $h(\mathbf{x})$ becomes diagonal in this representation:

$$h(\mathbf{x})|\psi_n^{\text{ad}}(\mathbf{x})\rangle = W_n(\mathbf{x})|\psi_n^{\text{ad}}(\mathbf{x})\rangle \tag{52}$$

where $W_n(\mathbf{x})$ is the nth adiabatic potential-energy surface. The corresponding electronic matrix elements of the partial Wigner transformed density operator $\rho(\mathbf{x}, \mathbf{p}, t)$ are given by

$$\rho_{nm}(\mathbf{x}, \mathbf{p}, t) = \langle\psi_n^{\text{ad}}(\mathbf{x})|\rho(\mathbf{x}, \mathbf{p}, t)|\psi_m^{\text{ad}}(\mathbf{x})\rangle \tag{53}$$

For notational convenience, in the following we will usually drop the dependency on the nuclear variables $\mathbf{x}, \mathbf{p}$.

By inserting Eqs. (51)–(53) into the QCL equation (50a), the equations of motion for the QC density matrix in the adiabatic representation can be written in the following suggestive form [34]:

$$\frac{\partial}{\partial t}\rho_{nn'} = \sum_{m,m'}\left(\mathcal{L}^{Q}_{nn',mm'} + \mathcal{L}^{C}_{nn',mm'} + \mathcal{L}^{QC}_{nn',mm'}\right)\rho_{mm'} \tag{54}$$

where

$$\mathcal{L}^{Q}_{nn',mm'} = -i/\hbar(W_n - W_{n'})\delta_{nm}\delta_{n'm'} \tag{55a}$$

$$\mathcal{L}^{C}_{nn',mm'} = \sum_j\left(-\frac{p_j}{m_j}\frac{\partial}{\partial p_j} + \frac{\partial}{\partial x_j}\frac{1}{2}(W_n + W_{n'})\frac{\partial}{\partial p_j}\right)\delta_{nm}\delta_{n'm'} \tag{55b}$$

account for the pure quantum-mechanical and the pure classical time evolution of the density matrix, respectively. Hereby, the quantum part $\mathcal{L}^Q$ introduces the phase $(W_n - W_{n'})/\hbar$ to the off-diagonal matrix elements of the density matrix, while the classical part $\mathcal{L}^C$ describes standard Newtonian dynamics with respect to the adiabatic potential energy $(W_n + W_{n'})/2$. It is noted that the evaluation of the equations of motion (54) requires the propagation of classical trajectories in diagonal and off-diagonal matrix elements of the electronic density matrix. This is in contrast to usual surface-hopping methods based on a wave-function ansatz [81–83]. These formulations do not explicitly account for the coherences $\rho_{n\neq m}(t)$ and therefore only require the propagation of classical trajectories associated with diagonal density-matrix elements ρ_{nn}.

The quantum-classical part of the Liouville operator

$$\mathscr{L}^{QC}_{nn',mm'} = -\sum_j \frac{p_j}{m_j} F^{(j)}_{nm} \left(1 + \sum_i S^{(i)}_{nm} \frac{\partial}{\partial p_i} \right) \delta_{n'm'} (1 - \delta_{nm}\delta_{n'm'}) + \text{h.c.} \tag{56a}$$

$$S^{(i)}_{nm} = \frac{(W_n - W_m)}{2\sum_j F^{(j)}_{nm} p_j / m_j} F^{(i)}_{nm} \tag{56b}$$

accounts for the non-Born–Oppenheimer transitions of the system which are induced by the first-order nonadiabatic couplings $F^{(j)}_{nm} = \langle \psi^{\text{ad}}_n | \partial / \partial x_j | \psi^{\text{ad}}_m \rangle$. Note that in a standard quantum-mechanical formulation, only the first part of $\mathscr{L}^{QC}$ appears—that is, the term $\propto pF_{nm}$. In this case, p and F_{nm} represent nonlocal quantum operators with respect to the nuclear degrees of freedom; however, in the QC case, p and F_{nm} are classical functions. Here the nonlocal nature of the nonadiabatic transition is accounted for by the term $\propto S_{nm}\partial/\partial p$, which affects a momentum change of the classical trajectory. This shift of nuclear momenta associated with the transition between adiabatic potential-energy surfaces has been discussed in various semiclassical theories of nonadiabatic processes [55–59, 61, 62, 64, 65].

It is noted that the momentum derivatives of the coupling elements (56a) represent a problem for a practical trajectory-based evaluation of the QCL equation. These terms require the knowledge of the function in question not only at a particular point in phase space but at the same time also at nearby points. A possible solution is to change from trajectories to a grid of Gaussian wave packets—as, for example, in the approach developed by Horenko et al. [38, 39]. Wan and Schofield [40] presented an alternative approach, which seems to circumvent the problem by instantaneously diagonalizing the force matrix, thus eliminating the off-diagonal forces. However, this approach has been applied only to a one-dimensional system, where the force vectors have a single component. MacKernan et al. [209] have recently proposed a multi-dimensional generalization of this idea.

A general yet approximate ansatz has been suggested by Kapral and Cicotti [34], who considered the limit of small momentum changes:

$$\delta p_j \equiv \frac{S^{(j)}_{nm}}{p_j} \ll 1 \tag{57}$$

where p_j refers to the nuclear momenta before the electronic transition. In this case, one may approximate the derivative term in Eq. (56a) as

$$1 + \sum_j S^{(j)}_{nm} \frac{\partial}{\partial p_j} \approx \exp\left(\sum_j S^{(j)}_{nm} \frac{\partial}{\partial p_j} \right) \tag{58}$$

Since $e^{S\frac{\partial}{\partial p}} f(p) = f(p+S)$, the approximation reduces the action of the differential operator to a simple shift of momenta. The nonadiabatic transition operator assumes the simple form

$$\mathscr{L}^{QC}_{nn',mm'} = -\sum_j \frac{p_j}{m_j} d^{(j)}_{nm} \exp\left(\sum_i S^{(i)}_{nm} \frac{\partial}{\partial p_i}\right) \delta_{n'm'}(1 - \delta_{nm}\delta_{n'm'}) + \text{h.c.} \tag{59}$$

Note that, apart from the magnitude of δp_j, the accuracy of this approximation also depends on the specific shape of the density functions $\rho_{nm}(t)$ as compared to the spatial regions of strong nonadiabatic coupling. The numerical studies presented below will employ this approximation and also check its validity.

The approximation (58) resembles the usual momentum-jump ansatz employed in various surface-hopping methods [55, 57, 58, 61, 66, 67, 82]. In order to determine the momentum shift of a trajectory, however, the latter formulations typically require the conservation of nuclear energy:

$$W_1 + T_1 = W_2 + T_2 \tag{60}$$

where T_1 and T_2 denote the kinetic energy of the trajectory on the adiabatic potential-energy surfaces W_1 and W_2, respectively. This ansatz gives rise to the problem of classically forbidden transitions—that is, transitions that are required by the hopping algorithm but have to be rejected because of a lack of kinetic energy [78]. The QCL formulation, on the other hand, does not conserve the energy of an *individual* trajectory but the energy of the trajectory *average*. Only in the limiting case of vanishing momentum changes, the approximation (58) satisfies condition (60). It therefore may be expected that the present formulation does not suffer from the problem of classically forbidden transitions, which may represent a serious consistency problem for standard surface-hopping methods [78].

For the numerical implementation of the QCL equation, the Liouville operator $\mathscr{L}$ is decomposed into a zero-order part $\mathscr{L}_0$ which is easy to evaluate and a nonadiabatic transition part $\mathscr{L}^{QC}$ whose evaluation is difficult. The splitting suggests that we (a) employ a short-time expansion of the full exponential by use of a first-order Trotter formula

$$\begin{aligned}\rho(t+\delta t) &= e^{-i/\hbar(\mathscr{L}_0 + \mathscr{L}^{QC})\delta t}\rho(t) \\ &= e^{-i/\hbar\mathscr{L}_0\delta t}(1 - i/\hbar\mathscr{L}^{QC}\delta t)\rho(t) + \mathscr{O}(\delta t^2)\end{aligned} \tag{61}$$

and (b) solve the equations of motion for a small time step δt by imposing two basic processes on $\rho(t)$ to obtain $\rho(t+\delta t)$: (i) classical propagation and phase accumulation on the corresponding potential-energy surface described by $\mathscr{L}^0$ and (ii) transitions between coupled potential-energy surfaces induced by $\mathscr{L}^{QC}$.

While this scheme is straightforward to evaluate using a grid in phase space [31], its implementation on a grid becomes prohibitive in the case of multidimensional systems. Following previous work [210–214], we therefore introduce a stochastic algorithm that uses a Monte Carlo sampling of local classical trajectories. To this end, the density matrix is represented as an ensemble of phase-space points described by weight functions $W_{nm}(\mathbf{x}, \mathbf{p}, t)$, phase functions $\sigma_{nm}(\mathbf{x}, \mathbf{p}, t)$, and the density $P_{nm}(\mathbf{x}, \mathbf{p}, t)$:

$$\rho_{nm}(\mathbf{x}, \mathbf{p}, t) = W_{nm}(\mathbf{x}, \mathbf{p}, t)\sigma_{nm}(\mathbf{x}, \mathbf{p}, t)P_{nm}(\mathbf{x}, \mathbf{p}, t) \tag{62}$$

The density is assumed to be positive and normalized

$$\sum_{n,m} \int d\mathbf{x}\, d\mathbf{p}\; P_{nm}(\mathbf{x}, \mathbf{p}, t) = 1 \tag{63}$$

and we have $|\sigma_{nm}(\mathbf{x}, \mathbf{p}, t)| = 1$.

Let us investigate the change of the various components of the density matrix under propagation from time t to $t + \delta t$. We first consider the action of $\mathscr{L}_0$. Due to this operator, the density matrix element $\rho_{nm}(t)$ attains a phase factor

$$\exp(-i/\hbar(W_n - W_m)\delta t) \tag{64}$$

and is subject to propagation in phase space driven by

$$\exp\left(\sum_j \tfrac{1}{2}\left(F_{nn}^{(j)} + F_{mn'}^{(j)}\right)\frac{\partial}{\partial p_j} - \frac{p_j}{m_j}\frac{\partial}{\partial x_j}\right)\delta t \tag{65}$$

While the weights $W_{nm}(\mathbf{x}, \mathbf{p}, t)$ do not change under these operations, the densities $P_{nm}(\mathbf{x}, \mathbf{p}, t)$ are affected by the propagation (65). Furthermore, the sign functions $\sigma_{nm}(\mathbf{x}, \mathbf{p}, t)$ accumulate a phase (64) according to the (adiabatic) energy difference at the spatial position $\mathbf{x}$. In the absence of nonadiabatic transition (i.e., for $\mathscr{L}^{QC} = 0$), a stochastic realization of the propagation process is thus straightforward: Assuming that the initial phase-space distributions $\rho_{nm}(\mathbf{x}, \mathbf{p}, t = 0)$ are given in analytical form, initial positions and momenta are sampled from this distribution, thus determining the densities $P_{nm}(\mathbf{x}, \mathbf{p}, 0)$. Furthermore, the initial weights and phases are chosen to correctly represent the initial state of the density matrix. The time evolution of the density matrix is then described by Eqs. (64) and (65) as explained above.

Let us now consider the stochastic realization of the off-diagonal operator $\mathscr{L}^{QC}$, which represents the main difficulty of the numerical implementation.

$\mathscr{L}^{QC}$ can be viewed as a branching or hopping term that causes transitions between the different components of the density matrix. Apart from acting on the densities and the sign functions, this operator therefore also modifies the weights $W_{nm}(\mathbf{x}, \mathbf{p}, t)$ of the density matrix. To introduce a statistical interpretation of this process, let us consider a general matrix $\mathbf{M}$. If we write

$$M_{ab} = \left(\sum_{i,j} |M_{ij}|\right) \frac{|M_{ab}|}{\left(\sum_{i,j} |M_{ij}|\right)} \frac{M_{ab}}{|M_{ab}|} \tag{66}$$

the matrix $\mathbf{M}$ will be decomposed into a scalar weight factor $w = \sum_{ij} |M_{ij}|$, a phase $\phi_{ab} = M_{ab}/|M_{ab}|$, and a matrix with elements

$$T_{ab} = \frac{|M_{ab}|}{\left(\sum_{ij} |M_{ij}|\right)} \tag{67}$$

Since

$$\sum_{a,b} T_{ab} = 1, \qquad T_{ab} \geq 0 \tag{68}$$

$\mathbf{T}$ may be regarded as a matrix of probabilities.

We now assume that $\mathbf{M} = \mathbf{M}(t)$ is a Liouville operator acting on the Liouville vector $\rho(t)$ to give $\rho(t + \delta t)$. If we employ Eq. (62), $\rho(t + \delta t)$ can be calculated as

$$\begin{aligned} \rho_a(t + \delta t) &= \sum_b M_{ab}(t)\rho_b(t) \\ &= \sum_b M_{ab}(t)P_b(t)W_b(t)\sigma_b(t) \\ &= \sum_b T_{ab}(t)P_b(t) \cdot w(t)W_b(t) \cdot \phi_{ab}(t)\sigma_b(t) \end{aligned} \tag{69}$$

Thus, $\rho(t + \delta t)$ is a sum of contributions with the phase factors $(\phi_{ab}(t)\sigma_b(t))$, the weight factors $(w(t)W_b(t))$, and phase-space distributions $P_b(t)$ which appear with a probability $T_{ab}(t)$. Associating $\mathbf{M}(t)$ with the Liouville operator $1 - i/\hbar \mathscr{L}^{QC}\delta t$, the scheme becomes somewhat more involved. In addition to the hopping probabilities, $\mathscr{L}^{QC}$ also includes a momentum shift. Hence the expression $T_{ab}(t)P_b(t)$ describes a transition from component b to component a, accompanied by an instantaneous change of momenta according to Eq. (58). A further subtlety should be mentioned. Since the momentum correction (5.10b) itself explicitly depends on the momentum $\mathbf{p}$, the application of the momentum shift operator (58) can cause contractions (or expansions) of phase space, thereby modifying the weight. For the model systems considered, however, these contractions have been found to be negligible.

Combining the steps described above, a simple stochastic algorithm for a trajectory-based propagation of the QCL equation can be constructed as follows:

1. Sample the initial distribution function $P_{nm}(\mathbf{x}, \mathbf{p}, t = 0)$ by a set of "random walkers." The ith random walker carries the information of its phase-space position $(\mathbf{x_i}, \mathbf{p_i})$, density-matrix component $(nm)_i$, weight $W(\mathbf{x_i}, \mathbf{p_i}, t)$, and phase $\sigma(\mathbf{x_i}, \mathbf{p_i}, t)$. Divide the whole time interval into N small enough pieces such that the desired accuracy of the Trotter scheme is guaranteed.
2. During a single time step, do the following:
   (a) Propagate the random walker according to the local forces on the potential-energy surface with index nm.
   (b) Multiply the phase of the random walker by $e^{-i/\hbar(W_n - W_m)\delta t}$.
   (c) Make a nonadiabatic transition $|\psi_a\rangle \rightarrow |\psi_b\rangle$ if required: To this end, generate a random number $0 < \zeta < 1$. Stay on the current surface a if $\zeta < T_{aa}$. Otherwise, jump to the surface b that satisfies the condition $\sum_{k=1}^{b-1} T_{ak} < \zeta < \sum_{k=1}^{b} T_{ak}$. Change the momenta according to $p_j^b = p_j^a + S_{ab}^{(j)}(\mathbf{x}^a, \mathbf{p}^a)$.
   (d) Multiply the weight with w and the phase with ϕ_{ab}.
3. Calculate the molecular observables of interest.
4. Iterate from step (1) and run as many trajectories as needed to reach statistical convergence.

Within the momentum-jump approximation (58), the algorithm provides an in principle *exact stochastic realization* of the QCL equation. Apart from practical problems to be discussed below, the trajectory implementation therefore represents a well-defined computational scheme.

Similar to path-integral [208] and semiclassical [110] propagation methods, however, the Monte Carlo evaluation of the QCL equation is plagued by the so-called sign problem, which is caused by exponentially increasing weights W accompanied by rapidly varying phases. As a consequence, the signal-to-noise ratio of the sampling deteriorates in the cause of propagation time, or, in other words, the sampling effort increases exponentially in time. As shown in the QCL calculations below, this problem manifests itself in the fact that most of the random walkers generated by the algorithm do not contribute but cancel each other in the phase average.

A well-known strategy to overcome this fundamental problem is to reorganize the sum over complex paths in a way that facilitates this canceling.

In our example, this effect can be achieved by generating new random walkers at each surface hop: Whenever in step 2c a hop is requested, a copy of the random walker is made. While the original walker stays on the potential-energy surface, the copy starts out on the new surface, whereby its phase and momenta are updated. Finally, the copy is added to the collection of random walkers already present, where the weights are adjusted accordingly. Proceeding this way, one may expect that at any time a sufficiently large number of random walkers is present to yield a reliable description of the density matrix. Although it is clear that the gain of accuracy has to be paid through an exponential rise of newly generated random walkers, it is found that the convergence of this scheme is considerably better than simply increasing the number of initially starting random walkers (see below).

The concept of generating new random walkers at curve crossings is quite similar to the "spawning" scheme suggested by Martinez and co-workers [151, 153]. In the "full multiple spawning" algorithm, trajectories are generated on the fly in order to serve as an efficient adaptive basis set. While the underlying theories are quite different (QCL equation versus Gaussian wave-packet approximation [215] to Schrödinger's equation), both methods employ the same idea to improve convergence. The spawning concept is also utilized in the "localized threads" algorithm suggested to solve the QCL equation [40]. It should be emphasized, however, that the latter scheme differs from the procedure outlined above in that new random walkers are generated at every time step *and* for every matrix element, thus requiring a much larger computational effort. In both methods, however, the increasing number of walkers represents a significant problem if longer propagation times are considered.

C. Results

In order to get a first impression on the performance of the QC Liouville approach, it is instructive to start with a simple one-mode spin–boson model, that is, Model IVa [205]. In what follows, the QCL calculations used the first-order Trotter scheme (61) with a time step $\delta t = 0.05$ fs. If not noted otherwise, we have employed the momentum-jump approximation (59) and the initial number of random walkers employed was $N = 50\,000$.

Figure 13 shows (a) the adiabatic population $P^{\text{ad}}(t)$, (b) the diabatic population $P^{di}(t)$, and (c) the adiabatic coherence $C^{\text{ad}}(t)$ for the one-mode two-state model under consideration. It is seen that for short times the QCL calculations are in quantitative agreement with the quantum reference data, whereas for larger times the agreement deteriorates. Similar to the quality of these results, the norm of the QC density matrix (dotted line in Fig. 13a) is seen to deviate from unity with increasing time. This finding indicates that the loss of accuracy may be a problem of numerical convergence rather than an intrinsic problem of the QCL formulation. Compared to the standard surface-hopping

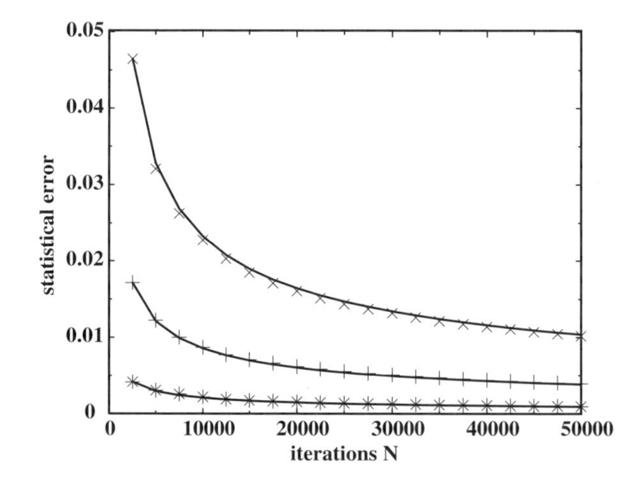

Figure 14. Statistical error of the adiabatic excited-state population at times $t = 10$ fs (***), 30 fs $(+++)$, and 50 fs $(\times \times \times)$ for Model IVa [205], plotted as a function of the number of iterations N. The full lines represent fits to a $1/\sqrt{N}$ dependence.

calculations also shown in Fig. 13, the QCL results represent a clear improvement, particularly for the coherence.

In order to study the origin of the deviations observed, we first consider the statistical convergence of the QCL data. As a representative example, Fig. 14 shows the absolute error of the adiabatic population as a function of the number of iterations N—that is, the number of initially starting random walkers. The data clearly reveal the well-known $1/\sqrt{N}$ convergence expected for Monte Carlo sampling. We also note the occurrence of the "sign problem" mentioned above. It manifests itself in the fact that the number of iterations increases almost exponentially with propagation time: While at time $t = 10$ fs only 200 iterations are sufficient to obtain an accuracy of 2%, one needs $N = 10\,000$ at $t = 50$ fs.

As explained above, the calculations employ an algorithm that generates new random walkers when a nonadiabatic transition is requested. To illustrate the performance of this scheme, Fig. 15 shows the average number of random walkers that have been generated from a single initial random walker during the propagation (i.e., for a single iteration). Here, the full and short dashed lines represent the numbers pertaining to the upper (ρ_{22}) and lower (ρ_{11}) electronic state, respectively, while the long dashed line represents the number of random walkers being in the coherences (ρ_{12}, ρ_{21}) of the electronic density matrix. As has been anticipated, the number of random walkers increases exponentially in time: The first step in the decay of the adiabatic population at time ~ 20 fs

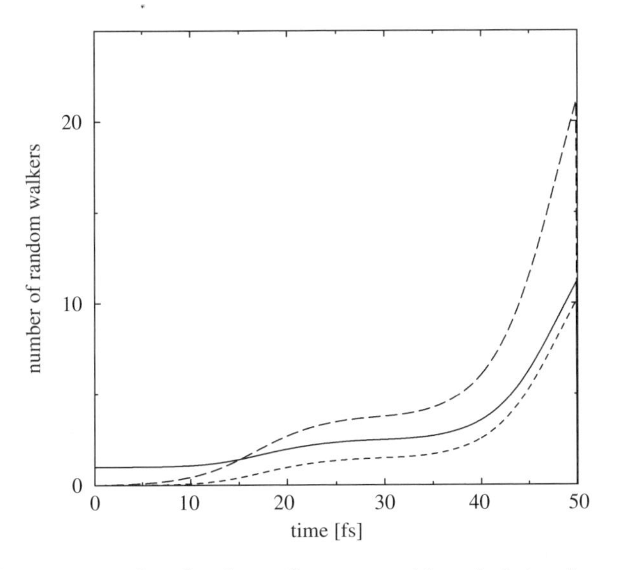

Figure 15. Average number of random walkers generated for a single iteration as obtained for Model IVa [205]. The full and short dashed lines correspond to the upper and lower electronic populations, respectively, while the long dashed line corresponds to the sum of the coherences of the electronic density matrix.

(cf. Fig. 13a) requires on average only the generation of about three more random walkers in total, while the second step at $t \approx 50$ fs already requires about 40 more random walkers. Furthermore, it is noted that already after $\sim$30 fs the random walkers are on average equally distributed between the four elements of the electronic density matrix. This distribution is a consequence of the fact that the probabilities to change from a coherence to the upper and to the lower state are the same [216]. Compared to the alternative scheme of simply increasing the number of initially starting random walkers, the present algorithm provides a speedup of about an order of magnitude [217].

Let us turn to the other main difficulty of a Monte Carlo implementation of the QCL approach—that is, the representation of the nonlocal operator (56) in terms of local classical trajectories. In the calculations presented above, the action of the derivative operator has been approximated by a local momentum jump (58) of the trajectory. To check the quality of this approximation, it is instructive to monitor the conservation of total energy in the QCL calculation. As shown in Fig. 16a, the total energy of the system is conserved up to $t \approx 30$ fs, where it suddenly starts to oscillate. For $t \gtrsim 60$ fs, the conservation of energy breaks down completely. To explain this somewhat surprising flaw of the QCL calculation, Fig. 16b shows the expectation value of the nuclear momentum. Starting at $\langle p(0) \rangle = 0$, the mean momentum oscillates with a period of $\sim$40 fs

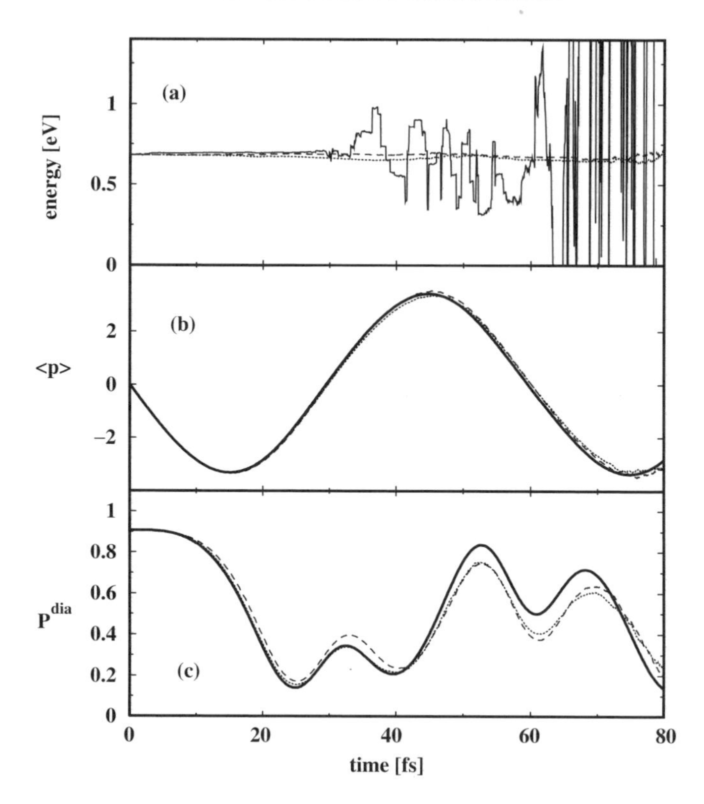

Figure 16. (*a*) Conservation of total energy as obtained in QCL calculations for Model IVa [205] using the standard (full line), the momentum-restricted (dashed line), and the energy-conserving (dotted line) scheme. (*b*) In all cases, the QCL results for the mean nuclear momentum $\langle p(t)\rangle$ are in excellent agreement with the quantum reference data (thick full line). The diabatic population probability $P^{\text{dia}}(t)$ shown in (*c*) is reproduced at least qualitatively.

corresponding to the vibrational frequency $\omega = 0.1$ eV employed in the model. Comparing Figs. 16a and 16b, it is found that the onset of the energy fluctuations at $t \sim 30$ and 60 fs coincides with the times the mean momentum approaches zero—that is, the turning points of the motion. This finding elucidates a problem of the momentum-jump approximation (58), which assumes a small relative momentum transfer $\delta p = S_{nm}/p$. For vanishing momentum p, however, δp may diverge. To prove this conjecture, we have changed the algorithm and excluded all trajectories with $\delta p > 1$. Figure 16a shows that the modified algorithm (dashed line) does conserve total energy. This reconfirms our assumption that the large deviations of total energy in the original QCL scheme is indeed caused by nonadiabatic transitions at turning points. Figure 16 furthermore shows quantum and QCL results obtained for the mean nuclear momentum and the diabatic population probability. While the momenta are in excellent agreement in all cases, the diabatic population obtained for the momentum-restricted QCL scheme appears to be somewhat less accurate than the results of the standard QCL calculation shown in Fig. 13b. We have also checked the performance of the QCL method in the case that the energy conservation of each *individual* trajectory is requested [cf. Eq. (60)]. The results obtained are found to be rather similar to the momentum-restricted QCL scheme.

Finally, it is interesting to investigate how the virtues of the QCL approach found for the one-mode model transfer to the general case of multimode dynamics. Due to the ultrafast decay of quantum coherences, one may in some sense expect that multidimensional problems are actually easier to handle for a QC description [19]. Furthermore, a multidimensional molecular system typically undergoes only a few electronic transitions and then remains on the lower adiabatic potential-energy surface. This is advantageous for the QCL description, whose problems only occur during nonadiabatic transitions. On the other hand, true intersections rather than avoided crossings may pose a problem for calculations using the adiabatic representation, since the nonadiabatic couplings are singular on the crossing surface. Furthermore, the sampling of phase space and the convergence of the QCL scheme may be more tedious in many dimensions [39, 41]. In the following, Model I is employed as a representative example.

Let us first consider the population probability of the initially excited adiabatic state of Model I depicted in Fig. 17. Within the first 20 fs, the quantum-mechanical result is seen to decay almost completely to zero. The result of the QCL calculation matches the quantum data only for about 10 fs and is then found to oscillate around the quantum result. A closer analysis of the calculation shows that this flaw of the QCL method is mainly caused by large momentum shifts associated with the divergence of the nonadiabatic couplings $F_{nm}^{(j)} = \langle \psi_n^{\mathrm{ad}} | \partial/\partial x_j | \psi_m^{\mathrm{ad}} \rangle$. We therefore chose to resort to a simpler approximation and require that the energy of each individual trajectory is conserved. Employing this scheme, the corresponding QCL result for the adiabatic population matches at least qualitatively the quantum data.

Due to the large-level density of the lower-lying adiabatic electronic state, the chances of a back transfer of the adiabatic population are quite small for a multidimensional molecular system. To a good approximation, one may therefore assume that subsequent to an electronic transition a random walker will stay on the lower adiabatic potential-energy surface [175]. This observation suggests a physically appealing computational scheme to calculate the time evolution of the system for longer times. First, the initial decay of the adiabatic population is calculated within the QCL approach up to a time t_0, when the

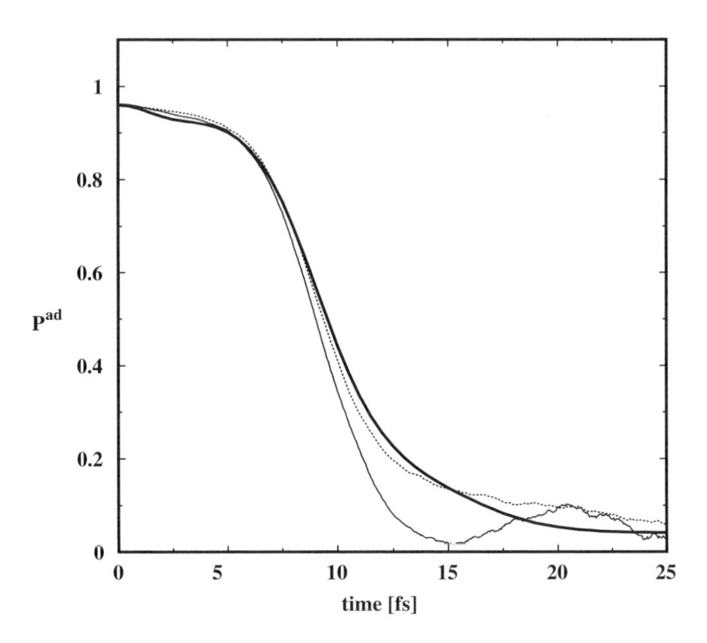

Figure 17. Initial decay of the adiabatic population probability obtained for Model I. Compared are quantum results (thick line) and standard (thin full line) and energy-conserving (dotted line) quantum-classical Liouville results.

system is (almost) completely localized in the lower adiabatic electronic state $|\psi_1^{ad}\rangle$. In a second step, we take the resulting phase-space distribution ρ_{11} $(\mathbf{x}, \mathbf{p}, t_0)$ as initial state for a standard quasi-classical trajectory propagation on the lower adiabatic potential-energy surface. While the adiabatic electronic population is constant by construction, the scheme yields diabatic electronic quantities as well as observables of the nuclear motion such as the time-dependent mean positions and momenta.

As an example, Fig. 18 shows the diabatic electronic population probability for Model I. The quantum-mechanical results (thick line) are reproduced well by the QCL calculations, which have assumed a "localization time" of $t_0 =$ 20 fs. The results obtained for the standard QCL (thin full line) and the energy-conserving QCL (dotted line) are of similar quality, thus indicating that the phase-space distribution $\rho_{11}(\mathbf{x}, \mathbf{p})$ at $t_0 = 20\,\text{fs}$ is similar for the two schemes. Also shown in Fig. 18 are the results obtained for a standard surface-hopping calculation (dashed line), which largely fail to match the beating of the quantum reference.

We conclude that the QCL description represents a promising approach to the treatment of multidimensional curve-crossing problems. The density-matrix

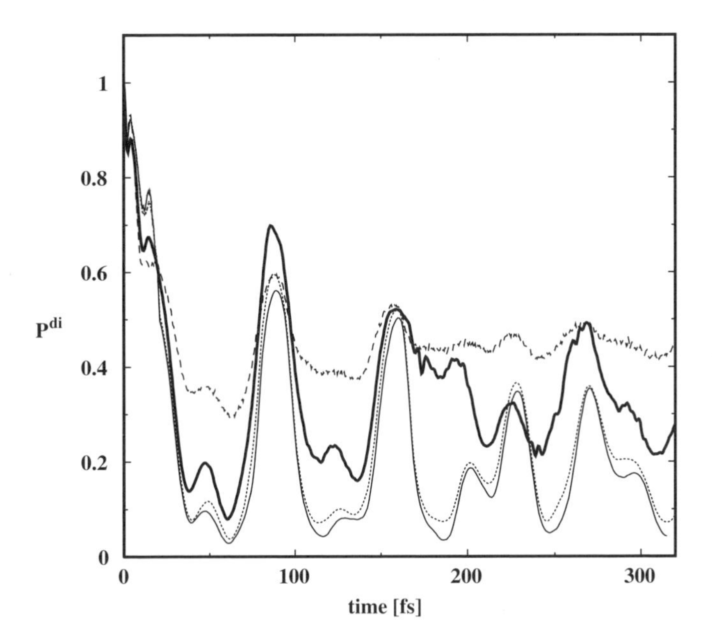

Figure 18. Diabatic electronic population probability obtained for Model I. Compared are quantum results (thick line) and standard (thin full line) and energy-conserving (dotted line) quantum-classical Liouville results.

formulation yields a consistent treatment of electronic populations *and* coherences, and the momentum changes associated with an electronic transition can be directly derived from the formalism without the need of *ad hoc* assumptions. Employing a Monte Carlo sampling scheme of local classical trajectories, however, we have to face two major complications—that is, the representation of nonlocal phase-space operators and the sampling problem caused by rapidly varying phases. At the present time, the QCL calculations performed for conical intersections cannot yet compete in accuracy and efficiency with more established methods. Recalling, though, that all numerical implementations of the QCL equation have been suggested within the last few years [32, 35, 38–42], it is clear that the QCL approach still holds a large potential to be discovered.

VI. MAPPING APPROACH

By definition, a mixed quantum-classical method treats the various degrees of freedom (DoF) of a system on a different dynamical footing—for example, quantum mechanics for the electronic DoF and classical mechanics for the nuclear DoF. As was discussed above, some of the problems with these methods are related to inconsistencies inherent in this mixed quantum-classical ansatz. To avoid these shortcomings, recently a conceptually different way to incorporate quantum mechanical DoF into a semiclassical or quasi-classical theory has been proposed, the so-called mapping approach [99, 100, 102]. In this formulation, the problem of a classical treatment of discrete DoF such as electronic states is bypassed by transforming the discrete quantum variables to continuous variables. In this section we introduce the general concept of the mapping approach and discuss the quasi-classical implementation of this method as well as applications to the five models introduced above. The semiclassical version of the mapping approach is discussed in Section VIII.

A. Theoretical Formulation

1. Spin Systems

There are a variety of formalisms that allow for a mapping of a discrete quantum system onto a continuous analog (for reviews see Refs. 218 and 219). The most prominent examples are Schwinger's theory of angular momentum [98] and the Holstein–Primakoff transformation [97], both of which allow a continuous representation of spin degrees of freedom. To discuss these two theories, we consider a spin DoF that is described by the spin operators S_1, S_2, S_3 with commutation relations

$$[S_j, S_k] = i\epsilon_{jkl} S_l \tag{70}$$

as well as the corresponding raising and lowering operators

$$S_\pm = S_1 \pm i S_2 \tag{71}$$

These operators act in the Hilbert space of spin states $|sm\rangle$ with

$$S_3|sm\rangle = m|sm\rangle, \qquad -s \le m \le s, \quad s = 1/2, 1, 3/2, \ldots \tag{72}$$

Consider, furthermore, a $(2s+1)$-dimensional subspace of the Hilbert space with fixed s. Then, according to Schwinger's theory of angular momentum [98], this discrete spin DoF can be represented by two bosonic oscillators described by creation and annihilation operators $a_n^\dagger, a_m$ with commutation relations

$$[a_n, a_m^\dagger] = \delta_{n,m} \tag{73}$$

and basis states $|n_1, n_2\rangle$ that satisfy

$$a_1^\dagger a_1 |n_1, n_2\rangle = n_1 |n_1, n_2\rangle \tag{74a}$$

$$a_2^\dagger a_2 |n_1, n_2\rangle = n_2 |n_1, n_2\rangle \tag{74b}$$

The mapping relations read

$$S_+ \rightarrow a_2^\dagger a_1 \tag{75a}$$

$$S_- \rightarrow a_1^\dagger a_2 \tag{75b}$$

$$S_3 \rightarrow (a_2^\dagger a_2 - a_1^\dagger a_1)/2 \tag{75c}$$

$$|sm\rangle \rightarrow \frac{(a_1^\dagger)^{s-m}(a_2^\dagger)^{s+m}}{\sqrt{(s+m)!(s-m)!}}|0_1, 0_2\rangle = |(s-m)_1, (s+m)_2\rangle \tag{75d}$$

The mapping preserves the commutation relations (70) of the spin operators. As can be seen from (75d), the image of the $(2s+1)$-dimensional spin Hilbert space is the subspace of the two-oscillator Hilbert space with $2s$ quantum of excitation—the so-called physical subspace [218, 220]. This subspace is invariant under the action of any operator which results by the mapping (75a)–(75d) from an arbitrary spin operator $A(S_1, S_2, S_3)$. Thus, starting in this subspace the system will always remain in it. As a consequence, the mapping yields the following identity for the matrix elements of an operator A:

$$\langle sm|A|sm'\rangle = \langle (s-m)_1, (s+m)_2|\mathscr{A}|(s-m')_1, (s+m')_2|\rangle \tag{76}$$

where $\mathscr{A}$ denotes the transformed operator which results from A through the mapping relations (75a)–(75d). In particular, if $\mathscr{A}$ denotes the time-evolution operator of the spin system, we have the the following exact identity for the propagator:

$$\langle sm|e^{-iHt}|sm'\rangle = \langle (s-m)_1, (s+m)_2|e^{-i\mathscr{H}t}|(s-m')_1, (s+m')_2\rangle \tag{77}$$

In the Schwinger representation the identity operator in the spin Hilbert space is mapped onto the constant of motion $a_1^\dagger a_1 + a_2^\dagger a_2$. The existence of this constant of motion is utilized by the Holstein–Primakoff transformation to eliminate one boson DoF, thus representing the spin DoF by a single oscillator [97]:

$$S_+ \rightarrow \sqrt{2s}\, a^\dagger \sqrt{1 - a^\dagger a/2s} \tag{78a}$$

$$S_- \rightarrow \sqrt{2s}\sqrt{1 - a^\dagger a/2s}\; a \tag{78b}$$

$$S_3 \rightarrow a^\dagger a - s \tag{78c}$$

$$|sm\rangle \rightarrow \frac{(a^\dagger)^{s+m}}{\sqrt{(s+m)!}}|0\rangle = |s+m\rangle \tag{78d}$$

The Holstein–Primakoff transformation also preserves the commutation relations (70). Due to the square-root operators in Eqs. (78a)–(78d), however, the mutual adjointness of S_+ and S_- as well as the self-adjointness of S_3 is only guaranteed in the physical subspace $\{|0\rangle, \ldots, |s+m\rangle\}$ of the transformation [219]. This flaw of the Holstein–Primakoff transformation outside the physical subspace does not present a problem on the quantum-mechanical level of description. This is because the physical subspace again is invariant under the action of any operator which results from the mapping (78) of an arbitrary spin operator $A(S_1, S_2, S_3)$. As has been discussed in Ref. 100, however, the square-root operators may cause serious problems in the semiclassical evaluation of the Holstein–Primakoff transformation.

Let us briefly mention some formal aspects of the above-introduced formalism, which have been discussed in detail by Blaizot and Marshalek [218]. First, it is noted that the both the Schwinger and the Holstein–Primakoff representations are not unitary transformations in the usual sense. Nevertheless, a transformation may be defined in terms of a formal mapping operator acting in the fermionic–bosonic product Hilbert space. Furthermore, the interrelation of the Schwinger representation and the Holstein–Primakoff representation has been investigated in the context of quantization of time-dependent self-consistent fields. It has been shown that the representations are related to each other by a nonunitary transformation. This lack of unitarity is a consequence of the nonexistence of a unitary polar decomposition of the creation and annihilation operators a_n and $a_n^\dagger$ [221] and the resulting difficulties in the definition of a proper phase operator in quantum optics [222].

Finally, it is noted that there exist alternative mappings of spin to continuous DoF. For example, Ref. 219 discusses various mappings that (like the Holstein–Primakoff transformation) represent a spin system by a single-boson DoF. The possibility of utilizing spin coherent states for this purpose is discussed in Section IX.

2. *N-Level Systems*

Let us next consider an N-level system with basis states $|\psi_n\rangle$ $(n = 1, \ldots, N)$ and the Hamiltonian

$$H = \sum_{n,m} h_{nm} |\psi_n\rangle\langle\psi_m| \tag{79}$$

In obvious analogy to Schwinger's theory of angular momentum, this N-level system can be represented by N oscillators, whereby the mapping relations for the operator and the basis states read [99]

$$|\psi_n\rangle\langle\psi_m| \to a_n^\dagger a_m \tag{80a}$$

$$|\psi_n\rangle \to |0_1, \ldots, 1_n, \ldots, 0_N\rangle \tag{80b}$$

Here a_n and $a_m^\dagger$ are the usual oscillator creation and annihilation operators with bosonic commutation relations (73), and $|0_1, \ldots, 1_n, \ldots, 0_N\rangle$ denotes a harmonic-oscillator eigenstate with a single quantum excitation in the mode n. According to Eq. (80a), the bosonic representation of the Hamiltonian (79) is given by

$$\mathcal{H} = \sum_{n,m} h_{nm} a_n^\dagger a_m \tag{81}$$

and the mapping preserves the commutation relations of the operator basis

$$[|\psi_n\rangle\langle\psi_m|, |\psi_k\rangle\langle\psi_l|] = |\psi_n\rangle\langle\psi_l|\delta_{m,k} - |\psi_k\rangle\langle\psi_m|\delta_{l,n} \tag{82}$$

The image of the N-level Hilbert space is the subspace of the N-oscillator Hilbert space with a single quantum excitation. Again, this (physical) subspace is invariant under the action of any operator which results by the mapping (80a) from an arbitrary N-level system operator. As a consequence we obtain the propagator identity

$$\langle\psi_n|e^{-iHt}|\psi_m\rangle = \langle 0_1, \ldots, 1_n, \ldots, 0_N|e^{-i\mathcal{H}t}|0_1, \ldots, 1_m, \ldots, 0_N\rangle \tag{83}$$

For a two-level system the mapping (80b) obviously coincides with Eq. (75d) for $s = 1/2$.

It is interesting to note that different bosonic Hamiltonians $\mathcal{H}$ may correspond to the same original Hamiltonian H. This ambiguity reflects the fact that a transformation of the bosonic Hamiltonian $\mathcal{H} \rightarrow \mathcal{H}'$ which corresponds to the identity transformation in the physical subspace does not change the dynamics in this subspace. For example, the two bosonic Hamiltonians

$$\mathcal{H} = E + V(a_1^\dagger a_2 + a_2^\dagger a_1) \tag{84}$$

$$\mathcal{H}' = E(a_1^\dagger a_1 + a_2^\dagger a_2) + V(a_1^\dagger a_2 + a_2^\dagger a_1) \tag{85}$$

are equivalent in the physical subspace $\{|0_1, 1_2\rangle, |1_1, 0_2\rangle\}$ and correspond to the same "discrete" two-level system Hamiltonian

$$H = E + V(|\psi_1\rangle\langle\psi_2| + |\psi_2\rangle\langle\psi_1|) \tag{86}$$

Since $\mathcal{H}$ and $\mathcal{H}'$ are equivalent in the physical subspace, both Hamiltonians generate the same quantum dynamics in this subspace. However, this is not necessarily true if approximations are employed in the evaluation of the dynamics. For example, adopting a semiclassical approximation, the quantum-mechanically equivalent Hamiltonians H and H' may yield different results. Experience shows that it is useful first to transform the Hamiltonian on the quantum-mechanical level to the simplest possible form (which in the present case would be $\mathscr{H}$) and then to apply the semiclassical approximation.

As discussed in Section VI.A for the case of spin systems, the formalism described above is not the only way to construct a mapping of a N-level system. First of all, it is clear that one may again eliminate one boson DoF by exploiting the operator $\sum_n a_n^\dagger a_n$ (which corresponds to the identity operator in the physical subspace) as a constant of motion. Furthermore, one may express the N^2-dimensional operator basis in terms of powers of the spin matrices S_1, S_2, and S_3. Employing the Holstein–Primakoff (or Schwinger) transformation, the entire N-level system can thus be represented by one (or two) oscillator(s). On a classical level, this strategy has been followed by Meyer and Miller in their spin-matrix mapping method [90]. Unfortunately, the higher powers of the spin operators result in highly nonlinear equations of motion, which are unfavorable for a semiclassical treatment.

3. *Vibronically Coupled Molecular Systems*

To apply the mapping formalism to vibronically coupled systems, we identify the $|\psi_n\rangle$ with electronic states and the h_{nm} with operators of the nuclear dynamics. Hereby, the adiabatic as well as a diabatic electronic representation may be employed. In a diabatic representation, we have [cf. Eq. (1)]

$$h_{nm} = T(\mathbf{p})\delta_{nm} + V_{nm}(\mathbf{x}) \tag{87}$$

When we also introduce "Cartesian" electronic variables $X_n = (a_n^\dagger + a_n)/\sqrt{2}, P_n = i(a_n^\dagger - a_n)/\sqrt{2}$, the molecular Hamiltonian in the diabatic oscillator representation can be written as

$$\mathscr{H} = T(\mathbf{p}) - \tfrac{1}{2}\sum_n V_{nn}(\mathbf{x}) + \tfrac{1}{2}\sum_{n,m}(X_nX_m + P_nP_m)V_{nm}(\mathbf{x}) \tag{88}$$

Here, for notational convenience, we have assumed that $V_{nm} = V_{nm}^\dagger$. We would like to emphasize that the mapping to the continuous Hamiltonian (88) does not involve any approximation, but merely represents the discrete Hamiltonian (1) in an extended Hilbert space. The quantum dynamics generated by both Hamilton operators is thus equivalent. The Hamiltonian (88) describes a general vibronically coupled molecular system, whereby both electronic and nuclear DoF are represented by continuous variables. Contrary to Eq. (1), the quantum-mechanical system described by Eq. (88) therefore has a well-defined classical analog.

B. Classical Dynamics

Since the mapping Hamiltonian (88) has a well-defined classical analog, the semiclassical or quasiclassical evaluation of general nonadiabatic problems is in principle a straightforward matter. The transition from quantum to classical mechanics can be performed by changing from the Heisenberg *operators* obeying Heisenberg's equations to the corresponding classical *functions* obeying Hamilton's equations

$$\dot{p}_n = -\frac{\partial \mathcal{H}}{\partial x_n}, \qquad \dot{x}_n = \frac{\partial \mathcal{H}}{\partial p_n} \tag{89a}$$

$$\dot{P}_j = -\frac{\partial \mathcal{H}}{\partial X_j}, \qquad \dot{X}_j = \frac{\partial \mathcal{H}}{\partial P_j} \tag{89b}$$

where now $\mathcal{H}$ denotes the classical Hamiltonian function corresponding to the Hamiltonian operator (88).

In addition to the equations of motion, one needs to specify a procedure to evaluate the observables of interest. Within a quasi-classical trajectory approach, the expectation value of an observable A is given by Eq. (16). For example, the expression for the diabatic electronic population probability, which is defined as the expectation value of the electronic occupation operator, reads

$$\begin{aligned} P_n^{\text{dia}}(t) &= \langle \tfrac{1}{2}(X_n^2 + P_n^2 - 1) \rangle_C \\ &= \int d\mathbf{x} \int d\mathbf{p} \int d\mathbf{X} \int d\mathbf{P}\; \rho_{\text{el}}(\mathbf{X}, \mathbf{P}) \rho_{\text{vib}}(\mathbf{x}, \mathbf{p}) \tfrac{1}{2}(X_n^2 + P_n^2 - 1) \end{aligned} \tag{90}$$

As has been discussed in Section III, the initial phase-space distribution ρ_{vib} for the nuclear DoF x_j and p_j may be chosen from the action-angle (18) or the Wigner (17) distribution of the initial state of the nuclear DoF. To specify the electronic phase-space distribution ρ_{el}, let us assume that the system is initially in the electronic state $|\psi_n\rangle$. According to Eq. (80b), the electronic state $|\psi_n\rangle$ is mapped onto N_{el} harmonic oscillators, whereby the nth oscillator is in its first excited state while the remaining $N_{\text{el}} - 1$ oscillators are in their ground state. The initial density operator is thus given by

$$\rho_{\text{el}} = |0_1 \ldots 1_n \ldots 0_{N_{\text{el}}}\rangle\langle 0_1 \ldots 1_n \ldots 0_{N_{\text{el}}}| \tag{91}$$

The initial electronic distribution $\rho_{\text{el}}(\mathbf{X}, \mathbf{P})$ thus factorizes in N_{el} harmonic-oscillator distributions, which may be sampled, for example, from the Wigner distribution of ρ_{el}. While this is known to work well for the ground state of the harmonic oscillator, the Wigner distribution pertaining to the first excited state may become negative, which can give rise to unphysical results. To avoid this well-known problem, it has proven advantageous to again change to classical action-angle variables N_k and Q_k [cf. Eq. (18)] and assume a constant initial action N_k for all trajectories [103].

C. Relation to Other Formulations

The mapping approach outlined above has been designed to furnish a well-defined classical limit of nonadiabatic quantum dynamics. The formalism applies in the same way at the quantum-mechanical, semiclassical (see Section VIII), and quasiclassical level, respectively. Most important, no additional assumptions but the standard semiclassical and quasi-classical approximations are needed to get from one level to another. Most of the established mixed quantum-classical methods such as the mean-field-trajectory method or the surface-hopping approach do invoke additional assumptions. The comparison of the mapping approach to these formulations may therefore (i) provide insight into the nature of these additional approximation and (ii) indicate whether the conceptual virtues of the mapping approach may be expected to result in practical advantages.

Let us first consider the relation to the mean-field trajectory method discussed in Section III. To make contact to the classical limit of the mapping formalism, we express the complex electronic variables d_n [cf. Eq. (28)] in terms of their real and imaginary parts, that is, $d_n = (X_n + iP_n)/\sqrt{2}$. Furthermore, we introduce the mean-field Hamiltonian function $\mathscr{H}_{\rm MF}$, which may be defined as

$$\begin{aligned}\mathscr{H}_{\rm MF} &= \langle \Psi(t)|H|\Psi(t)\rangle \\ &= T(\mathbf{p}) + \tfrac{1}{2}\sum_{n,m}(X_n X_m + P_n P_m)V_{nm}(\mathbf{x})\end{aligned} \tag{92}$$

As first noted by Dirac [85], the canonical equations of motion for the real variables X_n and P_n with respect to $\mathscr{H}_{\rm MF}$ are completely equivalent to Schrödinger's equation (28) for the complex variables d_n. Moreover, it is clear that the time evolution of the nuclear DoF [Eq. (32)] can also be written as Hamilton's equations with respect to $\mathscr{H}_{\rm MF}$. Similarly to the equations of motion for the mapping formalism [Eqs. (89a) and (89b)], the mean-field equations of motion for both electronic and nuclear DoF can thus be written in canonical form.

There is, however, an important conceptional difference between the two approaches. On the quasi-classical level, this difference simply manifests itself in the initial conditions chosen for the electronic DoF. Let us consider an electronic two-level system that is initially assumed to be in the electronic state $|\psi_1\rangle$. In the mean-field formulation, the initial conditions are $|d_1(0)| = 1$, $|d_2(0)| = 0$. Thus, when we change to action-angle variables [cf. Eq. (18)], the electronic initial distribution in (90), is given by $\rho_{\rm el} = \delta(N_1 - 1)\delta(N_2)$. In the mapping formalism, on the other hand, the initial electronic state $|\psi_1\rangle$ is represented by the first oscillator being in its first excited state and second oscillator being in its ground state [cf. Eq. (91)]. This corresponds to the electronic action-angle initial distribution $\rho_{\mathrm{el}} = \delta(N_1 - \frac{3}{2})\delta(N_2 - \frac{1}{2})$, stating that, just like the nuclear DoF, the electronic DoF hold zero-point energy. Since the two-level system to be described, of course, does not hold zero-point energy, the zero-point energy $\frac{1}{2}(V_{11} + V_{22})$ of the two oscillators needs to be subtracted from the Hamiltonian. In fact, when we compare the corresponding Hamiltonian functions $\mathscr{H}$ [Eq. (88)] and $\mathscr{H}_{\mathrm{MF}}$ [Eq. (92)], it is found that

$$\mathscr{H} = \mathscr{H}_{\mathrm{MF}} - \tfrac{1}{2}\textstyle\sum_n V_{nn}(\mathbf{x}) \tag{93}$$

thus assuring that the total energy is the same in both formulations.

What is the origin for the difference between the two formulations? In the mapping approach, we perform a quantum-mechanically exact transformation and subsequently employ the classical approximation to the *complete* system. As explained above, this results in harmonic-oscillator initial conditions and in the zero-point energy correction (93), which originates from nonvanishing commutators $[X_n, P_n] = i\hbar$ [99, 100]. As a consequence, the classical limit of the mapping formalism accounts for the dynamics of both quantum and classical DoF in a completely equivalent way. The mean-field trajectory method, on the other hand, is based on a partial classical limit for the nuclear DoF and, therefore, treats electronic and nuclear DoF differently with respect to their initial conditions (quantum-like initial conditions for the electronic DoF, quasi-classical initial conditions for the nuclear DoF). As shown in Section VI.E, this seemingly minor aspect may in fact completely determine the outcome of the classical modeling of nonadiabatic dynamics.

The general idea of an equivalent classical treatment of electronic and nuclear DoF was first suggested in the "classical electron-analog" models of McCurdy, Meyer, and Miller [87–91]. Exploiting various quantum-classical analogies, these authors constructed classical-path-like Hamiltonian functions similar to (92), thus treating the dynamics of electrons and nuclei on the same dynamical footing of classical mechanics. There exists a close connection between these classical models and the classical limit of the mapping approach. In particular, the classical electron analog model may be considered as the classical limit of the mapping formalism. The main difference between the two theories is of conceptual nature: While the formulation of Meyer and Miller is a *classical model*, the mapping approach is an exact quantum-mechanical formulation that allows for a well-defined classical limit. This difference becomes evident at the semiclassical level of the theory. Starting from an exact quantum-mechanical formulation, the mapping approach unambiguously defines the classical Hamiltonian as well as the boundary (or initial) conditions of the semiclassical propagator. The classical electron analog model, on the other hand, is based on a model ansatz and is therefore not unique. For example, the theory does not determine the boundary conditions of the semiclassical propagator, and "Langer-like modifications" need to be invoked to the off-diagonal elements of the Hamiltonian function to achieve meaningful semiclassical quantization conditions [89, 101]. In recent works on the semi-classical description of nonadiabatic dynamics [112–114, 116, 117], Miller and co-workers use a modified formulation of their original model [89], which is identical to the classical limit of the mapping approach proposed in Ref. 99. For a detailed discussion of the common and diverse features of the two formulations, we refer to Refs. 100 and 102.

D. Zero-Point Energy Problem and Level Density

Within the mapping approach, the electronic states are represented by products of harmonic-oscillator states, where one of the oscillators is in the first excited state and all others are in the ground state. Quantum-mechanically, the dynamics is restricted to this subspace with one quantum of excitation. In the classical limit of the mapping approach, however, this dynamical constraint is not necessarily fulfilled. The explanation for this failure of the classical formulation is given by the well-known zero-point energy (ZPE) problem of classical mechanics [223]. In quantum mechanics, each oscillator mode must hold an amount of energy that is larger or equal to the ZPE of this mode. In a classical trajectory calculation, on the other hand, energy can flow among the modes without this restriction. Since according to Eq. (90) the population probability of an electronic state is directly proportional to the mean energy content of the corresponding electronic oscillator, $P_n^{\text{dia}}(t) \propto \langle \frac{1}{2} V_{nn}(X_n^2 + P_n^2 - 1) \rangle$, this unphysical flow of ZPE may result in negative population probabilities. To illustrate the problem, Fig. 19 shows the adiabatic and diabatic population for Model I obtained with the mapping approach. It is seen that the overall agreement with the quantum results is of the same quality for the mean-field trajectory method and the mapping approach. Due to the ZPE problem, however, the latter may result in negative values of the adiabatic population.

The problem of an unphysical flow of ZPE is not a specific feature of the mapping approach, but represents a general flaw of quasi-classical trajectory methods. Numerous approaches have been proposed to fix the ZPE problem [223]. They include a variety of "active" methods [i.e., the flow of ZPE is controlled and (if necessary) manipulated during the course of individual trajectories] and several "passive" methods that, for example, discard trajectories not satisfying predefined criteria. However, most of these techniques share the problem that they manipulate *individual* trajectories, whereas the conservation of ZPE should correspond to a virtue of the *ensemble average* of trajectories.

Recently, Stock and Müller have proposed an alternative strategy to tackle the ZPE problem [102, 103, 224]. The theory is based on the observation that the unphysical flow of ZPE is a consequence of the fact that the classical phase-space distribution may enter regions of phase space that correspond to a

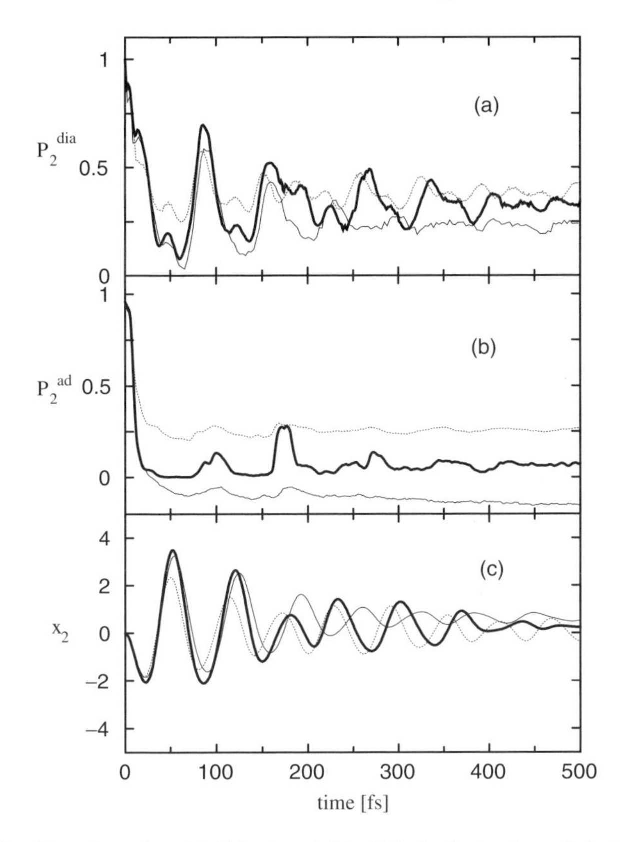

Figure 19. Time-dependent (*a*) diabatic and (*b*) adiabatic electronic excited-state populations and (*c*) vibrational mean positions as obtained for Model I. Shown are results of the mean-field trajectory method (dotted lines), the quasi-classical mapping approach (thin full lines), and exact quantum calculations (thick full lines).

violation of the uncertainty principle. To restrict the classically accessible phase space according to the rules of quantum mechanics, it has been proposed to invoke quantum corrections to the classical calculation. At the simplest level of the theory, these corrections have been shown to correspond to including only a fraction $\gamma(0 \leq \gamma \leq 1)$ of the full ZPE into the classical calculation. To study the effect of this modification on the mapping description, we note that in this formalism the electronic ZPE appears (i) in the quasi-classical description of the initial electronic state, [e.g., $\rho_{el} = \delta[N_1 - (1+\gamma/2)]\delta(N_2 - \gamma/2)]$, (ii) in the electronic projectors $\frac{1}{2}(P_n^2 + X_n^2 - \gamma)$, and (iii) in the Hamiltonian

$$\mathscr{H} = T(\mathbf{p}) + \sum_n \tfrac{1}{2}(P_n^2 + X_n^2 - \gamma)V_{nn}(\mathbf{x}) + \sum_{n>m}(X_nX_m + P_nP_m)V_{nm}(\mathbf{x}) \quad (94)$$

A value of $\gamma = 1$ corresponds to the original mapping formulation, which takes into account the full amount of ZPE. If, on the other hand, all electronic ZPE is neglected (i.e., $\gamma = 0$), the mapping approach becomes equivalent to the mean-field trajectory method.

To determine the optimal value of quantum correction γ, several criteria have been proposed, all of which are based on the idea that an appropriate classical theory should correctly reproduce long-time limits of the electronic populations. (Since the populations are proportional to the mean energy of the corresponding electronic oscillator, this condition also conserves the ZPE of this oscillator.) Employing phase-space theory, it has been shown that this requirement leads to the condition that the state-specific level densities

$$N_k(E) = \mathrm{tr}\{|\psi_k\rangle\langle\psi_k|\Theta(E - \mathscr{H})\} \tag{95}$$

should be the same for the quantum-mechanical description and its classical approximation [103, 224]. Although this is a rigorous criterion, unfortunately, it is not a very practical one since the quantum-mechanical state-specific level density in general is not easy to calculate. On a somewhat more approximate level, one may also require that the total integral quantum-mechanical level density (or level number)

$$N(E) = \mathrm{tr}\{\Theta(E - \mathscr{H})\} \tag{96}$$

should be equivalent to its classical approximation

$$N_C(E) = \frac{1}{(2\pi\hbar)^f} \int d\mathbf{x}\, d\mathbf{p} \int d\mathbf{X}\, d\mathbf{P}\; \Theta[E - \mathscr{H}(\mathbf{x}, \mathbf{p}, \mathbf{X}, \mathbf{P})] \tag{97}$$

where f is the number of electronic and vibrational DoF [102]. While this criterion may not yield the optimal quantum correction, nonetheless, it has been shown to cure the ZPE problem. We note in passing that there also have been several attempts to obtain analytical expressions for the quantum correction which, however, mostly apply to idealized model systems [224–226].

Considering the practical application of the mapping approach, it is most important to note that the quantum correction can also be determined in cases where no reference calculations exist. That is, if we *a priori* know the long-time limit of an observable, we can use this information to determine the quantum correction. For example, a multidimensional molecular system is for large times expected to completely decay in its adiabatic ground state, that is,

$$P_k^{\mathrm{ad}}(\infty) = \delta_{k0} \tag{98}$$

It is interesting to note that the latter criterion implies that the ground-state level density completely dominates the total level density—that is, that $N_0(E) \approx N(E)$. Hence the assumption (98) of complete decay into the adiabatic ground state is equivalent to the criterion that the classical and quantum total level densities should be equivalent. Furthermore, it is clear that this criterion determines an upper limit of γ. This is because larger values of the quantum correction would result in ground-state population larger than one (or negative excited-state populations).

To obtain the optimal quantum correction from the requirement (98), in general several trajectory calculations with varying values of γ need to be performed. It turns out, however, that often a *single* simulation is already sufficient. To explain this, we note that the quantum correction γ affects the adiabatic population probability $P_k^{\text{ad}}(\gamma, t) = \frac{1}{2}\langle X_k^2(\gamma, t) + P_k^2(\gamma, t) - \gamma\rangle$ directly as well as via the electronic variables X_k and P_k, which depend on γ via their initial conditions. Considering two mapping calculations that use the quantum corrections γ_1 and γ_2, respectively, one finds at long times $(t \to \infty)$ that [183]

$$
\begin{aligned}
P_0^{\text{ad}}(\gamma_1) - P_0^{\text{ad}}(\gamma_2) &= \tfrac{1}{2}(\gamma_2 - \gamma_1) + \tfrac{1}{2}\langle X_0^2(\gamma_1) - X_0^2(\gamma_2) + P_0^2(\gamma_1) - P_0^2(\gamma_2)\rangle \\
&\approx \tfrac{1}{2}(\gamma_2 - \gamma_1)
\end{aligned} \tag{99}
$$

Hence, using some first-guess correction γ_1 in a calculation that yields $P_0^{\text{ad}}(\gamma_1)$, the optimal correction γ_2 that satisfies (98) is readily obtained through $\gamma_2 = \gamma_1 + 2[P_0^{\text{ad}}(\gamma_1) - 1]$. Relation (99) has been found to be valid in systems undergoing ultrafast electronic relaxation, where at long times the electronic variables do not memorize their initial conditions [183].

E. Results

It is instructive to first consider how the classical approximation of the total level density, $N_C(E)$, depends on the amount of electronic ZPE included. Let us begin with a one-mode problem—that is, Model IVa, for which $N_C(E)$ can be evaluated analytically for high enough energies [226]. One obtains

$$
N_C(E) = \frac{(1+\gamma)}{\omega}\left[E + \frac{(1+\gamma)^2}{3}\lambda\right] \tag{100}
$$

where $\lambda = \kappa^2/2\omega$ represents the reorganization energy of the electron-transfer system. Figure 20 compares this approximation to the exact level density of the one-dimensional spin–boson model introduced above. The mapping calculation with $\gamma = 1$ is seen to match the quantum reference data almost perfectly, while the results for $\gamma = 0$ corresponding to the mean-field trajectory method underestimate the correct level density considerably. To understand this finding,

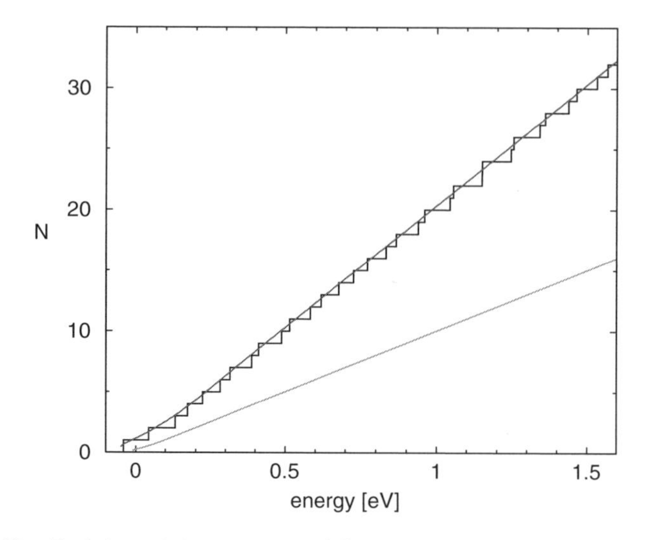

Figure 20. Total integral level density $N(E)$ as obtained for Model IVa. The mapping calculation ($\gamma = 1$, upper line) is seen to match the quantum staircase function almost perfectly, while the mean-field trajectory results ($\gamma = 0$, lower line) underestimate the correct level density considerably.

we note that for energies high enough that both electronic states are populated, the second term in Eq. (100) can be neglected, thus yielding the standard harmonic-oscillator result E/ω times the factor $1+\gamma$. In the mapping formulation ($\gamma = 1$), we therefore obtain twice the harmonic level density for the two-state system, which is in agreement with physical intuition. In the mean-field trajectory method ($\gamma = 0$), on the other hand, only a single oscillator is occupied, even at high energies. This result nicely explains the general finding that the mean-field method underestimates the level density. Considering Fig. 5, which displays the electronic populations and coherences of Model IVa, the comparison of classical and quantum calculations reveals that the mapping results are clearly superior to the mean-field results, in that they can track the quantum reference to much longer times.

Although the classical mapping formulation yields the correct quantum-mechanical level density in the special case of a one-mode spin–boson model, the classical approximation deteriorates for multidimensional problems, since the classical oscillators may transfer their ZPE. As a first example, Fig. 21a compares $N_C(E)$ as obtained for Model I in the limiting cases $\gamma = 0$ and 1 (thin solid lines) to the exact quantum-mechanical density $N(E)$ (thick line). The classical level density is seen to be either much higher (for $\gamma = 1$) or much lower (for $\gamma = 0$) than the quantum result. Since the integral level density can be

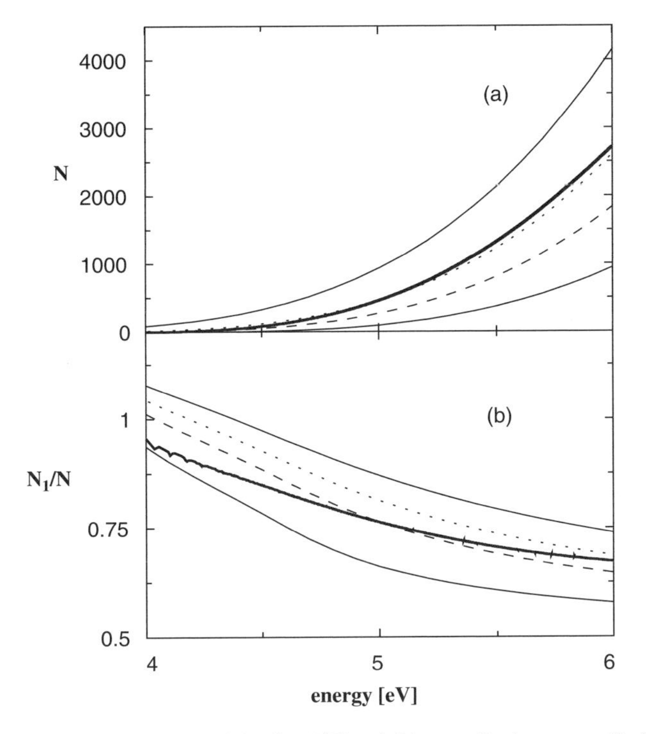

Figure 21. (*a*) Total integral level density $N(E)$ and (*b*) normalized state-specific level density $N_1(E)/N(E)$ as obtained for Model I. Thick lines show exact quantum-mechanical results, while thin lines display classical results for the limiting cases $\gamma = 0$, 1 (full lines) and the intermediate cases $\gamma = 0.44$ (dashed lines) and $\gamma = 0.68$ (dotted lines).

considered as a simple measure of the overall phase-space volume occupied by the system, this finding reflects directly the ZPE problem: Including no ZPE in the classical calculation, we have $N_C(E) \ll N(E)$ and therefore insufficient relaxation (see Fig. 19). Including the full ZPE, we have $N_C(E) \gg N(E)$ and therefore exaggerated relaxation due to unphysical flow of ZPE. Also shown are two intermediate cases, $\gamma = 0.44$ (dashed line) and $\gamma = 0.68$ (dotted line). The latter value of γ has been chosen to reproduce the quantum level density around $E = 5$ eV and indeed provides a quite accurate approximation.

The corresponding normalized state-specific level densities $N_1(E)/N(E)$ are displayed in Fig. 21b. Again, the limiting cases $\gamma = 0$ and 1 under- and overestimate the quantum data, respectively, while the intermediate value of $\gamma = 0.44$ (dashed line) provides a good approximation of the quantum results. It is noted, however, that requiring optimal agreement of the state-specific level density results in a lower ZPE correction than requiring optimal agreement of the total level density.

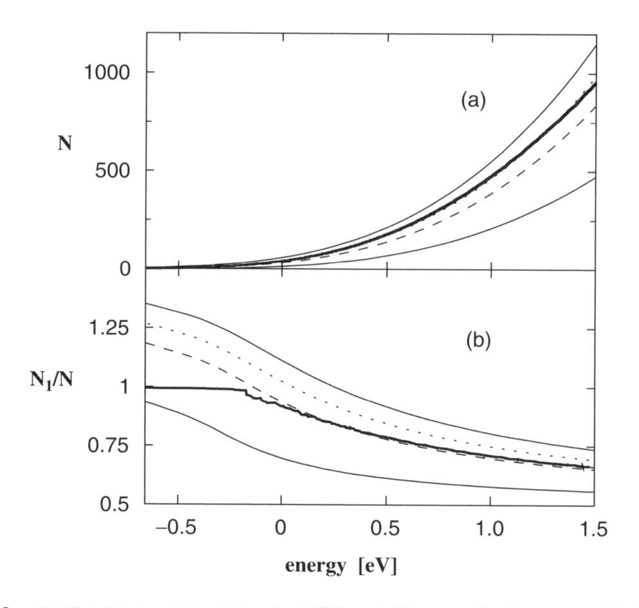

Figure 22. (a) Total integral level density $N(E)$ and (b) normalized state-specific level density $N_1(E)/N(E)$ as obtained for Model IVa. Thick lines show exact quantum-mechanical results, while thin lines display classical results for the limiting cases $\gamma = 0$, 1 (full lines) and the intermediate cases $\gamma = 0.6$ (dashed lines) and $\gamma = 0.8$ (dotted lines).

Figure 22 shows the same quantities for the intramolecular electron-transfer Model IVb. Similar to what occurs in the pyrazine model, the classical level density obtained with $\gamma = 1$ overestimates the total and state-specific level density while for $\gamma = 0$ the classical level densities are too small. Employing a ZPE correction of $\gamma = 0.8$ results in a very good agreement with the total quantum mechanical level density, while the criterion to reproduce the state-specific level density results in a ZPE correction of $\gamma = 0.6$.

Let us now study to what extent the ZPE-corrected mapping formulation is able to describe the nonadiabatic relaxation dynamics of our model problems. Beginning with Model I, Fig. 23 again shows (a) the diabatic population, (b) the adiabatic population, and (c) the mean position of the totally symmetric vibrational modes ν_{6a}. The ZPE-corrected mapping results for $\gamma = 0.44$ (thin solid lines) and the corresponding quantum results (thick lines) are seen to be in quite good overall agreement. In particular, it is found that for large times both the classical and the quantum diabatic population fluctuate around the same value of $P_2^{\text{dia}}(\infty) \approx 0.3$. This reconfirms our assertion that the agreement of the classical and the quantum state-specific level densities result in equivalent long-time values of the classical and quantum diabatic population. Most importantly,

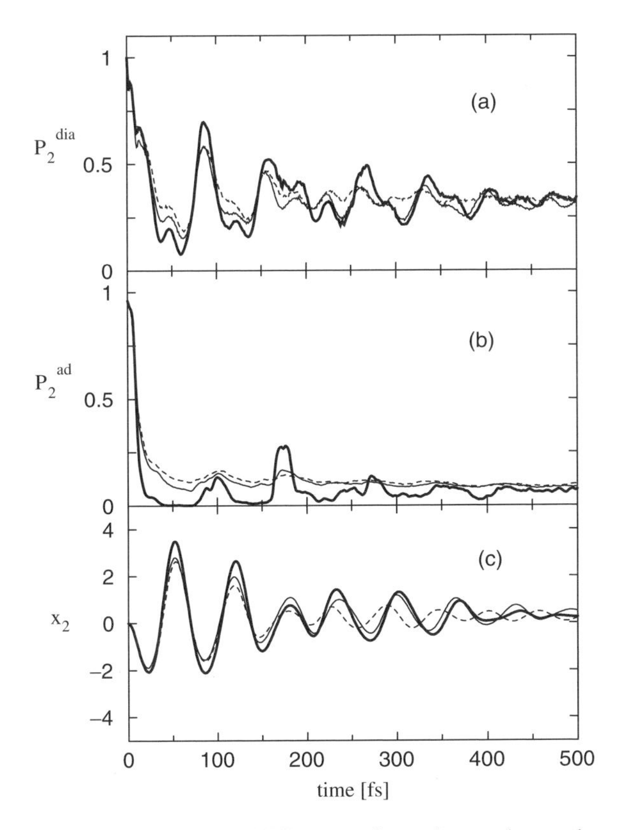

Figure 23. Same as Fig. 19, except ZPE-corrected mapping results are shown for $\gamma = 0.44$, whereby action-angle (full lines) and Wigner (dashed lines) initial conditions have been employed.

it is seen that the ZPE-corrected results also reproduce the entire time evolution of the electronic and vibrational observable under consideration. In particular, the data nicely match the coherent transients of the diabatic population and the vibrational motion.

We note in passing that the results above have been obtained with action-angle initial conditions for the electronic DoF. Employing Wigner initial conditions with $\gamma = 0.44$, similar results are obtained (dashed lines in Fig. 23), although the coherent beating of the signals is reproduced only qualitatively. Within the mapping approach, action-angle initial conditions generally have been found to be superior to Wigner initial conditions [103].

Employing the alternative criterion that requires the agreement of the classical and the quantum-mechanical total level density, we have also calculated the time-dependent observables for the quantum correction $\gamma =$

0.68 (data not shown). The overall quality of the data is quite similar to that of the data shown above, although the overall relaxation is somewhat exaggerated. As expected from the discussion above, we obtained $P_2^{\mathrm{ad}}(\infty) = 0$; that is, the requirement of a correct total level density coincides with the criterion (98).

Having determined the appropriate value of the quantum correction from the comparison of classical and quantum level densities, it is interesting to study the accuracy of the simple approximation (99). Extracting from Fig. 19 the long-time limits of the adiabatic ground-state populations as $P_0^{\mathrm{ad}}(\gamma = 0, \infty) = 0.75$ and $P_0^{\mathrm{ad}}(\gamma = 1, \infty) = 1.25$, the difference of the two populations yields $\frac{1}{2}(\gamma_2 - \gamma_1) = 0.5$, just as predicted by Eq. (99). Furthermore, we may employ the approximation to estimate the optimal quantum correction. Assuming that $P_0^{\mathrm{ad}}(\infty) = 1$, we obtain $\gamma = 0.5$, which is in qualitative agreement with the results obtained above.

We proceed with the discussion of Model II, representing the $\widetilde{C} \to \widetilde{B} \to \widetilde{X}$ internal-conversion process in the benzene cation. Figure 24 shows quantum-mechanical results, as well as various mapping results obtained for different values of the ZPE. As was discussed in Section III, the mean-field trajectory method, corresponding to the mapping approach without ZPE (i.e., $\gamma = 0$), essentially fails to describe the long-time relaxation of the population of the electronic $\widetilde{C}$- and $\widetilde{B}$-states (see panels B and E). Taking into account the full ZPE (i.e., $\gamma = 1$) shown in panels C and F, on the other hand, results in a quite good agreement with the quantum results for short times. For longer times, however, the overall relaxation is exaggerated and both the adiabatic and the diabatic population of the $\widetilde{C}$ state have negative values, indicating an unphysical flow of ZPE. It is noted that, due the higher number of electronic states and the larger excess energy of the initial state, the ZPE problem in Model II is considerably more severe than in Model I. For example, a detailed analysis showed that the mean-field trajectory method ($\gamma = 0$) predicts a level number that is about a factor of 50 smaller than the corresponding level number of the ZPE-corrected mapping result [102].

In the case of Model II, neither the state-specific nor the total quantum-mechanical level densities are available. To determine the optimal value of the ZPE correction, therefore criterion (98) was applied, which yielded $\gamma = 0.6$. The mapping results thus obtained (panels D and G) are seen to reproduce the quantum result almost quantitatively. It should be noted that this ZPE adjustment ensures that the adiabatic population probabilities remain within [0, 1] and at the same time also yields the best agreement with the quantum diabatic populations.

The excellent performance of the mapping formulation for this model encouraged us to consider an extended model of the benzene cation, for which no quantum reference calculations are available [227]. The model comprises 16 vibrational DoF and five coupled potential-energy surfaces, thus accounting for

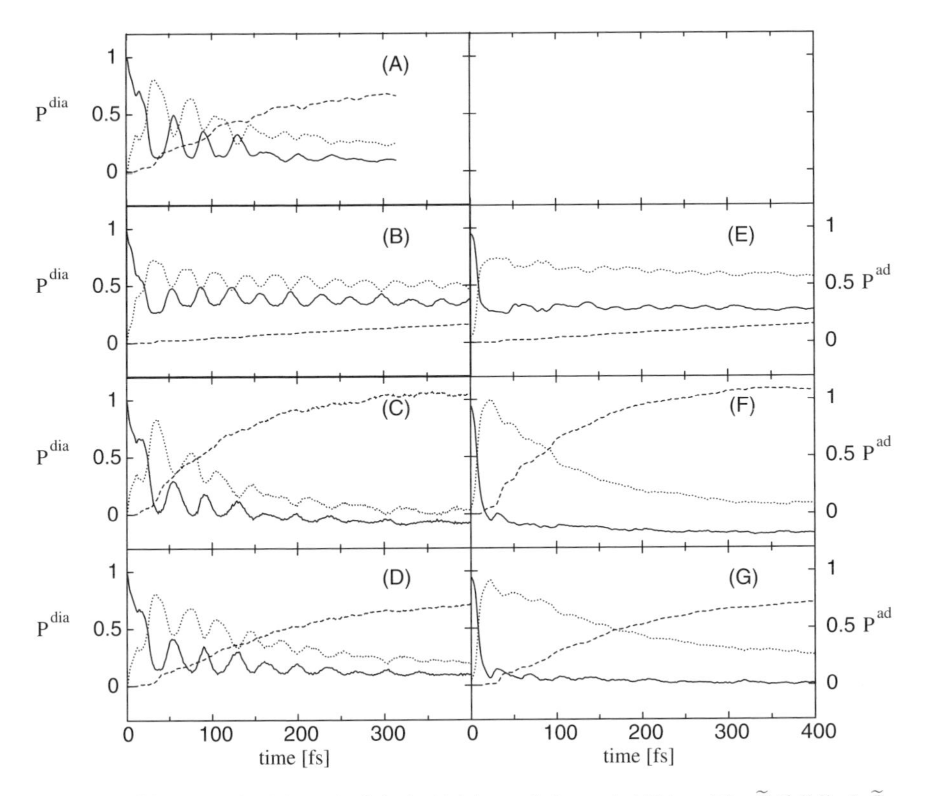

Figure 24. Diabatic (left) and adiabatic (right) population probabilities of the $\widetilde{C}$ (full line), $\widetilde{B}$ (dotted line), and $\widetilde{X}$ (dashed line) electronic states as obtained for Model II, which represents a three-state five-mode model of the benzene cation. Shown are: (A) exact quantum calculations of Ref. 180; mean-field trajectory results [panels (B),(E)]; and quasi-classical mapping results including the full [panels (C),(F)] and 60% [panels (D),(G)] of the electronic zero-point energy, respectively.

the degeneracy of the electronic states $\widetilde{X}$ and $\widetilde{B}$ [180]. Assuming again a complete decay into the adiabatic ground state, a ZPE correction of $\gamma = 0.33$ was determined. Figure 25 displays the diabatic and adiabatic population dynamics of this model. As may be expected, the inclusion of the additional DoF results in a faster dephasing of the coherent beating and in an even more efficient internal-conversion process. Recent quantum-mechanical studies of Köppel et al. [228] employing a refined five-state model including eight vibrational modes reconfirm these findings.

Next, we consider Model III, which describes an ultrafast photoinduced isomerization process. Figure 26 shows quantum-mechanical results as well as results of the ZPE-corrected mapping approach for three different observables:

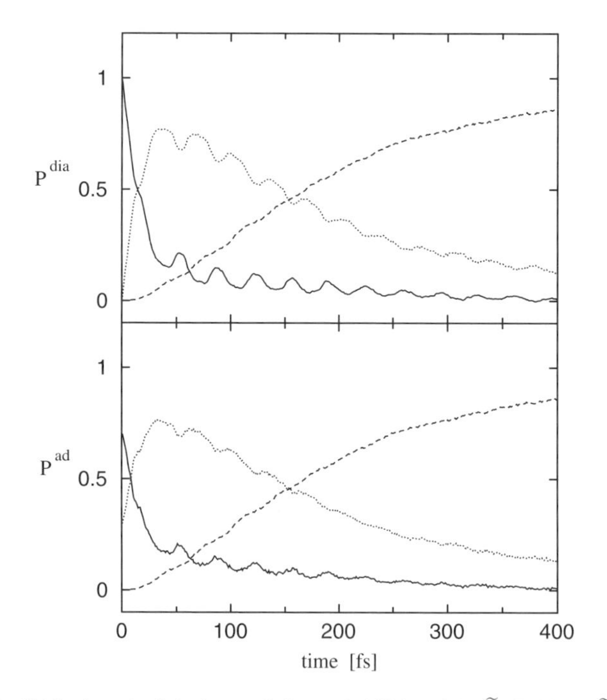

Figure 25. Diabatic and adiabatic population probabilities of the $\widetilde{C}$ (full line), $\widetilde{B}$ (dotted line), and $\widetilde{X}$ (dashed line) electronic states as obtained for a five-state 16-mode model of the benzene cation.

the adiabatic and diabatic population of the excited state, and P_{cis}, the probability that the system remains in the initially prepared *cis* conformation. A ZPE correction of $\gamma = 0.5$ has been used in the mapping calculation, based on the criterion to reproduce the quantum mechanical long-time limit of the adiabatic population. It is seen that the ZPE-corrected mapping approach represents an improvement compared to the mean-field trajectory method (cf. Fig. 3), in particular for the adiabatic population. The influence of the ZPE correction on the dynamics of the observables is, however, not as large as in the first two models.

Finally, we discuss applications of the ZPE-corrected mapping formalism to nonadiabatic dynamics induced by avoided crossings of potential energy surfaces. Figure 27 shows the diabatic and adiabatic electronic population for Model IVb, describing ultrafast intramolecular electron transfer. As for the models discussed above, it is seen that the MFT result ($\gamma = 0$) underestimates the relaxation of the electronic population while the full mapping result ($\gamma = 1$) predicts a too-small population at longer times. In contrast to the models

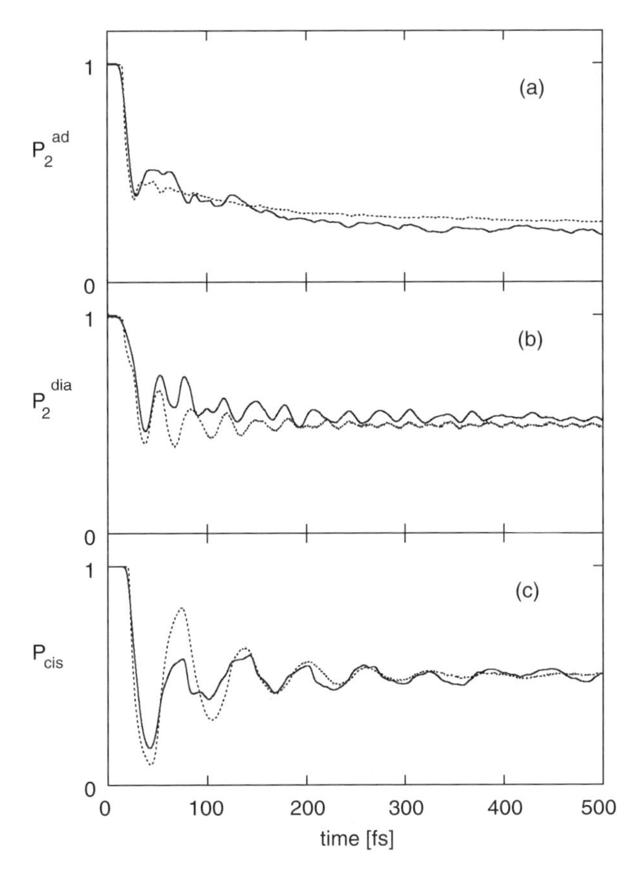

Figure 26. Time-dependent simulations of the nonadiabatic photoisomerization dynamics exhibited by Model III, comparing results of the ZPE-corrected classical mapping approach (dotted lines) and exact quantum calculations (full lines). Shown are the population probabilities $P_2^{\text{ad}}(t)$ and $P_2^{\text{dia}}(t)$ of the initially prepared adiabatic (*a*) and diabatic (*b*) electronic state, respectively, as well as the probability $P_{cis}(t)$ that the system remains in the initially prepared *cis* conformation (*c*).

discussed above, the adiabatic population does not attain an *a priori* known value at long times, and thus criterion (98) cannot be used to determine the ZPE-correction. The result obtained for a value of $\gamma = 0.6$ (which has been determined by requiring optimal agreement between the total classical and quantum mechanical level densities, see Fig. 22) is seen to reproduce the average long-time limit of the diabatic and adiabatic population rather well. A value of $\gamma = 0.8$, which was determined to reproduce the state-specific level density, results in an exaggerated relaxation of the electronic populations. Again the simple approximation (99) is found to correctly predict the dependence of

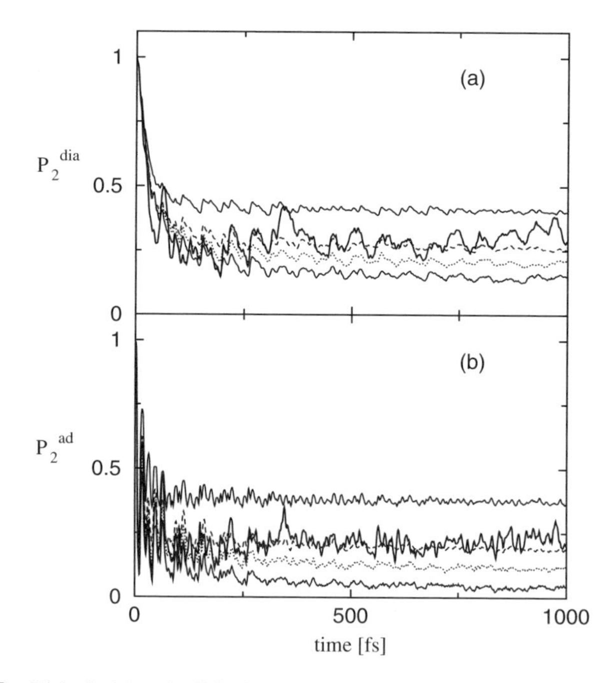

Figure 27. Diabatic (*a*) and adiabatic (*b*) population probabilities for Model IVb. Shown are exact quantum results (thick full lines), mean-field-trajectory results (upper thin full line), quasi-classical mapping results including the full zero-point energy (i.e. $\gamma = 1$, lower thin full line), as well as ZPE-corrected mapping results corresponding to $\gamma = 0.6$ (dashed line) and $\gamma = 0.8$ (dotted line), respectively.

the adiabatic long-time limit $P_2^{\text{ad}}(\gamma)$ on the quantum correction γ. For example, $P_2^{\text{ad}}(1) - P_2^{\text{ad}}(0.8) \approx 0.1 \approx P_2^{\text{ad}}(0.8) - P_2^{\text{ad}}(0.6)$.

Next, we consider Model Vb, which describes electron transfer in solution. Because the characteristic frequency of the (outer sphere) vibrational modes ($\omega_c = 7.5g_{12}$) in this model is larger than the electronic bias ($E_1 - E_2 = 2g_{12}$), the ZPE-corrections are employed to both the electronic and nuclear DoF. Figure 28 shows quantum (big dots) and classical (thin lines) results of the (a) adiabatic and (b) diabatic population of the donor state for this model. The classical results have been obtained for the limiting cases $\gamma = 0$ (dashed lines) and $\gamma = 1$ (dotted lines) as well as for the optimal ZPE correction $\gamma_{\text{opt}} = 0.6$ (full lines). While the latter results for $P_2^{\text{ad}}(t)$ decay to zero by construction, the uncorrected ($\gamma = 1$) results for the adiabatic population probability are seen to assume negative values, thus clearly exhibiting spurious flow of ZPE. Interestingly, the correct adiabatic behavior imposed on the classical system is directly reflected in the diabatic relaxation dynamics. As shown in Fig. 28b,

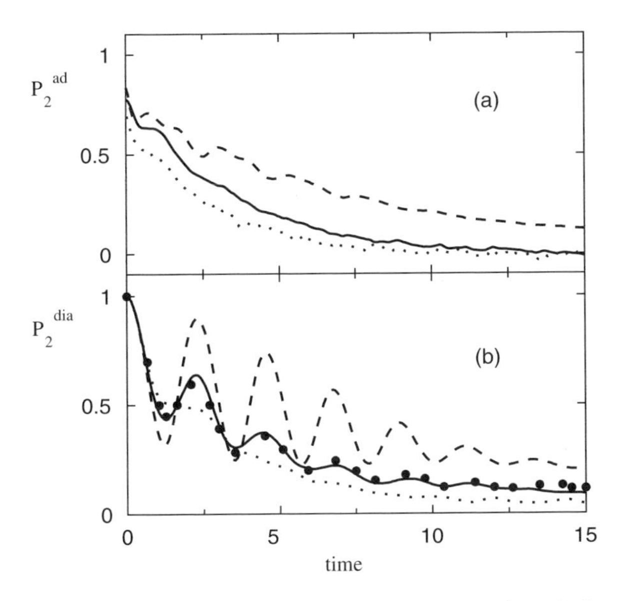

Figure 28. Time-dependent (*a*) adiabatic and (*b*) diabatic electronic excited-state populations as obtained for Model Vb describing electron transfer in solution. Quantum path-integral results [199] (big dots) are compared to mapping results for the limiting cases $\gamma = 0$ (dashed lines) and $\gamma = 1$ (dotted lines) as well as ZPE-adjusted mapping results for $\gamma_{\text{opt}} = 0.3$ (full lines).

the ZPE-adjusted results for the diabatic electronic population are in excellent agreement with quantum reference data for $\gamma_{\text{opt}} = 0.6$, while the limiting cases $\gamma = 0$ and $\gamma = 1$ considerably under- and overestimate the diabatic population probability, respectively. Note that because of the ZPE corrections for the nuclear DoF, the case $\gamma = 0$ corresponds to Boltzmann's initial condition for the nuclear DoF. The complete lack of ZPE excitation in the bath modes is seen to result in a significant underestimation of the damping of the coherent oscillations.

Let us finally consider an example that demonstrates the limits of the proposed procedure. To model charge-transfer dynamics in photosynthetic reaction centers, Sim and Makri [229] have recently reported numerically exact long-time path-integral simulations. Employing various spin–boson-type systems comprising the initial donor state $|\psi_1\rangle$, the intermediate bridge state $|\psi_2\rangle$, and the final acceptor state $|\psi_3\rangle$ of the electron transfer reaction, they investigated the effects of sequential (i.e., $E_1 > E_2 > E_3$) and superexchange (i.e., $E_2 > E_1 > E_3$) electron-transfer mechanisms. As a representative example, Fig. 29 shows results obtained for (a) the proposed sequential model of wild-type reaction centers and (b) the superexchange model for a mutant

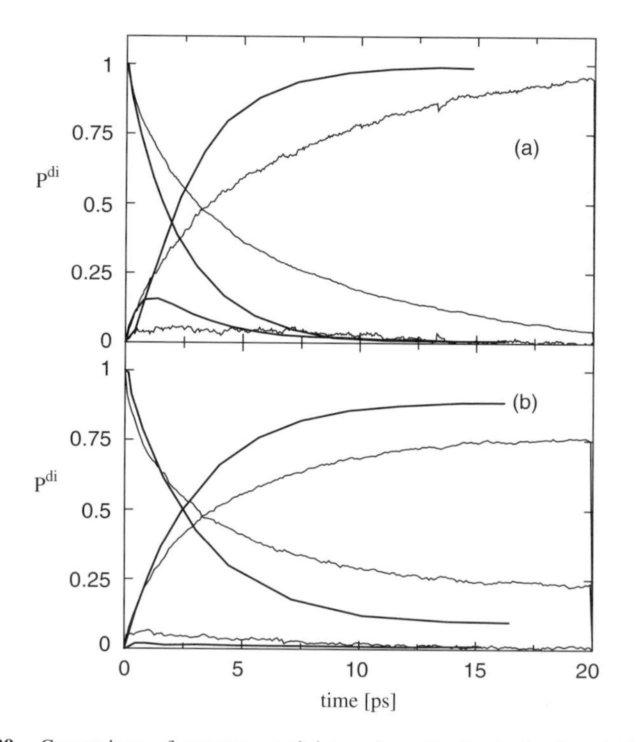

Figure 29. Comparison of quantum path-integral results (thick lines) and ZPE-corrected mapping results (thin lines) for the diabatic electronic populations of a three-state electron transfer model describing (*a*) sequential and (*b*) superexchange electron transfer.

reaction center. Both models are simulated at room temperature $T = 300$ K and employ an Ohmic spectral density with a Kondo parameter $\alpha = 1.67$ and a cutoff frequency $\omega_c = 600\,\mathrm{cm}^{-1}$. The sequential model is characterized by the vertical excitation energies $E_1 = 0, E_2 = -400\,\mathrm{cm}^{-1}$, and $E_3 = -2000\,\mathrm{cm}^{-1}$ and the diabatic couplings $g_{12} = 22\,\mathrm{cm}^{-1}$ and $g_{23} = 135\,\mathrm{cm}^{-1}$. The parameters of the superexchange model are $E_1 = 0, E_2 = 2000\,\mathrm{cm}^{-1}, E_3 = -630\,\mathrm{cm}^{-1}$, and $g_{12} = g_{23} = 240\,\mathrm{cm}^{-1}$.

The quantum results (thick lines) of the sequential model (a) are seen to completely decay in the diabatic electronic ground state of the system. For short times, furthermore, the model predicts transient electronic population in the intermediate state. In the superexchange model (b), on the other hand, there is a nonvanishing long-time population of the initial diabatic state, whereas the intermediate state is hardly ever populated. We have performed classical mapping simulations with ZPE correction (a) $\gamma = 0.6$ and (b) $\gamma = 0.34$. The ZPE corrections have been calculated via Eq. (98) and are only employed to the electronic variables. In general, the classical simulations are seen to predict a somewhat too-slow kinetics. In the case of the sequential model, the long-time limits of the diabatic populations are correctly reproduced, while this is only approximately the case for the superexchange model. Increasing the initial ZPE excitation results in faster kinetics but also in negative diabatic and adiabatic population probabilities. For example, in the case of the sequential model, the limiting cases $\gamma = 0$ and $\gamma = 1$ result in $\langle P_2^{\rm di} \rangle \approx 0.2$ and $\langle P_2^{\rm di} \rangle \approx -0.2$, respectively. In that sense, the ZPE-adjusted results can be regarded as the optimal classical results.

There are several reasons for the relatively poor performance of the classical description in Fig. 29. First, as was discussed above for Model II, the ZPE problem becomes more serious when the number of quantum states increases, because each mapped quantum state will contribute its ZPE to the system [224]. However, as has been analyzed in detail in Ref. 103, the main problem seems to be that the classical description may be nonergodic, while the quantum description is ergodic. As a consequence, the assumptions underlying our strategy are not met, and the simple quantum correction proposed does not necessarily guarantee a similar exploration of phase space in classical and quantum mechanics. Apart from issues associated with the ZPE problem, one might also suspect that the presence of nuclear tunneling represents a difficulty for the classical approach. It should be stressed, however, that the effects of electronic tunneling are, at least in principle, included in the formulation. Due to the mapping of the electronic states onto harmonic oscillators, tunneling between electronic states corresponds to energy transfer between coupled oscillators, which is readily described by the classical description.

F. Discussion

Although the mapping approach provides an in principle consistent classical description of nonadiabatic dynamics, the formulation has been shown to suffer seriously from the ZPE problem. This finding should not come as a surprise, since in the mapping formalism the electronic oscillators are constrained to the ground and first excited state [cf. Eq. (80b)], therefore representing a hard challenge for a classical description. In contrast to a semiclassical theory discussed in Section VIII, there is no rigorous solution to the ZPE problem on the quasi-classical level. Nevertheless, based on the phase-space theory developed in Refs. 103 and 224, a quantum correction γ was introduced which affects a reduction of the electronic ZPE included in the calculation. Employing this correction, the quasi-classical mapping formulation was shown to provide an accurate description of the electronic and vibrational relaxation dynamics of all five model problems considered. Generally speaking, the mapping calculations represent a significant improvement over the mean-field trajectory results and are of quality similar to that of the surface-hopping calculations. Since all these methods in some sense overstress the classical limit (e.g., ZPE excitation in the mapping approach, classically forbidden transitions in surface-hopping calculations), however, it is not easy to predict which method will perform better for a given problem.

We have discussed several ways to determine the optimal value of the quantum correction γ. Apart from the rigorous conditions associated with the state-specific and total level densities of the system, we have also established the quite useful criterion (98), which assumes that a multidimensional system at large times localizes in its adiabatic ground state. Experience shows that in most two-state problems a value of $\gamma = 0.5$ provides a fairly good description of the nonadiabatic relaxation dynamics. It is interesting to note that this rule of thumb was first established in an empirical analysis of the classical electron analog model [19]—that is, without knowing anything about the exact mapping relations or the crucial role of ZPE in this approach.

VII. PHASE-SPACE ANALYSIS AND VIBRONIC PERIODIC ORBITS

Because the mapping approach treats electronic and nuclear dynamics on the same dynamical footing, its classical limit can be employed to study the phase-space properties of a nonadiabatic system. With this end in mind, we adopt a one-mode two-state spin–boson system (Model IVa), which is mapped on a classical system with two degrees of freedom (DoF). Studying various Poincaré surfaces of section, a detailed phase-space analysis of the problem is given, showing that the model exhibits mixed classical dynamics [123]. Furthermore, a number of periodic orbits (i.e., solutions of the classical equation of motion that return to their initial conditions) of the nonadiabatic system are identified and discussed [125]. It is shown that these vibronic periodic orbits can be used to analyze the nonadiabatic quantum dynamics [126]. Finally, a three-mode model of non-adiabatic photoisomerization (Model III) is employed to demonstrate the applicability of the concept of vibronic periodic orbits to multidimensional dynamics [127].

A. Surfaces of Section

The starting point of our studies is the classical mapping Hamiltonian (88), which in the case of Model IVa is given by

$$\mathscr{H} = \frac{\omega}{2}(p^2 + x^2) + \tfrac{1}{2}\sum_{n=1,2}(X_n^2 + P_n^2 - 1)\kappa_n x + g(X_1X_2 + P_1P_2) \tag{101}$$

representing an electronic two-state system ($n = 1, 2$) with constant interstate coupling g and a single linearly displaced vibrational oscillator with diabatic potentials $V_n = \frac{1}{2}\omega x^2 + \kappa_n x$. As discussed in Section VI.A.2, the identity operator in the discrete Hilbert space yields the classical constant of motion

$$\sum_k \tfrac{1}{2}(X_k^2 + P_k^2 - 1) = N_{\text{tot}} \tag{102}$$

where $N_{\text{tot}} = $ const. for each individual trajectory. Introducing electronic action-angle variables (N_k, Q_k) via $\sqrt{2N_k + 1}\, e^{iQ_k} = X_k + iP_k$, one may eliminate one DoF by introducing the variables $N = N_2 = N_{\text{tot}} - N_1$, $Q = Q_2 - Q_1$ [89]. Assuming furthermore that $N_{\text{tot}} = \langle N_{\text{tot}} \rangle = 1$, we obtain the classical spin–boson Hamiltonian

$$\mathscr{H} = \frac{\omega}{2}(p^2 + x^2) + \kappa x(X^2 + P^2 - 2) + gX\sqrt{4 - X^2 - P^2} \tag{103}$$

where $\kappa = \kappa_2 = -\kappa_1$ and X and P denote the normal-mode coordinates corresponding to the action-angle variables N and Q. Equation (103) describes a nuclear oscillator (x, p) that is nonlinearly coupled to an electronic oscillator (X, P) representing the two-state system.

First, we wish to analyze the classical phase space of the mapped two-state problem (103) by examining the Poincaré surfaces of section of the problem [230]. Similar studies have been reported for the Ehrenfest mean-field formulation [231] and the classical electron analog model [232–234]. The surfaces of section are obtained by fixing the total energy E and the electronic phase Q and plotting the points of intersection of the classical trajectories with the x–p surface, where x and p are the nuclear phase-space variables. For fixed values of E, Q, x, and p the initial electronic action N is determined and, employing these initial conditions, a long trajectory is run. As a representative example, Fig. 30 shows the obtained surfaces of section for $Q = 0$ ($\dot{Q} > 0$) and various energies between 0.1 and 1.5 eV [123]. Except for very low or very high energies, the surfaces of section in Fig. 30 reveal that the two-state model exhibits mixed classical dynamics: Most of the area of the energetically available phase space belongs to chaotic motion, but there are also some islands of integrability. Furthermore, Fig. 30 demonstrates that the phase space accessible to the dynamics increases with the available energy. As the energy is increased, we observe two opposite effects: First, the resulting phase space projects onto the surfaces of section in the approximate form of an ellipse because at high energies the motion is governed by the oscillator potential. Second, the ellipse consists of a narrow band of irregular motion. Decreasing the energy, on the other hand, leads to a higher ratio of integrable islands compared to the chaotic part of the surfaces of section. For very low energies, $E \leq 2g = 0.2$ eV, the qualitative picture changes: The motion is mostly integrable and the forbidden area in the middle of the ellipse has vanished.

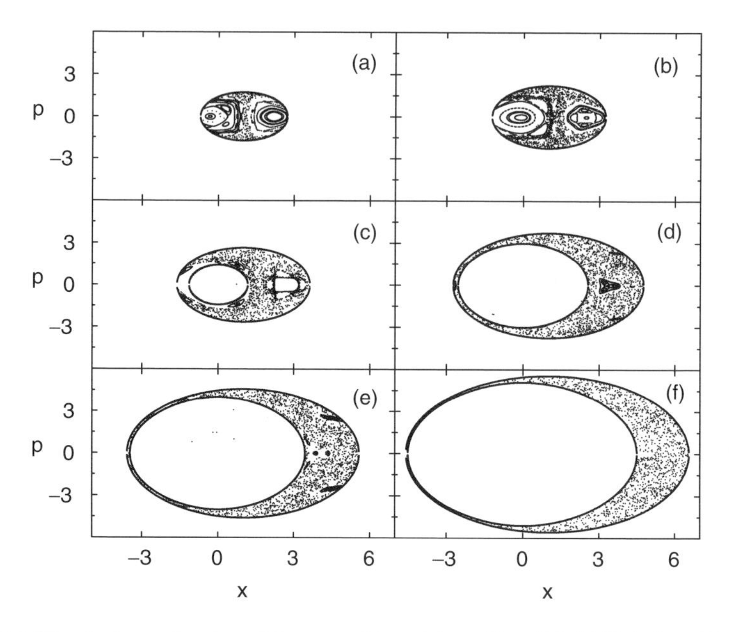

Figure 30. Surfaces of section of the mapped two-state problem (Model IVa) plotted for $Q = 0$ and the energies $E = (a)$ 0.1 eV, (b) 0.2 eV, (c) 0.3 eV, (d) 0.65 eV, (e) 1.0 eV, and (f) 1.5 eV.

We note that the integral over the energetically allowed phase space—that is, the classical level density (97)—was found in Fig. 20 to be in excellent agreement with the quantum-mechanical level density. This finding indicates that there is a valid correspondence between the quantum-mechanical two-state system and its classical mapping representation. A similar conclusion was drawn in a recent study of a mapped two-state problem, which focused on the Lyapunov exponents and the energy level statistics of the system [124, 235].

B. Vibronic Periodic Orbits

In order to identify the periodic orbits (POs) of the problem, we need to extract the periodic points (or fixed points) from the Poincaré map. Adopting the energy $E = 0.65$ eV, Fig. 31 displays the periodic points associated with some representative POs of the mapped two-state system. The properties of the orbits are collected in Table VI. The orbits are labeled by a Roman numeral that indicates how often trajectory intersects the surfaces of section during a cycle of the periodic orbit. For example, the two orbits that intersect only a single time are labeled Ia and Ib and are referred to as orbits of period I. The corresponding periodic points are located on the $p = 0$ axis at $x = 3.330$ and $x = -2.725$, respectively. Generally speaking, most of the short POs are stable and located in

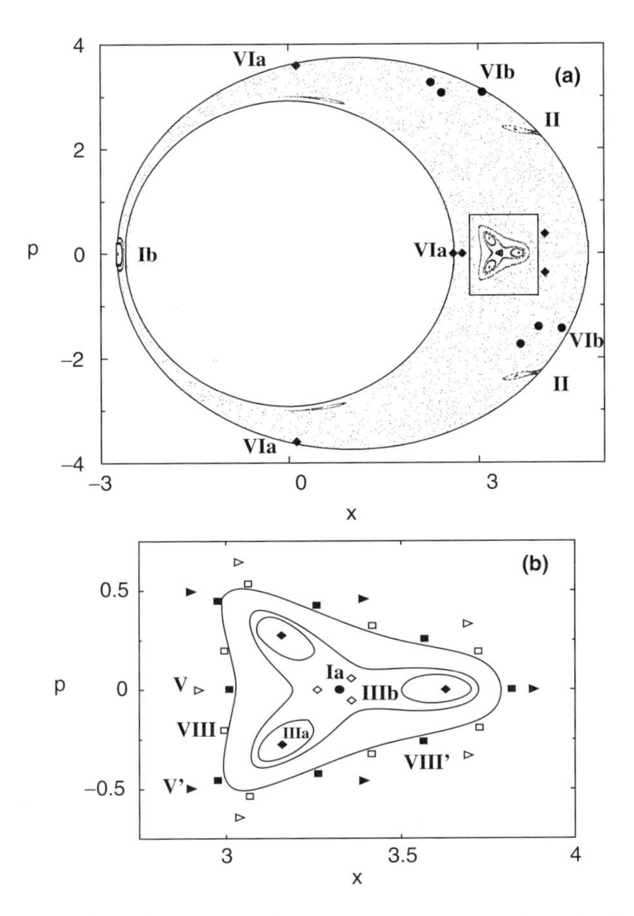

Figure 31. Periodic points of the Poincaré map at the energy $E = 0.65$ eV. The Roman numerals indicate how often the corresponding orbit intersects the surface of section. Panel (b) shows an enlargement of the main regular island around $(x, p) = (3.3, 0)$. The thin lines represent various tori of the system.

the isolated islands of phase space. Except for the orbits Ib and II, the majority of the stable POs are found in the main island around $(x, p) = (3.3,\ 0)$, which is shown as enlargement in Fig. 31b. In this island, we have found stable POs of period up to 13. Furthermore, we have identified a number of unstable POs. Except for the orbit IIIb, which is heteroclinic [230] and the shortest unstable PO found, the fixed points associated with the unstable orbits (e.g., VIa,b) are located in the chaotic region of phase space.

To represent the POs for a vibronically coupled system in a intuitively clear way, we consider the classical electronic population variable $N_{\text{dia}} =$

TABLE VI
Starting Points (x, p) for $Q = 0$, Periods T, Symmetry (yes/no), and Ljapunov Exponents λ of Some Representative Vibronic Periodic Orbits of the Mapped Two-State System, Assuming a Total Energy of 0.65 eV

Orbit	x	p	T[fs]	Symmetry	λ
Ia	3.330	0	39.2	yes	0
Ib	−2.725	0	36.4	yes	0
II	3.599	−2.352	46.3	no	0
IIIa	3.163	−0.275	117.8	yes	0
IIIb	3.261	0	117.8	yes	0.19
V	2.9216	0	196.6	no	0
VIa	4.061	−0.373	157.6	no	1.63
VIb	3.663	−1.736	135.1	yes	2.99
VIII	2.995	0.2	314.4	no	0

$(X^2 + P^2)/4$, which by construction varies between 0 (system is in $|\psi_1\rangle$) and 1 (system is in $|\psi_2\rangle$). Describing, as usual, the nuclear motion through the position x, the vibronic PO can then be drawn in the (N_{dia}, x) plane. Here, the subscript "dia" emphasizes that we refer to the population of the diabatic states $|\psi_n\rangle$, which are used to define the molecular Hamiltonian H. For interpretational purposes, on the other hand, it is often advantageous to change to the adiabatic electronic representation. Introducing the adiabatic population N_{ad}, where $N_{\text{ad}} = 0$ corresponds to the lower and $N_{\text{ad}} = 1$ to the upper adiabatic electronic state, the vibronic PO can be viewed in the (N_{ad}, x) plane. Alternatively, one may represent the vibronic PO as a curve $N_{\text{ad}} W_2 + (1 - N_{\text{ad}}) W_1$ between the adiabatic potential-energy curves $W_{1/2} = \frac{1}{2}(V_1 + V_2) \mp \frac{1}{2}\sqrt{(V_2 - V_1)^2 + 4g^2}$. That is, for $N_{\text{ad}} = 0$ the orbit moves along the lower adiabatic curve W_1, whereas for $N_{\text{ad}} = 1$ the orbit moves along the upper adiabatic curve W_2.

Figure 32 shows the diabatic (left panel) and adiabatic (right panel) representations of the most important vibronic POs of our system. All orbits are self-retracing and reflect the symmetry of the potentials with respect to the $x = 0$ axis. The shortest PO, orbit A (referred to as Ib in Table VI), is seen to vibrate between two turning points at $x \approx \pm 2.75$. Hereby, the PO oscillates between the diabatic states $|\psi_1\rangle$ and $|\psi_2\rangle$, while it is localized on the upper adiabatic electronic state. The latter finding explains the period time of 36.4 fs for the orbit, which is comparable to the period $T_2^{\text{ad}} = 35$ fs obtained by a quadratic fit of the upper adiabatic potential. Similarly, orbit B (referred to as II in Table VI) is localized on the lower adiabatic surface, thus resulting in a period of 46.3 fs, which is in good agreement with $T_1^{\text{ad}} = 50$ fs. Reflection with respect to the $x = 0$ axis generates a further PO of this type with the same period.

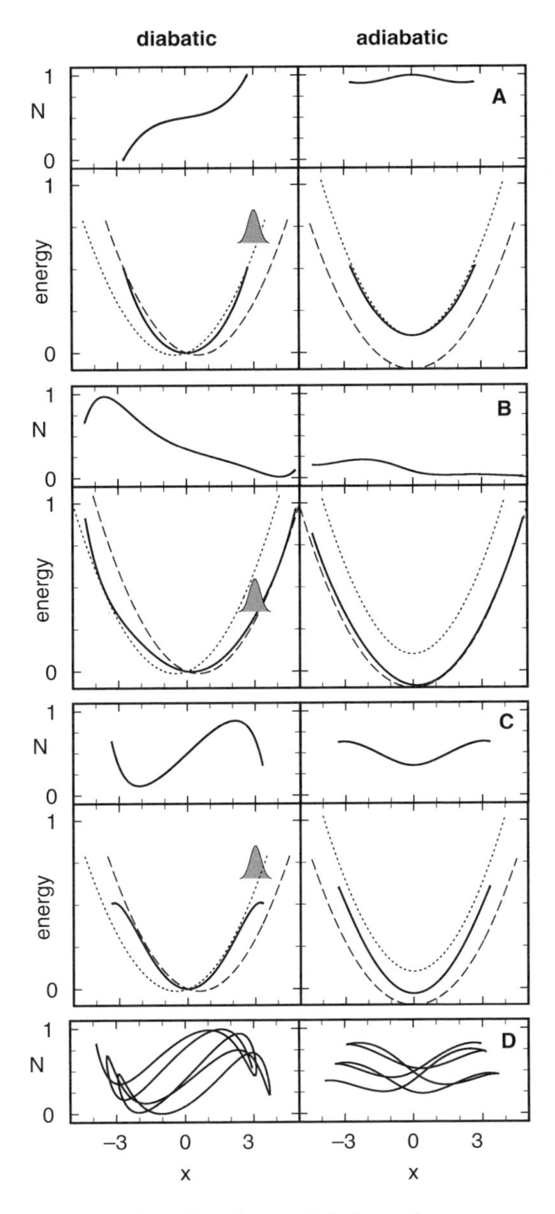

Figure 32. Vibronic periodic orbits of a coupled electronic two-state system with a single vibrational mode (Model IVa). All orbits are displayed as a function of the nuclear position x and the electronic population N, where $N = N_{\text{dia}}$ (left) and $N = N_{\text{ad}}$ (right), respectively. As a further illustration, the three shortest orbits have been drawn as curves in between the diabatic potentials V_1 and V_2 (left) as well as in between the corresponding adiabatic potentials W_1 and W_2 (right). The shaded Gaussians schematically indicate that orbits A and C are responsible for the short-time dynamics following impulsive excitation of V_2 at $(x_0, p_0) = (3, 0)$, while orbit B and its symmetric partner determine the short-time dynamics after excitation of V_1 at $(x_0, p_0) = (3, -2.45)$.

Apart from POs that virtually evolve on a single adiabatic potential-energy surface, there are numerous orbits that propagate on several or "in between" adiabatic surfaces. Orbit C (referred to as Ia in Table VI) is the shortest PO of this type, with a period of 39.2 fs. While the adiabatic population stays around $N_{\text{ad}} = 0.5$, the PO oscillates between the diabatic states with a Rabi-type frequency $\omega(x) = \sqrt{(V_2 - V_1)^2 + 4g^2}$. Averaging $\omega(x)$ over orbit C, we obtain ≈ 12 fs for the period of the diabatic oscillation; that is, within one cycle the orbit oscillates about three times between the diabatic states. It turns out that orbits of type C are of particular importance because there are numerous POs that are composed of such orbits with slightly shifted turning points. Orbit D represents an example with five repetitions and a period of 196.6 fs.

C. Periodic Orbit Analysis of Electron-Transfer Dynamics

To perform a PO analysis of nonadiabatic quantum dynamics, we employ a quasi-classical approximation that expresses time-dependent quantities of a vibronically coupled system in terms of the vibronic POs of the system [123]. Considering the quasi-classical expression (16) for the time-dependent expectation value of an observable A, this approximation assumes that the integrable islands in phase space represent the most significant contributions to the dynamics of the observables considered [236]. As a consequence, the short-time dynamics of the system is determined by its shortest POs and can be approximated by a time average over these orbits. Denoting the kth PO with period T_k by $[q_k(t), p_k(t)]$, we obtain [123]

$$A_C(t) \simeq \sum_k w_k \int_0^{T_k} d\tau \rho_0[q_k(\tau), p_k(\tau)] A[q_k(t+\tau), p_k(t+\tau)] \equiv \sum_k w_k A_k \quad (104)$$

where A_k denotes the single-orbit contribution to the observable A along the kth PO. The contribution of each PO is weighted by the integration over the initial distribution ρ_0 and by the factor w_k, which accounts for the phase-space weight of the integrable islands the respective orbit belongs to.

Let us investigate to what extent this simple classical approximation is able to describe the nonadiabatic dynamics exhibited by our model. To this end, we consider the diabatic electronic population probability $P^{\text{dia}}(t)$ defined in Eq. (21), which directly reflects the non-Born–Oppenheimer dynamics of the system. Assuming that the system is initially prepared at $x = 3$ in the diabatic state $|\psi_2\rangle$, the corresponding initial distribution ρ_0 mainly overlaps with orbits A and C, since at $x = 3$ these orbits do occupy the state $|\psi_2\rangle$. Similarly, excitation of $|\psi_1\rangle$ mainly overlaps with orbits of type B, which at $x = 3$ occupy the state $|\psi_1\rangle$ (see Fig. 32). In a first approximation, the electronic population probability $P(t)$ may therefore be calculated by including these orbits in the

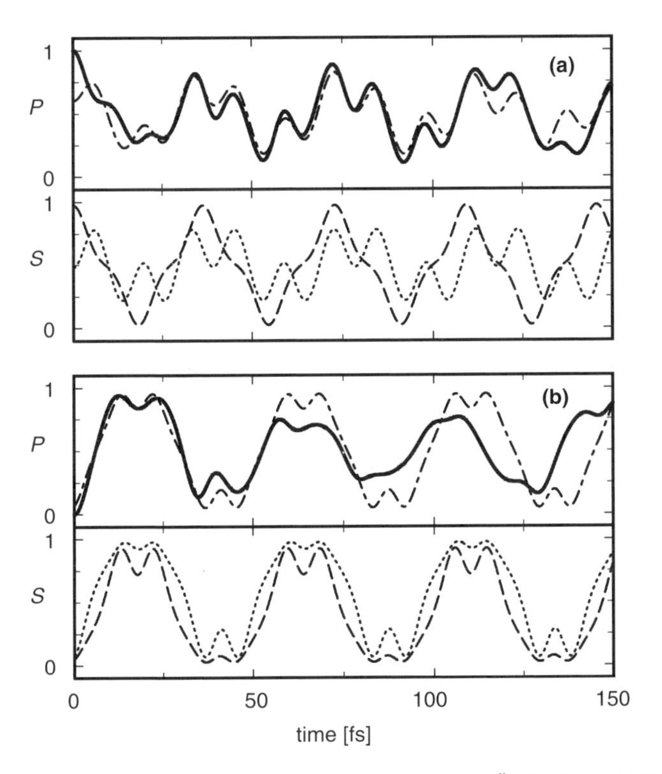

Figure 33. Time-dependent electronic population probability $P_2^{\mathrm{dia}}(t)$ of Model IVa. Compared are exact quantum results (thick lines), the classical approximation of Eq. (104) (dashed-dotted lines), and the single-orbit contribution $\mathcal{S}_k$ to the electronic populations (dashed and dotted lines). (*a*) When we excite the diabatic state $|\psi_2\rangle$, only orbits of type A (dashed line) and C (dotted line) contribute. (*b*) When we excite the diabatic state $|\psi_1\rangle$, only orbit B (dashed line) and its symmetric partner (dotted line) contribute.

evaluation of Eq. (104). Comparing this approximation $P^{\mathrm{dia}}(t) = \sum_k w_k \mathcal{S}_k(t)$ to exact quantum results, Fig. 33 reveals that in both cases already two vibronic PO are sufficient to describe the short-time dynamics of the electronic population. The deviation in the time evolution of $P(t)$ can therefore be explained as a direct consequence of the fact that different vibronic POs contribute to the nonadiabatic dynamics when the $|\psi_1\rangle$ or $|\psi_2\rangle$ state is excited.

Furthermore, all four periods found in the quantum calculation can be readily explained in terms of the single-orbit contributions $\mathcal{S}_k$ to the electronic population probability. In accordance with the discussion above, the 36-fs period of $\mathcal{S}_A$ and the 46-fs period of $\mathcal{S}_B$ reflect quasi-harmonic motion on the upper and lower adiabatic potential, respectively. The 12-fs period of $\mathcal{S}_C$ can be attributed to Rabi-type oscillations between the two diabatic states. The 8-fs beating of $\mathscr{S}_B$ is caused by the fact that due to the reversal of orbit B at the turning point $x \approx -4.5$, the PO goes within 8 fs twice through its diabatic maximum around $x \approx -3.7$ (see Fig. 32B).

The POs identified above can also be used for the analysis of other observables of our simple electron-transfer model. For example, it has been shown that a calculation employing two POs qualitatively reproduce the short-time evolution of the probability distribution $P(x,t) = \langle \Psi(t)|\psi_2\rangle|x\rangle\langle x|\langle\psi_2|\Psi(t)\rangle$ [126]. Describing complex wave-packet motion on the two coupled potential energy surfaces, this quantity is also of interest since it can be monitored in femtosecond pump-probe experiments [163]. In fact, it has been shown in Ref. 126 employing again the quasi-classical approximation (104) that the time- and frequency-resolved stimulated emission spectrum is nicely reproduced by the PO calculation. Hence vibronic POs may provide a clear and physically appealing interpretation of femtosecond experiments reflecting coherent electron transfer. We note that POs have also been used in semiclassical trace formulas to calculate spectral response functions [3].

D. Quasi-periodic Orbit Analysis of Photoisomerization Dynamics

So far, we have restricted our studies to a highly idealized one-mode two-state model of electron transfer. In an attempt to extend our investigations to multidimensional and anharmonic vibronic-coupling systems, in the following we are concerned with nonadiabatic *cis–trans* photoisomerization dynamics exhibited by Model III [127]. The model represents an ideal test case for a classical periodic-orbit analysis because preliminary studies [182] have shown that the classical mapping method is able to account for the nonadiabatic dynamics at least qualitatively, and because the wave-packet dynamics exhibited by the model is quite complex and does not possess an obvious and simple physical interpretation.

To obtain a first impression of the nonadiabatic wave-packet dynamics of the three-mode two-state model, Fig. 34 shows the quantum-mechanical probability density $P_k^{\text{ad}}(\varphi,t) = \langle\Psi(t)|\psi_k^{\text{ad}}\rangle|\varphi\rangle\langle\varphi|\langle\psi_k^{\text{ad}}|\Psi(t)\rangle$ of the system, plotted as a function of time t and the isomerization coordinate φ. To clearly show the $S_1 \to S_0$ internal-conversion process, the wave function has been projected on the S_1 (upper panel) and S_0 (lower panel) adiabatic electronic state, respectively. Starting at time $t = 0$ as a Gaussian wave packet localized at $\varphi_0 = 0$ with $p_0 = 0$, the system follows the gradient of the potential-energy surface (PES) to reach the conical intersections at $\varphi \approx \pm\pi/2$ and decays almost completely into the electronic ground state on a timescale of 100 fs. While the initial excited-state dynamics of the system is described by a localized wave-packet, the subsequent time evolution in the electronic ground state is much harder to characterize. Reaching the lower adiabatic electronic state via the "photochemical funnel" represented by the intersection, the wave function rapidly

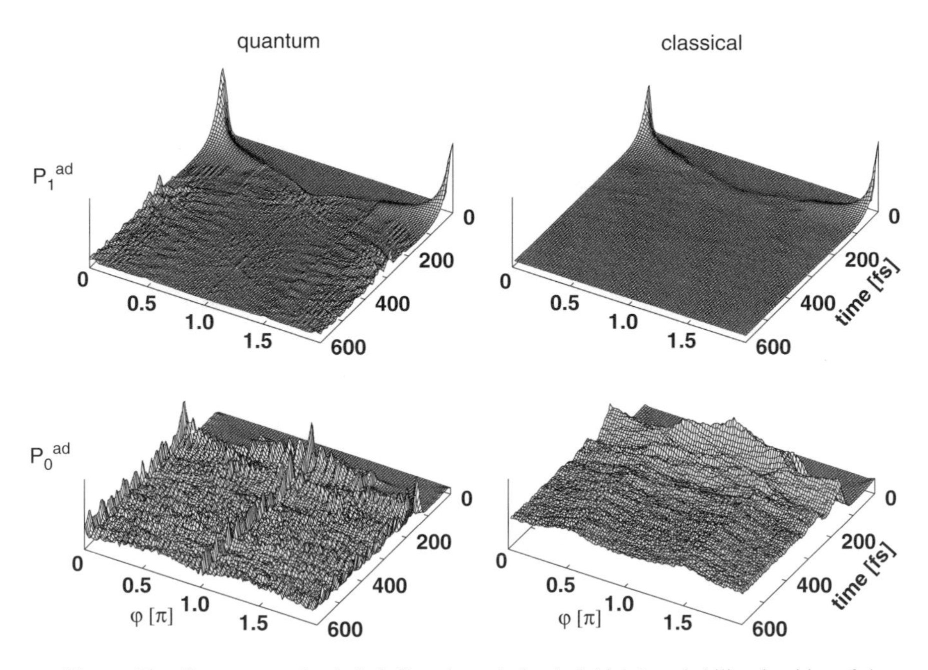

Figure 34. Quantum-mechanical (left) and quasi-classical (right) probability densities of the nonadiabatic *cis–trans* photoisomerization system (Model III) for the case $E = 0$, plotted as a function of time t and the isomerization coordinate φ. To clearly show the $S_1 \rightarrow S_0$ internal-conversion process, the wave function has been projected on the S_1 (upper panel) and S_0 (lower panel) adiabatic electronic state, respectively.

spreads out over the entire PES and remains completely delocalized for larger times. Obviously, the diffuseness of the vibrational wave function hampers a simple intuitive understanding of the dynamics in terms of specific nuclear motion. We thus have the curious situation that, even though the complete quantum-mechanical wave function of the system is available, it is far from obvious how to characterize the nuclear motion on coupled PESs in simple physical terms.

From a general experience with wave-packet motion in periodic potentials [237], it may be expected that the complexity of the dynamics is partially caused by the symmetric excitation of the system (i.e., at $\varphi_0 = 0$ and with zero initial momentum), which results in a bifurcation of the wave function right from the beginning. To simplify the analysis, it is therefore helpful to invoke an initial preparation that results in a preferred direction of motion of the system. With this end in mind, we next assume that the initial wave packet contains a dimensionless average momentum of $p_0 = 23.24$, corresponding to an

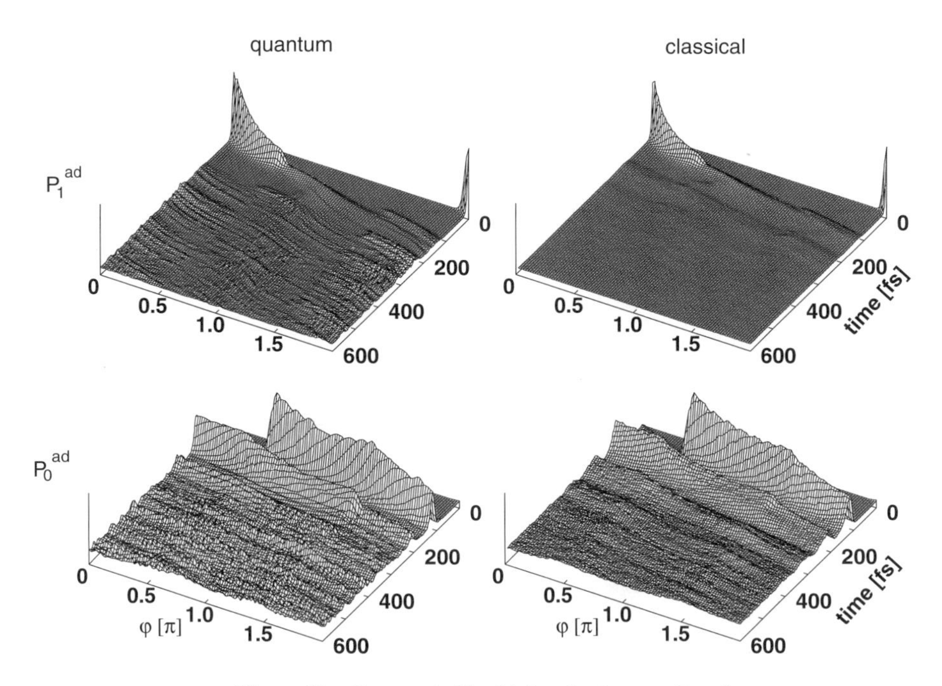

Figure 35. Same as in Fig. 34, but for the case $E > 0$.

additional kinetic energy of $T \approx 0.15$ eV. We refer to this case as "$E > 0$" and to the preparation with no kinetic energy as "$E = 0$." Figure 35 shows the time-dependent probability densities $P_0^{\text{ad}}(\varphi, t)$ and $P_1^{\text{ad}}(\varphi, t)$ for $E > 0$. As a consequence of the initial momentum, the wave packet is seen to move in one direction only. Within ~100 fs, the system reaches the conical intersection at $\varphi \approx \pi/2$ and decays almost completely into the electronic ground state. The next two encounters with conical intersections at $\varphi \approx 3/2\pi$ and $5/2\pi$ only in minor way affect the wave packet; hence its unidirectional motion on the lower adiabatic PES is continued for almost two periods. For larger times, however, smaller parts of the population recur via the intersections to the upper PES, thus rendering the overall appearance of the wave function again rather diffuse. Nevertheless, the splitting of the wave packet at conical intersections is much clearer to see than in the symmetric case discussed above, although the effect is expected there as well.

To study to what extent the mapping approach is able to reproduce the quantum results of Model III, Figs. 34 and 35 show the quasi-classical probability densities $P_k^{\text{ad}}(\varphi, t)$ for the two cases. The classical calculation for $E = 0$ is seen to accurately match the initial decay of the quantum-mechanical wave packet as well as the appearance of the wave function on the lower PES. Although the classical ground-state probability density does not include the finer details of the quantum calculation, it nevertheless reproduces the overall structure of the wave function remarkably well. Since we use a simple quasi-classical approach that does not account for possible quantum effects such as interference and tunneling, this finding indicates that—within the mapping formulation—the dynamics underlying the nonadiabatic photoisomerization is essentially of classical nature. The situation is quite similar in the case including initial kinetic energy shown in Fig. 35. Here, the quasi-classical calculation is almost in quantitative agreement with the quantum result, with the exception that the quantum calculation shows some minor recurrences of the population to the excited state also for larger times.

To apply the concept of vibronic POs established for a simple one-mode two-state model of nonadiabatic photoisomerization, we face the following problems: (i) The mapping Hamiltonian of Model III comprises two electronic and three nuclear degrees of freedom, thus resulting in a 10-dimensional phase space. (ii) The molecular system contains several timescales, ranging from the torsional period of ~ 150 fs to ~ 2 fs for the electronic oscillators. Hence the shortest possible vibronic PO with a single period in the torsional mode therefore contains almost 100 oscillations of the electronic oscillators. (iii) The phase space of Model III is found to be largely irregular. Only for energies below the surface crossing, stable orbits could be found. For energies corresponding to photoexcitation, however, most trajectories were found to sample phase space in an ergodic manner and lose the memory of their initial conditions within a few picoseconds. As a representative example, Fig. 36A shows the time evolution of the torsional coordinate φ of such a trajectory, obtained for the case $E = 0$. Starting out to propagate along the torsional potentials in positive direction, the trajectory is trapped in the *trans* potential for $0.5\,\text{ps} \lesssim t \lesssim 1.5\,\text{ps}$, before it continues to propagate back the torsional potentials in negative direction. Apart from a few oscillations in the *cis* potential, this direction of motion is kept up to $t \approx 3$ ps. During the next two picoseconds the trajectory exhibits a somewhat irregular behavior, which is followed by a quite regular oscillation in the *cis* potential.

Although the phase space of the nonadiabatic photoisomerization system is largely irregular, Fig. 36A demonstrates that the time evolution of a long trajectory can be characterized by a sequence of a few types of "quasi-periodic" orbits. The term quasi-periodic refers here to orbits that are close to an unstable periodic orbit and are, over a certain timescale, exactly periodic in the slow torsional mode and approximately periodic in the high-frequency vibrational and electronic degrees of freedom. In Fig. 36B, these orbits are schematically drawn as lines in the adiabatic potential-energy curves W_0 and W_1. The first class of quasi-periodic orbits we wish to consider are orbits that predominantly

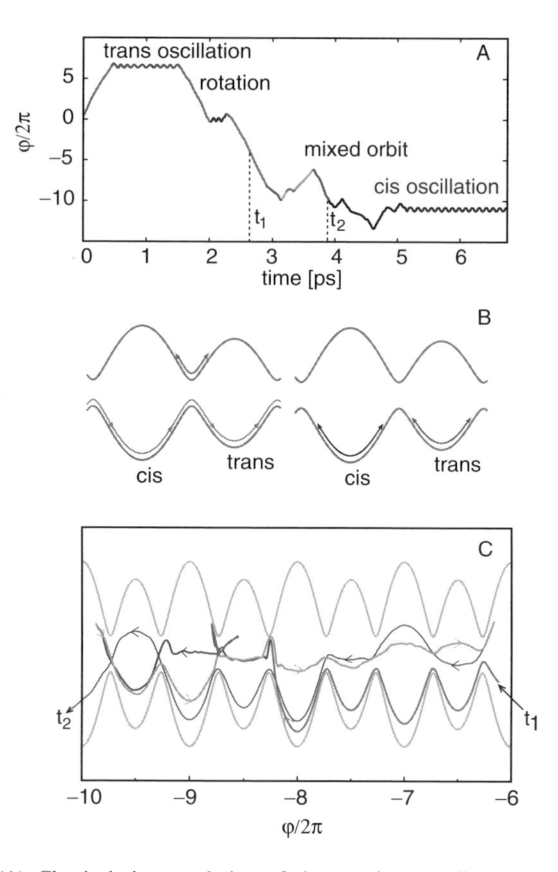

Figure 36. (A) Classical time evolution of the reaction coordinate φ as obtained for a representative trajectory describing nonadiabatic photoisomerization. (B) The vibronic motion of the system can be characterized by various quasi-periodic orbits, which are schematically drawn here as lines in the adiabatic potentials W_0 and W_1. (C) As an example of a mixed orbit, the time evolution between $t_1 = 2.7$ ps and $t_2 = 3.9$ ps is shown here in more detail.

evolve on the ground-state adiabatic PES. For example, there are orbits that oscillate in the ground-state potential well of the *cis* configuration and the *trans* configuration, respectively. These orbits will be referred to as "*cis* oscillations" and "*trans* oscillations." Furthermore, there are orbits that propagate along the ground-state torsional potential in a single direction. As this motion corresponds to a rotation of the molecule, these orbits will be referred to as "rotating orbits." The next class of quasi-periodic orbits to consider are orbits that predominantly propagate on the excited-state potential-energy surface. The most common kind here is an orbit that oscillates in the excited-state potential well between the *cis* and *trans* configuration. Interestingly, we hardly found rotating orbits on the upper PES for the model system and energies considered. The last class of quasi-periodic orbits consists of various "mixed orbits," which are trajectories that propagate "in between" two potentials—that is, on a superposition of the adiabatic PESs (see Fig. 36C). In this case, it is found that the system may "spontaneously" reverse the direction of torsional motion, whenever it approaches a surface crossing.

We are now in a position to analyze the wave-packet dynamics shown in Figs. 34 and 35 in terms of the quasi-periodic orbits. To investigate the types of orbits contributing to the wave-packet dynamics, we have performed a statistical analysis of all trajectories that contribute in the quasi-classical sampling procedure over a timescale of 600 fs. Let us begin with the case $E > 0$ shown in Fig. 35. Due to the initial mean momentum, the classical motion is clearly dominated by rotating orbits. Indeed, the analysis reveals that a large part (46%) of the trajectories perform continuous rotating motion (i.e., no turns) during the first 600 fs. About 30% of the trajectories exhibit a single turn, 14% two turns, and only 10% three and more turns. As may be expected, the rotating orbits show up as diagonally moving wave trains with a recurrence time of $\sim$100 fs. The trajectories with one or more turns, on the other hand, mainly show up in the excited-state distribution, which indicates a reflection of the orbits at the conical intersections at $\varphi = 0.5\pi$ and 1.5π, occurring at times $\sim$100 and $\sim$150 fs. At larger times, the system predominately evolves on the lower PES. Again, the turning points of the orbits at the conical intersections can be seen clearly. Furthermore, it is noted that these "turning orbits" result in quite diffuse structures, which may explain the increasing diffuseness of the quantum-mechanical wave function at larger times.

In the case $E = 0$, the quantum-mechanical wave function shown in Fig. 34 was found to be quite delocalized and diffuse. Interestingly, the statistical analysis reveals [127] that the prominent broad appearance of the wave packet is already caused by continuously rotating orbits (32% relative contribution). In other words, the clear signature of the rotating motion, which was found for the preparations $E > 0$ and $E < 0$, is hardly visible in this case. This somewhat surprising effect arises because the energy $E = 0$ corresponds to a preparation of the system right at the turning point—that is, on top of the torsional barrier. As a consequence, we obtain a large contribution of irregular orbits—in particular, mixed rotating orbits. The latter exhibit a broad distribution of periods and thus lead to a rapid delocalization of the resulting wave trains. Adding the orbits containing one or more turns, moreover, yields the quite structureless appearance of the total distribution. Again, the turns mostly occur in the vicinity of the conical intersections at $\varphi = 0.5\pi$ and 1.5π.

To summarize, we have characterized the nonadiabatic photoisomerization dynamics in terms of a few quasi-periodic orbits. These orbits are close to an unstable periodic orbit but are exactly periodic only with respect to the slow reaction coordinate of the system and only over a certain timescale. Although the phase space of the nonadiabatic photoisomerization system was found to be largely irregular, only a few types of quasi-periodic orbits were shown to be of importance for the time evolution of the system, including rotating orbits and orbits that oscillate in the *cis* or *trans* configuration of the electronic ground state. While the majority of orbits evolve on the lower adiabatic PES, there also are mixed rotating orbits, which evolve in between the coupled PESs. Since the latter are mostly irregular orbits that exhibit a broad distribution of periods, mixed orbits lead to a rapid delocalization of the resulting wave trains. The wave-packet dynamics associated with a nonadiabatic photoreaction therefore represents a beautiful example of the well-known correspondence of diffuse quantum-mechanical wave functions and irregular classical trajectories [3, 25, 238–243]. In the case of nonadiabatic photoisomerization, the chaotic behavior is mainly caused by (i) the relatively high energies corresponding to photoexcitation, (ii) the large anharmonicity of the isomerization potentials, and (iii) the reflection of the trajectory at surface crossings.

VIII. SEMICLASSICAL MAPPING FORMULATION

All approaches for the description of nonadiabatic dynamics discussed so far have used the simple quasi-classical approximation (16) to describe the dynamics of the nuclear degrees of freedom. As a consequence, these methods are in general not able to account for processes or observables for which quantum effects of the nuclear degrees of freedom are important. Such processes include nuclear tunneling, interference effects in wave-packet dynamics, and the conservation of zero-point energy. In contrast to quasi-classical approximations, semiclassical methods take into account the phase $\exp(iS_t/\hbar)$ of a classical trajectory and are therefore capable—at least in principle—of describing quantum effects.

A semiclassical description is well established when both the Hamilton operator of the system and the quantity to be calculated have a well-defined classical analog. For example, there exist several semiclassical methods for calculating the vibrational autocorrelation function on a single excited electronic surface, the Fourier transform of which yields the Franck–Condon spectrum [108, 109, 150, 244]. In particular, semiclassical methods based on the initial-value representation of the semiclassical propagator [104–111, 245–248], which circumvent the cumbersome root-search problem in boundary-value-based semiclassical methods, have been successfully applied to a variety of systems (see, for example, Refs. 110, 111, 161, and 249 and references therein). The mapping procedure introduced in Section VI results in a quantum-mechanical Hamiltonian with a well-defined classical limit, and therefore it extends the applicability of the established semiclassical approaches to nonadiabatic dynamics. The thus obtained semiclassical version of the mapping approach, as well as the equivalent formulation that is obtained by requantizing the classical electron analog model of Meyer and Miller [112], have been applied to a variety of systems with nonadiabatic dynamics in recent years [99, 100, 112–122, 130, 131].

Following a brief introduction of the basic concepts of semiclassical dynamics, in particular of the semiclassical propagator and its initial value representation, we discuss in this section the application of the semiclassical mapping approach to nonadiabatic dynamics. Based on numerical results for the $S_1 - S_2$ conical intersection in pyrazine as well as several spin–boson models, we discuss advantages and problems of this semiclassical method.

A. Semiclassical Propagator

To introduce the basic concept of a semiclassical propagator, let us consider an n-dimensional quantum system with Hamiltonian H, which is assumed to possess a well-defined classical analog. In order to obtain the semiclassical approximation to the transition amplitude $\langle \Psi_f | e^{-iHt/\hbar} | \Psi_i \rangle$ between the initial state $|\Psi_i\rangle$ and the final state $|\Psi_f\rangle$, the amplitude is expressed in terms of the coordinate-dependent propagator $\langle \mathbf{q}_f | e^{-iHt/\hbar} | \mathbf{q}_i \rangle$ (to illustrate the semiclassical approximation, we include $\hbar$ in the formulas of this subsection explicitly)

$$\langle \Psi_f | e^{-iHt/\hbar} | \Psi_i \rangle = \int d\mathbf{q}_f \int d\mathbf{q}_i \langle \Psi_f | \mathbf{q}_f \rangle \langle \mathbf{q}_f | e^{-iHt/\hbar} | \mathbf{q}_i \rangle \langle \mathbf{q}_i | \Psi_i \rangle \tag{105}$$

which then is evaluated within the semiclassical Van Vleck–Gutzwiller approximation [3]

$$\langle \mathbf{q}_f | e^{-iHt/\hbar} | \mathbf{q}_i \rangle_{\mathrm{VVG}} = \sum_{\mathrm{traj}} \frac{e^{iS_t/\hbar - i\pi\nu/2}}{\sqrt{\det(2\pi i\hbar \partial \mathbf{q}_f / \partial \mathbf{p}_i)}} \tag{106}$$

Here the sum includes all trajectories that start from point $\mathbf{q}_0 = \mathbf{q}_i$ at time 0 and end up at point $\mathbf{q}_t = \mathbf{q}_f$ at time t, $S_t = \int_0^t d\tau (\mathbf{p} \cdot \dot{\mathbf{q}} - H)$ is the classical action along such a trajectory, and the monodromy matrix elements $\partial \mathbf{q}_f / \partial \mathbf{p}_i$ account for the dependency of the trajectory $\mathbf{q}_t$ with respect to its initial momentum $\mathbf{p}_0$. The Maslov index ν counts the zeroes of the Van Vleck determinant. The Van Vleck–Gutzwiller propagator (106) can be obtained by invoking the stationary-phase approximation to the Feynman path-integral representation of the quantum propagator [250, 251]. Its physical significance is that for a system in the semiclassical limit, where a typical action is large compared to $\hbar$, the major contribution to the propagator $\langle \mathbf{q}_f | e^{-iHt/\hbar} | \mathbf{q}_i \rangle$ comes from the classical trajectories satisfying the correct boundary condition.

The evaluation of the semiclassical Van Vleck–Gutzwiller propagator (106) amounts to the solution of a boundary-value problem. That is, given a trajectory characterized by the position $\mathbf{q}(t) = \mathbf{q}_t$ and momentum $\mathbf{p}(t) = \mathbf{p}_t$, we need to find the roots of the equation $\mathbf{q}_t = \mathbf{q}_t(\mathbf{q}_0, \mathbf{p}_0)$. To circumvent this cumbersome root search, one may rewrite the semiclassical expression for the transition amplitude (105) as an initial-value problem [104–111]

$$\langle \Psi_f | e^{-iHt/\hbar} | \Psi_i \rangle_{\text{VVG}} = \int \frac{d\mathbf{q}_0 d\mathbf{p}_0}{(2\pi i\hbar)^{N/2}} \langle \Psi_f | \mathbf{q}_t \rangle \left[\det\left(\frac{\partial \mathbf{q}_t}{\partial \mathbf{p}_0} \right) \right]^{1/2} e^{iS_t(\mathbf{q}_0,\mathbf{p}_0)/\hbar} \langle \mathbf{q}_0 | \Psi_i \rangle \tag{107}$$

As a consequence, the semiclassical propagator is given as a phase-space integral over the initial conditions $\mathbf{q}_0$ and $\mathbf{p}_0$, which is amenable to a Monte Carlo evaluation. For this reason, semiclassical initial-value representations are regarded as the key to the application of semiclassical methods to multidimensional systems.

In the past two decades, a variety of semiclassical initial-value representations have been developed [105–111], which are equivalent within the semiclassical approximation (i.e., they solve the Schrödinger equation to first order in $\hbar$), but differ in their accuracy and numerical performance. Most of the applications of initial-value representation methods in recent years have employed the Herman–Kluk (coherent-state) representation of the semiclassical propagator [105, 108, 187, 245, 252–255], which for a general n-dimensional system can be written as

$$\left(e^{-iHt/\hbar} \right)_{HK} = \int \frac{d\mathbf{q}_0 d\mathbf{p}_0}{(2\pi\hbar)^n} |\mathbf{q}_t \mathbf{p}_t \rangle C_t e^{iS_t/\hbar} \langle \mathbf{q}_0 \mathbf{p}_0 | \tag{108}$$

where $(\mathbf{p}_0, \mathbf{q}_0)$ are initial momenta and coordinates for classical trajectories, $\mathbf{p}_t = \mathbf{p}_t(\mathbf{p}_0, \mathbf{q}_0)$ and $\mathbf{q}_t = \mathbf{q}_t(\mathbf{p}_0, \mathbf{q}_0)$ are the classically time-evolved phase-space variables and S_t is the classical action integral along the trajectory. The preexponential factor C_t is given by

$$C_t(\mathbf{p}_0, \mathbf{q}_0) = \left\{ \det\left[\frac{1}{2} \left(\gamma^{1/2} \frac{\partial \mathbf{q}_t}{\partial \mathbf{q}_0} \gamma^{-1/2} + \gamma^{-1/2} \frac{\partial \mathbf{p}_t}{\partial \mathbf{p}_0} \gamma^{1/2} - i\hbar \gamma^{1/2} \frac{\partial \mathbf{q}_t}{\partial \mathbf{p}_0} \gamma^{1/2} \right.\right.\right.$$
$$\left.\left.\left. + \frac{i}{\hbar} \gamma^{-1/2} \frac{\partial \mathbf{p}_t}{\partial \mathbf{q}_0} \gamma^{-1/2} \right) \right] \right\}^{1/2} \tag{109}$$

As a result of the balanced treatment of position and momentum, it involves a combination of all elements of the monodromy matrix

$$\mathbf{M}_t = \begin{pmatrix} \frac{\partial \mathbf{p}_t}{\partial \mathbf{p}_0} & \frac{\partial \mathbf{p}_t}{\partial \mathbf{q}_0} \\ \frac{\partial \mathbf{q}_t}{\partial \mathbf{p}_0} & \frac{\partial \mathbf{q}_t}{\partial \mathbf{q}_0} \end{pmatrix} \tag{110}$$

In Eq. (109), γ denotes an n-dimensional diagonal matrix, with element γ_j being the width parameter for the coherent state of the jth dimension. The coordinate space representation of an n-dimensional coherent state is the product of n one-dimensional minimum uncertainty wave packets

$$\langle \mathbf{x}|\mathbf{pq}\rangle = \prod_{j=1}^{n} \left(\frac{\gamma_j}{\pi}\right)^{1/4} e^{-\frac{\gamma_j}{2}(x_j-q_j)^2 + ip_j(x_j-q_j)/\hbar} \tag{111}$$

Within the applicability of the semiclassical approximation, the propagator (108) is rather insensitive to the particular value of the width parameters γ_j, but this parameter can of course affect the numerical efficiency of the calculation. In the numerical studies presented below, we have chosen the width γ_j as the width of the harmonic ground state of the jth vibrational mode. In the dimensionless units used here, this choice corresponds to $\gamma_j = 1$ for all degrees of freedom.

The calculation of the Herman–Kluk propagator for multidimensional nonseparable systems is a challenging task because it involves a multidimensional integral over an oscillating integrand. In addition, the preexponential factor C_t can become large for chaotic trajectories. Therefore, it is rather difficult to converge the integral for longer times using simple Monte Carlo integration schemes. Several smoothing techniques have been proposed to overcome this well-known problem of semiclassical propagators [108, 109, 122, 256–259]. The basic idea of these techniques is to integrate out the local oscillations analytically, using a linearization of the integrand over a small phase-space cell. In some of the numerical results reported below, we have adopted the method of Walton and Manolopoulos [109]. This method combines the Herman–Kluk propagator with the cellular dynamics algorithm of Heller [106]. It is based on the Filinov [260, 261] or stationary-phase Monte Carlo method [262].

B. Nonadiabatic Dynamics

The mapping procedure introduced in Section VI results in the quantum-mechanical Hamiltonian

$$\hat{\mathscr{H}} = T(\hat{\mathbf{p}}) - \tfrac{1}{2}\sum_{n=1}^{N_{\rm el}} V_{nn}(\hat{\mathbf{x}}) + \tfrac{1}{2}\sum_{n,m=1}^{N_{\rm el}} (\hat{X}_n\hat{X}_m + \hat{P}_n\hat{P}_m)V_{nm}(\hat{\mathbf{x}}) \tag{112}$$

describing a vibrationally coupled electronic $N_{\rm el}$-state system including $N_{\rm vib}$ vibrational degrees of freedom via $\mathbf{x} = (x_1, \ldots, x_{N_{\rm vib}})$. Because, in contrast to the original Hamiltonian (1), the mapping Hamiltonian (112) includes only continuous degrees of freedom, the problem allows for a straightforward semiclassical evaluation. To illustrate the concept, let us consider the transition amplitude $\langle\psi_f|\langle\mathbf{v}_f|e^{-iHt}|\mathbf{v}_i\rangle|\psi_i\rangle$ as defined—for example, in Eq. (24). The semiclassical mapping approach consists of two steps: First, one uses the quantum mechanical identity (83) between the discrete and the continuous representation

$$\langle\psi_f|\langle\mathbf{v}_f|e^{-i\hat{H}t}|\mathbf{v}_i\rangle|\psi_i\rangle = \langle 0_1,\ldots,1_f,\ldots,0_{N_{el}}|\langle\mathbf{v}_f|e^{-i\hat{\mathscr{H}}t}|\mathbf{v}_i\rangle|0_1,\ldots,1_i,\ldots,0_{N_{el}}\rangle \tag{113}$$

Subsequently, any of the well-established semiclassical approximations for the quantum propagator $e^{-i\hat{\mathscr{H}}t}$ can be applied. Employing, for example, the Herman–Kluk propagator, the semiclassical approximation for this transition amplitude is given by

$$\begin{aligned}\langle\psi_f|\langle\mathbf{v}_f|e^{-i\hat{H}t}|\mathbf{v}_i\rangle|\psi_i\rangle_{HK} = &\int\frac{d\mathbf{X}_0 d\mathbf{X}_0}{(2\pi)^{N_{\mathrm{el}}}}\int\frac{d\mathbf{x}_0 d\mathbf{p}_0}{(2\pi)^{N_{\mathrm{vib}}}}\langle 0_1,\ldots,1_f,\ldots,0_{N_{\mathrm{el}}}|\mathbf{X}_t\mathbf{P}_t\rangle\\ &\times\langle\mathbf{v}_f|\mathbf{x}_t\mathbf{p}_t\rangle C_t e^{iS_t}\langle\mathbf{x}_0\mathbf{p}_0|\mathbf{v}_i\rangle\langle\mathbf{X}_0\mathbf{P}_0|0_1,\ldots,1_i,\ldots,0_{N_{\mathrm{el}}}\rangle\end{aligned} \tag{114}$$

In contrast to the quasi-classical approaches discussed in the previous chapters of this review, Eq. (114) represents a description of nonadiabatic dynamics which is "semiclassically exact" in the sense that it requires only the basic semiclassical Van Vleck–Gutzwiller approximation [3] to the quantum propagator. Therefore, it allows the description of electronic and nuclear quantum effects.

In what follows, we wish to discuss some properties of the mapping formalism which are particularly relevant to its semiclassical implementation. The mapping procedure results in Hamiltonian (112), which—compared to a Hamiltonian that describes nuclear dynamics on a single potential-energy surface—has a rather peculiar structure. For example, it involves kinetic coupling terms for the electronic degrees of freedom and a nontrivial metric. Furthermore, the classical limit $\mathscr{H}$ of the mapping Hamiltonian (112) gives rise to the following equations of motion for the nuclear degrees of freedom:

$$\ddot{\mathbf{x}} = -\sum_n \tfrac{1}{2}(P_n^2 + X_n^2 - 1)\frac{\partial}{\partial\mathbf{x}}V_{nn}(\mathbf{x}) - \sum_{n>m}(X_n X_m + P_n P_m)\frac{\partial}{\partial\mathbf{x}}V_{nm}(\mathbf{x}) \tag{115}$$

Depending on the actual value of the electronic coordinates and momenta, the expression $\left(P_n^2 + X_n^2 - 1\right)$ may become negative and, thus, the propagation for the nuclear trajectories might take place on inverted potential-energy surfaces. This is due to the term $-\frac{1}{2}\sum_n V_{nn}(\mathbf{x})$ in the mapping Hamiltonian, which, as was discussed in Section VI.B, describes the zero-point energy of the electronic oscillators and represents (at the purely classical level) the major difference to the mean-field trajectory method. From a numerical point of view, the propagation of trajectories on an inverted potential-energy surface might result in instabilities. While this represents no problem in the case of the model systems studied in this chapter (which describe bound-state dynamics with rather smooth potentials), it might result in numerical difficulties for "hard-wall" potentials such as, for example, a Morse or Lennard-Jones potential [118].

To alleviate this problem, Bonella and Coker [118, 119] have proposed a different variant of the semiclasscial mapping approach, where the trajectories are calculated according to the mean-field Hamiltonian

$$\mathscr{H}_{\mathrm{MF}} = \mathscr{H} + \tfrac{1}{2}\sum_{n} V_{nn}(\mathbf{x}) \tag{116}$$

which was introduced in Eq. (92). The term $-\frac{1}{2}\sum_n V_{nn}(\mathbf{x})$, which is neglected in the mean-field Hamiltonian, enters the expression for the semiclassical transition amplitude as a weighting function in the integrand; that is, the classical action S_t in Eq. (114) is replaced by $\widetilde{S}_t + \frac{1}{2}\int_0^t d\tau \sum_n V_{nn}[\mathbf{x}(\tau)]$, where $\widetilde{S}_t$ denotes the action for the mean-field Hamiltonian $\mathscr{H}_{\mathrm{MF}}$. Computational studies have shown that, in the case of models plagued by the above-mentioned numerical instabilities, the modified mapping approach indeed may represent an improvement over the original formulation [118, 119]. For the problems under consideration in this chapter, however, the original semiclassical treatment appears to be superior. Moreover, the classical mapping Hamiltonian $\mathscr{H}$ leads to the correct quantum-mechanical level density if the zero-point energy problem is of minor importance (see, for example, Sections VI.B and VII, as well as Fig. 20).

It is instructive to consider the origin of the differences in the semiclassical formulations. As discussed in Ref. 100, the modified mapping approach can be obtained from the quantum-mechanical mapping Hamiltonian (112) by employing a coherent-state representation (or Q function) for the electronic degrees of freedom. This corresponds to a normal ordering of the quantum Hamiltonian, while the classical mapping Hamiltonian (88) corresponds to the Wigner function of the quantum Hamiltonian—that is, symmetric $X-P$ ordering. Within the semiclasscial approximation, we then obtain for symmetric ordering the standard Van Vleck–Gutzwiller propagator, while for normal ordering we additionally obtain a correction term in the action [100, 263], which in the special case of the mapping Hamiltonian gives exactly the term introduced by Bonella and Coker.

Alternatively, in Bonella and Coker's derivation [118,119] the difference between the two classical Hamiltonians in Eq. (116) arises from a different $\hbar$-dependence in the semiclassical limit $\hbar \to 0$. To illustrate the idea, the quantum-mechanical mapping Hamiltonian (112) is rewritten by introducing position and momentum operators $\widetilde{X}_n = \sqrt{\hbar}\hat{X}_n$ and $\widetilde{P}_n = \sqrt{\hbar}\hat{P}_n$ with $[\widetilde{X}_n, \widetilde{P}_m] = i\hbar\delta_{nm}$ and the "frequencies" $\hbar\Omega_{nm} = V_{nm}$. This yields

$$\hat{\mathscr{H}} = T(\hat{\mathbf{p}}) - \tfrac{1}{2}\sum_n \hbar\Omega_{nn}(\hat{\mathbf{x}}) + \tfrac{1}{2}\sum_{n,m}(\widetilde{X}_n\widetilde{X}_m + \widetilde{P}_n\widetilde{P}_m)\Omega_{nm}(\hat{\mathbf{x}}) \tag{117}$$

In order to derive the classical limit of this Hamiltonian, one may employ the stationary-phase approximation to the corresponding path-integral representation of the quantum propagator [250]. In the action of the path integral, the term $-\frac{1}{2}\sum_n \hbar\Omega_{nn}(\mathbf{x})$ gives a contribution that is formally independent on $\hbar$ and is therefore treated as a slowly varying part in the stationary-phase approximation. As a consequence, the resulting classical Hamiltonian misses the term $-\frac{1}{2}\sum_n \hbar\Omega_{nn}(\mathbf{x})$ and is simply given by the mean-field Hamiltonian (116).

It should be noted, however, that the limit $\hbar \to 0$ is only a formal procedure, which does not necessarily lead to a unique or correct semiclassical limit. In the case of the mapping formulation, this is because of the following reasons: (i) For a given molecule, the frequencies $\Omega_{nm}(\mathbf{x})$ will in general also depend in a nontrivial way on $\hbar$. (ii) A slowly varying term may as well be included in the stationary phase treatment [147]. (iii) As indicated by the term $-\frac{1}{2}\sum_n V_{nn}(\mathbf{x})$ resulting from the commutator $\left[a_n, a_m^\dagger\right] = \delta_{nm}$, the effective "action constant" of the electronic degrees of freedom actually is 1 and thus constant even in the formal limit $\hbar \to 0$. This suggests that the applicability of a semiclassical treatment of the mapping formulation may rather be rooted in the fact that the semiclassical mapping description of a pure N-level system is exact. Hence, the semiclassical treatment should still be valid in the case of a vibronically coupled N-level system, as long as the nonadiabatic coupling is not too strong. The discussion above suggests that the validity of a specific choice of the stationary-phase approximation, and thus the semiclassical approximation, has to be checked for the specific molecular system under consideration. This validity may depend, for example, on the separability of the electronic and nuclear degrees of freedom and on the various timescales of the system [264].

Another ambiguity in defining the classical mapping Hamiltonian is related to the fact that different bosonic quantum Hamiltonians $\mathscr{H}$ may correspond to the same original quantum Hamiltonian H. This problem was already discussed in Section VI.A.2 for N-level systems. In the context of nonadiabatic dynamics, a different version of the mapping Hamiltonian is given by

$$\mathscr{H} = T(\mathbf{p}) + \bar{V}(\mathbf{x}) + \tfrac{1}{2}\sum_{n,m}(X_nX_m + P_nP_m)(V_{nm}(\mathbf{x}) - \bar{V}(\mathbf{x})\delta_{nm}) \tag{118}$$

where $\bar{V}(\mathbf{x}) = \frac{1}{N_{\text{el}}}\sum_n V_{nn}(\mathbf{x})$ denotes the average nuclear potential. While the quantum dynamics generated by the Hamiltonians (112) and (118) is equivalent (in the physical subspace), this is not necessarily true within the semiclassical approximation. For most of the systems considered in this work, we have found that the mapping Hamiltonian (118) gives results that are in better agreement with quantum mechanical reference calculations. Employing the mapping Hamiltonian (118) instead of (112) also alleviates to some extent the above-mentioned numerical problems associated with the propagation of nuclear trajectories on an inverted potential-energy surface.

To conclude this section, we would like to mention some technical details of the semiclassical calculation. The various elements of the semiclassical expression (114) are obtained in the following way: The phase-space integral is evaluated using standard importance-sampling Monte Carlo schemes. Thereby, as was mentioned above, filtering techniques can be employed to facilitate the integration over the oscillating integrand (for details see Refs. 116 and 161). The classical action and the monodromy matrix for each trajectory $(\mathbf{q}_t, \mathbf{p}_t) = (\mathbf{x}_t, \mathbf{X}_t, \mathbf{p}_t, \mathbf{P}_t)$ are obtained by solving the differential equations

$$\frac{dS_t}{dt} = \mathbf{p}_t \cdot \dot{\mathbf{q}}_t - \mathcal{H} \tag{119}$$

$$\frac{d}{dt}\mathbf{M}_t = \begin{pmatrix} -\frac{\partial^2 \mathcal{H}}{\partial \mathbf{q} \partial \mathbf{p}} & -\frac{\partial^2 \mathcal{H}}{\partial \mathbf{q} \partial \mathbf{q}} \\ \frac{\partial^2 \mathcal{H}}{\partial \mathbf{p} \partial \mathbf{p}} & \frac{\partial^2 \mathcal{H}}{\partial \mathbf{p} \partial \mathbf{q}} \end{pmatrix} \mathbf{M}_t \tag{120}$$

along the trajectory. The calculation of the preexponential factor C_t involves a complex square root, Eq. (109), the correct branch of which is determined by requiring continuity of C_t [108]. In particular for larger systems, it is thereby advantageous to combine some of the phase factors of C_t and e^{iS_t}, in order to avoid the calculation of the preexponential factor for very small time intervals [120, 265].

A further simplification of the semiclassical mapping approach can be obtained by introducing electronic action-angle variables and performing the integration over the initial conditions of the electronic DoF within the stationary-phase approximation [120]. Thereby the number of trajectories required to obtain convergence is reduced significantly [120]. A related approach is discussed below within the spin-coherent state representation.

C. Results

The semiclassical mapping approach outlined above, as well as the equivalent formulation that is obtained by requantizing the classical electron-analog model of Meyer and Miller [112], has been successfully applied to various examples of nonadiabatic dynamics including bound-state dynamics of several spin–boson-type electron-transfer models with up to three vibrational modes [99, 100], a series of scattering-type test problems [112, 118, 120], a model for laser-driven population transfer between two adiabatic potential-energy surfaces [115], a study of quantum-classical correspondence [124], the photodissociation dynamics of ozone [114] and ICN [121, 266], the decay dynamics of the $^2\Pi_g$ *d*-wave shape resonance in electron–N_2 scattering [267], and a model of nonadiabatic dynamics with multiple surface crossings [117], as well as the photoinduced nonadiabatic dynamics of pyrazine at the S_1–S_2 conical intersection [116, 122, 268]. As representative examples, we discuss here the application of the semiclassical mapping approach to describe the photoinduced nonadiabatic dynamics of pyrazine at the S_1–S_2 conical intersection [116, 268] and to the Models IVa and IVb describing intramolecular electron transfer processes.

While the simulations of the pyrazine system discussed in the previous sections of this chapter have employed a three-mode model (Model I), the semiclassical simulations we will present here are based on two different models: a four-mode model and a model including all 24 normal modes of the pyrazine molecule. Let us first consider the four-mode model of the S_1–S_2 conical intersection in pyrazine which was developed by Domcke and co-workers [269]. In addition to the three modes considered in Model I, it takes into account another Condon-active mode (ν_{9a}). Figure 37 shows the modulus of the autocorrelation function [cf. Eq. (24)] of this model after photoexcitation to the S_2 electronic state. The exact quantum results (full line) are compared to the

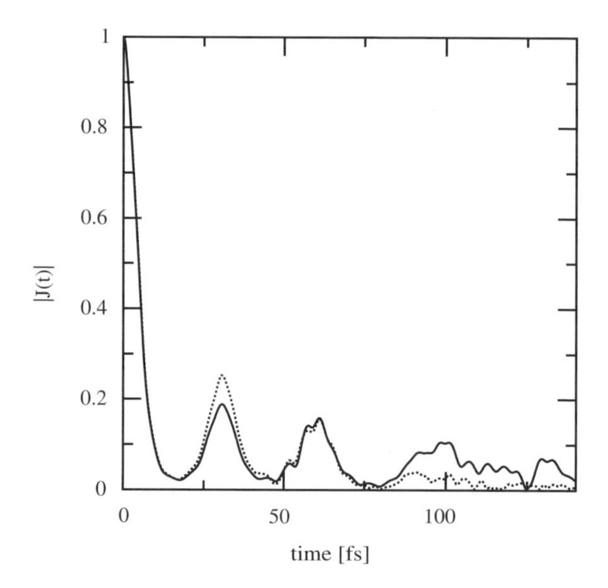

Figure 37. Modulus of the autocorrelation function for the four-mode model of pyrazine. The full line is the quantum result, and the dotted line is the semiclassical result.

semiclassical results (dotted line). The autocorrelation function exhibits a fast initial decay, reflecting the initial displacement of the wave packet on the S_2 surface. The suppression of the ensuing recurrences (which is absent for the uncoupled system, i.e., $V_{12} = 0$ [174]) reflects the ultrafast electronic dephasing in the S_2 electronic state of pyrazine. This dephasing process is incomplete due to the limited density of states of the four-mode model. It is seen that the semiclassical result reproduces all essential features of the autocorrelation function up to 100 fs. Both the first two recurrences and the high-frequency modulations are well-described. Upon closer inspection, one recognizes that the fine structure of the autocorrelation function is better reproduced than the overall damping of the amplitude; for example, the semiclassical result underestimates the damping of the first recurrence and has too small an amplitude for times $t > 80$ fs. The former deviation is related to the nonunitarity of the semiclassical approximation [270] and can be corrected to some extent by normalizing the data (see below). The latter deviation is presumably a result of the filtering technique used in the calculation, which tends to exaggerate the damping of the autocorrelation function for longer times.

The S_2 absorption spectrum is displayed in Fig. 38. Figure 38a compares the semiclassical and the quantum result, and Fig. 38b shows the experimental data from Ref. 271. Both theoretical results have been obtained by Fourier transformation of the autocorrelation function. Thereby a phenomenological dephasing constant $T_2 = 30$ fs has been included to reproduce the homogeneous width of the experimental spectrum. The absorption spectrum shows a diffuse S_2 band with irregularly spaced structures, which cannot be assigned in terms of harmonic modes in the S_2 state [163, 174]. The weak tail in the energy region of the S_1 state represents the well-known phenomenon of vibronic-intensity borrowing. It is seen that the semiclassical and the exact quantum results are in very good agreement in both parts of the spectrum. It is interesting to compare this semiclassical result with a calculation performed by Stock and Miller some time ago for the same model using a quasi-classical approach (i.e.. without semiclassical phase information) based on the classical electron analog model [94]. Although this classical method was able to reproduce the global features of the absorption spectrum, it was not capable of reproducing the finer structure. In contrast, the semiclassical method describes these fine structures very well, demonstrating that the inclusion of phase information (and hence quantum interference) is important to describe the absorption spectrum in this system correctly.

The ultrafast initial decay of the population of the diabatic S_2 state is illustrated in Fig. 39 for the first 30 fs. Since the norm of the semiclassical wave function is only approximately conserved, the semiclassical results are displayed as rough data (dashed line) and normalized data (dotted line) [i.e., $P_2^{\text{norm}} = P_2/(P_1 + P_2)$]. The normalized results for the population are seen to match the

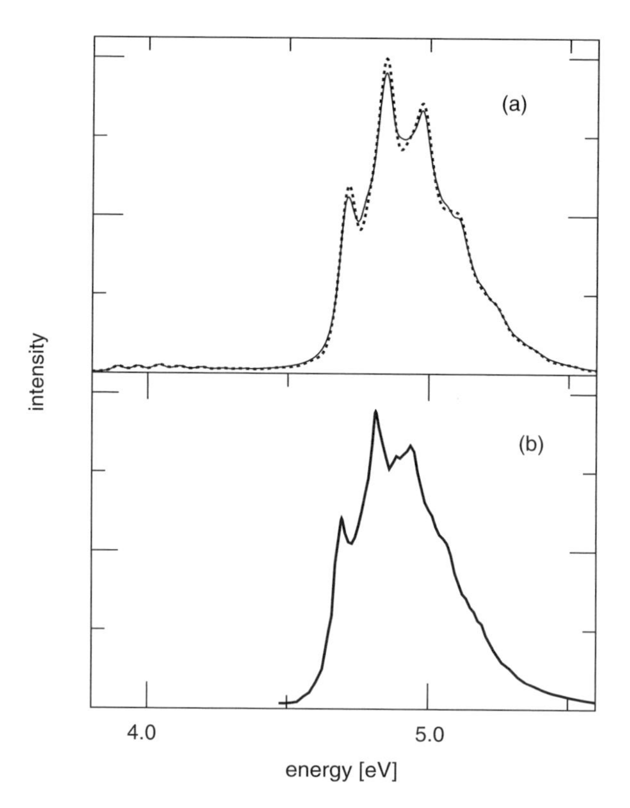

Figure 38. Absorption spectrum of pyrazine in the energy region of the S_1–S_2 conical intersection. Shown are (*a*) quantum mechanical (full line) and semiclassical (dotted line) results for the four-mode model (including a phenomenological dephasing constant of $T_2 = 30$ fs) and (*b*) the experimental data [271].

quantum reference data quantitatively. It should be emphasized that the deviation of the norm shown in Fig. 39 is not a numerical problem, but rather confirms the common wisdom that a two-level system as well as its bosonic representation is a prime example of a quantum system and therefore difficult to describe within a semiclassical theory. Nevertheless, besides the well-known problem of norm conservation, the semiclassical mapping approach clearly reproduces the nonadiabatic quantum dynamics of the system. It is noted that the semiclassical results displayed in Fig. 39 have been obtained without using filtering techniques. Therefore, due to the highly chaotic classical dynamics of the system, a very large number of trajectories ($\sim 10^7$) is needed to achieve convergence, even over the relatively short timescale of 30 fs. To improve the convergence and to facilitate the calculation of the population for longer times,

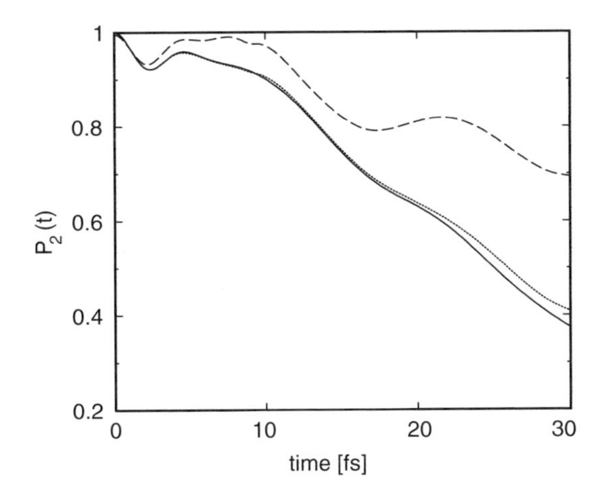

Figure 39. Initial decay of the diabatic population of the S_2 state for the four-mode pyrazine model. Compared are quantum (full line), semiclassical (dashed line), and normalized semiclassical (dotted line) results.

effective filtering techniques or more sophisticated semiclassical techniques such as the forward–backward initial-value representation [272–274] need to be employed.

Within the four-mode model, the experimental absorption spectrum can only be obtained by including a rather large phenomenological dephasing parameter $T_2 = 30$ fs. Recently, Raab et al. have reported multiconfiguration time-dependent Hartree calculations [275, 276] based on a model Hamiltonian that takes into account all 24 normal modes of the pyrazine molecule [277]. Figure 40 displays the modulus of the autocorrelation function for the 24-mode model. The semiclassical results are compared with the results of Raab et al. [277]. The comparison with the four-mode model (Fig. 39) shows that the inclusion of the remaining 20 modes leads to a damping of the recurrences of the autocorrelation function. The semiclassical result is seen to reproduce the quantum result rather well up to 70 fs. In particular, the calculation is seen to well reproduce the damping of the first recurrence due to the inclusion of the additional modes. This is in contrast to more classical methods, such as the quasi-classical implementation of the mapping approach. As we have discussed in Section VI, this failure to describe the correct relaxation behavior is related to an incorrect treatment of the zero-point energy in the classical implementation. The results in Fig. 40 demonstrate that the semiclassical method is capable of describing this effect correctly without requiring further zero-point energy modifications.

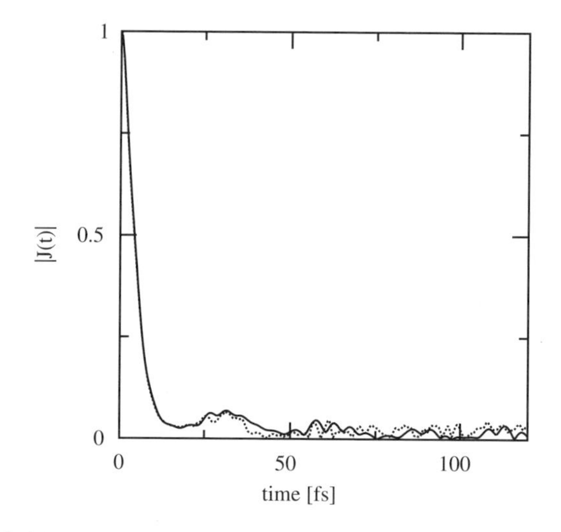

Figure 40. Modulus of the autocorrelation function for the 24-mode pyrazine model. The full line is the quantum result [277], and the dotted line is the semiclassical result.

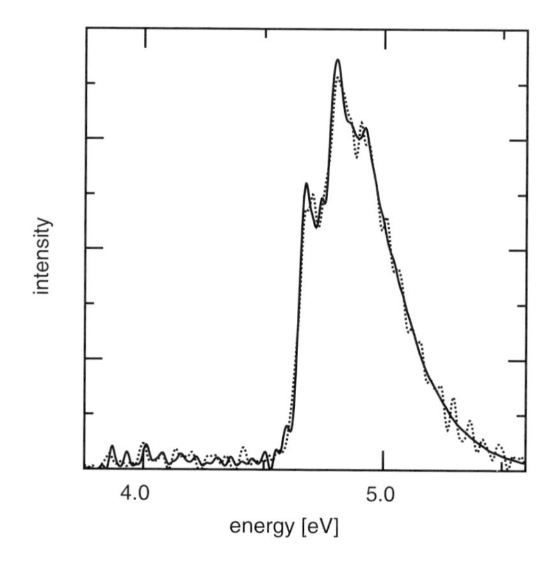

Figure 41. Absorption spectrum for the 24-mode pyrazine model. The full line is the quantum result [277], and the dotted line is the semiclassical result. In both spectra a phenomenological dephasing constant of $T_2 = 150$ fs was used.

Figure 41 shows the absorption spectrum for the 24-mode model of pyrazine. As was done by Raab et al. [277], we have included a phenomenological dephasing time of $T_2 = 150\,\text{fs}$ to model the experimental broadening due to finite resolution and rotational motion. It can be seen that the inclusion of all 24 normal modes of the pyrazine molecule leads to a shape of the spectrum which is in good agreement with the experimental result (Fig. 38b). The semiclassical result is seen to be in fairly good agreement with the quantum result. The spurious structure in the semiclassical spectrum is presumably due to the statistical error.

We next consider the application of the semiclassical mapping approach to describe ultrafast intramolecular electron transfer reactions. Figure 42 displays results of the semiclassical mapping approach for the single-mode Model IVa. Since this problem encompasses only moderately chaotic dynamics (cf. the discussion in in Section VII), no filtering techniques need to employed and converged results are obtained with a relatively moderate number of trajectories. Both the population of the initially prepared electronic state (Fig. 42a) and the autocorrelation function (Fig. 42b) are seen to be in excellent agreement with the quantum-mechanical results if normalized semiclassical data are used. Also shown in Fig. 42 is the norm of the semiclassical wave function, indicating a

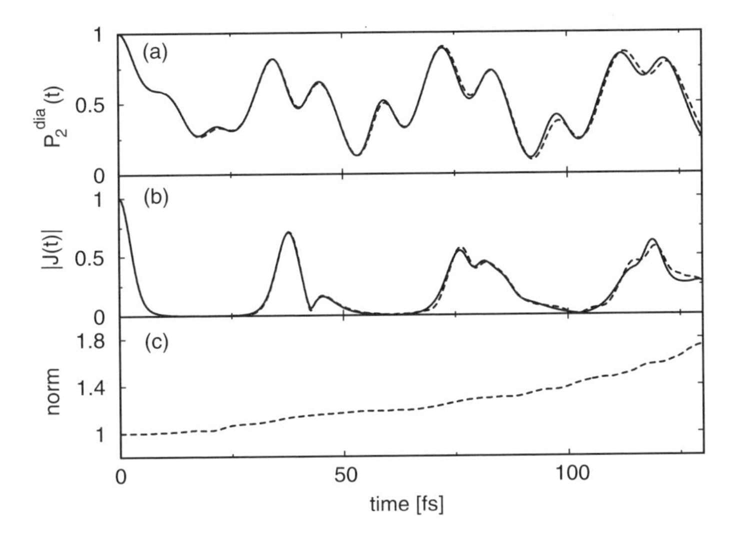

Figure 42. Diabatic population (*a*) and modulus of the autocorrelation function (*b*) of the initially prepared state for Model IVa. The full line is the quantum result, and the dashed line depicts the semiclassical mapping result. The semiclassical data have been normalized. Panel (*c*) shows the norm of the semiclassical wave function.

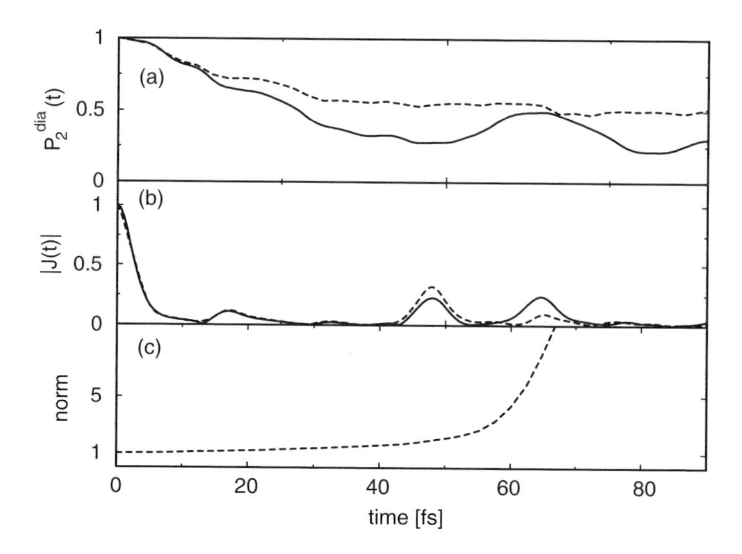

Figure 43. Diabatic population (*a*) and modulus of the autocorrelation function (*b*) of the initially prepared state for Model IVb. The full line is the quantum result, and the dashed line depicts the semiclassical mapping result. The semiclassical data have been normalized. Panel (*c*) shows the norm of the semiclassical wave function.

deviation from unity, which is for this single-mode problem not as severe as for the highly chaotic pyrazine model discussed above.

Finally we consider the more challenging electron-transfer Model IVb, which comprises three vibrational modes with strong electronic–vibrational coupling. Figure 43 displays results of the semiclassical mapping approach for the population of the initially prepared electronic state as well as the corresponding autocorrelation function. The results have been obtained without using filtering methods. It is seen that for short times (<40 fs) the semiclassical results for the autocorrelation function are in very good agreement with the quantum-mechanical reference data. The population dynamics, on the other hand, is only reproduced qualitatively. Similar to the pyrazine problem, the dynamics of Model IVb is strongly chaotic. As a result, the norm of the semiclassical wave function increases drastically after about 50 fs, and more sophisticated sampling schemes would be needed to converge the semiclassical calculation.

To conclude, the results presented in this section demonstrate that the semiclassical implementation of the mapping approach is able to describe rather well the ultrafast dynamics of the nonadiabatic systems considered. In particular, it is capable of describing the correct relaxation dynamics of the autocorrelation function as well as the structures of the absorption spectrum of pyrazine. The former is related to a better treatment of the zero-point energy, and the latter reflects the correct inclusion of quantum interference effects. The ability to describe these quantum effects is in contrast to the quasi-classical implementation of the mapping approach discussed in Section VI.

Although this result is quite encouraging, it should be mentioned that the required numerical effort for the semiclassical calculation is rather large. In case of the pyrazine system, for example, we have found that even though the required CPU time seems to have a better scaling (with respect to the number of nuclear degrees of freedom) in the semiclassical approach than in the quantum multiconfiguration time-dependent Hartree method, it is still comparable in the case of the 24-mode model. There are two main reasons for the rather large numerical effort in the semiclassical calculation: (i) Although an integral conditioning (filtering) technique was used in the calculation, the oscillatory nature of the integrand still requires a large number ($\sim 10^7$) of trajectories to converge the Monte Carlo integration for longer times [278]. (ii) Due to the calculation of the preexponential factor, the numerical effort per trajectory has an unfavorable $(N_{\rm el} + N_{\rm vib})^3$ scaling. Both problems need to be addressed further to make the semiclassical approach a practical method for the treatment of complex molecular systems.

IX. SPIN-COHERENT STATE METHOD

The mapping approach relates a quantum system with discrete DoF to a system with continuous DoF which is amenable to a semiclassical treatment. Another possibility to obtain a continuous description of discrete DoF is based on the coherent states of the particular system under investigation [279, 280]. In particular, the spin-coherent state path integral has been used to investigate the semiclassical description of spin systems [137–147]. This approach can also be applied to N-level systems. A general semiclassical formulation based on a stationary-phase approximation of the coherent-state path integral is given, for example, in Ref. 138. Following a brief introduction into spin-coherent state theory, we discuss in this section the semiclassical propagator in the spin-coherent state representation as well as its close connection to the semiclassical mapping approach.

A. Notation

Let us start with a brief review of spin-coherent state theory. For simplicity we focus on a two-level (or spin 1/2) system. The coherent states for a two-level system with basis states $|\psi_1\rangle, |\psi_2\rangle$ can be written as [136, 139]

$$|\mu\rangle = \frac{\mu}{\sqrt{1+|\mu|^2}}|\psi_1\rangle + \frac{1}{\sqrt{1+|\mu|^2}}|\psi_2\rangle \tag{121}$$

where μ is a complex parameter. As with the coherent states of the harmonic oscillator, these states are nonorthogonal

$$\langle\mu|\mu'\rangle = \frac{1+\mu^*\mu'}{\sqrt{1+|\mu|^2}\sqrt{1+|\mu'|^2}} \tag{122}$$

and overcomplete

$$\mathbf{1} = \frac{2}{\pi}\int \frac{d^2\mu}{(1+|\mu|^2)^2}|\mu\rangle\langle\mu| \tag{123}$$

Here the integration is over the complex plane:

$$d^2\mu = d(\mathrm{Re}\,\mu)d(\mathrm{Im}\,\mu) \tag{124}$$

Sometimes it is useful to parameterize the complex parameter μ by polar and azimuthal angles θ and ϕ $(0 \leq \theta \leq \pi, 0 \leq \phi \leq 2\pi)$:

$$\mu = e^{i\phi}\tan\left(\frac{\theta}{2}\right) \tag{125}$$

B. Semiclassical Propagator

To discuss the semiclassical spin-coherent state propagator, we consider a general transition amplitude $\langle\psi'|e^{-iHt}|\psi\rangle$, which can be expressed as an integral over the spin-coherent state propagator,

$$\langle\psi'|e^{-iHt}|\psi\rangle = \left(\frac{2}{\pi}\right)^2\int\frac{d^2\mu_f}{(1+|\mu_f|^2)^2}\int\frac{d^2\mu_i}{(1+|\mu_i|^2)^2}\langle\psi'|\mu_f\rangle\langle\mu_f|e^{-iHt}|\mu_i\rangle\langle\mu_i|\psi\rangle \tag{126}$$

The path integral representation of the spin-coherent state propagator is formally given by [139]

$$\langle\mu_f|e^{-iHt}|\mu_i\rangle = \int_{\mu_i}^{\mu_f^*}\frac{\mathscr{D}\mu}{\sqrt{(1+|\mu_f|^2)(1+|\mu_i|^2)}}e^{iS_\mu} \tag{127}$$

with the action (for details, see, e.g., Ref. 143)

$$S_\mu = \ln(1+\mu_f^*\mu(t)) + \ln(1+\mu_i\bar{\mu}(t)) + \int_0^t d\tau\left(\frac{i}{2}\frac{\bar{\mu}\dot{\mu}-\dot{\bar{\mu}}\mu}{1+\bar{\mu}\mu} - \langle\bar{\mu}|H|\mu\rangle\right) \tag{128}$$

Employing the stationary-phase approximation to the path integral, the semiclassical spin-coherent state propagator is obtained [139, 140, 143, 281, 282]:

$$\langle \mu_f | e^{-iHt} | \mu_i \rangle_{SC} = \left((1 + \mu_f^* \mu(t))(1 + \mu_i \bar{\mu}(t)) \frac{\partial^2 S_\mu}{\partial \mu_i \partial \mu_f^*} \right)^{1/2} e^{iS_\mu + \frac{i}{2}\phi} \tag{129}$$

where S_μ is the action (128) along the classic trajectory (see below) and ϕ denotes the Solari–Kochetov phase [143, 283, 284],

$$\phi = \tfrac{1}{2} \int_0^t d\tau \left(\frac{\partial}{\partial \bar{\mu}} (1 + \mu\bar{\mu})^2 \frac{\partial \langle \bar{\mu} | H | \mu \rangle}{\partial \mu} + \frac{\partial}{\partial \mu} (1 + \mu\bar{\mu})^2 \frac{\partial \langle \bar{\mu} | H | \mu \rangle}{\partial \bar{\mu}} \right) \tag{130}$$

The trajectories in the semiclassical propagator (129) are determined by the equations of motion

$$\dot{\mu} = -i(1 + \mu\bar{\mu})^2 \frac{\partial \langle \bar{\mu} | H | \mu \rangle}{\partial \bar{\mu}} \tag{131a}$$

$$\dot{\bar{\mu}} = i(1 + \mu\bar{\mu})^2 \frac{\partial \langle \bar{\mu} | H | \mu \rangle}{\partial \mu} \tag{131b}$$

and the boundary conditions $\mu(0) = \mu_i$, $\bar{\mu}(t) = \mu_f^*$. It is important to note that in general $\bar{\mu}(\tau)$ is not the complex conjugate of $\mu(\tau)$ because this would result in an overdetermined boundary-value problem [139, 140, 143].

Although the action (128) is nonquadratic in μ and $\bar{\mu}$, the semiclassical approximation (129) to the path integral (127) yields the exact quantum-mechanical result for a two-level (spin 1/2) system [141, 143, 284]. As has been discussed by several workers [139, 285], this somewhat surprising result relies on the linearity of Heisenberg's equation of motion for the spin operators. In light of this fact, generalized coherent state theory appears as a promising alternative for a semiclassical treatment of quantum systems which involve discrete states. The semiclassical spin-coherent state propagator and variants thereof have been applied to various systems, where a spin (or two-level system) interacts with a continuous degree of freedom, in particular the dynamics of the Jaynes–Cummings Model [144] and a trace-formula description of spin–orbit interaction [145–147]. In its present form, however, the theory is not suited for a computational evaluation of multidimensional problems. This is because the evaluation of the semiclassical propagator (129) requires the solution of a boundary value problem. While this problem could be circumvented by rewriting the propagator in terms of an initial value problem [similar to the transformation of the Van Vleck–Gutzwiller propagator to an initial-value problem, cf. Eq. (107)], the resulting trajectories are generally complex-valued thus rendering a numerical solution difficult. In the next section we discuss two attempts to define an initial-value representation of the semiclasscial spin-coherent state propagator.

C. Initial-Value Representation

Within the theoretical framework of time-dependent Hartree–Fock theory, Suzuki has proposed an initial-value representation for a spin-coherent state propagator [286]. When we adopt a two-level system with quantum Hamiltonian H, this propagator reads

$$\langle \psi_f | e^{-iHt} | \psi_i \rangle_{\rm SU} = N_t^{-1} \frac{2}{\pi} \int \frac{d^2\mu_0}{(1+|\mu_0|^2)^2} \langle \psi_f | \mu_t \rangle e^{iS_\mu} \langle \mu_0 | \psi_i \rangle \tag{132}$$

where N_t is a time-dependent factor guaranteeing the preservation of the norm of the semiclassical propagator. The dynamics of the classical trajectories is determined by the equation of motion,

$$\dot{\mu} = -i(1+|\mu|^2)^2 \frac{\partial \langle \mu | H | \mu \rangle}{\partial \mu^*} \tag{133}$$

For a two-level system (or, more generally, for spin systems with a Hamiltonian that depends linearly on the spin operators S_1, S_2, and S_3) Suzuki's propagator gives the exact quantum-mechanical result and the normalization factor $N_t = 1$. A comparison of Eq. (132) with Eq. (108) reveals that Suzuki's propagator for spin-coherent states appears to be very similar to the propagator of Herman and Kluk employing harmonic-oscillator coherent states. It is important to note, however, that missing the determinant factor C_t, Suzuki's expression resembles Heller's "frozen Gaussian approximation" [150]. In the case of a general nonlinear Hamiltonian, it should therefore be considered as a quasi-classical rather than a rigorous semiclassical approximation.

Another possibility to introduce a semiclasscial initial value representation for the spin-coherent state propagator is to exploit the close relation between Schwinger's representation of a spin system and the spin-coherent state theory [100, 133–135]. To illustrate this approach, we consider an electronic two-level system coupled to $N_{\rm vib}$ nuclear DoF. Within the mapping approach the semiclassical propagator for this system is given by

$$\begin{aligned} \langle \psi_f | \langle \mathbf{v}_f | e^{-iHt} | \mathbf{v}_i \rangle | \psi_i \rangle_{HK} &= \langle n_1', n_2' | \langle \mathbf{v}_f | e^{-i\mathcal{H}t} | \mathbf{v}_i \rangle | n_1, n_2 \rangle_{HK} \\ &= \int \frac{d\mathbf{x}_0 d\mathbf{p}_0}{(2\pi)^{N_{\rm vib}}} \int \frac{d\mathbf{X}_0 d\mathbf{P}_0}{(2\pi)^2} \langle \mathbf{v}_f | \mathbf{x}_t \mathbf{p}_t \rangle \langle n_1', n_2' | \mathbf{X}_t \mathbf{P}_t \rangle \\ &\quad \times C_t e^{iS_t} \langle \mathbf{X}_0 \mathbf{P}_0 | n_1, n_2 \rangle \langle \mathbf{x}_0 \mathbf{p}_0 | \mathbf{v}_i \rangle \end{aligned} \tag{134}$$

Note that because of the mapping relation (80b), the quantum numbers in Eq. (134) fulfill the identity $n_1' + n_2' = 1 = n_1 + n_2$.

In order to express the propagator (134) in terms of spin-coherent states, we introduce the following parameterization of the complex electronic variables [133]:

$$X_{1_t} + iP_{1_t} = \sqrt{2I_t}\sin\left(\frac{\theta_t}{2}\right)e^{i(\phi_t-\chi_t)/2} \tag{135a}$$

$$X_{2_t} + iP_{2_t} = \sqrt{2I_t}\cos\left(\frac{\theta_t}{2}\right)e^{-i(\phi_t+\chi_t)/2} \tag{135b}$$

The equations of motion for the new electronic variables are given by

$$\dot{I} = 0 \tag{136a}$$

$$\dot{\chi} = V_{11} + V_{22} + 2V_{12}\frac{\cos(\phi)}{\sin(\theta)} \tag{136b}$$

$$\dot{\phi} = V_{22} - V_{11} - 2V_{12}\cot(\theta)\cos(\phi) \tag{136c}$$

$$\dot{\theta} = -2V_{12}\sin(\phi) \tag{136d}$$

Due to the change of variables in Eqs. (135a) and (135b), two of these equations of motion can readily be solved: The electronic population $I = (X_1^2 + P_1^2 + X_2^2 + P_2^2)/2$ is a constant of motion, and Eq. (136) can formally be integrated to give

$$\chi_t = \int_0^t ds\left\{V_{11} + V_{22} + 2V_{12}\frac{\cos(\phi)}{\sin(\theta)}\right\} + \chi_0 \tag{137}$$

When we express the electronic DoF in Eq. (134) in terms of the new variables (135a) and (135b), perform the integration over χ_0, and employ the complex notation introduced in Eq. (125), the Herman–Kluk propagator can be written in the following form [100]:

$$\begin{aligned}(e^{-iHt})_{HK} = \int_0^\infty dI\frac{I^2}{2}e^{-I}\int\frac{d\mathbf{x}_0\,d\mathbf{p}_0}{(2\pi)^{N_{\text{vib}}}}\frac{2}{\pi}\int\frac{d^2\mu_0}{(1+|\mu_0|^2)^2}\times|\mathbf{x}_t\mathbf{p}_t\rangle|\mu_t\rangle \\ \times C_t\exp\{-i(1-I)(\chi_t-\chi_0+\phi_t-\phi_0)/2\}e^{iS_t^I}\langle\mu_0|\langle\mathbf{x}_0\mathbf{p}_0|\end{aligned} \tag{138}$$

Thereby, the action is given by

$$S_t^I = \int_0^t d\tau(\mathbf{p}_\tau\dot{\mathbf{x}}_\tau - T) + I\int_0^t d\tau\left[\frac{i}{2}(\mu_\tau^*\dot{\mu}_\tau - \dot{\mu}_\tau^*\mu_\tau) - h_{\text{el}}\right] \tag{139}$$

where

$$h_{\mathrm{el}} = \frac{V_{11} + V_{22}\mu^*\mu + V_{12}\mu + V_{21}\mu^*}{1 + \mu^*\mu} \tag{140}$$

denotes the part of the Hamiltonian which involves electronic variables.

Up to now the Herman–Kluk propagator (113) has merely been rewritten without any further approximations. A comparison reveals, however, that the form of the propagator (138) is already very similar to Suzuki's expression (132). The most essential difference between the two formulations is the sampling over the initial value of the electronic population I in the Schwinger representation which is absent in Suzuki's propagator. Replacing the sampling by a fixed value $I = 1$ (which corresponds to the quantum-mechanical value of the electronic population), we obtain an initial-value representation of the semiclassical spin-coherent state propagator:

$$(e^{-iHt})_{SCS} = \frac{2}{\pi} \int \frac{d\mathbf{x}_0 d\mathbf{p}_0}{(2\pi)^{N_{\mathrm{vib}}}} \int \frac{d^2\mu_0}{(1 + |\mu_0|^2)^2} |\mathbf{x}_t\mathbf{p}_t\rangle|\mu_t\rangle C_t e^{iS_t} \langle\mu_0|\langle\mathbf{x}_0\mathbf{p}_0| \tag{141}$$

where the action S_t is given by Eq. (139) with $I = 1$. It is noted that the condition $I = 1$ agrees with the steepest descend analysis of Bonella and Coker [120]. Approximating the preexponential factor C_t (which contains the stability-matrix information) by the normalization factor N_t^{-1}, the spin-coherent state propagator (141) reduces to Suzuki's expression (132) extended to nonadiabatic dynamics [100, 142].

D. Application to Nonadiabatic Dynamics

To illustrate the computational performance and accuracy of the spin-coherent state propagators discussed above, we consider again the four-mode model for pyrazine and Models IVa and IVb describing intramolecular electron transfer. We start the discussion with Model IVa, which exhibits relatively simple and regular dynamics. Figure 44 depicts results for the population of the initially prepared diabatic electronic state as well as the autocorrelation function employing Suzuki's propagator (132) and the semiclassical spin-coherent state propagator (141), respectively. All semiclasscial results are depicted as normalized data. The norm of the respective semiclasscial wave function (which corresponds for Suzuki's propagator to the normalization factor N_t) is shown in Fig. 44c. Overall, it is seen that both spin-coherent state propagators give results that are in rather good agreement with the quantum-mechanical reference data. For longer times, the semiclassical propagator (141) reproduces the autocorrelation function better than Suzuki's propagator. Another difference concerns the norm conservation: The norm of Suzuki's propagator [i.e., the normalization

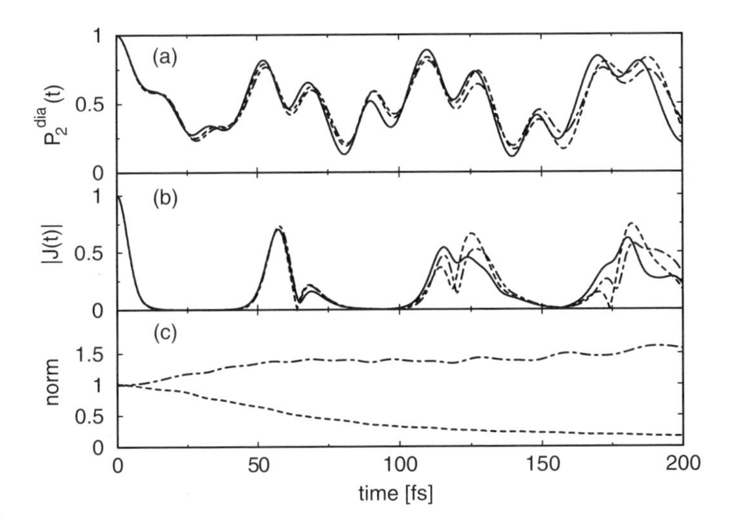

Figure 44. Diabatic population (*a*) and modulus of the autocorrelation function (*b*) of the initially prepared state for Model IVa. The full line is the quantum result, the dashed–dotted line is the result of the semiclassical spin-coherent state propagator, and the dashed line depicts the result of Suzuki's propagator. The semiclassical data have been normalized. Panel (*c*) shows the norm of the semiclassical wave functions.

factor N_t in Eq. (132)] decreases rapidly to rather small values. This is because this propagator neglects the preexponential factor C_t. Compared to the semiclassical mapping approach (cf. Fig. 42), both spin-coherent state propagators give slightly worse results; however, the overall difference between all three semiclassical results is rather small. The numerical effort for the three semiclassical calculations is, however, quite different. While the semiclassical mapping approach requires about $\sim 5 \times 10^5$ trajectories for convergence, this number reduces to $\approx 2 \times 10^4$ for the spin-coherent state propagator (141). This improved numerical performance is related to the fact that the spin-coherent state propagator (141) only includes trajectories with $I = 1$ which are typically more regular than trajectories with higher values of I. The number of trajectories required for convergence reduces by almost another order of magnitude for the propagator of Suzuki; this doesn't include the preexponential factor C_t, which is one source of the oscillations of the integrand in the initial-value representation of the semiclassical propagator.

The results obtained for the three-mode Model IVb are depicted in Fig. 45. As was found for the semiclassical mapping approach, the spin-coherent state propagators can only reproduce the short-time dynamics for the electronic population. The autocorrelation function, on the other hand, is reproduced at least qualitatively correctly by the semiclassical spin-coherent state propagator.

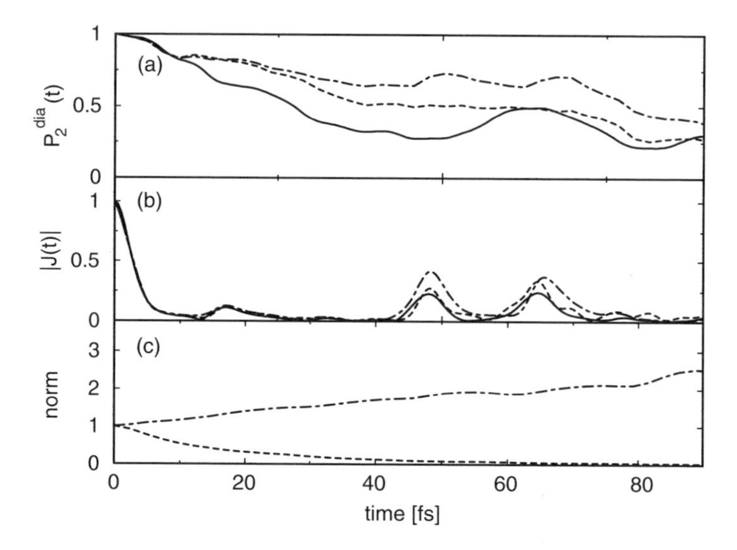

Figure 45. Diabatic population (*a*) and modulus of the autocorrelation function (*b*) of the initially prepared state for Model IVb. The full line is the quantum result, the dashed–dotted line is the result of the semiclassical spin-coherent state propagator, and the dashed line depicts the result of Suzuki's propagator. The semiclassical data have been normalized. Panel (*c*) shows the norm of the semiclassical wave functions.

Suzuki's propagator gives also quite good results for short times [287] but predicts spurious oscillations for longer times. It is noted that in contrast to the semiclassical mapping approach, the semiclassical spin-coherent state calculation could be converged without using filtering techniques. As was discussed above, this is due to the more regular behavior of trajectories with $I = 1$, which is also apparent in the relatively slow increase of the norm of the semiclassical wave function in Fig. 45c (as compared to the semiclassical mapping result in Fig. 43c).

Finally, we consider in Fig. 46 the application to the four-mode model of pyrazine, which, due to the strongly chaotic dynamics, is the most challenging of the three systems studied in this section. Similar to what occurs with the three-mode electron-transfer system considered above, the semiclassical spin-coherent state propagators can only describe the short-time population dynamics. The more approximate Suzuki propagator gives results that are in better overall agreement with the quantum-mechanical results. However, both propagators essentially miss the pronounced recurrence at time $t \approx 85$ fs. The autocorrelation function predicted by the semiclassical spin-coherent state propagator is, on the other hand, in rather good agreement with the quantum-mechanical result. Suzuki's propagator predicts—similar as for Model IVb—

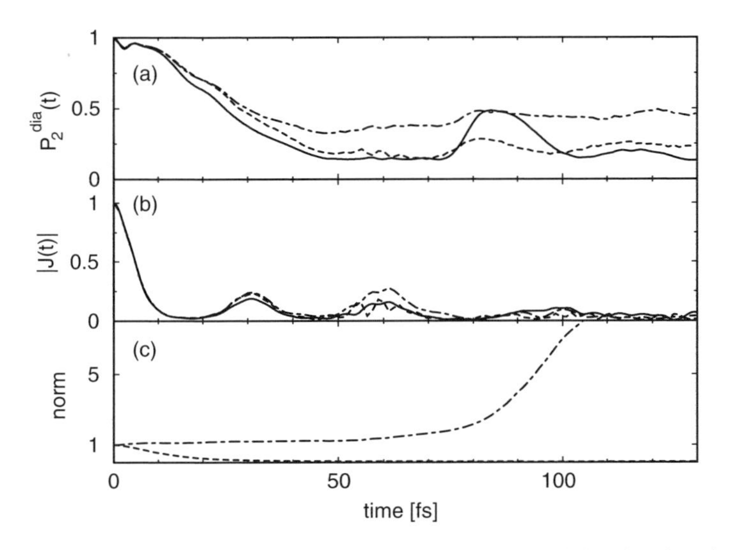

Figure 46. Diabatic population (a) and modulus of the autocorrelation function (b) of the initially prepared state for a four-mode model describing the $S_2 \rightarrow S_1$ internal-conversion process in pyrazine. The full line is the quantum result, the dashed–dotted line is the result of the semiclassical spin-coherent state propagator, and the dashed line depicts the result of Suzuki's propagator. The semiclassical data have been normalized. Panel (c) shows the norm of the semiclassical wave functions.

spurious oscillations for longer times. In contrast to the semiclassical mapping approach, the results shown in Fig. 46 have been obtained without the use of filtering methods. The number of trajectories required for convergence (e.g., $\sim 10^6$ for the autocorrelation function) is nevertheless smaller compared to the semiclassical mapping approach, thus demonstrating the better numerical performance of the spin-coherent state propagator.

In conclusion, it has been shown that the spin-coherent state propagator (141) represents a valuable formulation for the semiclassical description of non-adiabatic dynamics. While for some systems with predominantly chaotic classical dynamics the accuracy of this propagator has been found to be inferior compared to the semiclassical propagator (138) based on Schwinger's representation [100], for the systems considered here, the results are of similar quality. Since its numerical performance is advantageous, the spin-coherent state propagator thus represents a promising approach to the semiclassical description of nonadiabatic dynamics in complex molecular systems with many DoF. The generalization of Suzuki's propagator to nonadiabatic dynamics was found to give quite good results for regular to moderately chaotic dynamics (see also Ref. 142). For the two multidimensional systems with predominantly chaotic dynamics considered, the relatively simple method was shown to predict spurious oscillations in the autocorrelation function. Since this "quasi-classical" propagator doesn't require the calculation of the preexponential factor C_t, it nevertheless represents an interesting alternative to the other quasi-classical methods discussed in this chapter. Because both spin-coherent state propagators have so far only been applied to a relatively small number of test systems, further studies are required to obtain a comprehensive picture of the validity of the semiclassical approximation in the spin-coherent state representation.

X. CONCLUSIONS

In this chapter we have considered mixed quantum-classical (MQC) methods to describe ultrafast nonadiabatic molecular dynamics. We have started with an attempt to classify existing MQC strategies in formulations resulting from a partial classical limit, from a connection ansatz, and from a mapping formalism. Focusing on methods that have been applied to multidimensional problems, we have restricted the discussion to the mean-field trajectory method, the surface-hopping model, the quantum-classical Liouville equation, the quasi-classical and semiclassical version of the mapping approach, and the spin-coherent state representation. Since many of the existing works on MQC theory are closely related, we have tried to point out some of these connections. In particular, we have discussed the relation of the mean-field trajectory method with the quantum-mechanical time-dependent self-consistent field ansatz (Section III.D), the connection of empirical surface-hopping models with the stochastic implementation of the quantum-classical Liouville equation (Section V.B), the relation of the mapping approach with the mean-field trajectory method and the classical electron analog model (Section VI.C), and the connection between the semiclassical mapping approach and the semiclassical spin-coherent state propagator (Section IX.C).

To obtain a comprehensive picture of the performance of the various MQC approaches, we have considered five model problems, where each model represents a specific challenge for an approximate description. The detailed computational studies have shown that the practical applicability of MQC methods depends largely on whether the MQC formulation merely requires an simple quasiclassical average to calculate the observables of interests, or whether phase-space integrals involving rapidly oscillating phases need to be evaluated for this purpose. The first group comprises the mean-field trajectory method, the surface-hopping model, and the quasi-classical version of the mapping approach. The second group includes the quantum-classical Liouville description, the semiclassical mapping formulation, and the semiclassical spin-coherent state method. While the latter formulations suffer from the dynamical sign problem, the first group of methods may employ standard Monte Carlo sampling techniques. This yields converged results for typically 10^2–10^4 trajectories and allows for the treatment of molecular systems with a large number of degrees of freedom.

At present, the surface-hopping method is the most popular approach to describe nonadiabatic dynamics at conical intersections. The model employs the simple and physically appealing picture in which a molecular system always evolves on a *single adiabatic* potential-energy surface. Nonadiabatic transitions are realized in this approach via instantaneous hops of the trajectories between the coupled potential-energy surfaces according to some hopping criterion. As discussed in Section IV.B, there is no algorithm that provides both a practical and dynamically consistent stochastic realization of the Schrödinger equation in the classical-path approximation. Nevertheless, it has been found that the surface-hopping method gives an at least qualitative description of the dynamics in all cases considered. We have proposed a modified version of Tully's algorithm which was shown to give significantly better results than the standard method. In particular, it has been shown that it is important to use "binned" electronic population probabilities. There have been numerous suggestions to further improve the hopping algorithm [70–74]; however, the performance of all these variants seems to depend largely on the problem under consideration.

The mean-field trajectory method represents the simplest and in general most widely used MQC approach. Applied to models involving a single conical intersection, the method has been shown to afford a qualitative description of the nonadiabatic dynamics. Although the overall quality of the MQC results deteriorates at longer times, the mean-field trajectory method appropriately describes the branching of trajectories at a surface crossing as well as the electronic and vibrational coherences associated with this process. On the other hand, it was found that the method largely fails to account for the correct branching of trajectories in the case of multiple conical intersections. To explain these findings, a detailed discussion has been given in Sections III.D and IV.C, which also tries to clarify several common misconceptions concerning the theory and the performance of the mean-field trajectory method.

The mapping approach consists of (i) a quantum-mechanically exact transformation of the electronic states of the system to harmonic oscillators and (ii) a subsequent classical (or semiclassical) approximation to the *complete* system. As a consequence, the classical limit of the mapping formalism accounts for the dynamics of both quantum and classical degrees of freedom in a completely equivalent way. The mean-field trajectory method, on the other hand, is based on a *partial* classical limit for the nuclear degrees of freedom. Therefore, it treats electronic and nuclear degrees of freedom differently with respect to their initial conditions—that is, quantum-like initial conditions for the electronic degrees of freedom and quasi-classical initial conditions for the nuclear degrees of freedom. Correcting for this inconsistency, the mapping formulation is able to describe the nonadiabatic dynamics of all model problems considered rather well. Since the equations of motion of both methods are quite similar, the mapping formulation furthermore explains some of the failures of the mean-field trajectory method.

Although the mapping approach provides an in principle consistent classical description of nonadiabatic dynamics, the formulation has been shown to suffer seriously from the zero-point energy problem. This is because in the mapping formalism the electronic oscillators are constrained to the ground and first excited state, therefore representing a hard challenge for a classical description. Although on a quasi-classical level there is no rigorous solution to the zero-point energy problem, a quantum correction was introduced which affects a reduction of the electronic zero-point energy included in the calculation. Employing this correction, the quasi-classical mapping formulation was shown to provide an accurate description of the electronic and vibrational relaxation dynamics of all model problems considered. Generally speaking, the zero-point energy-corrected mapping calculations represent a significant improvement over the mean-field trajectory results and are of similar quality as the surface-hopping calculations. Since both methods in some sense overstress the classical limit (e.g., zero-point energy excitation in the mapping approach, classically forbidden transitions in surface-hopping calculations), however, it is not easy to predict which method will perform better for a given problem.

A conceptually more consistent treatment of nonadiabatic quantum dynamics can be obtained by either the quantum-classical Liouville formulation, the semiclassical version of the mapping approach, or the semiclassical spin-coherent state method. In particular, it has been shown that the quantum-classical Liouville equation is actually exact for the linear vibronic-coupling model considered here. The numerical deviations of this MQC description and the quantum-mechanical reference calculations therefore merely point out the numerical problems of a trajectory-based Monte Carlo implementation of the Liouville equation. While the quantum-classical Liouville equation is based on a partial classical limit for the nuclear degrees of freedom, the semiclassical version of the mapping approach as well as the spin-coherent state propagator employ a semiclassical treatment of both nuclear and electronic degrees of freedom. Hence these latter approaches account for electronic and nuclear quantum phenomena in a natural manner, including, for example, effects of quantum coherence and zero-point energy. Unlike the quasi-classical version of the mapping approach, its semiclassical version is therefore not plagued by the zero-point energy problem and is well-suited for the simulation of stationary and time-resolved optical spectra. At present the practical applicability of the quantum-classical Liouville description, the semiclassical version of the mapping approach, and the semiclassical spin-coherent state method is limited because of the dynamical sign problem. Because all three methods have been proposed only in the last few years, however, they still hold a great potential for improvement.

Apart from beautiful insights to the elusive world of quantum phenomena, the methodological development discussed in this chapter has also given rise to a wealth of applications to complex molecular system. By interfacing the dynamical methods with a quantum-mechanical and/or molecular-mechanics calculation of the potential energy, first attempts have been made to simulate quantum-mechanical processes in a microscopically described condensed-phase environment. Representative examples include the description of quantum-mechanical properties of liquids via a quantum-classical evaluation of real-time correlation functions [288–291], the modeling of photoinduced conformational dynamics in peptides [292, 293], the quantum-classical simulation of hydrogen transfer in enzymes [294], and the modeling of nonadiabatic *cis–trans* photoisomerization in rhodopsin [295] and photoactive yellow protein [296]. While this list is certainly not complete, it nonetheless reveals the potential of quantum-classical methods for future applications.

Acknowledgments

A major part of this chapter is based on results obtained by former graduate students—in particular, Stefan Dilthey, Susanne Hahn, Uwe Müller, Mark Santer, Birgit Strodl, and Igor Uspenskiy—to whom we wish to express our gratitude for a fruitful and pleasant collaboration. Furthermore, we want to thank our academic teachers Wolfgang Domcke and William H. Miller for introducing us to the exciting field of nonadiabatic dynamics and semiclassical theories. MT would like to thank Haobin Wang for many stimulating interactions on semiclassical dynamics. Continuous and generous support by the Deutsche Forschungsgemeinschaft is also gratefully acknowledged.

References

1. R. P. Feynman and A. R. Hibbs, *Quantum Mechanics and Path Integrals*, McGraw-Hill, New York, 1965.
2. W. P. Schleich, *Quantum Optics in Phase Space*, Wiley-VCH, Berlin, 2001.
3. M. C. Gutzwiller, *Chaos in Classical and Quantum Mechanics*, Springer, New York, 1990.
4. F. Haake, *Quantum Signatures of Chaos*, Springer, Berlin, 2000.
5. U. Weiss, *Quantum Dissipative Systems*, 2nd ed., World Scientific, Singapore, 1999.
6. N. F. Mott, *Proc. Cambridge Philos. Soc.* **27**, 553 (1931).
7. M. Born and R. Oppenheimer, *Ann. Phys. Leipzig* **84**, 457 (1927).
8. P. Ehrenfest, *Z. Phys.* **45**, 455 (1927).
9. W. Thorson, J. Delos, and S. Boorstein, *Phys. Rev. A* **4**, 1052 (1971).
10. G. D. Billing, *Chem. Phys. Lett.* **30**, 391 (1975).
11. R. B. Gerber, V. Buch, and M. A. Ratner, *J. Chem. Phys.* **77**, 3022 (1982).
12. D. Diestler, *J. Chem. Phys.* **78**, 2240 (1983).
13. D. A. Micha, *J. Chem. Phys.* **78**, 7138 (1983).
14. R. Kosloff and A. Hammerich, *Faraday Discuss.* **91**, 239 (1991).
15. R. Alimi, A. García-Vela, and R. B. Gerber, *J. Chem. Phys.* **96**, 2034 (1992).

16. H. J. C. Berendsen and J. Mavri, *J. Chem. Phys.* **99**, 8637 (1993).
17. P. Bala, B. Lesyng, and J. A. McCammon, *Chem. Phys. Lett.* **219**, 259 (1994).
18. G. Stock, *J. Chem. Phys.* **103**, 1561 (1995).
19. G. Stock, *J. Chem. Phys.* **103**, 2888 (1995).
20. J. Fang and C. C. Martens, *J. Chem. Phys.* **104**, 3684 (1996).
21. F. A. Bornemann, P. Nettesheim, and C. Schütte, *J. Chem. Phys.* **105**, 1074 (1996).
22. C. Zhu, A. W. Jasper, and D. G. Truhlar, *J. Chem. Phys.* **120**, 5543 (2004).
23. L. M. Raff and D. L. Thompson, in *Theory of Chemical Reaction Dynamics*, Vol. 3, M. Baer, ed., Chemical Rubber Company, Boca Raton, FL, 1985.
24. W. L. Hase, *Advances in Classical Trajectory Methods*, Vol. 1, Jai Press, London, 1992.
25. R. Schinke, *Photodissociation Dynamics*, University Press, Cambridge, 1993.
26. *Quantum and Classical Dynamics in Condensed Phase Simulations*, B. J. Berne, G. Ciccotti, and D. F. Coker, eds., World Scientific, Singapore, 1998.
27. I. V. Aleksandrov, *Z. Naturforsch.* **36a**, 902 (1981).
28. W. Boucher and J. Traschen, *Phys. Rev. D* **37**, 3522 (1988).
29. A. Anderson, *Phys. Rev. Lett.* **74**, 621 (1995).
30. O. V. Prezhdo and V. V. Kisil, *Phys. Rev. A* **56**, 162 (1997).
31. C. C. Martens and J. Fang, *J. Chem. Phys.* **106**, 4918 (1997).
32. A. Donoso and C. C. Martens, *J. Phys. Chem. A* **112**, 4291 (1998).
33. J. Caro and L. L. Salcedo, *Phys. Rev. A* **60**, 842 (1999).
34. R. Kapral and G. Cicotti, *J. Chem. Phys.* **110**, 8919 (1999).
35. S. Nielsen, R. Kapral, and G. Cicotti, *J. Chem. Phys.* **112**, 6543 (2000).
36. A. Sergi and R. Kapral, *J. Chem. Phys.* **118**, 8566 (2003).
37. A. Sergi, D. Mac Kernan, G. Ciccotti, and R. Kapral, *Theor. Chem. Acc.* **110**, 49 (2003).
38. I. Horenko, C. Salzmann, B. Schmidt, and C. Schütte, *J. Chem. Phys.* **117**, 11075 (2002).
39. I. Horenko, M. Weiser, B. Schmidt, and C. Schütte, *J. Chem. Phys.* **120**, 8913 (2004).
40. C.-C. Wan and J. Schofield, *J. Chem. Phys.* **112**, 4447 (2000).
41. C.-C. Wan and J. Schofield, *J. Chem. Phys.* **116**, 494 (2002).
42. M. Santer, U. Manthe, and G. Stock, *J. Chem. Phys.* **114**, 2001 (2001).
43. K. Ando and M. Santer, *J. Chem. Phys.* **118**, 10399 (2003).
44. Q. Shi and E. Geva, *J. Chem. Phys.* **121**, 3393 (2004).
45. P. Pechukas, *Phys. Rev.* **181**, 174 (1969).
46. J. Burant and J. Tully, *J. Chem. Phys.* **112**, 6097 (2000).
47. E. Gindensperger, C. Meier, and J. Beswick, *J. Chem. Phys.* **113**, 9369 (2000).
48. O. Prezhdo and C. Brooksby, Phys. Rev. Lett. **86**, 3215 (2001).
49. E. R. Bittner, *J. Chem. Phys.* **119**, 1358 (2003).
50. I. Burghardt and G. Parlant, *J. Chem. Phys.* **120**, 3055 (2004).
51. R. E. Wyatt, *Quantum Dynamics with Trajectories*, Springer, Berlin, 2005.
52. L. D. Landau, *Z. Sowjetunion U.R.S.S.* **2**, 46 (1932).
53. C. Zener, *Proc. R. Soc.* **A137**, 696 (1932).
54. E. C. G. Stückelberg, *Helv. Phys. Acta* **5**, 369 (1932).
55. W. H. Miller and T. F. George, *J. Chem. Phys.* **56**, 5637 (1972).

56. E. E. Nikitin, *Theory of Elementary Atomic and Molecular Processes in Gases*, Clarendon, Oxford, 1974.
57. M. F. Herman, *J. Chem. Phys.* **81**, 754, 764 (1984).
58. F. J. Webster, P. J. Rossky, and R. A. Friesner, *Comput. Phys. Commun.* **63**, 494 (1991).
59. R. G. Littlejohn and W. G. Flynn, *Phys. Rev. Lett.* **66**, 2839 (1991).
60. R. G. Littlejohn and W. G. Flynn, *Phys. Rev. A* **45**, 7697 (1992).
61. D. F. Coker and L. Xiao, *J. Chem. Phys.* **102**, 496 (1995).
62. J. Bolte and S. Keppler, *Phys. Rev. Lett.* **81**, 1987 (1998).
63. S. Keppeler, *Ann. Phys.* **304**, 40 (2003).
64. C. Zhu and H. Nakamura, *J. Chem. Phys.* **106**, 2599 (1997).
65. C. Zhu, Y. Teranishi, and H. Nakamura, *Adv. Chem. Phys.* **117**, 127 (2001).
66. J. C. Tully and R. K. Preston, *J. Chem. Phys.* **55**, 562 (1971).
67. J. C. Tully, *J. Chem. Phys.* **93**, 1061 (1990).
68. S. Hammes-Schiffer and J. Tully, *J. Chem. Phys.* **101**, 4657 (1994).
69. N. C. Blais, D. Truhlar, and C. A. Mead, *J. Chem. Phys.* **89**, 6204 (1988).
70. P. Kuntz, *J. Chem. Phys.* **95**, 141 (1991).
71. O. Prezhdo and P. J. Rossky, *J. Chem. Phys.* **107**, 825 (1997).
72. D. Sholl and J. Tully, *J. Chem. Phys.* **109**, 7702 (1998).
73. Y. L. Volobuev, M. Hack, M. Topaler, and D. Truhlar, *J. Chem. Phys.* **112**, 9716 (2000).
74. A. W. Jasper, S. N. Stechmann, and D. G. Truhlar, *J. Chem. Phys.* **116**, 5424 (2002).
75. C. Zhu, S. Nangia, A. W. Jasper, and D. G. Truhlar, *J. Chem. Phys.* **121**, in press (2004).
76. A. Ferretti, G. Grannucci, A. Lami, M. Persico, and G. Villani, *J. Chem. Phys.* **104**, 5517 (1996).
77. J.-Y. Fang and S. Hammes-Schiffer, *J. Phys. Chem. A* **103**, 9399 (1999).
78. U. Müller and G. Stock, *J. Chem. Phys.* **107**, 6230 (1997).
79. A. I. Krylov and R. B. Gerber, *J. Chem. Phys.* **105**, 4626 (1996).
80. V. Batista and D. Coker, *J. Chem. Phys.* **106**, 6923 (1997).
81. S. Chapman, *Adv. Chem. Phys.* **82**, 423 (1992).
82. D. F. Coker, in *Computer Simulation in Chemical Physics*, M. P. Allen and D. J. Tildesley, eds., Kluwer Academic Publishers, Dordecht, 1993, p. 315.
83. J. C. Tully, *Faraday Discuss.* **110**, 407 (1998).
84. M. Hack and D. Truhlar, *J. Phys. Chem.* **104**, 7917 (2000).
85. P. A. M. Dirac, *Proc. Roy. Soc. (London) A* **114**, 243 (1927).
86. F. Strocchi, *Rev. Mod. Phys.* **38**, 36 (1966).
87. W. H. Miller and C. W. McCurdy, *J. Chem. Phys.* **69**, 5163 (1978).
88. C. W. McCurdy, H.-D. Meyer, and W. H. Miller, *J. Chem. Phys.* **70**, 3177 (1979).
89. H.-D. Meyer and W. H. Miller, *J. Chem. Phys.* **70**, 3214 (1979).
90. H.-D. Meyer and W. H. Miller, *J. Chem. Phys.* **71**, 2156 (1979).
91. H.-D. Meyer and W. H. Miller, *J. Chem. Phys.* **72**, 2272 (1980).
92. W. H. Miller, *Adv. Chem. Phys.* **25**, 69 (1974).
93. H.-D. Meyer, *Chem. Phys.* **82**, 199 (1983).
94. G. Stock and W. H. Miller, *Chem. Phys. Lett.* **197**, 396 (1992).
95. E. M. Goldfield, P. L. Houston, and G. S. Ezra, *J. Chem. Phys.* **84**, 3120 (1986).

96. I. Benjamin and K. R. Wilson, *J. Chem. Phys.* **90**, 4176 (1989).
97. T. Holstein and H. Primakoff, *Phys. Rev.* **58**, 1098 (1940).
98. J. Schwinger, in *Quantum Theory of Angular Momentum*, L. C. Biedenharn and H. V. Dam, eds., Academic Press, New York, 1965.
99. G. Stock and M. Thoss, *Phys. Rev. Lett.* **78**, 578 (1997).
100. M. Thoss and G. Stock, *Phys. Rev. A* **59**, 64 (1999).
101. M. F. Herman and R. Currier, *Chem. Phys. Lett.* **114**, 411 (1985).
102. U. Müller and G. Stock, *J. Chem. Phys.* **108**, 7516 (1998).
103. U. Müller and G. Stock, *J. Chem. Phys.* **111**, 77 (1999).
104. W. H. Miller, *J. Chem. Phys.* **53**, 3578 (1970).
105. M. F. Herman and E. Kluk, *Chem. Phys.* **91**, 27 (1984).
106. E. J. Heller, *J. Chem. Phys.* **94**, 2723 (1991).
107. G. Campolieti and P. Brumer, *J. Chem. Phys.* **96**, 5969 (1992).
108. K. G. Kay, *J. Chem. Phys.* **100**, 4377 (1994).
109. A. R. Walton and D. E. Manolopoulos, *Mol. Phys.* **87**, 961 (1996).
110. M. A. Sepúlveda and F. Grossmann, *Adv. Chem. Phys.* **96**, 191 (1996).
111. W. H. Miller, *J. Phys. Chem. A* **105**, 2942 (2001).
112. X. Sun and W. H. Miller, *J. Chem. Phys.* **106**, 6346 (1997).
113. X. Sun, H. Wang, and W. H. Miller, *J. Chem. Phys.* **109**, 7064 (1998).
114. V. S. Batista and W. H. Miller, *J. Chem. Phys.* **108**, 498 (1998).
115. F. Grossmann, *Phys. Rev. A* **60**, 1791 (1999).
116. M. Thoss, W. H. Miller, and G. Stock, *J. Chem. Phys.* **112**, 10282 (2000).
117. E. A. Coronado, J. Xing, and W. Miller, *Chem. Phys.* **349**, 521 (2001).
118. S. Bonella and D. Coker, *J. Chem. Phys.* **114**, 7778 (2001).
119. S. Bonella and D. F. Coker, *Chem. Phys.* **268**, 189 (2001).
120. S. Bonella and D. F. Coker, *J. Chem. Phys.* **118**, 4370 (2003).
121. V. S. Batista and P. Brumer, *J. Phys. Chem. A* **105**, 2591 (2001).
122. H. Wang, D. E. Manolopoulos, and W. H. Miller, *J. Chem. Phys.* **115**, 6317 (2001).
123. S. Dilthey, B. Mehlig, and G. Stock, *J. Chem. Phys.* **116**, 69 (2002).
124. H. Fujisaki, *Phys. Rev. E* **69**, 037201 (2004).
125. S. Dilthey and G. Stock, *Phys. Rev. Lett.* **87**, 140404 (2001).
126. S. Dilthey and G. Stock, *J. Phys. Chem. A* **106**, 8483 (2002).
127. B. Balzer, S. Dilthey, S. Hahn, M. Thoss, and G. Stock, *J. Chem. Phys.* **119**, 4204 (2003).
128. J.-L. Liao and G. A. Voth, *J. Phys. Chem. B* **106**, 8449 (2002).
129. D. V. Shalashilin and M. S. Child, *J. Chem. Phys.* **121**, 3563 (2004).
130. A. Novikov, U. Kleinekathöfer, and M. Schreiber, *Chem. Phys.* **296**, 149 (2004).
131. E. Rabani, S. A. Egorov, and B. J. Berne, *J. Phys. Chem. A* **103**, 9539 (1999).
132. Q. Shi and E. Geva, *J. Phys. Chem. A* **108**, 6109 (2004).
133. Y. Takahashi and F. Shibata, *J. Phys. Soc. Japan* **38**, 656 (1975).
134. M. Bergeron, *Fortschr. Phys.* **40**, 119 (1992).
135. T. Boudjedaa, A. Bounames, L. Chetouani, T. F. Hammann, and K. Nouiser, *J. Math. Phys.* **36**, 1602 (1995).

136. J. M. Radcliffe, *J. Phys. A* **4**, 313 (1971).
137. J. R. Klauder, *Phys. Rev. D* **19**, 2349 (1979).
138. W.-M. Zhang and D. H. Feng, *Phys. Rep.* **252**, 1 (1995).
139. E. Ercolessi, G. Morandi, F. Napoli, and P. Pieri, *J. Math. Phys.* **37**, 535 (1996).
140. E. A. Kochetov, *J. Phys. A* **31**, 4473 (1998).
141. A. Alscher and H. Grabert, *J. Phys. A* **32**, 4907 (1999).
142. A. Lucke, C. Mak, and J. Stockburger, *J. Chem. Phys.* **111**, 10843 (1999).
143. M. Stone, K.-S. Park, and A. Garg, *J. Math. Phys.* **41**, 8025 (2000).
144. A. Alscher and H. Grabert, *Eur. Phys. J. D* **14**, 127 (2001).
145. M. Pletyukhov, C. Amann, M. Mehta, and M. Brack, *Phys. Rev. Lett.* **89**, 116601 (2002).
146. O. Zaitsev, *J. Phys. A* **35**, 721 (2002).
147. M. Pletyukhov and O. Zaitsev, *J. Phys. A* **36**, 5181 (2003).
148. B. Berne and G. D. Harp, *Adv. Chem. Phys.* **17**, 63 (1970).
149. S. Egorov, K. F. Everitt, and J. L. Skinner, *J. Phys. Chem. A* **103**, 9494 (1999).
150. E. J. Heller, *J. Chem. Phys.* **75**, 2923 (1981).
151. T. J. Martinez, M. Ben-Nun, and R. D. Levine, *J. Chem. Phys.* **100**, 7884 (1996).
152. M. Ben-Nun and T. J. Martinez, *Chem. Phys. Lett.* **298**, 57 (1998).
153. M. Ben-Nun, J. Quenneville, and T. J. Martinez, *J. Phys. Chem. A* **104**, 5161 (2000).
154. D. Shalashilin and M. Child, *J. Chem. Phys.* **115**, 5367 (2001).
155. T. Vreven, F. Bernardi, M. Garavelli, M. Olivucci, M. A. Robb, and H. B. Schlegel, *J. Am. Chem. Soc.* **119**, 12687 (1997).
156. M. Garavelli, F. Bernardi, M. Olivucci, M.J. Bearpark, S. Klein, and M. A. Robb, *J. Phys. Chem. A* **105**, 11496 (2001).
157. M. Hartmann, J. Pittner, and V. Bonačić-Koutecký, *J. Chem. Phys.* **114**, 2123 (2001).
158. G. Granucci, M. Persico, and A. Toniolo, *J. Chem. Phys.* **114**, 10608 (2001).
159. P. Cattaneo and M. Persico, *J. Am. Chem. Soc.* **123**, 7638 (2001).
160. N. L. Doltsinis and D. Marx, *Phys. Rev. Lett.* **88**, 166402 (2002).
161. M. Thoss and H. Wang, *Annu. Rev. Phys. Chem.* **55**, 299 (2004).
162. H. Köppel, W. Domcke, and L. S. Cederbaum, *Adv. Chem. Phys.* **57**, 59 (1984).
163. W. Domcke and G. Stock, *Adv. Chem. Phys.* **100**, 1 (1997).
164. L. Seidner, G. Stock, and W. Domcke, *Chem. Phys. Lett.* **228**, 665 (1994).
165. R. A. Marcus and N. Sutin, *Biochim. Biophys. Acta* **811**, 265 (1985).
166. P. F. Barbara, T. J. Meyer, and M. A. Ratner, *J. Phys. Chem.* **100**, 13148 (1996).
167. J. Jortner and M. Bixon, in *Electron Transfer: From Isolated Molecules to Biomolecules, Dynamics and Spectroscopy, Advances in Chemical Physics*, Vols. 106 and 107, Wiley, New York, 1999.
168. A. J. Leggett, S. Chakravarty, A. T. Dorsey, M. P. A. Fisher, A. Garg, and W. Zwerger, *Rev. Mod. Phys.* **59**, 1 (1987).
169. J. Michl and V. Bonačić-Koutecký, *Electronic Aspects of Organic Photochemistry*, Wiley, New York, 1990.
170. W. Domcke, D. R. Yarkony, and H. Köppel, eds., *Conical Intersections: Electronic Structure, Dynamics and Spectroscopy*, World Scientific, Singapore, 2004.
171. E. Wigner, *Phys. Rev.* **40**, 749 (1932).

172. L. Seidner and W. Domcke, *Chem. Phys.* **186**, 27 (1994).
173. R. Schneider and W. Domcke, *Chem. Phys. Lett.* **150**, 235 (1988).
174. R. Schneider, W. Domcke, and H. Köppel, *J. Chem. Phys.* **92**, 1045 (1990).
175. U. Manthe and H. Köppel, *J. Chem. Phys.* **93**, 345, 1658 (1990).
176. M. Durga Prasad, *Chem. Phys. Lett.* **194**, 27 (1992).
177. G. Stock and W. Domcke, *J. Chem. Phys.* **93**, 5496 (1990).
178. G. Stock, R. Schneider, and W. Domcke, *J. Chem. Phys.* **90**, 7184 (1989).
179. H. Köppel, L. Cederbaum, and W. Domcke, *J. Chem. Phys.* **89**, 2023 (1988).
180. H. Köppel, *Chem. Phys. Lett.* **205**, 361 (1993).
181. L. Seidner, G. Stock, and W. Domcke, *J. Chem. Phys.* **103**, 3998 (1995).
182. G. Stock, *J. Chem. Phys.* **103**, 10015 (1995).
183. I. Uspenskiy, B. Balzer, and G. Stock, to be published (2005).
184. S. Hahn and G. Stock, *J. Phys. Chem. B* **104**, 1146 (2000).
185. S. Hahn and G. Stock, *Chem. Phys.* **259**, 297 (2000).
186. S. Hahn and G. Stock, *J. Chem. Phys.* **116**, 1085 (2002).
187. F. Grossmann and M. F. Herman, *J. Phys. A* **35**, 9489 (2002).
188. B. Balzer, S. Hahn, and G. Stock, *Chem. Phys. Lett.* **379**, 351 (2003).
189. B. Wolfseder and W. Domcke, *Chem. Phys. Lett.* **259**, 113 (1996).
190. S. K. Dorn, R. B. Dyer, P. O. Stoutland, and W. H. Woodruff, *J. Am. Chem. Soc.* **115**, 6398 (1993).
191. K. Tominaga, D. A. V. Kliner, A. E. Johnson, N. E. Levinger, and P. F. Barbara, *J. Chem. Phys.* **98**, 1228 (1993).
192. C. Wang, B. K. Mohney, B. B. Akhremitchev, and G. C. Walker, *J. Phys. Chem. A* **104**, 4314 (2000).
193. W. Holzapfel, U. Finkele, W. Kaiser, D. Oesterhelt, H. Scheer, H. U. Stilz, and W. Zinth, *Proc. Natl. Acad. Sci. USA* **87**, 5168 (1990).
194. M. H. Vos, J. C. Lambry, S. J. Robles, D. C. Youvan, J. Breton, and J. L. Martin, *Proc. Natl. Acad. Sci. USA* **88**, 8885 (1991).
195. M. E. Michel-Beyerle, *The Reaction Center of Photosynthetic Bacteria: Structure and Dynamics*, Springer, Berlin, 1996.
196. N. Makri, *J. Phys. Chem. B* **103**, 2823 (1999).
197. Y. Georgievskii, C. P. Hsu, and R. A. Marcus, *J. Chem. Phys.* **110**, 5307 (1999).
198. C. H. Mak and D. Chandler, *Phys. Rev. A* **44**, 2352 (1991).
199. D. E. Makarov and N. Makri, *Chem. Phys. Lett.* **221**, 482 (1994).
200. For the applications considered, it was found that it virtually makes no difference whether the phase q is sampled from the intervall $[0, 2\pi]$ or chosen as $q = 0$ for each trajectory.
201. H. Wang, M. Thoss, and W. H. Miller, *J. Chem. Phys.* **115**, 2979 (2001).
202. H. Wang and M. Thoss, *Israel J. Chem.* **42**, 167 (2002).
203. H. Wang and M. Thoss, *J. Phys. Chem. A* **107**, 2126 (2003).
204. See, for example, the discussions in Ref. 18 and in L. E. Ballentine, Y. Yang, and J. P. Zibin, *Phys. Rev. A* **50**, 2854 (1994).
205. To be consistent with original work [42], the data shown in Figs. 13–16 were obtained for a slightly modified (but dynamically equivalent) version of Model IVa, where $g_{12} = 0.152\,\text{eV}$

(instead of $g_{12} = 0.1\,\text{eV}$) and (in agreement with Table IV) $\omega_1 = g_{12}$ and $\kappa_1^{(2)} = -\kappa_1^{(1)} = 0.5g_{12}$.

206. E. R. Bittner and P. J. Rossky, *J. Chem. Phys.* **103**, 8130 (1995).
207. O. V. Prezhdo, *J. Chem. Phys.* **111**, 8366 (1999).
208. C. H. Mak and R. Egger, *Adv. Chem. Phys.* **93**, 39 (1996).
209. D. MacKernan, personal communication.
210. V. P. Maslov and A. M. Chebotarev, *J. Sov. Math.* **13**, 315 (1980).
211. J. Bertrand and G. Rideau, *Lett. Math. Phys.* **7**, 327 (1983).
212. S. John and E. A. Remler, *Ann. Phys.* **180**, 152 (1987).
213. T. C. Schmidt and K. Möhring, *Phys. Rev. A* **48**, 3418 (1993).
214. V. S. Filinov, Y. V. Medvedev, and V. L. Kamsky, *Mol. Phys.* **85**, 711 (1995).
215. E. J. Heller, *J. Chem. Phys.* **62**, 1544 (1975).
216. This general feature of the stochastic scheme may cause convergence problems. For example, consider a situation in which the molecular system is predominately in a single state, say ρ_{11}. Although the expectation values of the population $P_2 = \text{tr}\rho_{22}$ and the corresponding coherences are zero, there are the same number of random walkers in these states which need to cancel themselves in the phase average.
217. Assuming that on average each random walker is propagated for half of the end time t, and requesting an accuracy of 2% at $t = 50$ fs, the present scheme requires a total number of $\sim 10^5$ random walkers to be propagated. Increasing simply the number of initially starting random walkers, on the other hand, $\sim 10^6$ random walkers would be required.
218. J. P. Blaizot and E. R. Marshalek, *Nucl. Phys. A* **309**, 422 (1978).
219. P. Garbaczewski, *Phys. Rep.* **36**, 65 (1978).
220. J. P. Blaizot and G. Ripka, *Quantum Theory of Finite Systems*, MIT Press, Cambridge, MA, 1986.
221. L. Susskind and J. Glogower, *Physics* **1**, 49 (1964).
222. A. Luis and L. L. Sánchez-Soto, *Phys. Rev. A* **48**, 752 (1993).
223. For a general discussion and an overview of existing methods, see, for example, Y. Guo, D. L. Thompson, and T. D. Sewell, *J. Chem. Phys.* **104**, 576 (1996).
224. G. Stock and U. Müller, *J. Chem. Phys.* **111**, 65 (1999).
225. A. A. Golosov and D. R. Reichman, *J. Chem. Phys.* **114**, 1065 (2001).
226. S. Dilthey and G. Stock, *Israel J. Chem.* **42**, 203 (2002).
227. U. Müller, Ph.D. thesis, University of Freiburg, unpublished (1999).
228. H. Köppel, M. Döscher, I. Baldea, H.-D. Meyer, and P. G. Szalay, *J. Chem. Phys.* **117**, 2657 (2002).
229. E. Sim and N. Makri, *J. Phys. Chem.* **101**, 5446 (1997).
230. A. J. Lichtenberg and M. A. Liebermann, *Regular and Stochastic Motion*, Springer, New York, 1983.
231. L. Müller, J. Stolze, H. Leschke, and P. Nagel, *Phys. Rev. A* **44**, 1022 (1991).
232. J. W. Zwanziger, E. R. Grant, and G. S. Ezra, *J. Chem. Phys.* **85**, 2089 (1986).
233. T. Zimmermann, H. Köppel, and L. S. Cederbaum, *J. Chem. Phys.* **91**, 3934 (1989).
234. D. M. Leitner, H. Köppel, and L. S. Cederbaum, *J. Chem. Phys.* **104**, 434 (1996).
235. H. Fujisaki and K. Takatsuka, *Phys. Rev. E* **63**, 066221 (2001).
236. I. C. Percival, *Adv. Chem. Phys.* **36**, 1 (1977).

237. B. Balzer, S. Dilthey, G. Stock, and M. Thoss, *J. Chem. Phys.* **119**, 5795 (2003).
238. E. J. Heller, *Chaos and Quantum Physics*, North-Holland, Amsterdam, 1991.
239. J. M. Gomez Llorente and E. Pollak, *Annu. Rev. Phys. Chem.* **43**, 91 (1992).
240. B. Eckhardt, in *Periodic Orbit Theory, Proceedings of the International School of Physics, Course CXIX*, G. Casati, I. Guarneri, and U. Smilansky, eds., North-Holland, Amsterdam, 1993.
241. O. Hahn and H. S. Taylor, *J. Chem. Phys.* **96**, 5915 (1992).
242. M. P. Jacobson, C. Jung, H. S. Taylor, and R. W. Field, *J. Chem. Phys.* **111**, 600 (1999).
243. M. Joyeux, S. C. Farantos, and R. Schinke, *J. Phys. Chem. A* **106**, 5407 (2002).
244. M. Ovchinnikov and V. A. Apkarian, *J. Chem. Phys.* **108**, 2277 (1998).
245. J. Ankerhold, M. Saltzer, and E. Pollak, *J. Chem. Phys.* **116**, 5925 (2002).
246. S. Zhang and E. Pollak, *Phys. Rev. Lett.* **91**, 190201 (2003).
247. E. Pollak and J. Shao, *J. Phys. Chem. A* **107**, 7112 (2003).
248. S. Zhang and E. Pollak, *J. Chem. Phys.* **121**, 3384 (2004).
249. S. Garashuk and D. J. Tannor, *Annu. Rev. Phys. Chem.* **51**, 553 (2000).
250. L. S. Schulman, *Techniques and Applications of Path Integration*, Wiley, New York, 1981.
251. H. Kleinert, *Path Integrals in Quantum Mechanics, Statistics and Polymer Physics*, World Scientific, Singapore, 1995.
252. F. Grossmann and A. L. Xavier, *Phys. Lett. A* **243**, 243 (1998).
253. M. Baranger, M. A. M. de Aguiar, F. Keck, H. J. Korsch, and B. Schellhaaß, *J. Phys. A* **34**, 7227 (2001).
254. M. Baranger, M. A. M. de Aguiar, F. Keck, H. J. Korsch, and B. Schellhaaß, *J. Phys. A* **35**, 9493 (2002).
255. W. H. Miller, *Mol. Phys.* **100**, 397 (2002).
256. B. W. Spath and W. H. Miller, *J. Chem. Phys.* **104**, 95 (1996).
257. B. E. Guerin and M. F. Herman, *Chem. Phys. Lett.* **286**, 361 (1998).
258. Y. Elran and K. G. Kay, *J. Chem. Phys.* **110**, 8912 (1999).
259. B. R. McQuarrie and P. Brumer, *Chem. Phys. Lett.* **319**, 27 (2000).
260. V. S. Filinov, *Nucl. Phys. B* **271**, 717 (1987).
261. N. Makri and W. H. Miller, *Chem. Phys. Lett.* **139**, 10 (1987).
262. J. D. Doll, D. L. Freeman, and T. L. Beck, *Adv. Chem. Phys.* **78**, 61 (1994).
263. V. Voros, *Phys. Rev. A* **40**, 6814 (1989).
264. For a nonadiabatic system with spin-orbit interaction, the validity of the semiclassical approximation (based on the spin-coherent state representation) has been discussed in detail in Ref. 147.
265. H. Wang, M. Thoss, K. Sorge, R. Gelabert, X. Gimenez, and W.H. Miller, *J. Chem. Phys.* **114**, 2562 (2001).
266. E. A. Coronado, V. S. Batista, and W. H. Miller, *J. Chem. Phys.* **112**, 5566 (2000).
267. M. Thoss, unpublished results.
268. G. Stock and M. Thoss, Mixed quantum-classical description of the dynamics at conical intersections, in *Canonical Intersections*, W. Domcke, D. R. Yarkony, and H. Köppel, eds., World Scientific, Singapore, 2004.
269. C. Woywod, W. Domcke, A. L. Sobolewski, and H.-J. Werner, *J. Chem. Phys.* **100**, 1400 (1994).

270. For a discussion of the nonunitarity of the semiclassical approximation in the context of nonadiabatic dynamics, see, for example, Ref. 100.
271. I. Yamazaki, T. Murao, T. Yamanaka, and K. Yoshihara, *Faraday Discuss.* **75**, 395 (1983).
272. N. Makri and K. Thompson, *Chem. Phys. Lett.* **291**, 101 (1998).
273. X. Sun and W. H. Miller, *J. Chem. Phys.* **110**, 6635 (1999).
274. M. Thoss, H. Wang, and W. H. Miller, *J. Chem. Phys.* **114**, 9220 (2001).
275. H.-D. Meyer, U. Manthe, and L. S. Cederbaum, *Chem. Phys. Lett.* **165**, 73 (1990).
276. H.-D. Meyer, *Phys. Rep.* **324**, 1 (2000).
277. A. Raab, G. A. Worth, H.-D. Meyer, and L. S. Cederbaum, *J. Chem. Phys.* **110**, 936 (1999).
278. Wang et al. have recently reperformed the calculation for the 24-mode pyrazine model employing an improved filtering technique, thereby reducing the number of trajectories required to obtain convergence by almost two orders of magnitude [122].
279. J. R. Klauder and B.-S. Skagerstam, *Coherent States*, World Scientific, Singapore, 1985.
280. A. M. Perelomov, *Generalized Coherent States and Their Applications*, Springer, Berlin, 1986.
281. T. Fukui, *J. Math. Phys.* **34**, 4455 (1993).
282. D. Ebert and V. Yarunin, *Fortschr. Phys.* **42**, 589 (1994).
283. H. G. Solari, *J. Math. Phys.* **28**, 1067 (1987).
284. E. A. Kochetov, *J. Math. Phys.* **36**, 4667 (1995).
285. E. Keski-Vakkuri, A. J. Niemi, G. Semenoff, and O. Tirkkonen, *Phys. Rev. D* **44**, 3899 (1991).
286. T. Suzuki, *Nucl. Phys. A* **398**, 557 (1983).
287. It is noted that in the results of Suzuki's propagator given in Fig. 7 of Ref. 100, an incorrect normalization factor was used, which resulted in a strong deviation of the overall magnitude of the autocorrelation function from the quantum-mechanical result.
288. Q. Shi and E. Geva, *J. Phys. Chem. A* **107**, 9070 (2003).
289. A. Nakayama and N. Makri, *J. Chem. Phys.* **119**, 8592 (2003).
290. J. A. Poulsen, G. Nyman, and P. J. Rossky, *J. Chem. Phys.* **119**, 12179 (2003).
291. I. R. Craig and D. E. Manolopoulos, *J. Chem. Phys.* **121**, 3386 (2004).
292. S. Spörlein, H. Carstens, C. Renner, R. Behrendt, L. Moroder, P. Tavan, W. Zinth, and J. Wachtveitl, *Proc. Natl. Acad. Sci. USA* **99**, 7998 (2002).
293. P. Nguyen and G. Stock, *J. Chem. Phys.* **119**, 11350 (2003).
294. S. Hammes-Schiffer, *Curr. Opin. Struct. Biol.* **14**, 192 (2004).
295. J. Saam, E. Tajkhorshid, S. Hayashi, and K. Schulten, *Biophys. J.* **83**, 3097 (2002).
296. G. Groenhof, M. Bouxin-Cademartory, B. Hess, S. de Visser, H. J. C. Berendsen, M. Olivucci, A. E. Mark, and M. A. Robb, *J. Am. Chem. Soc.* **126**, 4228 (2004).

NON-BORN–OPPENHEIMER VARIATIONAL CALCULATIONS OF ATOMS AND MOLECULES WITH EXPLICITLY CORRELATED GAUSSIAN BASIS FUNCTIONS

SERGIY BUBIN

Department of Physics and Department of Chemistry
University of Arizona, Tucson, AZ

MAURICIO CAFIERO

Department of Chemistry, Rhodes College, Memphis, TN

LUDWIK ADAMOWICZ

Department of Chemistry and Department of Physics
University of Arizona, Tucson, AZ

CONTENTS

Advances in Chemical Physics, Volume 131, edited by Stuart A. Rice

I. INTRODUCTION

Soon after the Schrödinger equation was introduced in 1926, several works appeared dealing with the fundamental problem of the nuclear motion in molecules. Very soon after, the relativistic equations were introduced for one- and two-electron systems. The experiments on the Lamb shift stimulated derivation of the expressions for fine energy corrections related to mass–velocity effects, radiative effects, and so on. Following this development, questions were raised whether an *ab initio* approach, in which only the values of fundamental constants are taken from experiments, is capable of reproducing experimental results with the precision which matches that of the experimental techniques. At first, the questions concerned the accuracy of the calculations, but soon they were extended to testing the model of the molecular electron structure provided by the nonrelativistic and relativistic quantum mechanics.

In order to answer these questions, accurate experimental and theoretical results were needed for representative molecular systems. Theoreticians, for obvious reasons, have favored very simple systems, such as the hydrogen molecular ion (H_2^+) for their calculations. However, with only one electron, this system did not provide a proper test case for the molecular quantum mechanical methods due to the absence of the electron correlation. Therefore, the two-electron hydrogen molecule has served as the system on which the fundamental laws of quantum mechanics have been first tested.

In an attempt to make the quantum-mechanical calculations on molecular systems practical and to provide a more intuitive interpretation of the computed results, it has long been a quest in the electronic structure theory of molecules to establish a solid base for separating the motion of light electrons from the motion of heavier nuclei. It is believed that the original work of Born and Oppenheimer [1] initiated the discussion by the analysis of the diatomic case. Further works of Cobes and Seiler [2], who managed, with the use of singular perturbation theory, to resolve the problem of the diverging series, which appeared in the Born–Oppenheimer (BO) expansion, and particularly of Klein et al. [3], who extended the formalism to polyatomic systems, have brought the consideration of the topic to a level of commonly accepted theory.

Apart from the further refinements of the BO approach, there has been a continuing interest in theoretically describing molecular systems with a method that treats the motions of both nuclei and electrons equivalently. This type of methodology has to entirely depart from the PES concept. It is particularly interesting how this type of approach describes the conventional notions of the molecular and electronic structures. In particular, the concept of chemical bonding, which at the BO level is an electronic phenomenon, has to be described in an approach departing from the BO approximation, as an effect derived from collective dynamical behavior of both electrons and nuclei.

Another motivation for considering molecular systems without assuming the BO approximation stems from the realization that in order to reach "spectroscopic" accuracy in quantum-mechanical calculations (i.e., error less than 1 microhartree), one needs to account for the coupling between motions of electrons and nuclei and, in some cases, also for the relativistic effects. Modern experimental techniques, such as gas-phase ion-beam spectroscopy, reach accuracy on the order of $0.001\,\text{cm}^{-1}$ (5 nanohartrees) [4]. In order for the molecular quantum mechanics to continue providing assistance in resolving and assigning experimental spectra and in studies of reaction dynamics, work has to continue on the development of more refined theoretical methodology, which accounts for nonadiabatic interactions. With such methodology, fundamental concepts of the molecular quantum mechanics can be explored and the basic theoretical framework of the high-resolution molecular spectroscopy can be tested. Recent advances in high-performance computing, especially in the area of massively parallel systems, has given momentum to proceed with the development of quantum-mechanical methods that depart from the BO approximation and describe the motions of the nuclei and electrons with a single wave function. In the context of the non-BO calculations, it is particularly interesting to study highly rho-vibrationally and electronically excited molecules and clusters, where more significant coupling between the two motions can occur. Essentially, whenever the spacings between electronic excitation levels become comparable to the spacings between the vibrational or rotational levels, nonadiabatic effects are likely to be found. Studies of these effects are relevant to astrophysical phenomena, molecule dynamics, and molecular behavior at high temperatures.

If one assumes the BO approximation and considers potential energy surfaces of a molecule, one can usually identify areas where there is a high density of electronic states and areas where the electronic states are more separated from each other. A strong nonadiabatic coupling can be expected to mainly occur in the areas with the high electronic state density. This simple realization has given rise to a considerable body of recent theoretical research on nonadiabatic phenomena, done in relation to conical intersections of molecular PESs [5–14], most notably by Yarkony and his group, among others. The reason for having large or even infinite nonadiabatic coupling terms in the conical areas is that fast-moving electrons may create exceptionally large forces causing the nuclei to strongly accelerate. The terms responsible for this accelerated motion cannot be ignored even in the first approximation. The consequences of the conical PES crossings to the dynamics of molecular reactions have also been considered by a number of groups (see, for example, the work of Hammes-Schiffer [15–26]). In those works, however, the non-BO effects are only considered at the conical intersection of two electronic PESs since the focus there has been more on determining the probability of the process splitting and following two different electronic PESs and less on very accurate global representation of coupled electronic–nuclear states, as it has been in our work.

The nonadiabatic coupling terms can quickly become large or even infinite (or singular) when two successive adiabatic states become degenarate. Such singular nonadiabatic coupling may not only lead to the breakdown of the

Born–Oppenheimer perturbation theory but may also make the application of the perturbation theory inadequate.

Though infrequent, fully non-Born–Oppenheimer high-accuracy calculations on atomic and molecular systems have been increasing in number in recent years. However, besides our group, there are only a few groups doing such calculations, particularly for molecular systems. One should mention the recent works of Frolov and Smith [27–29] and of Frolov [30–34] concerning some exotic systems involving muons and positrons, as well as one electron H_2 isotopomers. Our works, reviewed here, have concerned both atomic and molecular systems. Although for molecules most of our non-BO calculations concerned ground and excited states of diatomic systems, we have also recently extended the non-BO approach beyond molecules with two nuclei. The system with the largest number of particles we have calculated so far had two nuclei and five electrons (the LiH^- anion).

There are several elements in non-BO calculations that distinguish them from the conventional BO calculations. The first one concerns the Hamiltonian. If one neglects the relativistic effects and places the considered system in a laboratory Cartesian coordinate frame, the Hamiltonian has a simple form of a sum of one-particle kinetic energy operators for all particles involved in the system plus a sum representing all pair Coulombic interactions between the particles. It is convenient to separate the center-of-mass kinetic energy from the Hamiltonian. This is usually done by a coordinate transformation that involves choosing a new coordinate system whose first three coordinates are the Cartesian coordinates of the center of mass in the laboratory coordinate system and the remaining $3N - 3$ coordinates are internal coordinates. There are a number of ways to select the internal coordinates. In some works, these have been the Jacobi coordinates, the spherical coordinates, or the coordinates defined with respect with the center of mass of the system. In our approach we have used a cartesian coordinate system with the coordinate origin placed at one of the particles (usually the heaviest one). This will be described later in this chapter.

An important difference between the BO and non-BO internal Hamiltonians is that the former describes only the motion of electrons in the stationary field of nuclei positioned in fixed points in space (represented by point charges) while the latter describes the coupled motion of both nuclei and electrons. In the conventional molecular BO calculations, one typically uses atom-centered basis functions (in most calculations one-electron atomic orbitals) to expand the electronic wave function. The fermionic nature of the electrons dictates that such a function has to be antisymmetric with respect to the permutation of the labels of the electrons. In some high-precision BO calculations the wave function is expanded in terms of basis functions that explicitly depend on the interelectronic distances (so-called explicitly correlated functions). Such functions usually very effectively describe the electron correlation effects, which need to be included in any high-level BO calculation. An alternative to the explicitly correlated functions is to use expansions in terms of Slater determinants constructed using one electron functions (molecular orbitals). Such an approach is called the configuration interaction (CI) method, and it usually converges much slower than the approach using explicitly correlated basis functions.

Non-BO calculations usually need to be performed to very high precision since only then the non-BO effects, which are usually very small, can be adequately determined. This requires that not only the electron–electron correlation effects are described very accurately, but also the correlation effects due to the nucleus–nucleus interaction and due to the nucleus–electron interaction are accurately represented in the wave function. Since the electrons are light particles, their individual wave functions can strongly overlap, and the probability of finding two electrons (with opposite spins) simultaneously in the same point in space is much higher than for two nuclei, which are much heavier and avoid each other to a much higher degree. One can say that the nuclear correlation effects are much stronger than the electronic ones. Also, the correlation effects associated with the coupled motion of electrons and nuclei are significant because the electrons, particularly the core electrons, follow the nuclei very closely. In order to describe the three types of correlation effects simultaneously with high precision in the non-BO wave function, one needs to use basis functions, which not only depend on the interelectron distances, but also explicitly depend on electron–nucleus and nucleus–nucleus distances. The functions of this type that we have used in our non-BO calculations will be shown later in this chapter. We should mention that the basis set selection is the central point in the non-BO calculation.

After the separation of the kinetic energy operator due to the center-of-mass motion from the Hamiltonian, the Hamiltonian describes the internal motions of electrons and nuclei in the system. These in the BO approximation can be separated into the vibrational and rotational motions of the nuclear frame of the molecule and the electronic motion that only parametrically depends on the instantenous positions of the nuclei. When the BO approximation is removed, the electronic and nuclear motions become coupled and the only good quantum numbers, which can be used to quantize the stationary states of the system, are the principle quantum number, the quantum number quantizing the square of the total (nuclear and electronic) squared angular momentum, and the quantum number quantizing the projection of the total angular momentum vector on a selected direction (usually the z axis). The separation of different rotational states is an important feature that can considerably simplify the calculations.

If in the non-BO calculation one chooses a basis set of eigenfunctions of the operator representing the square of the total angular momentum and the operator representing the projection of the angular momentum on the selected axis, one can separate the calculation for different rotational states and perform them independently of each other. However, if one uses basis functions that are not rotational eigenfunctions, then the manifold of states that one gets includes all types of internal excitations (i.e., rotational, vibrational, and electronic). Since the rotational state spacing is usually much smaller in comparison with the spacings between the vibrational and electronic levels, the different vibrational and electronic levels (or, as we should more correctly call them, the vibro-electronic levels, because the vibrational and electronic motions are coupled in the non-BO calculation) are separated by a large number of rotational levels. This creates a problem if the calculations are done with the use of the variational method, because in order to determine, say, the vibrational spectrum of the molecule corresponding to the zero angular momentum, one needs to "fish" them out from the very high density spectrum of all internal states. There are two ways to overcome this obvious difficulty in the calculation. One, as mentioned, is to use in the calculation the basis functions of rotational eigenstates. The second is based on including in the Hamiltonian an operator that artificially shifts up the energies of states with the rotational quantum numbers different from the quantum number of the states targeted by the calculation. The first approach is what we have used in the calculations shown in this chapter. The second approach is currently being developed for calculating rotationally excited states using explicitly correlated basis functions that incorporate centers displaced away from the origin of the coordinate systems (due to this displacement, these functions are not rotational eigenfunctions).

We start this chapter by showing the coordinate transformation that allows us to separate the operator representing the kinetic energy of the center-of-mass motion from the total Hamiltonian expressed in terms of the Cartesian laboratory coordinates. Next we discuss the symmetry of the internal Hamiltonian and the spatial and permutational symmetry of the wave functions. In the following section we describe the algorithms involved in calculating Hamiltonian matrix elements and their derivatives with respect to the nonlinear parameters involved in the basis functions. We start the discussion of the numerical results by showing some atomic calculations. Next, we present some of our recent calculations on diatomic systems and we discuss their accuracy and the nonadiabatic effects they describe. In the following section we consider the interaction of a molecular system described without assuming the Born–Oppenheimer approximation with an external stationary electric field. In that section we also review some of our calculations concerning electrical properties of some small diatomic molecules. The field-dependent calculations have been done using basis functions whose centers are allowed to "float" away from the origin of the coordinate system to describe the polarization of the molecule along the direction of the field. In the following section we describe the use of the "floating" basis functions in field-independent non-BO calculations. The chapter is concluded with (a) a description of directions for our future works on the non-BO approach and (b) outstanding problems that, in our view, need to be addressed in further advancing the development and the implementation of the non-BO molecular quantum mechanics.

II. HAMILTONIAN, SEPARATION OF THE CENTER OF MASS, INTERACTION WITH ELECTRIC FIELD

A. Nonrelativistic Hamiltonian

A system of $n+1$ particles of masses M_i and charges Q_i may be described at any point in time by the $n+1$ vectors, R_i, describing the positions of the particles:

$$R_i = \begin{pmatrix} x_i \\ y_i \\ z_i \end{pmatrix} \tag{1}$$

and the $n+1$ vectors, P_i, describing the momenta of the particles:

$$P_i = \begin{pmatrix} P_{x,i} \\ P_{y,i} \\ P_{z,i} \end{pmatrix} \tag{2}$$

For convenience we collect the R_i vectors together:

$$R = \begin{pmatrix} R_1 \\ R_2 \\ \vdots \\ R_{n+1} \end{pmatrix} \tag{3}$$

and similarly collect the momenta together:

$$P = \begin{pmatrix} P_1 \\ P_2 \\ \vdots \\ P_{n+1} \end{pmatrix} \tag{4}$$

The kinetic energy of this system is given by

$$T = \sum_{i=1}^{n+1} \frac{P_i^2}{2M_i} \tag{5}$$

If we assume only Coulombic interactions between the particles, the potential energy is given by

$$V = \sum_{i=1}^{n+1} \sum_{j>i}^{n+1} \frac{Q_i Q_j}{r_{ij}} \tag{6}$$

where r_{ij} is the magnitude of the distance vector between particles i and j.

The total Hamiltonian for this system is thus

$$H = \sum_{i=1}^{n+1} \frac{P_i^2}{2M_i} + \sum_{i=1}^{n+1} \sum_{j>i}^{n+1} \frac{1}{4\pi\epsilon_o} \frac{Q_i Q_j}{r_{ij}} \tag{7}$$

We may transform this to the quantum-mechanical Hamiltonian operator by substitution of the configuration space operators

$$\hat{x} \rightarrow x \tag{8}$$

$$\hat{P}_x \rightarrow \frac{1}{i} \frac{\partial}{\partial x} \tag{9}$$

and we get

$$H = \sum_{i=1}^{n+1} \frac{-1}{2M_i} \nabla_i^2 + \sum_{i=1}^{n+1} \sum_{j>i}^{n+1} \frac{Q_i Q_j}{r_{ij}} \tag{10}$$

B. The Dipole Approximation

The interaction of matter particles and light calls for treating the light quantum-mechanically. It is often sufficient, though, to treat the matter quantum mechanically and the light classically via the semiclassical approach. The light, then, is treated as perpendicular oscillating electric and magnetic fields. The effect of the magnetic portion on matter is usually less than the effect of the electric portion, and so for the work presented here we may neglect the magnetic portion. If we further assume a static field, then we may express the interaction as the scalar product of the total dipole moment of the system and the field:

$$E = \mu \cdot \varepsilon \tag{11}$$

For a derivation of the above Hamiltonian, please see, for example, the book by Schatz and Ratner [71]. We may also express the total quantum-mechanical operator as

$$\mu \cdot \varepsilon = \sum_{i=1}^{n+1} \varepsilon \cdot R_i Q_i$$

where each term in the sum is the interaction of each particle with the electric field.

The Hamiltonian for the system in an electric field is thus

$$H = \sum_{i=1}^{n+1} \frac{P_i^2}{2M_i} + \sum_{i=1}^{n+1} \sum_{j>i}^{n+1} \frac{Q_i Q_j}{r_{ij}} - \sum_{i=1}^{n+1} \varepsilon \cdot R_i Q_i \tag{12}$$

C. Transformation to Center-of-Mass Coordinates

The number of internal degrees of freedom for any system may be reduced by a transformation to center-of-mass coordinates. For example, the system of $n + 1$ particles with $3(n + 1)$ degrees of freedom is reduced n pseudoparticles with $3n$ degrees of freedom, with the 3 leftover degrees of freedom describing the motion of the center of mass.

For the R vector described above, it may be shown that the appropriate transformation is

$$T = \begin{pmatrix} \frac{M_1}{M_T} & \frac{M_2}{M_T} & \frac{M_3}{M_T} & \frac{M_4}{M_T} & \cdots & \frac{M_{n+1}}{M_T} \\ -1 & 1 & 0 & 0 & 0 & \cdots \\ -1 & 0 & 1 & 0 & 0 & \cdots \\ -1 & 0 & 0 & 1 & 0 & \cdots \\ -1 & 0 & 0 & 0 & 1 & \cdots \\ \cdots & \cdots & \cdots & \cdots & \cdots & \cdots \end{pmatrix} \otimes I_3 \tag{13}$$

in the sense that T transforms R, the set of all $3(n + 1)$ coordinates, into r_0, the set of coordinates describing the position of the center of mass, and r, the set of $3n$ coordinates describing the positions of the n pseudoparticles:

$$TR = \begin{pmatrix} r_0 \\ r_1 \\ \vdots \\ r_n \end{pmatrix} \tag{14}$$

or

$$TR = \begin{pmatrix} r_0 \\ r \end{pmatrix} \tag{15}$$

where r is defined as

$$r = \begin{pmatrix} r_1 \\ r_2 \\ \vdots \\ r_n \end{pmatrix} \tag{16}$$

In the above, M_T is the sum of all $n+1$ masses in the original system.

The momenta are transformed by the inverse transformation:

$$T^{-1}P = \begin{pmatrix} p_0 \\ p_1 \\ \vdots \\ p_n \end{pmatrix} \tag{17}$$

or

$$T^{-1}P = \begin{pmatrix} p_0 \\ p \end{pmatrix} \tag{18}$$

where p_0 describes the momentum of the center of mass and p, the vector of momenta of the pseudoparticles, is defined as

$$p = \begin{pmatrix} p_1 \\ p_2 \\ \vdots \\ p_n \end{pmatrix} \tag{19}$$

The inverse transformation T^{-1} is given by

$$\begin{pmatrix} 1 & -\frac{M_2}{M_T} & -\frac{M_3}{M_T} & -\frac{M_4}{M_T} & \cdots & -\frac{M_{n+1}}{M_T} \\ 1 & \frac{M_T-M_2}{M_T} & -\frac{M_3}{M_T} & -\frac{M_4}{M_T} & -\frac{M_5}{M_T} & \cdots \\ 1 & -\frac{M_2}{M_T} & \frac{M_T-M_3}{M_T} & -\frac{M_4}{M_T} & -\frac{M_5}{M_T} & \cdots \\ 1 & -\frac{M_2}{M_T} & -\frac{M_3}{M_T} & \frac{M_T-M_4}{M_T} & -\frac{M_5}{M_T} & \cdots \\ 1 & -\frac{M_2}{M_T} & -\frac{M_3}{M_T} & -\frac{M_4}{M_T} & \frac{M_T-M_5}{M_T} & \cdots \\ \cdots & \cdots & \cdots & \cdots & \cdots & \cdots \end{pmatrix} \otimes I_3 \tag{20}$$

The charges map directly $Q_i \rightarrow q_{i-1}$, with the change on the particle at the center of mass mapping to a central potential.

These transformations result in the internal Hamiltonian (in atomic units):

$$\hat{H} = -\frac{1}{2}\left(\sum_{i}^{n}\frac{1}{m_i}\nabla_i^2 + \sum_{i\neq j}^{n}\frac{1}{M_0}\nabla_i'\nabla_j\right) + \sum_{i=1}^{n}\frac{q_0 q_i}{r_i} + \sum_{i<j}^{n}\frac{q_i q_j}{r_{ij}} - \sum_{i=1}^{n}\varepsilon\cdot r_i q_i \tag{21}$$

which may be written as

$$\hat{H} = -\nabla_r'\bar{M}\nabla_r + \sum_{i=1}^{n}\frac{q_0 q_i}{r_i} + \sum_{i<j}^{n}\frac{q_i q_j}{r_{ij}} - \sum_{i=1}^{n}\varepsilon\cdot r_i q_i$$

The mass matrix M enters the Hamiltonian for convenience of expression and is an $n \times n$ matrix with $\frac{1}{2m_i}$ on the diagonal elements and $\frac{1}{2M_0}$ on all of the off-diagonal elements; the $\bar{M}$ notation for any matrix will mean a Kronecker product with the 3×3 identity matrix, $\bar{M} = M \otimes I_3$.

III. PERMUTATIONAL SYMMETRY

Determination of a wave function for a system that obeys the correct permutational symmetry may be ensured by projection onto the irreducible representations of the symmetry groups to which the systems in question belong. For each subset of identical particles i, we can implement the desired permutational symmetry into the basis functions by projection onto the irreducible representation of the permutation group, $\mathcal{S}_{n_i}$, for total spin S_i using the appropriate projection operator $\hat{Y}_i$. The total projection operator would then be a product:

$$\hat{Y} = \prod_i \hat{Y}_i \tag{22}$$

For fermions, the projection operators are simply Young operators, derived from the appropriate Young tableau, as will be shown below.

A. Projection onto the Irreducible Representations of the *n*th-Order Symmetric Group

The energy of a quantum system is invariant to permutations of identical particles in the system. Thus, the Hamiltonian for a system with n identical particles can be said to commute with the elements of the nth-order symmetric group:

$$[\hat{H}, S_n] = 0 \tag{23}$$

This requires that the eigenfunctions of the Hamiltonian are simultaneously eigenfunctions of both the Hamiltonian and the symmetric group. This may be accomplished by taking the basis functions used in the calculations, which may be called primitive basis functions, and projecting them onto the appropriate irreducible representation of the symmetric group. After this treatment, we may call the basis functions symmetry-projected basis functions.

The projection operator takes the form of a sum of all of the possible permutations of the identical particles, $\hat{P}_i$, each multiplied by an appropriate constant, a_i:

$$\hat{Y} = \sum_{i=1}^{n!} a_i \hat{P}_i \qquad (24)$$

The format we will use to indicate the permutations is $(abc\ldots)$. For example, the permutation of particles 1 and 2 is (12). The permutation of particles 1, 3, and 4 is (134). The material below concerns how to obtain the coefficients, a_i.

The Pauli antisymmetry principle tells us that the wave function (including spin degrees of freedom), and thus the basis functions, for a system of identical particles must transform like the totally antisymmetric irreducible representation in the case of fermions, or spin $\frac{1}{2}k$ (for odd k) particles, and like the totally symmetric irreducible representation in the case of bosons, or spin k particles (where k may take on only integer values).

Projection operators for irreducible representations of the symmetric group are obtained easily from their corresponding Young tableaux. A Young tableau is created from a Young frame. A Young frame is a series of connected boxes such as

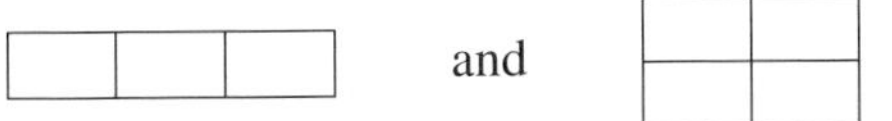

The shape of the Young frame corresponding to the desired irreducible representation of the symmetric group is obtained from the physics of the system. For example, for the totally antisymmetric representation of a group S_n we use a frame that is completely horizontal with n boxes. For example, for four particles we have

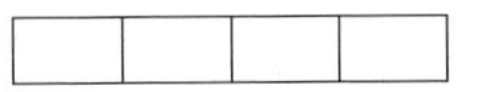

We then have to determine the dimension of the representation. This is done by filling in the boxes in the Young frame with the integers $1 \ldots n$ according to the following rules:

- Numbers must increase to the right.
- Numbers must increase down.

Obviously for the above frame, we can only fill in the numbers 1, 2, 3, and 4 according to the rules in one way:

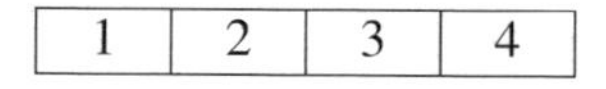

Thus this corresponds to a one-dimensional representation. The totally symmetric representation is created in a similar fashion, but with all of the boxes arranged vertically:

<table>
<tr><td>1</td></tr>
<tr><td>2</td></tr>
<tr><td>3</td></tr>
<tr><td>4</td></tr>
</table>

This is also a one-dimensional representation.

The above operators apply only to primitive basis functions that have the spin degree of freedom included. In the current work we follow the work of Matsen and use a spin-free Hamiltonian and spin-free basis functions. This approach is valid for systems wherein spin–orbit type perturbations are not considered. In this case we must come up with a different way of obtaining the Young tableaux, and thus the correct projection operators.

We explain here how to obtain the needed tableaux for fermions. We begin with the number of identical particles under consideration, n, and their total spin quantum number, s. We then calculate the symmetry quantum number, $p = \frac{n}{2} - s$. We then define a partition, μ, as

$$\mu = [2^p 1^{n-2p}] \tag{25}$$

This partition tells us to build a Young frame with

- 2 boxes in the first p rows
- 1 box in the remaining $n - 2p$ rows

We then fill in the numbers according to the above rules.

For example, for a system of four identical fermions (such as the four electrons in LiH) with all spins paired (i.e., $s = 0$), we have

$$\mu = [2^2 1^0] \tag{26}$$

and the Young frame:

<table>
<tr><td> </td><td> </td></tr>
<tr><td> </td><td> </td></tr>
</table>

Filling this in according to the rules, we have

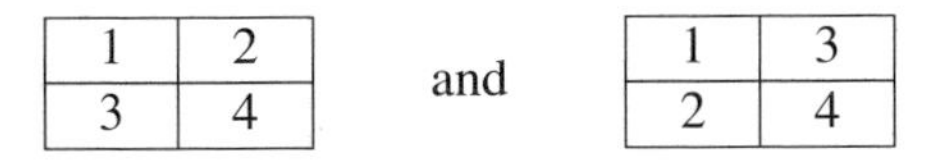

Since we have two ways of filling in the Young frame (i.e., we have two tableaux), we have a two-dimensional representation.

B. Projection Operators Obtained by the Operator Method

Now that we have established how to create the Young tableaux, we must outline a method of obtaining the desired projection operators from them. Please note that some of the following material has been adapted from Pauncz [73]. A simple way to do this for small n is described as follows. We define an operator A to be the antisymmetrizer for rows of the Young tableaux:

$$A = \prod^{rows} \sum_{i=1}^{n!} \delta \hat{P}_i, \tag{27}$$

where δ is positive for odd permutations and negative for even permutations. We also define an operator S to be the symmetrizer for columns of the Young tableaux:

$$S = \prod^{columns} \sum_{i=1}^{n!} \hat{P}_i. \tag{28}$$

The Young operator, or simply the projection operator, is then the product $\hat{Y} = AS$.

For example, for the four-electron case mentioned above, we take the first Young tableau

1	2
3	4

and generate the following operators:

$$\begin{aligned} A &= \{[E - (12)][E - (34)]\} \\ S &= \{[E + (13)][E + (24)]\} \end{aligned} \tag{29}$$

where we have used E to mean the identity operator. We can then simply build the projection operator using the rules of multiplication of permutations.

C. Projection Operators Obtained by the Representation Method

Another method that may be used to generate the projection operator involves the use a matrix representation of the operator. In particular, we will use the orthogonal representation. First we must assign a Yamanouchi symbol to each tableau we have created. This is done by going through the numbers from 1 to n in each tableau and writing down in which row the number occurs. Thus if we assign names to the above tableaux:

$$T(a) = \begin{array}{|c|c|}\hline 1 & 2 \\ \hline 3 & 4 \\ \hline \end{array} \qquad \text{and} \qquad T(b) = \begin{array}{|c|c|}\hline 1 & 3 \\ \hline 2 & 4 \\ \hline \end{array}$$

we have the Yamanouchi symbols: $YS[T(a)] = 1122$ and $YS[T(b)] = 1212$. We may use these Yamanouchi symbols to order the tableaux. We start with the last number in the symbol and see which tableau has the highest number. This will the be first tableau. If two or more Tableaux have the last numbers equal, then we go to the second to last numbers and use these as the criteria. This procedure is done recursively until all of the tableaux are assigned numbers. In the case of the tableaux we are using for examples, we must go the second to last numbers to find the ordering. Thus we have

Now for each tableaux i, we may define the axial distance between two adjacent numbers p and $p+1$, $d^i_{p,p+1}$ (for $1 \leq p < n$) to be the number of boxes that must be traversed to reach the box containing $p+1$ starting from the box containing p. The sign assigned to the axial distance will be positive if one had to go left and down, and negative if one had to go right and up.

Now we may find the matrix representation, U, of the operators. The dimensions of the matrices will be the same as the dimensions of the irreducible representation used. The matrix representation of the identity operator, $U(\hat{E})$, will of course be the identity matrix. If it is noted that any permutation may be written as a product of transpositions (permutations of order 2), and any transposition may be written as a product of elementary transpositions $(p\ p+1)$ [74], then it is only nessesary to find matrix representations of the elementary transpositions. The diagonal elements of the elementary transposition $(p\ p+1)$ are given by

$$U[(p\ p+1)]_{ii} = \frac{-1}{d^i_{p,p+1}} = D(i,k) \tag{30}$$

Thus one needs to know all of the i Young tableaux, as well as all of the axial distances in order to construct the matrix. The off-diagonal element is given by

$$U[(p\ p+1)]_{ij} = \sqrt{1 - D(i,k)^2} \tag{31}$$

if the tableaux i and j differ only by the transposition of p and $p+1$. Otherwise the off-diagonal elements are zero. The nonelementary transpositions may be obtained from the elementary transpositions with the use of the identity:

$$(i\ j+1) = (j\ j+1)(i\ j)(j\ j+1) \tag{32}$$

For the above tableaux and the associated symmetric group S_4, we only need the matrices $U[(12)], U[(23)]$, and $U[(34)]$. Thus for the two tableaux, we have only six axial distances: $d^1_{(12)} = -1, d^1_{(23)} = 2, d^1_{(34)} = -1, d^2_{(12)} = 1, d^2_{(23)} = -2$, and $d^2_{(34)} = 1$. Using these, one can find all of the representation matrices.

In the operator method described above, one may obtain a projection operator for each tableaux, so in the representation method, the same must be true. The projection operators are thus

$$\hat{Y}_{pp} = \sum_{i=1}^{n!} U[\hat{P}_i]_{pp} \hat{P}_i \tag{33}$$

where the sum is over the permutations, and $1 \leq p \leq$ dimension of the representation.

D. Decomposition of S_n in Cosets of S_{n-1}

The process of obtaining all of the representation matrices can become quite tedious as n increases, since the number of permutations increases as $n!$. This may be simplified by the decomposition of larger groups in cosets of smaller groups.

If we have a group, G of order $|g|$, and a subgroup, H of order $|h|$, where $|h| < |g|$, there is some set of elements g_i such that $g_i \in G$, but $g_i \notin H$. For the elements of H, h_j, it can be shown that $p_k = g_i \times h_j$ is not an element of H. Thus if we chose some g_i and multiply it by all of the elements of H, g_iH, this generates another subgroup, H' of order $|h|$, whose elements $h_j^{'}$ are not in H. If all of the elements of G are used up in H and H'—that is, if $|g| = 2\ |h|$—then we are done. If, for example, $|g|/|h| = 3$, then we can choose some other element g_k, where $g_k \notin H$ and $g_k \notin H'$ and form another subgroup $g_kH = H''$. We may continue this procedure until all of the elements of G are used up.

These subgroups of G generated by multiplication of some element of G by the subgroup H are called cosets of H in G. If the multiplication is carried out on the left, they are called left cosets, and vice versa for right multiplication.

If we wanted to generate the representation matrices of S_4, we would find the three elementary transpositions and the identity and then generate the other 20 matrices. On the other hand, we could find the representations of the group S_2 in S_4. This consists of two matrices, $U[(E)]$ and $U[(12)]$. We could then use $U[(23)]$ and $U[(13)]$ to generate all of the six elements in S_3. We could then use $U[(34)], U[(14)]$, and $U[(24)]$ to generate the rest of S_4. While this may seem at first more time-consuming, it is much more easily automated than the brute force approach.

E. Effects of Permutations on Basis Functions

The permutations discussed above act on the particle coordinates. In a less symbolic, more mathematical footing, we can consider the permutations as transformation matrices, $\hat{P}$, which act on the coordinate vector, R, turning them into the permuted coordinates. For example, if we consider the H_2^+ molecule with the coordinate vector

$$R = \begin{pmatrix} R_{H1} \\ R_{H2} \\ R_{e1} \end{pmatrix} \tag{34}$$

then the transformation matrix permuting the two protons would be

$$\hat{P} = \begin{pmatrix} 0 & 1 & 0 \\ 1 & 0 & 0 \\ 0 & 0 & 1 \end{pmatrix} \otimes I_3 \tag{35}$$

so that we have

$$\hat{P}R = \begin{pmatrix} R_{H2} \\ R_{H1} \\ R_{e1} \end{pmatrix} \tag{36}$$

Other permutations are done in a similar manner. When a transformation to the center of mass ($\hat{T}$) is performed, this also affects the permutations: $\bar{P} = \hat{T}^{-1}\hat{P}\hat{T}$. After this transformation, the new permutation acts on r: $\hat{P}R = \bar{P}r$. From now on we will refer to all permutations as $\hat{P}$, and center-of-mass transformation will have to be inferred from the context.

These permutations on coordinates are equivalent to operations on the basis functions. We will use shifted spherical Gaussians for this example (these functions will be discussed in a detailed way below in this chapter):

$$\begin{aligned} Pg &= \exp\{-(PR - s)'\bar{A}(PR - s)\} \\ &= \exp\{-(PR - PP^{-1}s)'\bar{A}(PR - PP^{-1}s)\} \\ &= \exp\{-[P(R - P^{-1}s)]'\bar{A}P(R - P^{-1}s)\} \\ &= \exp\{-(R - P^{-1}s)'P'\bar{A}P(R - P^{-1}s)\} \end{aligned} \tag{37}$$

Here we dropped the hat on the P for convenience. So in general we need not only the permutations, but also their inverses. The inverses are easy to obtain, however.

In the case of transpositions, such as that described above, it is obvious that $P = P^{-1} = P'$. Also, for products of transpositions, P_a and P_b, where $[P_a, P_b] = 0$, we also have $P = P^{-1}$. Furthermore, still for transpositions, $\bar{P} = \bar{P}^{-1}$, since

$$\begin{aligned} PP &= 1 \\ PTT^{-1}P &= 1 \\ T^{-1}PTT^{-1}PT &= T^{-1}T \\ \bar{P}\bar{P} &= 1 \end{aligned} \tag{38}$$

and thus $\bar{P} = \bar{P}^{-1}$.

Going from transpositions to permutations of higher order, we make use of the fact that any permutation of order n may be written as a product of $n - 1$ transpositions. For the permutation of order 3, we have $P = P_aP_b$ where P_a and P_b are transpositions that do not necessarily commute. We find the inverse easily:

$$\begin{aligned} P &= P_aP_b \\ P^{-1} &= (P_aP_b)^{-1} \\ P^{-1} &= P_b^{-1}P_a^{-1} \end{aligned}$$

but since P_a and P_b are transpositions,

$$\begin{aligned} P^{-1} &= P_bP_a \\ P^{-1} &= P_b'P_a' \\ P^{-1} &= P' \end{aligned} \tag{39}$$

The same can be shown to be true for higher-order permutations as well.

IV. ATOMIC NON-BO CALCULATIONS

The symmetry requirements and the need to very effectively describe the correlation effects have been the main motivations that have turned our attention to explicitly correlated Gaussian functions as the choice for the basis set in the atomic and molecular non-BO calculations. These functions have been used previously in Born–Oppenheimer calculations to describe the electron correlation in molecular systems using the perturbation theory approach [35–42]. While in those calculations, Gaussian pair functions (geminals), each dependent only on a single interelectron distance in the exponential factor, $\exp(-\beta r_{ij}^2)$, were used, in the non-BO calculations each basis function needs to depend on distances between all pairs of particles forming the system.

In our non-BO calculations performed so far, we have considered atomic systems with only s-electrons and molecular systems with only σ-electrons. The atomic non-BO calculations are much less complicated than the molecular calculations. After separation of the center-of-mass motion from the Hamiltonian and placing the atom nucleus in the center of the coordinate system, the internal Hamiltonian describes the motion of light pseudoelectrons in the central field on a positive charge (the charge of the nucleus) located in the origin of the internal coordinate system. Thus the basis functions in this case have to be able to accurately describe only the electronic correlation effect and the spherically symmetric distribution of the electrons around the central positive charge.

In our atomic calculations, the s-type explicitly correlated Gaussian functions have the following form:

$$\phi_k = \exp[-\mathbf{r}'(A_k \otimes I_3)\mathbf{r}] \tag{40}$$

The above function is a one-center correlated Gaussian with exponential coefficients forming the symmetric matrix A_k. ϕ_k are rotationally invariant functions as required by the symmetry of the problem—that is, invariant with respect to any orthogonal transformation. To show the invariance, let U be any 3×3 orthogonal matrix (any proper or improper rotation in 3-space) that is applied to rotate the $\mathbf{r}$ vector in the 3–D space. Prove the invariance:

$$((I_n \otimes U)\mathbf{r})'(A_k \otimes I_3)(I_n \otimes U)\mathbf{r} = \mathbf{r}'(I_n \otimes U')(A_k \otimes I_3)(I_n \otimes U)\mathbf{r} \tag{41}$$
$$= \mathbf{r}'(A_k \otimes U'U)\mathbf{r} \tag{42}$$
$$= \mathbf{r}'(A_k \otimes I_3)\mathbf{r} \tag{43}$$

The n-particle one-center correlated Gaussians, ϕ_k, can also be expressed in the more conventional form used in the electronic structure calculations as

$$\phi_k = \exp\big[-\alpha_{1k}r_1^2 - \alpha_{2k}r_2^2 - \cdots - \alpha_{nk}r_n^2 - \beta_{12,k}r_{12}^2 - \beta_{13,k}r_{13}^2 - \cdots - \beta_{nn-1,k}r_{nn-1}^2\big] \tag{44}$$

In this form, the n-particle correlated Gaussian is a product of n orbital Gaussians centered at the origin of the coordinate system and $n(n-1)/2$ Gaussian pair functions (geminals).

To describe bound stationary states of the system, the ϕ_k's have to be square-normalizable functions. The square-integrability of these functions may be achieved using the following general form of an n-particle correlated Gaussian with the negative exponential of a positive definite quadratic form in $3n$ variables:

$$\phi_k = \exp\left[-\mathbf{r}'\left(L_k L_k' \otimes I_3\right)\mathbf{r}\right] \tag{45}$$

Here $\mathbf{r}$ is a $3n \times 1$ vector of Cartesian coordinates for the n pseudoelectrons, and L_k is an $n \times n$ lower triangular matrix of rank n whose elements may vary in the range $[-\infty, \infty]$.

As mentioned, most calculations we have done so far have concerned molecular systems. However, prior to development of the non-BO method for the diatomic systems, we performed some very accurate non-BO calculations of the electron affinities of H, D, and T [43]. The difference in the electron affinities of the three systems is a purely nonadiabatic effect resulting from different reduce masses of the pseudoelectron. The pseudoelectrons are the heaviest in the T/T$^-$ system and the lightest in the H/H$^-$ system. The calculated results and their comparison with the experimental results of Lineberger and co-workers [44] are shown in Table I. The calculated results include the relativistic, relativistic recoil, Lamb shift, and finite nuclear size corrections labeled ΔE_{corr} calculated by Drake [45]. The agreement with the experiment for H and D is excellent. The 3.7-cm^{-1} increase of the electron affinity in going from H to D is very well reproduced by the calculations. No experimental EA value is available for T.

TABLE I
Electron Affinity of Hydrogen, Deuterium, and Tritium Atoms Obtained with 300 Explicitly Correlated Gaussian Functions[a]

	Hydrogen	Deuterium	Tritium
$E_H - E_{H^-}$	6083.4058 cm^{-1}	6087.0201 cm^{-1}	6088.2233 cm^{-1}
ΔE_{corr}[b]	0.307505 cm^{-1}	0.307589 cm^{-1}	0.307616 cm^{-1}
EA	6083.0983 cm^{-1}	6086.7126 cm^{-1}	6087.9157 cm^{-1}
Lykke et al. (experiment)[c]	6082.99 $\pm$ 0.15 cm^{-1}	6086.2 $\pm$ 0.6 cm^{-1}	

[a]The term ΔE_{corr} contains relativistic, relativistic recoil, Lamb shift, and finite nuclear size corrections [43].
[b]Reference 45.
[c]Reference 44.

The H/H^-, D/D^-, and T/T^- calculations were done using the variational method and 300 Gaussian functions per system. While these many functions ensure adequate convergence of the calculation for small atoms, it is usually far from adequate for even the smallest diatomic molecules. Later in this chapter, we will show calculations for HD^+ and H_2 systems where 2000 and even more basis functions were required.

V. DIATOMIC NON-BO CALCULATIONS

A. Correlated Gaussian Basis Set

The general form of an n-pseudoparticle correlated Gaussian function is given by

$$\phi_k = \exp[-\mathbf{r}' A_k \mathbf{r}] \tag{46}$$

Here, $\mathbf{r}$ is a $n \times 1$ vector of Cartesian coordinates of pseudoparticles, such that

$$\mathbf{r}' \equiv (\mathbf{r}_1', \mathbf{r}_2', \ldots, \mathbf{r}_n') \tag{47}$$

and A_k is a symmetric $n \times n$ matrix of nonlinear variational parameters,

$$A_k \equiv \begin{pmatrix} (A_k)_{1,1} & (A_k)_{1,2} & \cdots & (A_k)_{1,n} \\ (A_k)_{2,1} & (A_k)_{2,2} & \cdots & (A_k)_{2,n} \\ \vdots & \vdots & \ddots & \vdots \\ (A_k)_{n,1} & (A_k)_{n,2} & \cdots & (A_k)_{n,n} \end{pmatrix}$$

Effectively, vector $\mathbf{r}$ has $3 \times n \times 1$ components since each $\mathbf{r}_i$ in (47) is itself a three-dimensional vector. Technically speaking, in place of A_k in (46), one should write the Kronecker product $A_k \otimes I_3$ with I_3 being the 3×3 identity matrix. However, to simplify notations and avoid writing routinely this obvious Kronecker product, below in this section we will be using the following convention for matrix–vector multiplications involving such vectors:

$$\begin{aligned} \mathbf{r}' A_k \mathbf{r} &= (\mathbf{r}_1', \mathbf{r}_2', \ldots, \mathbf{r}_n') \begin{pmatrix} (A_k)_{1,1}\mathbf{r}_1 + (A_k)_{1,2}\mathbf{r}_2 + \cdots + (A_k)_{1,n}\mathbf{r}_n \\ (A_k)_{2,1}\mathbf{r}_1 + (A_k)_{2,2}\mathbf{r}_2 + \cdots + (A_k)_{2,n}\mathbf{r}_n \\ \vdots \\ (A_k)_{n,1}\mathbf{r}_1 + (A_k)_{n,2}\mathbf{r}_2 + \cdots + (A_k)_{n,n}\mathbf{r}_n \end{pmatrix} \\ &= (A_k)_{1,1}\mathbf{r}_1'\mathbf{r}_1 + (A_k)_{1,2}\mathbf{r}_1'\mathbf{r}_2 + \cdots + (A_k)_{n,n}\mathbf{r}_n'\mathbf{r}_n \end{aligned}$$

where $\mathbf{r}_i'\mathbf{r}_j = x_i x_j + y_i y_j + z_i z_j$. Thus, we first carry out all matrix–vector multiplications treating $\mathbf{r}_i$ as numbers and then, at the end, we replace each product $\mathbf{r}_i\mathbf{r}_j$ with a dot product. In other words, everywhere a product of two quantities denoting three-dimensional vectors appears, it should be considered as a dot product if no special notice is made.

A set of nonlinear parameters A_k, in general case, is unique for each function ϕ_k. To satisfy the requirement of square integrability of the wave function, each matrix A_k must be positively defined. It imposes certain restrictions on the values that the elements of matrix A_k may take. To ensure the positive definiteness and to simplify some calclations, it is very convenient to represent matrix A_k in a Cholesky factored form,

$$A_k = L_k L_k' \tag{48}$$

where the elements of lower triangular matrix L_k may take any real values.

Functions (46) have been succesfully used in numerous quantum-mechanical variational calculations of atomic and exotic systems where there is, at most, one particle (nuclei), which is substantially heavier than other constituents. However, as is well known, simple correlated Gaussian functions centered at the origin cannot provide a satisfactory convergence rate for nearly adiabatic systems, such as molecules, containing two or more heavy particles. In the diatomic case, which we we will mainly be concerned with in this section, one may introduce in basis functions (46) additional factors of powers of the internuclear distance. Such factors shift the peaks of Gaussians to some distance from the origin. This allows us to adequately describe the localization of nuclei around their equilibrium position.

If we label the two heavy particles as one and two, then the distance between these particles is given by $r_1 = |\mathbf{R_1} - \mathbf{R_2}|$ and the basis functions have the following form:

$$\varphi_k = r_1^{m_k} \exp[-\mathbf{r}' A_k \mathbf{r}] \tag{49}$$

$r_1^{m_k}$ can be conveniently written as the square root of a quadratic form in $\mathbf{r}$ using the matrix J_{11}, which is defined as an $n \times n$ matrix with 1 in the $1,1$ position and with 0's elsewhere:

$$r_1^m = [\mathbf{r}' J_{11} \mathbf{r}]^{m/2} \tag{50}$$

Similarly, any preudoparticle coordinate $\mathbf{r}_i$ or interpseudoparticle distance $\mathbf{r}_{ij}$ can be represented as

$$r_{ij} = \left[\mathbf{r}' J_{ij} \mathbf{r}\right]^{1/2} \tag{51}$$

$$J_{ij} = \begin{cases} E_{ii} & \text{if } i = j \\ E_{ii} + E_{jj} - E_{ij} - E_{ji} & \text{if } i \neq j \end{cases} \tag{52}$$

where E_{ij} is a matrix with a 1 in the i,jth position and with 0's elsewhere.

B. Implementation of the Permutational Symmetry

Permutations of real particles induce transformations on internal coordinates. Let P be a permutation of real particles; then P transforms basis functions (49) as

$$P\varphi_k = r_1^{m_k} \exp\left[-\mathbf{r}'\left(\tau'_P A_k \tau_P\right)\mathbf{r}\right] \tag{53}$$

Here, the $n \times n$ matrix τ_P is not an elementary permutation matrix. In case when P is a transposition corresponding to the interchange of 1st and jth particle ($P \equiv P_{1j}$), the matrix $\tau_{P_{1j}}$ is the identitity matrix with all elements in $(j-1)$th column replaced with -1. If P is a transposition corresponding to the interchange of ith and jth particle ($i \neq 1$ and $i \neq 1$), then $\tau_{P_{ij}}$ is the identity matrix whose $(i-1),(i-1)$th and $(j-1),(j-1)$th elements are replaced with 0's and $(i-1),(j-1)$th and $(j-1),(i-1)$th elements are 1's. All transformation matrices τ_P for permutations that are not transpositions can be represented as products of τ_P corresponding to a certain sequence of transpositions. The latter is due to the fact that any permutation can be represented as a product of certain transpositions.

Note that the factor $r_1^{m_k}$ is invariant under any permutation on systems of particles for which φ_k is a valid basis.

A symmetry projector, $\mathscr{P}$, for an irreducible representation of the permutational symmetry group of a system is given by

$$\mathscr{P} = \sum_P \chi_P P \mapsto \sum_P \chi_P \tau_P \tag{54}$$

Hence, $\mathscr{P}$ acts on φ_k as

$$\mathscr{P}\varphi_k = \sum_P \chi_P \; r_1^{m_k} \exp\left[-\mathbf{r}'\left(\tau'_P A_k \tau_P\right)\mathbf{r}\right] \tag{55}$$

The coefficients χ_P are from the matrix elements of the irreducible representation for the desired state.

Computational effort for computing matrix elements with symmetry-projected basis functions can be reduced by a factor equal to the order of the group by exploiting commutation of the symmetry projectors with the Hamiltonian and identity operators. In general,

$$\langle \mathscr{P}\varphi_k | H | \mathscr{P}\varphi_l \rangle = \langle \varphi_k | H | \mathscr{P}^\dagger \mathscr{P}\varphi_l \rangle \tag{56}$$

Thus, symmetry projection need only be performed on the ket. Typically, projection operators are Hermitian and essentially idempotent: $\mathscr{P}^\dagger \mathscr{P} \propto \mathscr{P}$ in any case, and we will simply write $\mathscr{P}$ for ket projector.

The matrix elements needed in calculations are of the form

$$O_{kl} = \langle \varphi_k | O | \mathscr{P} \varphi_l \rangle = \sum_P \chi_P \langle r_1^{m_k} \exp[-\mathbf{r}' A_k \mathbf{r}] | O | r_1^{m_l} \exp[-\mathbf{r}' (\tau_P' A_l \tau_P) \mathbf{r}] \rangle$$

where O is some operator that commutes with all the permutations from the symmetry group of the system under consideration. We will not write the summation over terms in the symmetry projector in the formulas that follow and simply note that these terms are accounted for by making the following substitution in the integral formulae:

$$A_l \mapsto \tau_P' A_l \tau_P = \tau_P' L_l (\tau_P' L_l)' \tag{57}$$

No other modification to the integral formulae need be made.

C. Integrals and Energy Derivatives

The evaluation of matrix elements for explicitly correlated Gaussians (46) and (49) can be done in a very elegant and relatively simple way using matrix differential calculus. A systematic description of this very powerful mathematical tool is given in the book by Magnus and Neudecker [105]. The use of matrix differential calculus allows one to obtain compact expressions for matrix elements in the matrix form, which is very suitable for numerical computations [116, 118] and perhaps facilitates a new theoretical insight. The present section is written in the spirit of Refs. 116 and 118, following most of the notation conventions therein. Thus, the reader can look for information about some basic ideas presented in these references if needed.

1. Some Notations

The vec operator transforms a matrix into a vector by stacking the columns of the matrix one underneath the other. Let B be an $m \times n$ matrix and let b_j be its jth column; then vec B is the $mn \times 1$ vector

$$\text{vec}\, B = \begin{bmatrix} b_1 \\ b_2 \\ \vdots \\ b_n \end{bmatrix} \tag{58}$$

An operator similar to vec is the vech, "vector half," operator. Let B be a square $n \times n$ matrix . Then vech B is the $n(n+1)/2 \times 1$ vector obtained by stacking the lower triangular elements of B. For example, if $n = 3$, then

$$\operatorname{vech} B = \begin{bmatrix} B_{11} \\ B_{21} \\ B_{31} \\ B_{22} \\ B_{32} \\ B_{33} \end{bmatrix} \tag{59}$$

For symmetric B, $\operatorname{vech} B$ contains the independent elements of B.

Other notation used: $\operatorname{diag} B$ is the diagonal $n \times n$ matrix consisting of the diagonal elements of the square matrix B. The trace of B is denoted $\operatorname{tr} B$, and the determinant of B is denoted $|B|$. The Kronecker product of two matrices is denoted by symbol $\otimes$. Other notation will be introduced as needed.

2. *Overlap Matrix Elements*

The following well-known integral is used in the derivations below:

$$\int_{-\infty}^{\infty} \exp[-x'Ax + b'x] dx_1 dx_2 \ldots dx_n = \frac{\pi^{n/2}}{|A|^{1/2}} \exp\left[\frac{1}{4} b'A^{-1}b\right] \tag{60}$$

Here, A is a positive definite $n \times n$ matrix, A^{-1} is the inverse of A, b is $n \times 1$ vector, and x is an $n \times 1$ vector variable.

The overlap of simple Gaussians (46) follows directly from (60). If $A_{kl} = A_k + A_l$,

$$\langle \phi_k | \phi_l \rangle = \frac{\pi^{3n/2}}{|A_{kl}|^{3/2}} \tag{61}$$

For further derivations we will need the matrix element of the Dirac delta function, $\delta(\mathbf{r}_{ij} - \xi)$. Using the following representation of the delta function,

$$\delta(\mathbf{r}_{ij} - \xi) = \lim_{\alpha \to \infty} \left(\frac{\alpha}{\pi}\right)^{3/2} \exp\left[-\alpha(\mathbf{r}_{ij} - \xi)^2\right]$$

we can obtain

$$\begin{aligned} \langle \phi_k | \delta(\mathbf{r}_{ij} - \xi) | \phi_l \rangle &= \lim_{\alpha \to \infty} \frac{\alpha^{3/2}}{\pi^{3/2}} \left\langle \phi_k \middle| \exp\left[-\alpha r_{ij}^2 + 2\alpha \mathbf{r}_{ij} \xi - \alpha \xi^2\right] \middle| \phi_l \right\rangle \\ &= \lim_{\alpha \to \infty} \frac{\alpha^{3/2}}{\pi^{3/2}} \left\langle \phi_k \middle| \exp\left[-\alpha \mathbf{r}' J_{ij} \mathbf{r} + 2\alpha (j_i' \mathbf{r} - j_j' \mathbf{r}) \xi - \alpha \xi^2\right] \middle| \phi_l \right\rangle \\ &= \lim_{\alpha \to \infty} \frac{\alpha^{3/2}}{\pi^{3/2}} \frac{\pi^{3n/2}}{|A_{kl} + \alpha J_{ij}|^{3/2}} \exp\left[\alpha^2 \xi^2 (j_i' - j_j')(A_{kl} + \alpha J_{ij})^{-1} (j_i - j_j) - \alpha \xi^2\right] \end{aligned} \tag{62}$$

In the second line we introduced an $n \times 1$ vector j_i whose ith component is 1 and all others are zeros. Since J_{ij} is a rank 1 matrix, we can rewrite the determinant in the last formula in the following form:

$$|A_{kl} + \alpha J_{ij}| = |A_{kl}||I + \alpha A_{kl}^{-1} J_{ij}| = |A_{kl}|(1 + \alpha \mathrm{tr}[A_{kl}^{-1} J_{ij}])$$

where I is $n \times n$ identity matrix. Because the limit of the preexponential part of (62) is a finite number, the limit of the exponent must be $-\beta\xi^2$, with β being some finite number:

$$\begin{aligned} \langle \phi_k | \delta(\mathbf{r}_{ij} - \xi) | \phi_l \rangle &= \frac{\pi^{3(n-1)/2}}{|A_{kl}|^{3/2}} \frac{1}{\mathrm{tr}[A_{kl}^{-1} J_{ij}]^{3/2}} \exp[-\beta\xi^2] \\ &= \langle \phi_k | \phi_l \rangle \frac{1}{\pi^{3/2}} \frac{1}{\mathrm{tr}[A_{kl}^{-1} J_{ij}]^{3/2}} \exp[-\beta\xi^2] \end{aligned} \tag{63}$$

Using the normalization condition,

$$\int \langle \phi_k | \delta(\mathbf{r}_{ij} - \xi) | \phi_l \rangle d\xi = \langle \phi_k | \phi_l \rangle$$

We can easily see that $\beta = \mathrm{tr}[A_{kl}^{-1} J_{ij}]^{-1}$. Thus,

$$\langle \phi_k | \delta(\mathbf{r}_{ij} - \xi) | \phi_l \rangle = \langle \phi_k | \phi_l \rangle \frac{1}{\pi^{3/2}} \frac{1}{\mathrm{tr}[A_{kl}^{-1} J_{ij}]^{3/2}} \exp\left[-\frac{\xi^2}{\mathrm{tr}[A_{kl}^{-1} J_{ij}]} \right] \tag{64}$$

The last relationship allows one to evaluate the matrix element of an arbitrary function $f(\mathbf{r}_{ij})$, which depends on a single pseudoparticle coordinate or a single interpseudoparticle coordinate,

$$\begin{aligned} \langle \phi_k | f(\mathbf{r}_{ij}) | \phi_l \rangle &= \int f(\mathbf{r}_{ij}) \langle \phi_k | \delta(\mathbf{r}_{ij} - \xi) | \phi_l \rangle d\xi \\ &= \langle \phi_k | \phi_l \rangle \frac{1}{\pi^{3/2}} \int f\left(\frac{\xi}{\mathrm{tr}[A_{kl}^{-1} J_{ij}]^{1/2}} \right) e^{-\xi^2} d\xi \end{aligned} \tag{65}$$

In the most important case, when f depends only on the absolute value of the interpseudoparticle distance, this formula becomes

$$\langle \phi_k | f(r_{ij}) | \phi_l \rangle = \langle \phi_k | \phi_l \rangle \frac{4}{\sqrt{\pi}} \int_0^\infty f\left(\mathrm{tr}[A_{kl}^{-1} J_{ij}]^{1/2} \xi \right) \xi^2 e^{-\xi^2} d\xi \tag{66}$$

Having this general expression we now can obtain the overlap matrix element for basis functions with premultipliers (49):

$$\begin{aligned}\langle\varphi_k|\varphi_l\rangle &= \langle\phi_k|r_1^{m_k+m_l}|\phi_l\rangle \\ &= \frac{2}{\sqrt{\pi}}\Gamma\left(\frac{m_{kl}+3}{2}\right)\mathrm{tr}[A_{kl}^{-1}J_{11}]^{m_{kl}/2}\langle\phi_k|\phi_l\rangle \\ &= \frac{2}{\sqrt{\pi}}\Gamma\left(\frac{m_{kl}+3}{2}\right)(A_{kl}^{-1})_{11}^{m_{kl}/2}\langle\phi_k|\phi_l\rangle \end{aligned} \tag{67}$$

Here, $\Gamma(x)$ is the Euler gamma function and $m_{kl} = m_k + m_l$.

In real calculations, it is advisable to use normalized basis functions in order to avoid problems with numerical instabilities. Therefore, the overlap matrix elements, S_{kl}, are defined using normalized basis functions. After simplification, we obtain

$$\begin{aligned}S_{kl} &= \frac{\langle\varphi_k|\varphi_l\rangle}{(\langle\varphi_k|\varphi_k\rangle\langle\varphi_l|\varphi_l\rangle)^{1/2}} \\ &= \gamma_1(m_k, m_l)2^{3n/2}\left[\left(\frac{(A_{kl}^{-1})_{11}}{(A_k^{-1})_{11}}\right)^{m_k}\left(\frac{(A_{kl}^{-1})_{11}}{(A_l^{-1})_{11}}\right)^{m_l}\left(\frac{||L_k||\;||L_k||}{|A_{kl}|}\right)^3\right]^{1/2}\end{aligned} \tag{68}$$

where $||L_k||$ denotes the absolute value of the determinant of the Cholesky factor of A_k, and

$$\gamma_1(m_k, m_l) = 2^{\frac{m_{kl}}{2}}\frac{\Gamma\left(\frac{m_{kl}+3}{2}\right)}{\left[\Gamma\left(m_k+\frac{3}{2}\right)\Gamma\left(m_l+\frac{3}{2}\right)\right]^{1/2}} \tag{69}$$

is a constant that can be precomputed for a range of m_k and m_l values to speed up matrix element calculations.

3. *Kinetic Energy Matrix Elements*

To evaluate the integral

$$\langle\varphi_k|-\nabla_{\mathbf{r}}'M\nabla_{\mathbf{r}}|\varphi_l\rangle = \langle\nabla_{\mathbf{r}}'r_1^{m_k}\phi_k|M|\nabla_{\mathbf{r}}r_1^{m_l}\phi_l\rangle$$

a few preliminary results will be needed. Recalling (50), we first evaluate the gradient of φ_k with respect to vector $\mathbf{r}$:

$$\begin{aligned}\nabla_{\mathbf{r}}\phi_k &= -2\phi_k A_k\mathbf{r} \\ \nabla_{\mathbf{r}}\varphi_k &= \nabla_{\mathbf{r}}[\mathbf{r}'J_{11}\mathbf{r}]^{m_k/2}\phi_k = [\mathbf{r}'J_{11}\mathbf{r}]^{m_k/2}\phi_k\left(\frac{m_k}{r_1^2}J_{11}\mathbf{r} - 2A_k\mathbf{r}\right)\end{aligned}$$

Hence,

$$\begin{aligned}
&\langle \varphi_k | -\nabla_{\mathbf{r}}' M \nabla_{\mathbf{r}} | \varphi_l \rangle \\
&= \left\langle \phi_k \left| [\mathbf{r}' J_{11} \mathbf{r}]^{m_k/2} \left(\frac{m_k}{r_1^2} J_{11}\mathbf{r} - 2A_k \mathbf{r} \right) M [\mathbf{r}' J_{11} \mathbf{r}]^{m_l/2} \left(\frac{m_l}{r_1^2} J_{11}\mathbf{r} - 2A_l \mathbf{r} \right) \right| \phi_l \right\rangle \\
&= \left\langle \phi_k \left| [\mathbf{r}' J_{11} \mathbf{r}]^{m_{kl}/2} \frac{m_k m_l}{r_1^4} \mathbf{r}' J_{11} M J_{11} \mathbf{r} \right| \phi_l \right\rangle - 2 \left\langle \phi_k \left| [\mathbf{r}' J_{11} \mathbf{r}]^{m_{kl}/2} \frac{m_k}{r_1^2} \mathbf{r}' J_{11} M A_l \mathbf{r} \right| \phi_l \right\rangle \\
&\quad - 2 \left\langle \phi_k \left| [\mathbf{r}' J_{11} \mathbf{r}]^{m_{kl}/2} \frac{m_l}{r_1^2} \mathbf{r}' A_k M J_{11} \mathbf{r} \right| \phi_l \right\rangle + 4 \left\langle \phi_k \left| [\mathbf{r}' J_{11} \mathbf{r}]^{m_{kl}/2} \mathbf{r}' A_k M A_l \mathbf{r} \right| \phi_l \right\rangle
\end{aligned}$$

Making use of the fact that for an arbitrary matrix B,

$$\mathbf{r}' B \mathbf{r} = \mathrm{tr}[B \mathbf{r}\mathbf{r}'] = (\mathrm{vec}\, B')' \mathrm{vec}(\mathbf{r}\mathbf{r}')$$

and replacing $\mathbf{r}' J_{11} M J_{11} \mathbf{r}$ with $r_1^2 M_{11}$, we now have

$$\begin{aligned}
&\langle \varphi_k | -\nabla_{\mathbf{r}}' M \nabla_{\mathbf{r}} | \varphi_l \rangle \\
&= m_k m_l M_{11} \left\langle \varphi_k \left| \frac{1}{r_1^2} \right| \varphi_l \right\rangle - 2m_k (\mathrm{vec}(A_l M J_{11}))' \left\langle \varphi_k \left| \frac{\mathrm{vec}(\mathbf{r}\mathbf{r}')}{r_1^2} \right| \varphi_l \right\rangle \\
&\quad - 2m_l (\mathrm{vec}(J_{11} M A_k))' \left\langle \varphi_k \left| \frac{\mathrm{vec}(\mathbf{r}\mathbf{r}')}{r_1^2} \right| \varphi_l \right\rangle + 4(\mathrm{vec}(A_l M A_k))' \langle \varphi_k | \mathrm{vec}(\mathbf{r}\mathbf{r}') | \varphi_l \rangle \\
&= m_k m_l M_{11} \left\langle \varphi_k \left| \frac{1}{r_1^2} \right| \varphi_l \right\rangle + 2m_k (\mathrm{vec}(A_l M J_{11}))' \frac{\partial}{\partial \mathrm{vec} A_{kl}} \left\langle \varphi_k \left| \frac{1}{r_1^2} \right| \varphi_l \right\rangle \\
&\quad + 2m_l (\mathrm{vec}(J_{11} M A_k))' \frac{\partial}{\partial \mathrm{vec} A_{kl}} \left\langle \varphi_k \left| \frac{1}{r_1^2} \right| \varphi_l \right\rangle - 4(\mathrm{vec}(A_l M A_k))' \frac{\partial}{\partial \mathrm{vec} A_{kl}} \langle \varphi_k | \varphi_l \rangle
\end{aligned}$$

In the last formula we used the relation $\langle \varphi_k | \mathrm{vec}(\mathbf{r}\mathbf{r}') | \varphi_l \rangle = -\frac{\partial}{\partial \mathrm{vec} A_{kl}} \langle \varphi_k | \varphi_l \rangle$. The gradient of the overlap with respect to $(\mathrm{vec} A_{kl})'$ is

$$\begin{aligned}
&\frac{\partial}{\partial (\mathrm{vec} A_{kl})'} \langle \varphi_k | \varphi_l \rangle = 2\pi^{(3n-1)/2} \Gamma\left(\frac{m_{kl}+3}{2} \right) \\
&\times \left\{ \mathrm{tr}[A_{kl}^{-1} J_{11}]^{m_{kl}/2} \frac{\partial}{\partial (\mathrm{vec} A_{kl})'} \frac{1}{|A_{kl}|^{3/2}} + \frac{1}{|A_{kl}|^{3/2}} \frac{\partial}{\partial (\mathrm{vec} A_{kl})'} \mathrm{tr}[A_{kl}^{-1} J_{11}]^{m_{kl}/2} \right\} \quad (70)
\end{aligned}$$

It is known from matrix differential calculus that for a matrix variable X and a constant matrix C the following is true:

$$d|X| = |X| \mathrm{tr}[X^{-1} dX] = |X| (\mathrm{vec}(X^{-1})')' d\mathrm{vec}(X)$$

$$\frac{\partial |X|}{\partial (\mathrm{vec} X)'} = |X| (\mathrm{vec}(X^{-1})')'$$

and

$$d\text{tr}[CX^{-1}] = (\text{vec}C')'\text{vec}(X^{-1}) = (\text{vec}C')'\Big[-(X')^{-1} \otimes X^{-1}\Big]d\text{vec}X,$$

$$\frac{\partial\text{tr}[CX^{-1}]}{\partial\text{vec}(X)'} = -(\text{vec}C')'\Big[(X')^{-1} \otimes X^{-1}\Big]$$

Applying these differentiation rules to the expression (70), one obtains

$$\frac{\partial}{\partial(\text{vec}A_{kl})'}\langle\varphi_k|\varphi_l\rangle = \langle\varphi_k|\varphi_l\rangle\left\{-\frac{3}{2}(\text{vec}(A_{kl}^{-1}))' - \frac{m_{kl}}{2}\frac{1}{\text{tr}[A_{kl}^{-1}J_{11}]}(\text{vec}J_{11})'\left[A_{kl}^{-1} \otimes A_{kl}^{-1}\right]\right\}$$

or

$$\frac{\partial}{\partial\text{vec}A_{kl}}\langle\varphi_k|\varphi_l\rangle = \langle\varphi_k|\varphi_l\rangle\left\{-\frac{3}{2}\text{vec}(A_{kl}^{-1}) - \frac{m_{kl}}{2}\frac{1}{(A_{kl}^{-1})_{11}}\left[A_{kl}^{-1} \otimes A_{kl}^{-1}\right]\text{vec}J_{11}\right\}$$

Hence, the matrix elements appearing in the kinetic energy are

$$\left\langle\varphi_k\left|\frac{1}{r_1^2}\right|\varphi_l\right\rangle = \frac{2}{m_{kl}+1}\frac{1}{(A_{kl}^{-1})_{11}}\langle\varphi_k|\varphi_l\rangle$$

$$\frac{\partial}{\partial\text{vec}A_{kl}}\langle\varphi_k|\varphi_l\rangle = \left\{-\frac{3}{2}\text{vec}(A_{kl}^{-1}) - \frac{m_{kl}}{2}\frac{1}{(A_{kl}^{-1})_{11}}\text{vec}(A_{kl}^{-1}J_{11}A_{kl}^{-1})\right\}\langle\varphi_k|\varphi_l\rangle$$

$$\frac{\partial}{\partial\text{vec}A_{kl}}\left\langle\varphi_k\left|\frac{1}{r_1^2}\right|\varphi_l\right\rangle = \left\{-\frac{3}{2}\text{vec}(A_{kl}^{-1}) - \frac{m_{kl}-2}{2}\frac{1}{(A_{kl}^{-1})_{11}}\text{vec}(A_{kl}^{-1}J_{11}A_{kl}^{-1})\right\} \times\frac{2}{m_{kl}+1}\frac{\langle\varphi_k|\varphi_l\rangle}{(A_{kl}^{-1})_{11}}$$

where we used the fact that $[X' \otimes Y]\text{vec}(Z) = \text{vec}(YZX)$. The kinetic energy matrix element is then

$$\begin{aligned}
&\langle\varphi_k|-\nabla_{\mathbf{r}}'M\nabla_{\mathbf{r}}|\varphi_l\rangle \\
&= m_k m_l M_{11}\frac{2}{m_{kl}+1}\frac{1}{(A_{kl}^{-1})_{11}}\langle\varphi_k|\varphi_l\rangle \\
&\quad - m_k(\text{vec}(A_l M J_{11}))'\left\{3\text{vec}(A_{kl}^{-1}) + \frac{m_{kl}-2}{(A_{kl}^{-1})_{11}}\text{vec}(A_{kl}^{-1}J_{11}A_{kl}^{-1})\right\}\frac{2}{m_{kl}+1}\frac{\langle\varphi_k|\varphi_l\rangle}{(A_{kl}^{-1})_{11}} \\
&\quad - m_l(\text{vec}(J_{11}MA_k))'\left\{3\text{vec}(A_{kl}^{-1}) + \frac{m_{kl}-2}{(A_{kl}^{-1})_{11}}\text{vec}(A_{kl}^{-1}J_{11}A_{kl}^{-1})\right\}\frac{2}{m_{kl}+1}\frac{\langle\varphi_k|\varphi_l\rangle}{(A_{kl}^{-1})_{11}} \\
&\quad + 2(\text{vec}(A_l M A_k))'\left\{3\text{vec}(A_{kl}^{-1}) + \frac{m_{kl}}{(A_{kl}^{-1})_{11}}\text{vec}(A_{kl}^{-1}J_{11}A_{kl}^{-1})\right\}\langle\varphi_k|\varphi_l\rangle \qquad (71)
\end{aligned}$$

After simplification and rearrangement the final expression for the kinetic energy takes the following form:

$$T_{kl} = 6\mathrm{tr}[A_k M A_l A_{kl}^{-1}]S_{kl} + \frac{2}{(A_{kl}^{-1})_{11}}\left[\frac{m_k m_l M_{11}}{m_{kl}+1} - m_k(A_{kl}^{-1}A_l M A_l A_{kl}^{-1})_{11} - m_l(A_{kl}^{-1}A_k M A_k A_{kl}^{-1})_{11}\right]S_{kl} \tag{72}$$

4. *Potential Energy Matrix Elements*

We will derive the potential energy components by finding the integral $R_{kl}^{ij} = \langle 1/r_{ij}\rangle$. There are two cases, m_{kl} even and m_{kl} odd.

Let us first introduce some simplifying definitions:

$$a = \mathrm{tr}\left[J_{11}A_{kl}^{-1}\right] \tag{73}$$

$$b = \mathrm{tr}\left[J_{ij}A_{kl}^{-1}\right] \tag{74}$$

$$c = \mathrm{tr}\left[J_{11}A_{kl}^{-1}J_{ij}A_{kl}^{-1}\right] \tag{75}$$

which for implementation purposes can be written as

$$a = \left(A_{kl}^{-1}\right)_{11} \tag{76}$$

$$b = \begin{cases} \left(A_{kl}^{-1}\right)_{ii}, & i = j \\ \left(A_{kl}^{-1}\right)_{ii} + \left(A_{kl}^{-1}\right)_{jj} - 2\left(A_{kl}^{-1}\right)_{ij}, & i \neq j \end{cases} \tag{77}$$

and

$$c = \begin{cases} \left(A_{kl}^{-1}\right)_{1i}^{2}, & i = j \\ \left(\left(A_{kl}^{-1}\right)_{1i} - \left(A_{kl}^{-1}\right)_{j1}\right)^{2}, & i \neq j \end{cases} \tag{78}$$

Let m_{kl} be even. If $p = m_{kl}/2$ with $p = 0, 1, 2, \ldots$, then

$$\langle \varphi_k | 1/r_{ij} | \varphi_l \rangle = \langle \phi_k | r_1^{2p}/r_{ij} | \phi_l \rangle \tag{79}$$

Using an integral transformation for $1/r_{ij}$,

$$\langle \varphi_k | 1/r_{ij} | \varphi_l \rangle = \frac{2}{\sqrt{\pi}}\int_0^{\infty} \langle \varphi_k | \exp\left[-v^2 \mathbf{r}' J_{ij}\mathbf{r}\right] | \varphi_l \rangle dv \tag{80}$$

followed by a differential transformation for r_1^{2p} and then integration over $\mathbf{r}$, we have

$$\begin{aligned}
&\langle \varphi_k | 1/r_{ij} | \varphi_l \rangle \\
&= \frac{2}{\sqrt{\pi}}(-1)^p \int_0^\infty \langle \phi_k | \frac{\partial^p}{\partial u^p} \exp[-u\mathbf{r}' J_{11}\mathbf{r}] \exp\left[-v^2 \mathbf{r}' J_{ij}\mathbf{r}\right] | \phi_l \rangle dv \Big|_{u=0} \\
&= \frac{2}{\sqrt{\pi}}(-1)^p \frac{\partial^p}{\partial u^p} \int_0^\infty \int_{-\infty}^\infty \exp\left[-\mathbf{r}'\left(A_{kl} + uJ_{11} + v^2 J_{ij}\right)\mathbf{r}\right] d\mathbf{r}\, dv \Big|_{u=0} \\
&= \frac{2}{\sqrt{\pi}}(-1)^p \frac{\partial^p}{\partial u^p} \int_0^\infty \frac{\pi^{3n/2}}{\left|A_{kl} + u J_{11} + v^2 J_{ij}\right|^{3/2}} dv \Big|_{u=0} \\
&= \frac{2}{\sqrt{\pi}} \langle \phi_k | \phi_l \rangle (-1)^p \frac{\partial^p}{\partial u^p} \int_0^\infty \frac{1}{\left|I_n + u J_{11} A_{kl}^{-1} + v^2 J_{ij} A_{kl}^{-1}\right|^{3/2}} dv \Big|_{u=0}
\end{aligned} \tag{81}$$

Now, since J_{11} and J_{ij} are rank-one matrices, we can write the determinant in the integral above as a polynomial in the traces we defined in Refs. 73–75 and then integrate over v, yielding

$$\begin{aligned}
&\langle \varphi_k | 1/r_{ij} | \varphi_l \rangle \\
&= \frac{2}{\sqrt{\pi}} \langle \phi_k | \phi_l \rangle (-1)^p \frac{\partial^p}{\partial u^p} \int_0^\infty \frac{1}{\left(1 + ua + v^2 b + uv^2(ab - c)\right)^{3/2}} dv \Big|_{u=0} \\
&= \frac{2}{\sqrt{\pi}} \langle \phi_k | \phi_l \rangle (-1)^p \frac{\partial^p}{\partial u^p} \frac{1}{(1 + au)\left(b + u(ab - c)\right)^{1/2}} \Big|_{u=0}
\end{aligned} \tag{82}$$

Then, differentiating p times with respect to u, setting u to zero, and simplifying gives the final result,

$$R_{kl}^{ij} = S_{kl} \frac{\gamma_2(p)}{\sqrt{b}} \sum_{q=0}^{p} \gamma_3(q) \left(1 - \frac{c}{ab}\right)^q \tag{83}$$

where

$$\gamma_2(p) = \frac{\Gamma(p+1)}{\Gamma(p+3/2)} \tag{84}$$

and

$$\gamma_3(q) = \frac{\Gamma(q+1/2)}{\Gamma(q+1)\Gamma(1/2)} \tag{85}$$

The case when m_{kl} is odd involves an additional integral transformation, which unfortunately makes these terms somewhat more complicated. The integral evaluation is similar to that for the even case and proceeds as follows.

Let $(m_{kl}+1)/2 = p$ with $p = 1, 2, 3, \ldots$, then

$$\langle \varphi_k | 1/r_{ij} | \varphi_l \rangle = \langle \phi_k | r_1^{2p-1}/r_{ij} | \phi_l \rangle = \left\langle \phi_k \left| \frac{r_1^{2p}}{r_1 r_{ij}} \right| \phi_l \right\rangle \tag{86}$$

Making transformations as for the even case, but with the addition of an extra integral transformation for $1/r_1$, we obtain

$$\begin{aligned}
&\langle \varphi_k | 1/r_{ij} | \varphi_l \rangle \\
&= \left(\frac{2}{\sqrt{\pi}}\right)^2 (-1)^p \int_0^\infty \int_0^\infty \langle \phi_k | \frac{\partial^p}{\partial u^p} \exp[-u\mathbf{r}' J_{11}\mathbf{r}] \exp\left[-w^2 \mathbf{r}' J_{11}\mathbf{r}\right] \exp\left[-v^2 \mathbf{r}' J_{ij}\mathbf{r}\right] | \phi_l \rangle dw dv \Big|_{u=0} \\
&= \frac{4}{\pi} (-1)^p \frac{\partial^p}{\partial u^p} \int_0^\infty \int_0^\infty \int_{-\infty}^\infty \exp\left[-\mathbf{r}'\left(A_{kl} + uJ_{11} + w^2 J_{11} + v^2 J_{ij}\right)\mathbf{r}\right] d\mathbf{r}\, dw\, dv \Big|_{u=0} \\
&= \frac{4}{\pi} \langle \phi_k | \phi_l \rangle (-1)^p \frac{\partial^p}{\partial u^p} \int_0^\infty \int_0^\infty \frac{1}{\left|I_n + (u+w^2) J_{11} A_{kl}^{-1} + v^2 J_{ij} A_{kl}^{-1}\right|^{3/2}} dw\, dv \Bigg|_{u=0}
\end{aligned} \tag{87}$$

Again, reducing the determinant to a polynomial in traces and integrating gives

$$\begin{aligned}
&\langle \varphi_k | 1/r_{ij} | \varphi_l \rangle \\
&= \frac{4}{\pi} \langle \phi_k | \phi_l \rangle (-1)^p \frac{\partial^p}{\partial u^p} \int_0^\infty \int_0^\infty \frac{1}{\left(1 + (u+w^2)a + v^2 b + uv^2(ab-c)\right)^{3/2}} dw\, dv \Bigg|_{u=0} \\
&= \frac{4}{\pi} \langle \phi_k | \phi_l \rangle (-1)^p \frac{\partial^p}{\partial u^p} \int_0^\infty \frac{1}{(1 + a(u+w^2))\left(b + (u+w^2)(ab-c)\right)^{1/2}} dw \Bigg|_{u=0} \\
&= \frac{4}{\pi} \langle \phi_k | \phi_l \rangle (-1)^p \frac{\partial^p}{\partial u^p} \frac{1}{(c(1+au))^{1/2}} \arcsin\left[\left(\frac{c}{a(b + u(ab-c))}\right)^{1/2}\right] \Bigg|_{u=0}
\end{aligned} \tag{88}$$

Differentiating and simplifying gives the final result for odd m_k values:

$$R_{kl}^{ij} = \frac{2}{\sqrt{\pi}} S_{kl} \sqrt{a} \left[\gamma_3(p) \frac{\arcsin\left[\sqrt{\frac{c}{ab}}\right]}{\sqrt{c}} + \frac{1}{2\sqrt{ab-c}} \sum_{q=1}^{p} \sum_{t=0}^{q-1} \frac{1}{q} \gamma_3(p-q) \gamma_3(t) \left(1 - \frac{c}{ab}\right)^{q-t} \right] \tag{89}$$

The inclusion of the arcsin and the double summation in this formula unfortunately complicates these odd power terms compared to the even power case. The implementation of odd powers m_k requires significantly more computer time due to the complexity of this formula. Furthermore, we found that variation of near optimal m_k by plus or minus one had negligeable effect on energy convergence. Therefore, in our calculations utilizing gradient formulas for energy optimization, we excluded the odd power case.

5. *Some Other Matrix Elements*

The matrix elements of r_1 and r_1^2, which in the case of a diatomic molecule are the internuclear distance and its square, can easily be obtained using relationship (66):

$$\begin{aligned}\langle \varphi_k | r_1 | \varphi_l \rangle &= \langle \phi_k | r_1^{m_{kl}+1} | \phi_l \rangle = \langle \phi_k | \phi_l \rangle \frac{2}{\sqrt{\pi}} \mathrm{tr}\left[J_{11} A_{kl}^{-1}\right]^{(m_{kl}+1)/2} \Gamma\left(\frac{m_{kl}+4}{2}\right) \\ &= \langle \varphi_k | \varphi_l \rangle \frac{m_{kl}+2}{2} \gamma_2 \left(\frac{m_{kl}}{2}\right) \sqrt{\left(A_{kl}^{-1}\right)_{11}} \end{aligned} \tag{90}$$

$$\langle \varphi_k | r_1^2 | \varphi_l \rangle = \langle \phi_k | r_1^{m_{kl}+2} | \phi_l \rangle = \langle \varphi_k | \varphi_l \rangle \frac{m_{kl}+3}{2} \left(A_{kl}^{-1}\right)_{11} \tag{91}$$

The evaluation of the expectation values of other than r_1 distances and their squares can be done by differentiating expressions (83) and (67), respectively. We will restrict ourselves to the case when m_k's are even, so that $p = m_{kl}/2$ with $p = 0, 1, 2, \ldots$. Let us first consider the simpler case of r_{ij}^2:

$$\begin{aligned}\langle \varphi_k | r_{ij}^2 | \varphi_l \rangle &= -\frac{\partial}{\partial u} \langle \varphi_k | \exp\left[-u\,\mathbf{r}' J_{ij} \mathbf{r}\right] | \varphi_l \rangle \Big|_{u=0} \\ &= -\frac{\partial}{\partial u} \frac{2}{\sqrt{\pi}}\, \Gamma\left(p + \frac{3}{2}\right) \mathrm{tr}[(A_{kl} + uJ_{ij})^{-1} J_{11}]^p \frac{\pi^{3n/2}}{\left|A_{kl} + uJ_{ij}\right|^{3/2}} \Bigg|_{u=0}\end{aligned} \tag{92}$$

Here, we will need some simple facts from matrix differential calculus. If X is a matrix variable and β is a parameter that X depends on, then

$$\frac{\partial |X|}{\partial \beta} = |X| \mathrm{tr}\left[X^{-1} \frac{\partial X}{\partial \beta}\right] \tag{93}$$

$$\frac{\partial \mathrm{tr}[X]}{\partial \beta} = \mathrm{tr}\left[\frac{\partial X}{\partial \beta}\right] \tag{94}$$

$$\frac{\partial (X^{-1})}{\partial \beta} = -X^{-1} \frac{\partial X}{\partial \beta} X^{-1} \tag{95}$$

Using these formulae, one can show that

$$\frac{\partial}{\partial u}\mathrm{tr}[(A_{kl} + uJ_{ij})^{-1}J_{11}] = -\mathrm{tr}[(A_{kl} + uJ_{ij})^{-1}J_{ij}(A_{kl} + uJ_{ij})^{-1}J_{11}] \tag{96}$$

and

$$\langle\varphi_k|r_{ij}^2|\varphi_l\rangle = \langle\varphi_k|\varphi_l\rangle\left(p\frac{c}{a} + b\right) \tag{97}$$

In the same manner for the first power of r_{ij}, one obtains

$$\begin{aligned}
&\langle\varphi_k|r_{ij}|\varphi_l\rangle \\
&= -\frac{\partial}{\partial u}\left\langle\phi_k\left|\frac{\exp[-u\mathbf{r}'J_{ij}\mathbf{r}]}{r_{ij}}\right|\phi_l\right\rangle\Bigg|_{u=0} \\
&= -\frac{\partial}{\partial u}\frac{2}{\sqrt{\pi}}\Gamma\left(p+\frac{3}{2}\right)\mathrm{tr}[(A_{kl}+uJ_{ij})^{-1}J_{11}]^p\frac{\pi^{3n/2}}{|A_{kl}+uJ_{ij}|^{3/2}}\frac{\gamma_2(p)}{\mathrm{tr}[(A_{kl}+uJ_{ij})^{-1}J_{ij}]^{1/2}} \\
&\quad\times\sum_{q=0}^{p}\gamma_3(q)\left(1-\frac{\mathrm{tr}[(A_{kl}+uJ_{ij})^{-1}J_{ij}(A_{kl}+uJ_{ij})^{-1}J_{11}]}{\mathrm{tr}[(A_{kl}+uJ_{ij})^{-1}J_{11}]\mathrm{tr}[(A_{kl}+uJ_{ij})^{-1}J_{ij}]}\right)^q\Bigg|_{u=0} \\
&= \langle\varphi_k|1/r_{ij}|\varphi_l\rangle\left\{-p\frac{c}{a}-\frac{3}{2}b+\frac{1}{2}\frac{c}{b}+\left(\frac{2t}{ab}-\frac{c^2}{a^2b}-\frac{ch}{ab^2}\right)\sum_{q=1}^{p}\gamma_3(q)\,q\left(1-\frac{c}{ab}\right)^{q-1}\right\}
\end{aligned} \tag{98}$$

where

$$\begin{aligned}
h &= \mathrm{tr}[A_{kl}^{-1}J_{ij}A_{kl}^{-1}J_{ij}] \\
t &= \mathrm{tr}[A_{kl}^{-1}J_{ij}A_{kl}^{-1}J_{ij}A_{kl}^{-1}J_{11}]
\end{aligned}$$

The last two traces are equal to

$$h=\begin{cases}(A_{kl}^{-1})_{ii}^2, & i=j \\ \left[(A_{kl}^{-1})_{ii}+(A_{kl}^{-1})_{jj}-2(A_{kl}^{-1})_{ij}\right]^2, & i\neq j\end{cases} \tag{99}$$

$$t=\begin{cases}0, & i\neq 1 \text{ and } j\neq 1 \\ \left[(A_{kl}^{-1})_{11}-(A_{kl}^{-1})_{1j}\right]^2\left[(A_{kl}^{-1})_{11}+(A_{kl}^{-1})_{jj}-2(A_{kl}^{-1})_{1j}\right], & i=1 \text{ and } i\neq j \\ (A_{kl}^{-1})_{11}^3, & i=j=1\end{cases} \tag{100}$$

In order to perform calculations of the correlation function of particles 1 and 2 (which is the same quantity as the density of the 1st pseudoparticle, $g_1(\xi)$) using basis (49), we need the matrix elements of the delta function, $\delta(\mathbf{r}_1 - \xi)$. They can be evaluated by replacing $A_{kl} \to A_{kl} + uJ_{11}$ in expression (64) and then differentiating p times with respect to u.

$$\begin{aligned}\langle\varphi_k|\delta(\mathbf{r}_1 - \xi)|\varphi_l\rangle \\ &= (-1)^p \frac{\partial^p}{\partial u^p} \frac{\pi^{3n/2}}{|A_{kl} + uJ_{11}|^{3/2}} \frac{1}{\pi^{3/2}} \frac{1}{\mathrm{tr}[(A_{kl} + uJ_{11})^{-1} J_{11}]^{3/2}} \\ &\times \exp\left[- \frac{\xi^2}{\mathrm{tr}[(A_{kl} + uJ_{11})^{-1} J_{11}]} \right]\Bigg|_{u=0}\end{aligned}$$

Applying formula (96) and using the fact that $\mathrm{tr}[XJ_{ii}XJ_{ii}] = \mathrm{tr}[XJ_{ii}]^2$ for an arbitrary matrix X reduces the final result to

$$\begin{aligned}\langle\varphi_k|\delta(\mathbf{r}_1 - \xi)|\varphi_l\rangle \\ &= \langle\varphi_k|\varphi_l\rangle \frac{1}{2\pi} \frac{1}{\Gamma\left(p + \frac{3}{2}\right)} \frac{1}{(A_{kl}^{-1})_{11}^{3/2}} \left(\frac{\xi^2}{(A_{kl}^{-1})_{11}} \right)^p \exp\left[- \frac{\xi^2}{(A_{kl}^{-1})_{11}} \right]\end{aligned} \tag{101}$$

The matrix elements of $\delta(\mathbf{r}_{ij})$ can be easily obtained by straightforward integration. The procedure is very similar to the evaluation of the overlap integral and yields

$$\begin{aligned}\langle\varphi_k|\delta(\mathbf{r}_{ij})|\varphi_l\rangle &= \frac{2}{\sqrt{\pi}} \Gamma\left(p + \frac{3}{2}\right) (D_{kl}^{-1})_{11}^p \frac{\pi^{3(n-1)/2}}{|D_{kl}|^{3/2}} \\ &= \langle\varphi_k|\varphi_l\rangle \frac{1}{\pi^{3/2}} \left(\frac{|A_{kl}|}{|D_{kl}|} \right)^{3/2} \left(\frac{(D_{kl}^{-1})_{11}}{(A_{kl}^{-1})_{11}} \right)^p\end{aligned} \tag{102}$$

where D_{kl} is an $(n-1) \times (n-1)$ matrix formed from A_{kl} by adding the jth row to the ith one, then adding and jth column to the ith column, and then crossing out the jth column and row.

6. *Energy Gradient*

The integral formulas above are sufficient for performing energy calculations and evaluating some expectation values. However, optimization of the many nonlinear parameters contained in the exponent matrices L_k demands excessive computational resources if an approximate numerical energy gradient is used. Several orders of magnitude of computational effort can be saved by utilizing analytic gradients. Additionally, more thorough optimization can be achieved due to increased accuracy of analytic gradients compared with numerical approximations. We feel that analytic gradient formulas are essential in many practical situations. To this end, the energy gradient formulas are now presented.

We begin with the derivative of the secular equation with respect to energy eigenvalues. For some background on matrix differential calculus, see the Refs. 116 and 117.

The secular equation,

$$(H - \epsilon S)c = 0 \tag{103}$$

defines the energy ϵ as an implicit function of the $N \times N$ matrices H and S, where N is the number of basis functions. H and S are themselves functions of the $Nn(n+1)/2 \times 1$ vector $L = \left[(\text{vech } L_1)', \cdots, (\text{vech } L_N)'\right]$ of nonlinear exponential parameters contained in the matrices L_k; recall that $A_k = L_k L_k'$. The energy gradient with respect to L is then

$$g = \nabla_L \epsilon = \frac{1}{c'Sc}\left(\frac{\partial \text{vech } H}{\partial L'} - \epsilon \frac{\partial \text{vech } S}{\partial L'}\right)'(\text{vech}[2cc' - \text{diag } cc']) \tag{104}$$

The matrix $(\partial \text{vech } H/\partial L' - \epsilon\, \partial \text{vech } S/\partial L') = \partial \text{vech}(H - \epsilon S)/\partial L'$ in the gradient above is sparse and has dimension $N(N+1)/2 \times Nn(n+1)/2$. This sparse matrix together with the eigenvector c can be collapsed to a dense partitioned form with dimension $N \times Nn(n+1)/2$,

$$G = \begin{pmatrix} c_1^2 \frac{\partial (H-\epsilon S)_{11}}{\partial (\text{vech } L_1)'} & 2c_1c_2 \frac{\partial (H-\epsilon S)_{12}}{\partial (\text{vech } L_2)'} & \cdots & 2c_1c_N \frac{\partial (H-\epsilon S)_{1N}}{\partial (\text{vech } L_N)'} \\ 2c_2c_1 \frac{\partial (H-\epsilon S)_{21}}{\partial (\text{vech } L_1)'} & c_2^2 \frac{\partial (H-\epsilon S)_{22}}{\partial (\text{vech } L_2)'} & \cdots & 2c_2c_N \frac{\partial (H-\epsilon S)_{2N}}{\partial (\text{vech } L_N)'} \\ \vdots & \vdots & \ddots & \vdots \\ 2c_Nc_1 \frac{\partial (H-\epsilon S)_{N1}}{\partial (\text{vech } L_1)'} & 2c_Nc_2 \frac{\partial (H-\epsilon S)_{N2}}{\partial (\text{vech } L_2)'} & \cdots & c_N^2 \frac{\partial (H-\epsilon S)_{NN}}{\partial (\text{vech } L_N)'} \end{pmatrix} \tag{105}$$

With G defined as above, the gradient can be computed by summing over the rows of G. That is

$$g_i = \frac{1}{c'Sc}\sum_j G_{ji}$$

The nonzero terms in the matrices $\partial \text{vech } H/\partial a'$ and $\partial \text{vech } S/\partial a'$ are contained in the $1 \times n(n+1)/2$ vectors $\partial H_{kl}/\partial(\text{vech } L_k)'$ and $\partial S_{kl}/\partial(\text{vech } L_k)'$ and also in $\partial H_{kl}/\partial(\text{vech } L_l)'$ and $\partial S_{kl}/\partial(\text{vech } L_l)'$. Matrix derivatives depend on the arrangement of elements in a matrix variable; therefore, since symmetry projection on kets effectively reorders elements of the exponent matrix L_l, the formulas for derivatives with respect to the exponent matrices in the ket, vech $[L_l]$, are different from those with respect to the exponent matrices in the bra, vech $[L_k]$. We assume that under particle permutations, A_l transforms as

$$\tau_P' A_l \tau_P = \tau_P' L_l L_l' \tau_P = \tau_P' L_l (\tau_P' L_l)'$$

where τ_P are some transformation matrices. Also, the matrix derivatives for the diagonal blocks of G are complicated somewhat by the symmetry projection on the kets. However, they can be computed using the following relationship, for example:

$$\frac{\partial H_{kk}}{\partial(\text{vech } L_k)'} = \left.\frac{\partial H_{kl}}{\partial(\text{vech } L_k)'}\right|_{l=k} + \left.\frac{\partial H_{kl}}{\partial(\text{vech } L_l)'}\right|_{l=k} \tag{106}$$

Thus, only two sets of formulas for the derivatives need be computed.

Matrix elements are scalar-valued matrix functions of the exponent matrices L_k. Therefore, the appropriate mathematical tool for finding derivatives is the matrix differential calculus [116, 118]. Using this, the derivations are nontrivial but straightforward. We will only present the final results of the derivations. The reader wishing to derive these formulas, or other matrix derivatives, is referred to the Ref. 116 and references therein.

We want to note that only one term in the symmetry projection will be represented. As was the case for the integral formulas, the symmetry terms require the substitution $A_l \mapsto \tau_P' A_l \tau_P = \tau_P' L_l (\tau_P' L_l)'$ or, more generally, $L_l \mapsto \tau_P' L_l$. This is required for derivatives with respect to both vech $[L_k]$ and vech $[L_l]$. The derivatives with respect to vech $[L_l]$ will require further modification, and this will be noted in the formulas below.

Using the normalized overlap formula, Eq. (68), the derivative with respect to the nonzero terms of the lower triangular matrix L_k is

$$\begin{aligned}\frac{\partial S_{kl}}{\partial(\text{vech } L_k)'} &= \frac{3}{2} S_{kl} \text{ vech } \left[(L_k^{-1})' - 2A_{kl}^{-1} L_k\right]' \\ &+ S_{kl} \frac{m_k}{\left(A_k^{-1}\right)_{11}} \text{vech}\left[A_k^{-1} J_{11} A_k^{-1} L_k\right]' \\ &- S_{kl} \frac{m_{kl}}{\left(A_{kl}^{-1}\right)_{11}} \text{vech}\left[A_{kl}^{-1} J_{11} A_{kl}^{-1} L_k\right]'\end{aligned} \tag{107}$$

For the derivative, with respect to the elements vech L_l, we account for the symmetry terms by making a multiplication by τ_P in addition to the substitutions described above. Thus,

$$\frac{\partial S_{kl}}{\partial(\text{vech } L_l)'} = \frac{3}{2} S_{kl} \text{ vech}\left[\tau_P \left(L_l^{-1}\right)' - 2\tau_P A_{kl}^{-1} L_l\right]' + S_{kl} \frac{m_l}{\left(A_l^{-1}\right)_{11}} \text{vech}\left[\tau_P A_l^{-1} J_{11} A_l^{-1} L_l\right]' - S_{kl} \frac{m_{kl}}{\left(A_{kl}^{-1}\right)_{11}} \text{vech}\left[\tau_P A_{kl}^{-1} J_{11} A_{kl}^{-1} L_l\right]' \tag{108}$$

The kinetic energy gradient components are obtained by differentiating Eq. (72) with respect to vech L_k and vech L_l:

$$\begin{aligned} \frac{\partial T_{kl}}{\partial(\text{vech } L_k)'} &= \frac{\partial S_{kl}}{\partial(\text{vech } L_k)'} \frac{T_{kl}}{S_{kl}} + 2S_{kl} \Bigg[6 \text{ vech}\left[A_{kl}^{-1} A_l M A_l A_{kl}^{-1} L_k\right]' \\ &+ 2\left(A_{kl}^{-1}\right)_{11}^{-2} \left(\frac{m_k m_l M_{11}}{m_{kl}+1} - m_k \left(A_{kl}^{-1} A_l M A_l A_{kl}^{-1}\right)_{11} - m_l \left(A_{kl}^{-1} A_k M A_k A_{kl}^{-1}\right)_{11} \right) \\ &\times \text{vech } \left[A_{kl}^{-1} J_{11} A_{kl}^{-1} L_k\right]' \\ &- 2\left(A_{kl}^{-1}\right)_{11}^{-1} m_l \text{ vech}[A_{kl}^{-1} J_{11} A_{kl}^{-1} A_k M A_l A_{kl}^{-1} L_k + A_{kl}^{-1} A_l M A_k A_{kl}^{-1} J_{11} A_{kl}^{-1} L_k] \\ &+ 2\left(A_{kl}^{-1}\right)_{11}^{-1} m_k \text{ vech}[A_{kl}^{-1} J_{11} A_{kl}^{-1} A_l M A_l A_{kl}^{-1} L_k + A_{kl}^{-1} A_l M A_l A_{kl}^{-1} J_{11} A_{kl}^{-1} L_k] \Bigg] \end{aligned} \tag{109}$$

and the derivative with respect to vech L_l including the symmetry projector term τ_P is given by

$$\begin{aligned} \frac{\partial T_{kl}}{\partial(\text{vech } L_l)'} &= \frac{\partial S_{kl}}{\partial(\text{vech } L_l)'} \frac{T_{kl}}{S_{kl}} + 2S_{kl} \Bigg[6 \text{ vech}\left[\tau_P A_{kl}^{-1} A_k M A_k A_{kl}^{-1} L_l\right]' \\ &+ 2\left(A_{kl}^{-1}\right)_{11}^{-2} \left(\frac{m_k m_l M_{11}}{m_{kl}+1} - m_k \left(A_{kl}^{-1} A_l M A_l A_{kl}^{-1}\right)_{11} - m_l \left(A_{kl}^{-1} A_k M A_k A_{kl}^{-1}\right)_{11} \right) \\ &\times \text{vech } \left[\tau_P A_{kl}^{-1} J_{11} A_{kl}^{-1} L_l\right]' \\ &- 2\left(A_{kl}^{-1}\right)_{11}^{-1} m_k \text{ vech}[\tau_P A_{kl}^{-1} J_{11} A_{kl}^{-1} A_l M A_k A_{kl}^{-1} L_l + \tau_P A_{kl}^{-1} A_k M A_l A_{kl}^{-1} J_{11} A_{kl}^{-1} L_l] \\ &+ 2\left(A_{kl}^{-1}\right)_{11}^{-1} m_l \text{ vech}[\tau_P A_{kl}^{-1} J_{11} A_{kl}^{-1} A_k M A_k A_{kl}^{-1} L_l + \tau_P A_{kl}^{-1} A_k M A_k A_{kl}^{-1} J_{11} A_{kl}^{-1} L_l] \Bigg] \end{aligned} \tag{110}$$

The purpose of the gradient formulas is to enhance optimization efforts, and we have found that well-optimized wave functions utilizing a mixture of even and odd powers, m_k, did not produce results any better than using even powers only. Hence, for this reason, along with reasons that were stated at the end of the section describing the potential energy matrix elements, we will not derive the gradient terms for the potential energy matrix elements that utilize odd m_k.

Using the even m_k potential energy integral formula, Eq. (83), the definitions for a, b, and c given in Eqs. (73)–(75), and the definitions for γ_2 and γ_3 given in Eqs. (84 and 85), the gradient terms for the potential, with $m_{kl} = 2p$, are given by

$$\begin{aligned}
&\frac{\partial R_{kl}^{ij}}{\partial(\mathrm{vech}\, L_k)'} \\
&= \frac{\partial S_{kl}}{\partial(\mathrm{vech}\, L_k)'} \frac{R_{kl}^{ij}}{S_{kl}} - \frac{\partial b}{\partial(\mathrm{vech}\, L_k)'} \frac{R_{kl}^{ij}}{2b} + \gamma_2(p) S_{kl} b^{-1/2} \\
&\times \left[\sum_{q=1}^{p} \gamma_3(q) q \left(1 - \frac{c}{ab}\right)^{q-1} \frac{c}{(ab)^2} \left(\frac{\partial a}{\partial(\mathrm{vech}\, L_k)'} b + a \frac{\partial b}{\partial(\mathrm{vech}\, L_k)'} - \frac{ab}{c} \frac{\partial c}{\partial(\mathrm{vech}\, L_k)'} \right) \right]
\end{aligned} \tag{111}$$

where

$$\frac{\partial a}{\partial(\mathrm{vech}\ L_k)'} = -2\ \mathrm{vech}\left[A_{kl}^{-1} J_{11} A_{kl}^{-1} L_k\right] \tag{112}$$

$$\frac{\partial b}{\partial(\mathrm{vech}\ L_k)'} = -2\ \mathrm{vech}\left[A_{kl}^{-1} J_{ij} A_{kl}^{-1} L_k\right] \tag{113}$$

and

$$\begin{aligned}
\frac{\partial c}{\partial(\mathrm{vech}\ L_k)'} &= -2\ \mathrm{vech}\left[A_{kl}^{-1} J_{11} A_{kl}^{-1} J_{ij} A_{kl}^{-1} L_k\right] \\
&\quad - 2\ \mathrm{vech}\left[A_{kl}^{-1} J_{ij} A_{kl}^{-1} J_{11} A_{kl}^{-1} L_k\right]
\end{aligned} \tag{114}$$

The derivative with respect to vech L_l has the same form as the above but with L_k replaced by $\tau_P' L_l$ and each of the expressions vech $\left[A_{kl}^{-1} \cdots\right]$ replaced by vech $\left[\tau_P A_{kl}^{-1} \cdots\right]$.

D. Variational Method and Minimization of the Energy Functional

The energy in variational calculations is obtained by minimizing the Rayleigh quotient. In the case of basis set (49), this quotient has the following form:

$$E(\{c_k\}, \{m_k\}, \{L_k\}) = \min_{\{\{c_k\},\{m_k\},\{L_k\}\}} \frac{c' H(\{m_k\}, \{L_k\}) c}{c' S(\{m_k\}, \{L_k\}) c} \tag{115}$$

where the minimization takes place with respect to the linear coefficients, $\{c_k\}$ of the wave function expansion in terms of the basis functions, and with respect to the nonlinear exponential parameters, $\{L_k\}$, and preexponential powers, $\{m_k\}$, of the basis functions. H and S are the Hamiltonian and overlap matrices of size $N \times N$ (N is the number of basis functions). The finding of the linear coefficients, $\{c_k\}$, is usually done by solving secular equation (103). However, in ground-state calculations one may minimize the above quotient directly, without diagonalization. In some cases such a procedure, when $\{c_k\}$ and $\{L_k\}$ are considered to be independent, may have certain advantages. But in many situations we do reduce the minimization with respect to $\{c_k\}$ to the generalized symmetric eigenvalue problem. In our calculations we used the inverse iteration method to solve this problem. The idea of the method consists in performing the iterations,

$$(H - \epsilon_{\text{appr}} S) c_{k+1} = S c_k \tag{116}$$

where ϵ_{appr} is an approximate value of the exact solution of the generalized symmetric eigenvalue problem. The starting vector c_0 can be chosen randomly. As long as the exact eigenvalue we need to obtain is closer to ϵ_{appr} than any other eigenvalue, the iterations (116) will converge. Typically, just a few iterations are needed to find ϵ with sufficient accuracy. Each iteration in (116) is performed in a few steps:

$$H - \epsilon_{\text{appr}} S = LDL' \tag{117}$$

$$Lx = Sc_k \tag{118}$$

$$Dy = x \tag{119}$$

$$L' c_{k+1} = y \tag{120}$$

Here, L is a lower triangular matrix (not to be confused with L_k, the Cholesky factor of the matrix of nonlinear parameters A_k), and D is a diagonal matrix. The scheme of the solution of the generalized symmetric eigenvalue problem above has proven to be very efficient and accurate in numerous calculations. But the main advantage of this scheme is revealed when one has to routinely solve the secular equation with only one row and one column of matrices H and S changed. In this case, the update of factorization (117) requires only $\propto N^2$ arithmetic operations while, in general, the solution "from scratch" needs $\propto N^3$ operations.

It is well known that the convergence of variational expansions in terms of correlated Gaussians, both the simple ones and those with premultipliers, strongly depends on how one selects the nonlinear parameters in the Gaussian exponentials. In order to get high-accuracy results in the calculations, one needs to perform optimizations of those parameters at some level. Due to a usually large number of basis functions in non-BO calculations and, consequently, a larger number of the exponential parameters, this task represents a serious computational problem. The two most commonly applied approaches to the parameter optimization are (a) a full optimization, which is very effective when the analytical gradient of the variational energy functional with respect to the parameters is available, and (b) the method based on a stochastic selection of the parameters.

We found that in many practical situations a hybrid method that combines the gradient-driven optimization with the stochastic selection method turns out to be very efficient. In this approach, we first generate a relatively small basis set for each of the studied systems using the full gradient optimization. This generates a good starting point for each system for the next step of the procedure. In this next step we apply the following strategy. We incrementally increase the size of the basis set by including additional basis functions, one by one, with randomly selected values of the nonlinear parameters and values of the pre-exponential powers. After including a function into the basis set, we first optimize the power of its preexponential factor using the finite difference approach and then optimize the nonlinear parameters in its exponent using the analytical gradient approach. After adding several new basis functions using this approach (the number varies depending on the desired degree of optimization), the whole basis is reoptimized by means of the gradient approach applied in sequence to each basis functions, one function at a time. This continues until the number of basis functions reached a certain limit or until the necessary accuracy is achieved. Although this procedure has been proven to be quite efficient in optimizations of large basis sets of correlated Gaussians, it still requires a lot of computational resources, especially for systems with a large number of particles and a large number of particle permutations in the Young symmetry operator. In addition to that, as we found from our experience, a full (simultaneous) optimization of all nonlinear parameters may still be very desirable for highly vibrationally excited states where the Gaussians tend to be very strongly coupled. The way to partially overcome this problem of high computational demands is extensive parallelization of the computer code for use on multinode computational systems. For this purpose we used the message passing interface (MPI), a widely used tool in the world of parallel computations, and were able to achieve a sufficient parallelization level of the code for runs with several processors. This development enabled us to significanly extend our capabilities of optimizing large basis sets.

To illustrate the capabilities of the variational method, we will present later the results and discuss the details of some diatomic non-BO calculations on small molecules, which were carried out by our group.

E. The Ground and Excited States of H_2

Due to a small number of electrons and nuclei, the H_2 molecule often serves as the first target for testing different methods in quantum chemistry. Despite the fact that this system has been very well studied since the early days of quantum mechanics and the pioneering work of Heitler and London [106], very few studies did not invoke Born–Oppenheimer approximation. Moreover, all of the non-BO calculations usually deal with the ground or few low-lying states only. In this part we present highly accurate, nonrelativistic, variational, non-BO calculations of the "vibrational spectrum" of the H_2 molecule [121]. Although we use the traditional term "vibrational spectrum," the states we have calculated can be better characterized as states with zero total angular momentum, which is the sum of the angular momenta of the electrons and the nuclei.

The non-BO wave functions of different excited states have to differ from each other by the number of nodes along the internuclear distance, which in the case of basis (49) is r_1. To accurately describe the nodal structure in all 15 states considered in our calculations, a wide range of powers, m_k, had to be used. While in the calculations of the H_2 ground state [119], the power range was 0–40, in the present calculations it was extended to 0–250 in order to allow pseudoparticle 1 density (i.e., nuclear density) peaks to be more localized and sharp if needed. We should notice that if one aims for highly accurate results for the energy, then the wave function of each of the excited states must be obtained in a separate calculation. Thus, the optimization of nonlinear parameters is done independently for each state considered.

Table II contains total variational energies of the lowest 15 states corresponding to the rotational ground state ($J = 0$) calculated with 3000 basis functions each. We also show the expectation values of the internuclear distance and its square calculated as average values using the optimized wave function of each state ($\langle r_1 \rangle$ and $\langle r_1^2 \rangle$). The energy for the ground state of -1.1640250300 hartree is noticeably lower than the previously reported upper bound [119] of -1.1640250232 hartree. We are certain that for at least a few lowest excited states, the quality of the results is very similar to that for the ground state. However, for the highest states, where the number of nodes in the wave function is much higher, the quality of the calculations decreases, but we believe that it still allows determination of the transition energies, with the accuracy similar to the experimental uncertainty, if not higher.

In Table II we also compare our total variational energies with the energies obtained by Wolniewicz. In his calculations Wolniewicz employed an approach wherein the zeroth order the adiabatic approximation for the wave function was used (i.e., the wave function is a product of the ground-state electronic wave function and a vibrational wave function) and he calculated the nonadiabatic effects as corrections [107, 108]. In general the agreement between our results

TABLE II
Nonadiabatic Variational Energies for 15 States of the H_2 Molecule with Zero Total Angular Momentum (the Ground Rotational States) Obtained with 3000 Basis Functions for Each State and Expectation Values of the Internuclear Distance and the Square of the Internuclear Distance, $\langle r_{\text{H-H}} \rangle$ $\langle r^2_{\text{H-H}} \rangle$[a]

v	E	$\langle r_{\text{H-H}} \rangle$	$\langle r^2_{\text{H-H}} \rangle$	E, Wolniewicz [107]
0	−1.1640250300	1.4487380	2.1270459	−1.1640250185
1	−1.1450653676	1.5453495	2.4739967	−1.1450653629
2	−1.1271779152	1.6460579	2.8568172	−1.1271779324
3	−1.1103404429	1.7517082	3.2814143	−1.1103404855
4	−1.0945391187	1.8634245	3.7556995	−1.0945391940
5	−1.0797693217	1.9827332	4.2905417	−1.0797694803
6	−1.0660370737	2.1117587	4.9013207	−1.0660372849
7	−1.0533604890	2.2535349	5.6105163	−1.0533608258
8	−1.0417726950	2.4125952	6.4525567	−1.0417731139
9	−1.0313249454	2.5958940	7.4823876	−1.0313254708
10	−1.0220917849	2.8149490	8.7946796	−1.0220924876
11	−1.0141782601	3.0901798	10.566922	−1.0141791536
12	−1.0077301951	3.4627010	13.181814	−1.0077311979
13	−1.0029493758	4.0342373	17.680148	−1.0029504633
14	−1.0001150482	5.2110181	28.919890	−1.0001159762
—[b]	−0.9994556794			

[a]Also, the nonrelativistic energies of Wolniewicz are presented for comparison. All quantities in atomic units.
[b]Nonrelativistic dissociation threshold.

and the results of Wolniewicz is very good. However, one notices that the agreement is much better for the lower energies than for the higher ones. While for the two lowest states our energies are lower than those obtained by Wolniewicz, the energies for the higher states are progressively higher.

Included in Table III is the comparison of the transition frequencies calculated from the energies obtained in our calculations with the experimental transition frequencies of Dabrowski [125]. To convert theoretical frequencies into wavenumbers, we used the factor of 1 hartree $= 219474.63137\,\text{cm}^{-1}$. For all the frequencies our results are either within or very close to the experimental error bracket of $0.1\,\text{cm}^{-1}$. We hope that the advances in high-resolution spectroscopy will inspire remeasurements of the vibrational spectrum of H_2 with the accuracy lower than $0.1\,\text{cm}^{-1}$. With such high-precision results, it would be possible to verify whether the larger differences between the calculated and the experimental frequencies for higher excitation levels, which now appear, are due to the relativistic and radiative effects.

Finally, in Table III we also compare our results for the transition energies with the results obtained by Wolniewicz [107,108]. Wolniewicz also calculated

TABLE III

Comparison of Vibrational Frequencies $E_{v+1} - E_v$ (in cm^{-1}) of H_2 Calculated from Non-Born–Oppenheimer Energies [121] with the Experimental Values of Dabrowski [125] and with the results of Wolniewicz, Obtained Using the Conventional Approach Based on the Potential Energy Curve[a]

v	Dabrowski	Work [121]	Wolniewicz[b] (Diff.)	Wolniewicz[c] (Diff.)
0	4161.14	4161.165 (+0.025)	4161.163 (+0.023)	4161.167 (+0.027)
1	3925.79	3925.842 (+0.052)	3925.837 (+0.047)	3925.836 (+0.046)
2	3695.43	3695.398 (−0.032)	3695.392 (−0.038)	3695.389 (−0.041)
3	3467.95	3467.990 (+0.040)	3467.983 (+0.033)	3467.976 (+0.026)
4	3241.61	3241.596 (−0.014)	3241.577 (−0.033)	3241.564 (−0.046)
5	3013.86	3013.880 (+0.020)	3013.869 (+0.009)	3013.851 (−0.009)
6	2782.13	2782.189 (+0.059)	2782.161 (+0.031)	2782.136 (+0.006)
7	2543.25	2543.227 (−0.023)	2543.209 (−0.041)	2543.175 (−0.075)
8	2292.93	2293.016 (+0.086)	2292.993 (+0.063)	2292.950 (+0.020)
9	2026.38	2026.445 (+0.064)	2026.406 (+0.026)	2026.351 (−0.029)
10	1736.66	1736.818 (+0.158)	1736.776 (+0.116)	1736.707 (+0.047)
11	1415.07	1415.187 (+0.117)	1415.163 (+0.093)	1415.076 (+0.006)
12	1049.16	1049.269 (+0.109)	1049.250 (+0.090)	1049.139 (−0.021)
13	622.02	622.063 (+0.043)	622.098 (+0.078)	621.956 (−0.064)

[a]Differences between the calculated and the experimental results are shown in parentheses.
[b]Private communications (nonrelativistic values).
[c]From Ref. 108 (includes relativistic and radiative corrections).

transition energies corrected for the relativistic effects, and these results are also shown in Table III. Upon comparing the results, one notices that our transition energies are, in general, very similar to Wolniewicz's nonrelativistic results. Both sets of results show higher positive discrepancies in comparison with the experimental values for the higher excitation levels. These discrepancies decrease somewhat when the relativistic effects are included. However, the insufficient accuracy of the experiment, as well as perhaps that of some of the theoretical results, does not allow us to carry out a more detail analysis of the remaining discrepancies.

F. Charge Asymmetry in HD^+ Molecular Ion

The lack of a center of symmetry in HD^+, due to the difference in nuclear masses, creates a particularly interesting situation that requires a theoretical approach that may differ from those used to describe the parent cation, H_2^+, and its symmetric isotopomer, D_2^+. The asymmetry of the HD^+ system has been investigated both experimentally [109,110] and theoretically [111–114]. In recent work, Ben-Itzhak et al. [109] studied the dissociation of the electronic ground state of HD^+ following ionization of HD by fast proton impact and found the $H^+ + D(1s)$ dissociation channel is more likely than the $H(1s) + D^+$ dissociation channel by about 7%. They attributed this asymmetry breakdown to the finite nuclear mass correction to the Born–Oppenheimer (BO) approximation, which makes the $1s\sigma$ state 3.7 meV lower than the $2p\sigma$ state at the dissociation limit.

Near the dissociation threshold the density of bound states in the HD^+ spectrum increases. If we remain in the ground rotational state manifold (i.e., consider only states with total angular momentum equal to zero), there is a place in the spectrum where the spacing between the consecutive levels becomes comparable to the difference between the binding energies of the H and D atoms (equal to 29.84 cm^{-1}). The D atom is energetically more stable because it has slightly larger reduced mass than H, which makes the electron slightly closer (on average) approach the nucleus, resulting in stronger Coulombic attraction and a lower binding energy. According to the previous calculations based on the adiabatic approximation [113], the H/D energy gap is approximately matched by the energy difference between the $v = 20$ and $v = 21$ levels. In this region the vibrational wave function that corresponds to, say, $v = 20$ level, combined with the ground-state electronic wave function that places the electron at the proton, has energy similar to that of thc wavc function, with the vibrational component corresponding to $v = 21$ and with the electronic component localizing the electron at *d*. Since such two-wave functions have the same symmetry, their mixing can occur. This nonadiabatic coupling must be included in the calculation of the dissociation of HD^+ that yields a proton plus a deuterium atom since the electron favors the heavier nucleus.

In nearly all theoretical calculations of H_2^+ and its isotopes' spectra reported in the literature, a body-fixed coordinate system with the origin at the geometric center of the nuclei has been used. For example, in the recent work of Esry and Sadeghpour [113], as well as the works of Moss [111, 112, 114], the starting point was the H_2^+ BO Hamiltonian in prolate spheroidal coordinates (PSC); and electronic wave functions and energies were first obtained as a function of the internuclear distance. Here we demonstrate the capability of the variational method to treat such systems fully nonadiabatically [124]. The very high powers m_k in the preexponential multipliers, which, as in case of H_2 molecule calculations, ranged from 0 to 250, allow one to describe very sharp peaks in the "vibrational" part of the wave function.

The effort in the first stage of the calculations has been focused on generating very accurate variational wave functions and energies for the rotationless vibrational states of the HD^+ ion. As mentioned, this system has been studied by many researchers and very accurate, virtually exact nonrelativistic energies have been published in the literature [112]. This includes the energy for the highest vibrational $v = 22$ state, which is only about 0.4309 cm^{-1} below the $D + H^+$ dissociation limit.

The basis set for each vibrational state was generated in a separate calculation. To achieve a similar level of accuracy as obtained in the best

TABLE IV
Comparison of the Total Nonrelativistic Nonadiabatic Variational Energies, E (in a.u.), and Dissociation Energies, D (in cm^{-1}), of HD^+ Vibrational States with Zero Total Angular Momentum, Obtained in Work [124] and the Corresponding Quantities Obtained by Moss [111][a]

v	E, Work [124]	D, Work [124]	E, Moss	D, Moss
0	−0.5978979685	21516.0096	−0.5978979686	21516.0096
1	−0.5891818291	19603.0382	−0.5891818295	19603.0382
2	−0.5809037001	17786.1989	−0.5809037006	17786.1989
3	−0.5730505464	16062.6308	−0.5730505469	16062.6309
4	−0.5656110418	14429.8483	−0.5656110424	14429.8484
5	−0.5585755200	12885.7298	−0.5585755213	12885.7300
6	−0.5519359482	11428.5122	−0.5519359493	11428.5124
7	−0.5456859137	10056.7882	−0.5456859158	10056.7886
8	−0.5398206394	8769.5092	−0.5398206420	8769.5098
9	−0.5343370110	7565.9919	−0.5343370137	7565.9925
10	−0.5292336317	6445.9296	−0.5292336357	6445.9305
11	−0.5245109059	5409.4111	−0.5245109104	5409.4121
12	−0.5201711374	4456.9421	−0.5201711482	4456.9444
13	−0.5162186988	3589.4820	−0.5162187105	3589.4846
14	−0.5126601767	2808.4767	−0.5126601926	2808.4802
15	−0.5095046270	2115.9136	−0.5095046516	2115.9190
16	−0.5067638344	1514.3792	−0.5067638779	1514.3887
17	−0.5044526466	1007.1321	−0.5044526992	1007.1436
18	−0.5025891815	598.1488	−0.5025892341	598.1603
19	−0.5011947323	292.1025	−0.5011947991	292.1172
20	−0.5002924017	94.0638	−0.5002924543	94.0754
21	−0.4999103339	10.2097	−0.4999103614	10.2157
22	−0.4998657692	0.4288	−0.4998657786	0.4309
$D + H^+$	−0.4998638152			

[a]Total energies were calculated using dissociation energies, mass values, and hartree to cm^{-1} conversion factor from Ref. 111.

previous calculations [112], we used 2000 basis functions for each state—except the ground and first excited states, where we limited ourselves to 1000-term expansions because the energies for those states were essentially converged with this number of functions.

In Table IV we compare our variational energies with the best literature values of Moss [112]. As one can see, the values agree very well. The agreement is consistently very good for all the states calculated.

After the wave functions for all 23 $(v = 0, \ldots, 22)$ states were generated, we calculated the expectation values of the internuclear d–p distance, $\langle r_1 \rangle$, the deuteron–electron (d–e) distance, $\langle r_2 \rangle$, and the proton–electron (p–e) distance, $\langle r_{12} \rangle$, for each state. The expectation values of the squares of the distances were also computed.

TABLE V
Expectation Values of the Deuteron–Proton Distance, r_{d-p}, the Deuteron–Electron Distance, r_{d-e}, and the Proton–Electron Distance, r_{p-e}, and Their Squares for the Vibrational Levels of HD^+ in the Rotational Ground State[a]

v	$\langle r_{d-p} \rangle$	$\langle r_{d-e} \rangle$	$\langle r_{p-e} \rangle$	$\langle r^2_{d-p} \rangle$	$\langle r^2_{d-e} \rangle$	$\langle r^2_{p-e} \rangle$
0	2.055	1.688	1.688	4.268	3.534	3.537
1	2.171	1.750	1.750	4.855	3.839	3.843
2	2.292	1.813	1.814	5.492	4.169	4.173
3	2.417	1.880	1.881	6.185	4.526	4.531
4	2.547	1.948	1.950	6.942	4.915	4.921
5	2.683	2.020	2.022	7.771	5.339	5.346
6	2.825	2.095	2.097	8.682	5.804	5.813
7	2.975	2.175	2.177	9.689	6.318	6.329
8	3.135	2.259	2.261	10.81	6.888	6.902
9	3.305	2.348	2.351	12.06	7.527	7.545
10	3.489	2.445	2.448	13.48	8.250	8.272
11	3.689	2.549	2.554	15.09	9.074	9.105
12	3.909	2.664	2.670	16.96	10.03	10.07
13	4.154	2.791	2.799	19.16	11.15	11.21
14	4.432	2.934	2.946	21.79	12.49	12.57
15	4.754	3.099	3.116	25.01	14.13	14.26
16	5.138	3.292	3.319	29.11	16.20	16.41
17	5.611	3.527	3.572	34.55	18.92	19.30
18	6.227	3.821	3.910	42.25	22.66	23.47
19	7.099	4.198	4.421	54.35	28.13	30.38
20	8.550	4.569	5.516	77.74	35.66	46.64
21	12.95	2.306	12.19	176.0	12.94	168.2
22	28.53	1.600	28.46	900.4	4.261	901.8
D atom[b]		1.500			3.002	

[a]All quantities are in atomic units.
[b]In the ground state.

The results are shown in Table V. As can be expected, the average internuclear distance increases with the rising level of excitation. This increase becomes more prominent at the levels near the dissociation threshold. For example, in going from $v = 21$ to $v = 22$ the average internuclear distance increases more than twofold from 12.95 a.u. to 28.53 a.u. In the $v = 22$ state the HD^+ ion is almost dissociated. However, the most striking feature that becomes apparent upon examining the results is a sudden increase of the asymmetry between the deuteron–electron and proton–electron average distances above the $v = 20$ excitation level. In levels up to $v = 20$, there is some asymmetry of the electron distribution with the *p–e* distance being slightly longer than the *d–e* distance. For example, in the $v = 20$ state the *d–e* average distance is 4.569 a.u. and the *p–e* distance is 5.516 a.u. The situation becomes completely different for the $v = 21$ state. Here the *p–e* distance of 12.19 a.u. is almost equal to the average value of the internuclear distance, but the d–e distance becomes much smaller and equals only 2.306 a.u. It is apparent that in this state the electron is essentially localized at the deuteron and the ion becomes highly polarized. An analogous situation also occurs for the $v = 22$ state. Here, again, the p–e average distance is very close to the internuclear distance while the d–e distance is close to what it is in an isolated D atom.

As has been mentioned above, the inclusion of basis functions (49) with high power values, m_k, is very essential for the calculations of molecular systems. It is especially important for highly vibrationally excited states where there are many highly localized peaks in the nuclear correlation function. To illustrate this point, we calculated this correlation function (it corresponds to the internuclear distance, $r_{d-p} = r_1$), which is the same as the probability density of pseudoparticle 1. The definition of this quantity is as follows:

$$g_1(\xi) = \langle \delta(\mathbf{r}_1 - \xi) \rangle$$

In essence, it is the probability density of the two nuclei to have relative separation ξ. Since the orientation of the molecule is not fixed (nuclei are not fixed any more if we deal with an non-BO approach), $g_1(\xi)$ is a spherically symmetric function. The plots of $g_1(\xi)$ are presented in Figs. 1–4. It should be noted that all the correlation functions shown are normalized in such a way that

$$4\pi \int_0^\infty g_1(\xi)\xi^2 \, d\xi = 1$$

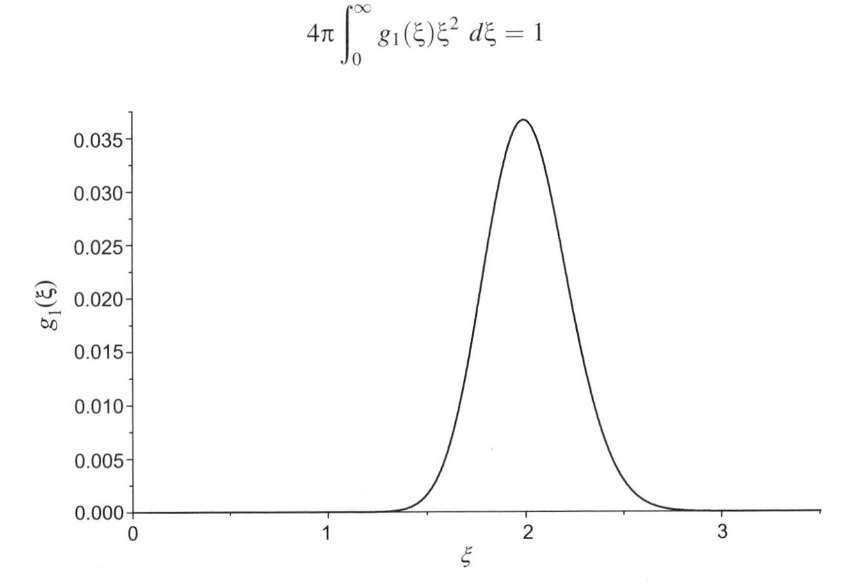

Figure 1. The correlation function $g_1(\xi)$ for $v = 0$ vibrational state of HD^+. All quantities are in a.u.

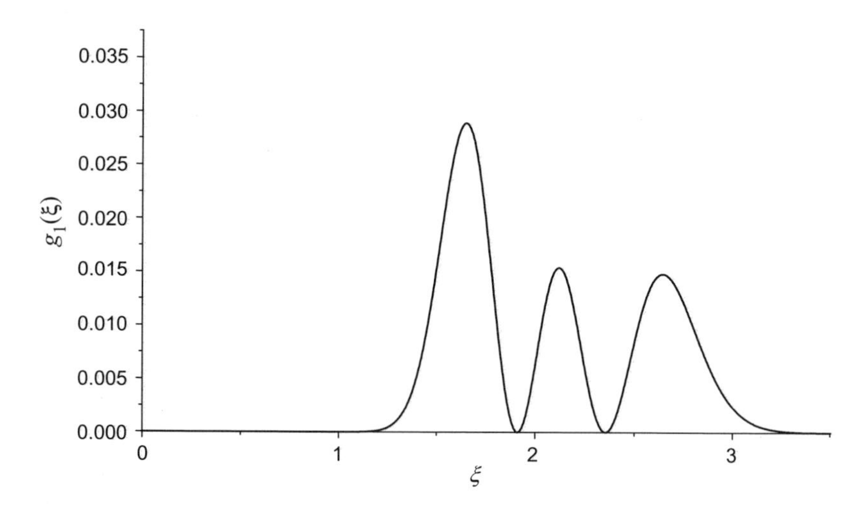

Figure 2. The correlation function $g_1(\xi)$ for $v = 2$ vibrational state of HD^+. All quantities are in a.u.

Upon looking at the graphs, one can clearly see how strongly localized the peaks of $g_1(\xi)$ are—in particular, for the case of the $v = 22$ state. It even surprises us, to some degree, how well a basis set consisting of just 2000 basis functions (49) is capable of fitting the wave function, so that it reproduces seven decimal figures in the total energy.

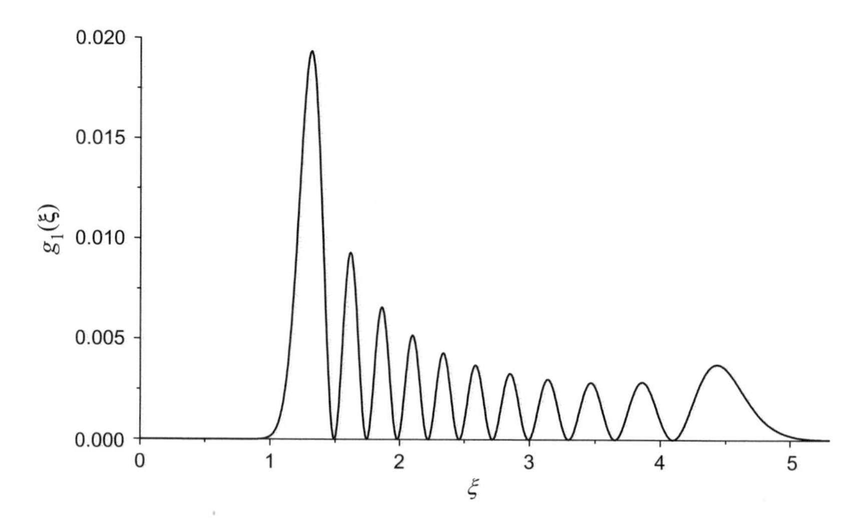

Figure 3. The correlation function $g_1(\xi)$ for $v = 10$ vibrational state of HD^+. All quantities are in a.u.

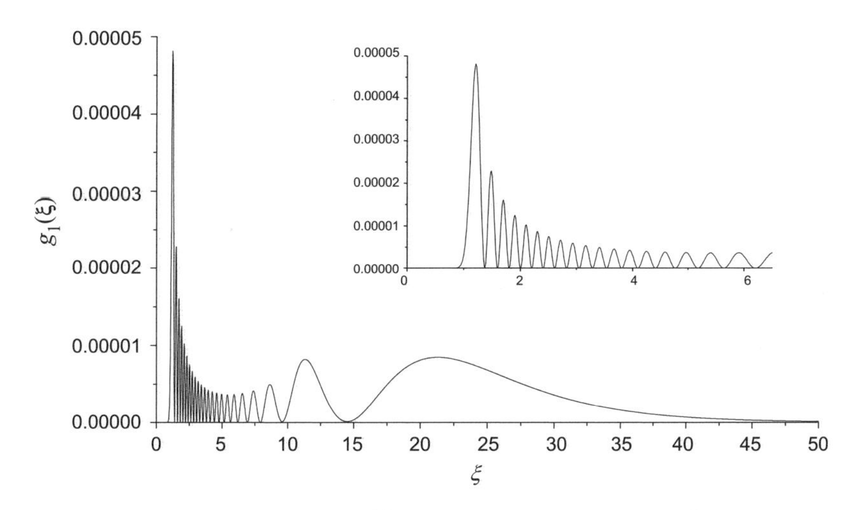

Figure 4. The correlation function $g_1(\xi)$ for $v = 22$ vibrational state of HD^+. All quantities are in a.u.

G. LiH and LiD Electron Affinity

The determination of electron affinities (EAs) is one of the most serious problems in quantum chemistry. While the Hartree–Fock electron affinity can be easily evaluated, most anions turn out to be unbound at this level of theory. Thus, the correlation effects are extremely crucial in evaluating EAs. At this point, lithium hydride and lithium hydride anion make up a very good benchmark system because they are still small enough yet exhibit features of more complicated systems. Four and five electrons, respectively, give rise to higher-order correlation effects that are not possible in H_2.

The theoretical interest in the LiH^- has increased since the electron affinity of LiH and its deuterated counterpart, LiD, were measured with the use of the photoelectron spectroscopy by Bowen and co-workers [126]. The adiabatic electron affinities of ^{7}LiH and ^{7}LiD determined in that experiment were 0.342 ± 0.012 eV for the former and 0.337 ± 0.012 eV for the latter system. The appearance of these data posed a challenge for theory to reproduce those values in rigorous calculations based on the first principles. Since the two systems are small, it has been particularly interesting to see if the experimental EAs can be reproduced in calculations where the BO approximation is not assumed [123].

Since the time that Bowen's and co-workers' article was published, the theory based on the BO approximation, except for one very recent multi-reference configuration interaction (MRCI) calculation by Chang et al. [127], has been unable to produce a value of the LiH adiabatic electron affinity that falls within the experimental uncertainty bracket. More than that, in many works where authors managed to obtain an EA that was relatively close to the experimentally determined one, the success was often due to fortuitous cancellation of errors in the total energies of LiH and LiH^- rather than the accuracy achieved in calculations. This inability seems somewhat odd, since there is nothing particularly unusual about and fundamentally difficult to describe in the LiH^-/LiH system. Since the LiH molecule is a polar system ($Li^{\delta+}$–$H^{\delta-}$), an excess electron in the process of attachment localizes on the electropositive alkali atom in a nonbonding orbital of the neutral molecule.

We would like to mention that one should not expect the nonadiabatic effects to play a noticeable role in the LiH and LiD electron affinity calculations. However, by applying the non-BO approach, we can directly determine the total energies of the anion and the neutral system in one-step calculations, and we do not need to resort to calculating the electronic potential energy curves for the anion and the neutral first and using them for calculating their nuclear vibration energies in the next step, as is done in the BO approach. Thus, using the non-BO method, we not only make the calculation free of any artifacts that may result from the two-step procedure used in the BO approximation, but we also obtain total and relative energies that, if the basis sets become more complete, approach the true nonrelativistic limits of those quantities, free of any approximations.

In comparison with H_2 and HD^+, LiH and LiH^- represent a significantly more challenging case from the point of view of computational demands, even if we restrict ourselves to the consideration of the ground state only. The total number of particles is six in the case of LiH and seven in the case of LiH^-, which means that effectively we deal with a 15- and an 18-dimensional problem, respectively. Considering that the number of permutations in the Young operator also increases as the number of identical particles increases (this number is 24 LiH and 120 for LiH^-), the amount of computational work required for performing sound calculations rises by a factor of several orders of magnitude. To attempt such calculations at the present time, one has to gain access to large-scale parallel computer systems. However, with advances in computer hardware, this kind of calculation may soon become quite ordinary and even routine.

The results of our calculations for the energy are shown in Table VI. In Table VII we show the values of the LiH and LiD electron affinities calculated as the difference of the energies of the anion and the neutral system for all lengths of the basis set for which the total energies are reported. A question can be raised whether it is appropriate to use the total energies obtained with the same length of the basis set for LiH^- and LiH (or LiD^- and LiD) in the electron affinity calculation. Since LiH^- has one more electron than LiH, it should require more basis functions for LiH^- than for LiH to achieve a similar level of

TABLE VI
Nonadiabatic Variational Ground-State Energies of the LiH^-, LiH, LiD^-, and LiD Molecules Obtained with Different Basis Set Sizes [123][a]

Basis Size	LiH	LiD	LiH^-	LiD^-
1000	−8.06632055	−8.06725433	−8.07778128	−8.07859843
1200	−8.06634454	−8.06728273	−8.07794715	−8.07875507
1400	−8.06636491	−8.06730593	−8.07807991	−8.07888651
1600	−8.06638295	−8.06732805	−8.07822084	−8.07903483
1800	−8.06639498	−8.06734216	−8.07830536	−8.07912238
2000	−8.06640408	−8.06735333	−8.07837566	−8.07919877
2200	−8.06641099	−8.06736128	−8.07842711	−8.07925549
2400	−8.06641554	−8.06736613	−8.07846014	−8.07929081
2600	−8.06642068	−8.06737208	−8.07848753	−8.07931985
2800	−8.06642353	−8.06737514	−8.07851089	−8.07934370
3000	−8.06642581	−8.06737749	−8.07852933	−8.07936248
3200	−8.06642787	−8.06737962	−8.07854406	−8.07937759
3400	−8.06642941	−8.06738123	−8.07855805	−8.07939162
3600	−8.06643070	−8.06738251	−8.07856887	−8.07940445

[a]All energies are in atomic units.

TABLE VII
Convergence of the Electron Affinities of LiH and LiD (in eV) in Terms of the Number of the Basis Functions [123] and the Corresponding Experimental Values of Sarkas et al.

Basis Size	LiH	LiD
1000	0.31186	0.30869
1200	0.31572	0.31218
1400	0.31878	0.31512
1600	0.32213	0.31856
1800	0.32410	0.32056
2000	0.32576	0.32233
2200	0.32698	0.32366
2400	0.32775	0.32449
2600	0.32836	0.32512
2800	0.32891	0.32568
3000	0.32935	0.32613
3200	0.32970	0.32648
3400	0.33004	0.32682
3600	0.33030	0.32713
Experiment [126]	0.342 ± 0.012	0.337 ± 0.012

accuracy of the results. This is indeed showing in the convergence of the results presented in Table VI. Certainly the 3600-term energy for LiH is better converged than the 3600-term energy for LiH^-. It should be mentioned that the LiH and LiD variational energies obtained in work [123] are the lowest non-BO ground-state energies ever obtained for these systems. It is significantly lower than the previous non-BO value of −8.06615576 hartree, which was also obtained in our group [120]. The same is true about the LiH^- and LiD^- energies, though in this case the final values are not as tightly converged as the energies for LiH and LiD. Also, work [123], to date, is the only one where a variational, non-BO calculation of the ground state of a five-electron system such as LiH^- or LiD^- has been attempted.

Since the variational approach has been used in this work and the LiH (LiD) energy at a particular basis set size is better converged than the LiH^- (LiD^-) energy, the calculated electron affinity is always lower than the result would be in the limit of the complete basis set. Thus, by calculating the EA at the basis set with the same size for the anion and for neutral system and by incrementally increasing the size of the basis, we could monitor the EA convergence and be sure that all the EA values we calculated including the final value (this being the 3600-term result) are lower bounds to the true EA.

As shown in Table VII, our best results for LiH and LiD electron affinities obtained with the 3600-term expansions of the wave functions for LiH^-/LiH and LiD^-/LiD are 0.33030 and 0.32713 eV, respectively. Even though, as stated, both values represent lower bounds to the true EAs, they both are within the uncertainty brackets of the experimental results of 0.342 ± 0.012 eV (LiH) and 0.337 ± 0.012 eV (LiD) obtained by Bowen and co-workers [126].

In the calculations, the powers, m_k, in the preexponential factors in the basis functions (49) were allowed to vary in the interval of 0–200. The obtained distributions of m_k's have the mean values of 70.0, 74.1, 67.6, and 67.9, along with the standard deviations of 49.7, 50.4, 49.3, and 49.7, for LiH, LiD, LiH^-, and LiD^-, respectively. A slightly higher mean power for LiD than for LiH can be explained by a more localized vibrational component of the wave function for the former than for the latter system. Lower mean powers for the anions than for the neutral systems result from two opposing effects. First, the bond lengths for the anions are slightly longer than for the neutral counterparts (see the next paragraph), which should require larger powers. Second, due the weakening of the bonds in the anions, the vibrational components of their wave functions become more delocalized, resulting in lowering of the powers. Apparently the second effect dominates over the first one.

Finally, Table VIII shows the expectation values of the internuclear distance and its square for LiH, LiD, LiH^-, and LiD^-. Here we see trends that can be easily understood considering that the attachment of an excess electron weakens slightly the Li–H bond and that increasing the mass of H by switching to D

TABLE VIII
Expectation Values of the Internuclear Distance, $r_1 \equiv r_{\mathrm{LiH}}$, and Its Square (in a.u.) for LiH, LiD, LiH^-, and LiD^-, Calculated with the 3600-Term Basis Sets [123]

System	$\langle r_1 \rangle$	$\langle r_1^2 \rangle$
LiH^-	3.214708	10.39391
LiH	3.061047	9.419733
LiD^-	3.199737	10.28362
LiD	3.049131	9.334349

results in making the vibrational component of the total wave function more localized, resulting in a slight contraction of the average length of the bond. This contraction is very similar in going from LiH and LiD and from LiH^- to LiD^-.

H. Molecules Containing Positron: e^+LiH

During the last several years, much advance has been made concerning the study of bound states of the positron with small systems. The ability of various atoms, ions, and molecules to bind a positron is now well established and represent a popular subject of research. However, most of the calculations performed were done on atomic systems (or those containing just one particle significanly heavier than the positron). For example, there has been considerable interest to the positronium hydride, HPs ($p^+e^+e^-e^-$), and its isotopomers (see Refs. 128–130 and references therein). Molecular systems containing a positron have been mainly treated with the use of the Quantum Monte Carlo [131, 132] or variational method [133–135] assuming BO approximation. An attempt of calculating e^+LiH was made in Ref. 136, where the authors used simple Gaussians (without premultipliers). However, the convergence of the total energy turned out to be very slow while the basis size was not large, which resulted in relatively low accuracy of the calculations.

In Ref. 122 we raised the question of whether one can use basis functions (49) in non-BO quantum mechanical calculations of molecular diatomic systems containing positrons and whether those functions are capable of providing a proper representation for the positron–nucleus and positron–electron correlation effects in a diatomic system. e^+LiH was chosen as a target system. Along with e^+LiH, we also performed the calculations of HPs and Li^+ because the total energies of these systems are needed for determining the dissotiation energy.

The convergence of the energy values for HPs, LiH, and e^+LiH in terms of the number of the basis functions is shown in Table IX. In the case of e^+LiH, as

TABLE IX
Total Non-Born–Oppenheimer Energies (in a.u.) of HPs, LiH, and e^+LiH as a Function of Basis Size [122]

Basis Size	HPs	LiH	e^+LiH
800	−0.7888705040	−8.066278419	−8.103075429
1000	−0.7888705983	−8.066320545	−8.103572816
1200	−0.7888706398	−8.066344535	−8.103905788
1400	−0.7888706611	−8.066364905	−8.104113213
1600	−0.7888706790	−8.066382950	−8.104256550
1800	−0.7888706877	−8.066394978	−8.104372009
2000	−0.7888706940	−8.066404077	−8.104478249
2200	−0.7888706984	−8.066410987	−8.104543434
2400	−0.7888707014	−8.066415542	−8.104598552
2600	−0.7888707036	−8.066420678	−8.104645276
2800	−0.7888707057	−8.066423527	−8.104683502
3000	−0.7888707062	−8.066425806	−8.104713922
3200	−0.7888707066	−8.066427866	−8.104739913

well as for LiH case, the powers m_k in (49) were selected from the interval 0–200. Although the positronium hydride wave function can be obtained with very high precision even without using powers of the hydrogen–positron distance in the preexponential factors in the basis functions, we did include some functions with small preexponential powers ranging from 0 to 10 to ensure better numerical stability in the calculations.

From the lowest energy values shown in Table IX, one can determine that the positron detachment energy of e^+LiH, PDE $= E(e^+\text{LiH}) - E(\text{LiH})$, is 0.038312 hartree. The lowest-energy fragmentation of e^+LiH corresponds to dissociation of the system into HPs + Li^+. To calculate the dissociation energy, DE $= E(e^+\text{LiH}) - E(i^+) - E(\text{HPs})$, one needs to determine the total energy of the Li^+ ion. Since the non-BO calculation of this quantity is very simple, rather than taking it from the literature, we recalculated it using our method. A 400-term expansion was sufficient to obtain a highly accurate result of −7.279321518 hartree, where, we believe, all the significant figures shown are exact.

Given the values of the HPs and Li^+ energies calculated in this work in addition to that of e^+LiH, our dissociation energy is 0.036548 hartree. This value qualitatively agrees with the value of 0.0382(2) hartree obtained in the Born–Oppenheimer calculations by Mella and et al. [131] using the Quantum Monte Carlo method. It also agrees with the Born–Oppenheimer result of 0.036936 obtained by Strasburger [135] with the use of explicitly correlated Gaussians and the variational method.

The lowest variational energy upper-bound for the Born–Oppenheimer LiH ground-state energy to date is −8.070538 hartree [137]. Assuming that the energies of Li and H with infinitely heavy nuclei are −7.4780603 [138] and −0.5 hartree, respectively, one obtains the infinite-mass dissociation energy of LiH of −0.0924777 hartree. The finite-mass energy of LiH can be estimated by subtracting this number from the sum of the finite-mass energies of Li and H atoms and by adding to the result the zero-point LiH energy. Using the finite-mass energy of Li of −7.4774519 hartree which one can calculate by using the expansion from the article of Yan and Drake [138], the corresponding value for the H atom of −0.4997278 hartree, and the zero-point LiH energy of 0.0031981 hartree (see Ref. 139), we obtain the LiH ground-state energy corrected for the finite nuclear masses equal to −8.066459 hartree. The use of experimentally determined zero-point energy of −0.0031799 [140] shifts this value to −8.066478 hartree. The difference between this value and our non-BO result of −8.066427866 hartree is larger than the estimated sum of their inaccuracies. Although, perhaps, this may partially be attributed to relatively low accuracy of the zero-point energy, it is clear that the nonadiabatic effect of the coupled electron–nuclear motion must play a role in the difference.

In Table X we present expectation values of the internuclear distance and its square for LiH and e^+LiH as well as the electron–positron contact densities for HPs and e^+LiH evaluated with the largest basis set of 3200 basis functions obtained in the calculations. It should be noted that the mean internuclear distance of LiH calculated here is slightly higher than the known value, 3.015 bohr, of the equilibrium nuclear distance—that is, the distance where the potential energy curve reaches its minimum. This is, obviously, an expected result since larger distances contribute more to the mean distance when one averages the internuclear distance over the "vibrational" part of the wave function, even if a purely Born–Oppenheimer calculation is carried out, and, hence, the mean distance is always larger than the distance corresponding the peak of the vibrational component of the wave function (minimum of the potential energy curve). Thus the discrepancy becomes more and more

TABLE X
Expectation Values of the Li–H Internuclear Distance, Its Square, and Electron–Positron Contact Densities Evaluated at 3200-Function Basis Size [122][a]

System	$\langle r_{\mathrm{LiH}} \rangle$	$\langle r_{\mathrm{LiH}}^2 \rangle$	$\langle \delta(\mathbf{r}_{e^-e^+}) \rangle$
HPs	—	—	2.44855×10^{-2}
LiH	3.06105	9.41977	—
e^+LiH	3.44470	11.9397	7.08879×10^{-3}

[a]All quantities are in atomic units.

noticeable as the nuclear masses decrease. For e^+LiH, our $r_{iH}(\equiv r_1)$ mean distance of 3.445 bohr agrees well with the the equilibrium internuclear separation of 3.348 bohr obtained by Strasburger [135] in the Born–Oppenheimer calculations.

An important characteristic of positronic systems relevant to the experiment is their lifetimes. The expectation value of the electron–positron contact density allows us to evaluate the two-photon annihilation rate for a positronic system using the expression

$$\Gamma_{2\gamma} = n\frac{\pi\alpha^4 c}{a_0}\langle\delta(\mathbf{r}_{e^-e^+})\rangle$$

where α is the fine structure constant, a_0 is the Bohr radius, c is the velocity of light, and n denotes the number of electron–positron pairs in the system (2 and 4 in the case of HPs and e^+LiH, respectively). The two-photon annihilation rates we obtained with 3200 basis functions are $2.4716 \times 10^9\ \mathrm{s}^{-1}$ for HPs and $1.4311 \times 10^9\ \mathrm{s}^{-1}$ for e^+LiH. This indicates that a positron attached to LiH survives much longer than in the HPs system. The HPs annihilation rate can be compared with the result of Yan and Ho obtained in a finite-mass calculation using Hylleraas coordinates [141] (HPs) and with the explicitly correlated Gaussian calculation [129] performed for $^\infty$HPs (HPs with infinetely heavy proton), both of which yielded the value of $2.4722 \times 10^9\ \mathrm{s}^{-1}$. In the case of e^+LiH, we can make a comparison with the Born–Oppenheimer Quantum Monte Carlo result of Mella et al. [132], which yielded $1.49 \times 10^9\ \mathrm{s}^{-1}$ (the vibrationally averaged result), and with the Born–Oppenheimer explicitly correlated Gaussian result of Strasburger [135] where the value of $1.375 \times 10^9\ \mathrm{s}^{-1}$ was obtained at the e^+LiH equilibrium distance of $R = 3.348$ bohr.

VI. NON-BO CALCULATIONS OF DIATOMIC MOLECULES IN ELECTRIC FIELD WITH SHIFTED GAUSSIANS

When a diatomic molecule enters an electric field, the rotational symmetry of the wave function is broken. The wave function is no longer an eigenfunction of the total angular momentum operator, $\hat{J}$, but, if we assume the electric field to be directed along the molecular z axis, is only an eigenfunction of $\hat{J}_z$. Thus, we cannot use the above spherically symmetric basis functions to expand the wave function. We find that expansion of the electric-field-perturbed wave function in terms of spherical Gaussians with floating centers provides a good description of the electric field effect.

It can be shown that the basis of spherical explicitly correlated Gaussian functions with floating centers (FSECG) form a complete set. These functions have the form

$$g_k(r) = \prod_{i=1}^{n} \exp(-\alpha_i^k (r_i - R_i^k)^2) \prod_{j>i}^{n} \exp(-\beta_{ij}^k (r_i - r_j)^2) \tag{121}$$

which may equivalently be expressed as

$$g_k(r) = \exp(-(r - s_k)'(A_k \otimes I_3)(r - s_k)) \tag{122}$$

The Kronecker product with the identity ensures rotational invariance (sphericalness); elliptical Gaussians could be obtained by using a full $n \times n$ A matrix. In the former formulation of the basis function, it is difficult to ensure the square integrability of the functions, but this becomes easy in the latter formulation. In this format, all that is required is that the matrix, A_k, be positive definite. This may be achieved by constructing the matrix from a Cholesky decomposition: $A_k = L_k' L_k$. Later in this work we will use the notation $\bar{A} = A \otimes I_3$ to indicate the Kronecker product with the 3×3 identity matrix.

The correct symmetry of the system may be ensured by projection onto the irreducible representations, as described above. Thus, when taking symmetry into account, the final form of the basis function is

$$\hat{Y} g_k(r) = \prod_i \hat{Y}_i \exp\{-(r - s_k)'[(L_k L_k') \otimes I_3](r - s_k)\} \tag{123}$$

and the spin-free spatial wave function has the form

$$\Psi = \sum_{k=1}^{m} c_k \hat{Y} g_k \tag{124}$$

where m is the size of the basis.

A. Integrals and Energy Derivatives

1. Born–Oppenheimer Integrals over Correlated Gaussians

Later we will discuss conventional Born–Oppenheimer calculations used in conjunction with the current work. Thus, for completeness, we will cover here the integrals needed for these calculations. These integrals are quite similar to the ones used in the non-Born–Oppenheimer calculaions, as will be shown below. We show first the integrals over the Born–Oppenheimer Hamiltonian:

$$\hat{H} = -\frac{1}{2}\nabla_r' \nabla_r + \sum_{j>i}^{n} \frac{1}{r_{ij}} - \sum_{i=1}^{n} \sum_{t=1}^{N} \frac{Z_t}{r_{it}} + \sum_{u>t}^{N} \frac{Z_t Z_u}{r_{tu}} \tag{125}$$

where explicit separation of the electronic and nuclear coordinates has been performed. In this case, n indicates only the number of electrons, rather than indicating the number of particles, while N indicates the number of nuclei. These molecular integrals have been published before in several formats [48, 52, 61, 75], but the format presented here is most useful for deriving the analytical gradients used in the optimization of the wave function.

Some definitions used in the integral formulas are

$$
\begin{aligned}
\bar{A}_{kl} &= \bar{A}_k + \bar{A}_l \\
e &= s_k'\bar{A}_k + s_l'\bar{A}_l \\
s &= \bar{A}_{kl}^{-1}e \\
p_{kl} &= \frac{|A_k|^{1/2}|A_l|^{1/2}}{|A_{kl}|}
\end{aligned}
$$

and

$$\gamma = -s_k'\bar{A}_k s_k - s_l'\bar{A}_l s_l + e'\bar{A}_{kl}^{-1}e$$

Also, we will use the J_{ij} matrix given by (52). We define a $3n$ vector t as containing the three coordinates of the tth nucleus repeated n times:

$$t = \begin{pmatrix} r_t \\ r_t \\ \vdots \\ r_t \end{pmatrix}$$

Repeated use is made of the following formula:

$$\int_{-\infty}^{\infty} \exp(-x'Mx + iz'x)dx = \pi^{n/2}|M|^{-1/2}\exp(-z'M^{-1}z/4) \tag{126}$$

where M is an $n \times n$ matrix, and x and z are vectors of size n.

The overlap integral is the simplest and serves as a basis for the rest of the integrals. The product of two functions is

$$
\begin{aligned}
g_k g_l &= \exp(-(r - s_k)'\bar{A}_k(r - s_k) - (r - s_l)'\bar{A}_l(r - s_l)) \\
&= \exp(-r'(\bar{A}_k + \bar{A}_l)r + 2(s_k'\bar{A}_k + s_l'\bar{A}_l)r) \times \exp(-s_k'\bar{A}_k s_k - s_l'\bar{A}_l s_l)
\end{aligned}
$$

Using (38), we get the following expression for the overlap integral:

$$\langle g_k|g_l\rangle = \pi^{3n/2}|A_{kl}|^{-3/2}\exp(-s_k'\bar{A}_k s_k - s_l'\bar{A}_l s_l + e'\bar{A}_{kl}^{-1}e) \tag{127}$$

Using the same method for the integrals $\langle g_k|g_k\rangle$ and $\langle g_l|g_l\rangle$, we find the overlap integral for the normalized basis functions to be

$$S_{kl} = 2^{3n}\left(\frac{|A_k|^{1/2}|A_l|^{1/2}}{|A_{kl}|}\right)^{3/2} \exp(\gamma) \tag{128}$$

The kinetic energy integral follows easily from the overlap integral. We write the kinetic energy operator

$$\hat{T} = -\frac{1}{2}\nabla_r \cdot \nabla_r \tag{129}$$

where the subscript r denotes the gradient with respect to all $3n$ coordinates of the n electrons. The integral is thus

$$\begin{aligned} T_{kl} &= \langle g_k|\hat{T}|g_l\rangle \\ &= -\frac{1}{2}\langle g_k|\nabla_r \cdot \nabla_r|g_l\rangle \\ &= \frac{1}{2}\langle \nabla'_r g_k|\nabla_r g_l\rangle \end{aligned}$$

Looking at (2), it is evident that $\nabla_r g_k = -\nabla_{s_k} g_k$, thus our integral becomes

$$\begin{aligned} T_{kl} &= \frac{1}{2}\langle \nabla'_{s_k} g_k|\nabla_{s_l} g_l\rangle \\ &= \frac{1}{2}\langle g_k|\nabla'_{s_k}\nabla_{s_l}|g_l\rangle \\ &= \frac{1}{2}\mathrm{tr}[\nabla_{s_l}\nabla'_{s_k}\langle g_k|g_l\rangle] \end{aligned}$$

The two derivatives of the overlap integral are easily evaluated directly from (40) and the result is

$$\begin{aligned} \nabla'_{s_k}\langle g_k|g_l\rangle &= S_{kl}(-2\bar{A}_k s_k + 2\bar{A}_k s) \\ \nabla_{s_l}\nabla'_{s_k}\langle g_k|g_l\rangle &= S_{kl}[(2\bar{A}_l s - 2\bar{A}_l s_l)'(2\bar{A}_k s - 2\bar{A}_k s_k) + 2\bar{A}_k\bar{A}_{kl}^{-1}\bar{A}_l] \end{aligned}$$

So the final expression for the kinetic energy integral is

$$T_{kl} = S_{kl}(2(s - s_k)'\bar{A}_k\bar{A}_l(s - s_l) + 3\mathrm{tr}[A_k A_{kl}^{-1} A_l]) \tag{130}$$

The two potential energy integrals, electron repulsion and nuclear attraction, are derived quite similarly. Both involve the Gaussian transformation

$$\frac{1}{r_{ij}} = \frac{2}{\pi^{1/2}} \int_0^\infty \exp(-u^2 r_{ij}^2)\, du \tag{131}$$

and the identity

$$\int_0^\infty (1 + \alpha u^2)^{-3/2} \exp(-\frac{au^2}{1 + \alpha u^2})\, du = \frac{\pi^{1/2}}{2a^{1/2}} \operatorname{erf}\left(\left(\frac{a}{\alpha}\right)^{1/2}\right) \tag{132}$$

The electron repulsion integral is the more straightforward of the two potential integrals. It should be recognized that r_{ij}^2 can be written as $r'\bar{J}_{ij}r$. Thus the integral, using (43), becomes

$$\begin{aligned} ER_{kl}^{ij} &= \langle g_k | \frac{1}{r_{ij}} | g_l \rangle \\ &= \frac{2}{\pi^{1/2}} \exp(\gamma) \int_{-\infty}^{\infty} dr \int_0^\infty du \ \exp(-(r - s)'\bar{A}_{kl}(r - s) - u^2 r' \bar{J}_{ij} r) \end{aligned}$$

which can be further rearranged to give

$$ER_{kl}^{ij} = \frac{2}{\pi^{1/2}} \exp(\gamma) \exp(-s'\bar{A}_{kl}s) \int_{-\infty}^{\infty} dr \int_0^\infty du \ \exp(-r'(\bar{A}_{kl} + u^2 \bar{J}_{ij})r + 2s'\bar{A}_{kl}r)$$

Applying (38) to the integral over r, we get

$$ER_{kl}^{ij} = \frac{2}{\pi^{1/2}} \exp(\gamma) \pi^{3n/2} \int_0^\infty du |\bar{A}_{kl} + u^2 \bar{J}_{ij}|^{-1/2} \exp\left(-\frac{u^2 s' \bar{J}_{ij} s}{1 + u^2 \mathrm{tr}[J_{ij} A_{kl}]}\right)$$

Following the approach given in the work of Kinghorn [116], we further get

$$ER_{kl}^{ij} = \frac{2}{\pi^{1/2}} S_{kl} \int_0^\infty du (1 + u^2 \mathrm{tr}[J_{ij} A_{kl}])^{-3/2} \exp\left(-\frac{u^2 s' \bar{J}_{ij} s}{1 + u^2 \mathrm{tr}[J_{ij} A_{kl}]}\right)$$

By applying (44), we get

$$ER_{kl}^{ij} = S_{kl} \left(\frac{1}{s'\bar{J}_{ij}s}\right)^{1/2} \operatorname{erf}\left(\left(\frac{s'\bar{J}_{ij}s}{\mathrm{tr}[J_{ij}A_{kl}]}\right)^{1/2}\right) \tag{133}$$

In order to solve the nuclear attraction integral, we use the t vector. We recognize that $r_{it}^2 = (r-t)'\bar{J}_{ii}(r-t)$. The integral can thus be written as

$$NA_{kl}^{it} = \langle g_k|\frac{1}{r_{it}}|g_l\rangle$$
$$= \frac{2}{\pi^{1/2}}\exp(\gamma)\int_{-\infty}^{\infty} dr \int_0^{\infty} du \; \exp(-(r-s)'\bar{A}_{kl}(r-s) - u^2(r-t)'\bar{J}_{ii}(r-t))$$

which, with a little rearrangement, can be put into the form

$$NA_{kl}^{it} = \frac{2}{\pi^{1/2}}\exp(\gamma)\exp(-(s-t)'\bar{A}_{kl}(s-t))$$
$$\times \int_{-\infty}^{\infty} dr \int_0^{\infty} du \exp(-(r-t)'(\bar{A}_{kl}+u^2\bar{J}_{ii})(r-t) + 2(s-t)'\bar{A}_{kl}(r-t))$$

By changing the integration variable from r to $(r-t)$ and applying (38), we get

$$NA_{kl}^{it} = \frac{2}{\pi^{1/2}}\exp(\gamma)\pi^{3n/2}\int_0^{\infty} du|\bar{A}_{kl}+u^2\bar{J}_{ii}|^{-1/2}\exp\left(-\frac{u^2(s-t)'\bar{J}_{ij}(s-t)}{1+u^2\mathrm{tr}[J_{ii}A_{kl}^{-1}]}\right)$$

the form of which is very similar to the electron repulsion integral. Using the same steps as in that integral, we get

$$NA_{kl}^{it} = S_{kl}\left(\frac{1}{(s-t)'\bar{J}_{ii}(s-t)}\right)^{1/2}\mathrm{erf}\left(\left(\frac{(s-t)'\bar{J}_{ii}(s-t)}{\mathrm{tr}[J_{ii}A_{kl}^{-1}]}\right)^{1/2}\right) \tag{134}$$

After converting the error function to the F_0 function, both of the potential energy integrals have the same general form

$$V_{kl} = \frac{2}{\pi^{1/2}}S_{kl}\alpha^{-1/2}F_0\left(\frac{\rho}{\alpha}\right) \tag{135}$$

For $V_{kl} = ER_{kl}^{ij}$, we have $\alpha = \mathrm{tr}[J_{ij}A_{kl}^{-1}]$ and $\rho = s'\bar{J}_{ij}s$; for $V_{kl} = NA_{kl}^{it}$, we have $\alpha = \mathrm{tr}[J_{ii}A_{kl}^{-1}]$ and $\rho = (s-t)'\bar{J}_{ii}(s-t)$.

The F_0 function is defined as

$$F_0(x) = \tfrac{1}{2}\sqrt{\tfrac{\pi}{x}}\,\mathrm{erf}(\sqrt{x})$$

2. *Non-Born–Oppenheimer Integrals over Correlated Gaussians*

Most of the integrals used in the non-BO calculations—the overlap integral and the potential energy integral—are similar to those presented above and will not be derived. A slight difference in the kinetic energy integral will be shown. In this case we find integrals over the operators in the non-BO Hamiltonian:

$$\hat{H} = -\nabla_r'\bar{M}\nabla_r + \sum_{i=1}^{n}\frac{q_0 q_i}{r_i} + \sum_{i<j}^{n}\frac{q_i q_j}{r_{ij}} \tag{136}$$

The kinetic energy integral is given by

$$T_{kl} = S_{kl}(4(s - s_k)'\bar{A}_k\bar{M}\bar{A}_l(s - s_l) + 6\mathrm{tr}[MA_lA_{kl}^{-1}A_k]) \tag{137}$$

This form differs from the Born–Oppenheimer form in that the constant $\frac{1}{2}$ (for electron masses) is replaced by the mass matrix, M (described above), but all other steps are similar.

3. *Electric Field Integrals*

The electric field term is most easily evaluated with one electric field component at a time. Taking the z component, we have

$$FT_{kl}^z = \langle g_k|\sum_{i=1}^{n}\mu_z^i\varepsilon_z|g_l\rangle$$

$$FT_{kl}^z = \langle g_k|\sum_{i=1}^{n}q_i r^{3i}\varepsilon_z|g_l\rangle$$

$$FT_{kl}^z = \sum_{i=1}^{n}\varepsilon_z q_i\langle g_k|r^{3i}|g_l\rangle \tag{138}$$

and the integral reduces to an evaluation of $\langle g_k|r_{3i}|g_l\rangle$. We will present the general case of this integral for all components, i:

$$r_{kl}^i = \langle g_k|r^i|g_l\rangle \tag{139}$$

This integral is solved by examining the expectation value of the operator:

$$o'E_{ii}\bar{A}_l^{-1}\nabla_r \tag{140}$$

where o is a vector of ones, and E_{ii} is a $3n \times 3n$ matrix as described above.

The gradient of the basis function is simply:

$$\nabla_r' g_l = g_l(2\bar{A}_l s_l - 2\bar{A}_l r) \tag{141}$$

so the expectation value of the operator is

$$\langle o'E_{ii}\bar{A}_l^{-1}\nabla_r\rangle = 2o'E_{ii}s_lS_{kl} - 2r_{kl}^i \tag{142}$$

Looking at the form of the basis function (2), it is obvious that

$$\nabla_r' g_l = -\nabla_{s_l}' g_l \tag{143}$$

so that

$$\langle o' E_{ii} \bar{A}_l^{-1} \nabla_r \rangle = -\langle o' E_{ii} \bar{A}_l^{-1} \nabla_{s_l} \rangle \tag{144}$$

The right-hand side of the above is easily evaluated by pulling the operator out of the integral:

$$\langle o' E_{ii} \bar{A}_l^{-1} \nabla_{s_l} \rangle = 2S_{kl}(o' E_{ii} s - o' E_{ii} s_l) \tag{145}$$

Setting one of the operators equal to the negative of the other, we get

$$r^i_{kl} = S_{kl} o' E_{ii} s = S_{kl} s(i) \tag{146}$$

4. *Energy Gradients*

The gradients of the molecular integrals with respect to the nonlinear variational parameters (i.e., the exponential parameters A_k and the orbital centers s_k) were derived using the methods of matrix differential calculus as introduced by Kinghorn [116]. It was shown there that the energy gradient with respect to all nonlinear variational parameters can be written as

$$\nabla_a E = \frac{1}{c' S c} \left(\frac{\partial \text{vech } H}{\partial a'} - E \frac{\partial \text{vech } S}{\partial a'} \right)' (\text{vech } [2cc' - \text{diag } cc']) \tag{147}$$

where a is a $m(n(n+1)/2 + 3n)$ vector of all the nonlinear parameters. Each function, g_k, contains the lower triangular matrix L_k with $n(n+1)/2$ nonzero elements and the $3n$ vector s_k. The vector a is made by stacking, function by function, the nonzero elements of L_k followed by the vector s_k. The matrices of derivatives $\partial\text{vech } H/\partial a'$ and $\partial\text{vech } S/\partial a'$ are sparse, since each row has at most $2(n(n+1)/2 + 3n)$ nonzero elements. The derivatives of the Hamiltonian are determined using the derivatives of the molecular integrals (Born–Oppenheimer):

$$\frac{\partial \text{vech } H}{\partial a'} = \frac{\partial \text{vech } T}{\partial a'} + \frac{\partial \text{vech } ER}{\partial a'} + \frac{\partial \text{vech } NR}{\partial a'} - \frac{\partial \text{vech } NA}{\partial a'} \tag{148}$$

where NR stands for the nuclear–nuclear repulsion energy, or (non-Born–Oppenheimer)

$$\frac{\partial \text{vech } H}{\partial a'} = \frac{\partial \text{vech } T}{\partial a'} + \frac{\partial \text{vech } V}{\partial a'} \tag{149}$$

The differential of the overlap integral with respect to s_k is straightforward,

$$\begin{aligned} d_{s_k} S_{kl} &= S_{kl} d_{s_k}(-s_k' \bar{A}_k s_k + e' \bar{A}_{kl}^{-1} e) \\ &= S_{kl}(-2s_k' \bar{A}_k ds_k + 2e' \bar{A}_{kl}^{-1} de) \\ &= S_{kl}(-2s_k' \bar{A}_k + 2s' \bar{A}_k) ds_k \\ &= 2S_{kl}((s - s_k)' \bar{A}_k) ds_k \end{aligned} \tag{150}$$

The kinetic energy differential is only slightly more complicated:

$$\begin{aligned} d_{s_k} T_{kl} &= \frac{T_{kl}}{S_{kl}} d_{s_k} S_{kl} + 2S_{kl} d_{s_k}((s - s_k)' \bar{A}_k \bar{A}_l (s - s_l)) \\ &= \frac{T_{kl}}{S_{kl}} d_{s_k} S_{kl} + 2S_{kl}((s - s_k)' \bar{A}_k \bar{A}_l ds + (s - s_l)' \bar{A}_k \bar{A}_l ds - (s - s_l)' \bar{A}_k \bar{A}_l \mathrm{d}s_k) \\ &= \frac{T_{kl}}{S_{kl}} d_{s_k} S_{kl} + 2S_{kl}((s - s_k) \bar{A}_k \bar{A}_l \bar{A}_{kl}^{-1} \bar{A}_k + (s - s_l) \bar{A}_k \bar{A}_l \bar{A}_{kl}^{-1} \bar{A}_k \\ &\quad - (s - s_l) \bar{A}_k \bar{A}_l)' ds_k \end{aligned} \tag{151}$$

In the general form of the potential energy integrals, only ρ has any dependence on s_k. The general differential is

$$d_{s_k} V_{kl} = \frac{V_{kl}}{S_{kl}} d_{s_k} S_{kl} + \frac{2}{\pi^{1/2}} S_{kl} \alpha^{-3/2} F_0' \left(\frac{\rho}{\alpha}\right) d\rho \tag{152}$$

For $V_{kl} = ER_{kl}^{ij}$:

$$d\rho = 2s' \bar{J}_{ij} \bar{A}_{kl}^{-1} \bar{A}_k ds_k \tag{153}$$

For $V_{kl} = NA_{kl}^{it}$:

$$d\rho = 2(s - t)' \bar{J}_{ij} \bar{A}_{kl}^{-1} \bar{A}_k ds_k \tag{154}$$

As shown in (4), the matrices A_k are actually products of triangular matrices $L_k L_k'$ and the derivatives reflect this. The derivatives we must find are

$$\frac{\partial O_{kl}}{\partial \text{vech } L_k}$$

In doing so, we make frequent use of the construct

$$v' \bar{M} v = \text{tr}[(v \cdot v) M] \tag{155}$$

to get rid of pesky Kronecker products.

The full differential of the overlap integral with respect to L_k is

$$d_k S_{kl} = 2^{3n}(d_k(p_{kl}^{3/2}) \exp(\gamma) + p_{kl}^{3/2} \exp(\gamma) \mathrm{d}_k \gamma) \tag{156}$$

We will examine first the part containing the determinants:

$$\begin{aligned}
d_k(p_{kl})^{3/2} &= \frac{3}{2}(p_{kl})^{1/2} d_k(p_{kl}) \\
&= \frac{3}{2}(p_{kl})^{1/2} \frac{1}{|A_{kl}|^2}\left(\frac{1}{2}|A_{kl}||A_l|^{1/2}|A_k|^{1/2}\mathrm{tr}[A_k^{-1}d_k(A_k)]\right. \\
&\quad \left. - |A_{kl}||A_l|^{1/2}|A_k|^{1/2}\mathrm{tr}[A_{kl}^{-1}d_k(A_k)]\right) \\
&= \frac{3}{2}(p_{kl})^{3/2}(\mathrm{tr}[L_k' A_k^{-1}\mathrm{d}L_k] - 2\mathrm{tr}[L_k' A_{kl}^{-1}dL_k]) \\
&= \frac{3}{2}(p_{kl})^{3/2}(\mathrm{vech}\,[A_k^{-1}L_k]' - 2\,\mathrm{vech}[A_{kl}^{-1}L_k]')\mathrm{vech}\,[dL_k].
\end{aligned} \tag{157}$$

Next, we need to find the differential of γ,

$$\begin{aligned}
d_k\gamma &= -d_k(s_k'\bar{A}_k s_k) + d_k(e'\bar{A}_{kl}^{-1}e) \\
&= -s_k' d(\bar{A}_k)s_k + 2e'\bar{A}_{kl}^{-1}d(\bar{A}_k)s_k - e'\bar{A}_{kl}^{-1}d(\bar{A}_k)\bar{A}_{kl}^{-1}e \\
&= -\mathrm{tr}[(s_k\cdot s_k)dA_k] + 2\mathrm{tr}[(s_k\cdot s)dA_k] - \mathrm{tr}[(s\cdot s)dA_k] \\
&= (2\mathrm{vech}\,[(s_k\cdot s)L_k] + 2\mathrm{vech}\,[(s_k\cdot s)'L_k] \\
&\quad - 2\mathrm{vech}\,[(s_k\cdot s_k)L_k] - 2\mathrm{vech}\,[(s\cdot s)L_k])'\mathrm{vech}\,[dL_k]
\end{aligned} \tag{158}$$

The differential of the kinetic energy integral is

$$d_kT_{kl} = \frac{T_{kl}}{S_{kl}}d_kS_{kl} + S_{kl}(2d_k((s-s_k)'\bar{A}_k\bar{A}_l(s-s_l)) + 3d_k\mathrm{tr}[A_kA_{kl}^{-1}A_l]) \tag{159}$$

Let us look at the first term in the parentheses:

$$\begin{aligned}
d_k((s-s_k)'\bar{A}_k\bar{A}_l(s-s_l)) &= d_k(s-s_k)'\bar{A}_k\bar{A}_l(s-s_l) \\
&\quad + (s-s_k)'d_k(\bar{A}_k)\bar{A}_l(s-s_l) + (s-s_k)'\bar{A}_k\bar{A}_l d_k(s-s_l) \\
&= -s'd_k(\bar{A}_k)\bar{A}_{kl}^{-1}\bar{A}_k\bar{A}_l(s-s_l) + s_k'd_k(\bar{A}_k)\bar{A}_{kl}^{-1}\bar{A}_k\bar{A}_l(s-s_l) \\
&\quad + (s-s_k)'d_k(\bar{A}_k)\bar{A}_l(s-s_l) - (s-s_k)'\bar{A}_k\bar{A}_l\bar{A}_{kl}^{-1}d_k(\bar{A}_k)s \\
&\quad + (s-s_k)'\bar{A}_k\bar{A}_l\bar{A}_{kl}^{-1}d_k(\bar{A}_k)s_k \\
&= -\mathrm{tr}[A_{kl}^{-1}A_kA_l((s-s_l)\cdot s)d_k(A_k)] + \mathrm{tr}[A_{kl}^{-1}A_kA_l((s-s_l)\cdot s_k)d_k(A_k)] \\
&\quad + \mathrm{tr}[A_l((s-s_l)\cdot(s-s_k))d_k(A_k)] - \mathrm{tr}[(s\cdot(s-s_k))A_kA_lA_{kl}^{-1}d_k(A_k)] \\
&\quad + \mathrm{tr}[(s_k\cdot(s-s_k))A_kA_lA_{kl}^{-1}d_k(A_k)] \\
&= (-\mathrm{vech}\,[A_{kl}^{-1}A_kA_l((s-s_l)\cdot s)L_k] - \mathrm{vech}\,[(A_{kl}^{-1}A_kA_l((s-s_l)\cdot s))'L_k] \\
&\quad + \mathrm{vech}\,[A_{kl}^{-1}A_kA_l((s-s_l)\cdot s_k)L_k] + \mathrm{vech}\,[(A_{kl}^{-1}A_kA_l((s-s_l)\cdot s_k))'L_k] \\
&\quad + \mathrm{vech}\,[A_l((s-s_l)\cdot(s-s_k))L_k] + \mathrm{vech}\,[(A_l((s-s_l)\cdot(s-s_k)))'L_k] \\
&\quad - \mathrm{vech}\,[(s\cdot(s-s_k))A_kA_lA_{kl}^{-1}L_k] - \mathrm{vech}\,[((s\cdot(s-s_k))A_kA_lA_{kl}^{-1})'L_k] \\
&\quad + \mathrm{vech}\,[(s_k\cdot(s-s_k))A_kA_lA_{kl}^{-1}L_k] \\
&\quad + \mathrm{vech}\,[((s_k\cdot(s-s_k))A_kA_lA_{kl}^{-1})'L_k])'\mathrm{vech}\,[dL_k]
\end{aligned} \tag{160}$$

And finally let's look at the second term in the parentheses:

$$\begin{aligned} d_k \text{tr}[A_k A_{kl}^{-1} A_l] &= \text{tr}[d_k A_k A_{kl}^{-1} A_l] - \text{tr}[A_k A_{kl}^{-1} d_k A_k A_{kl}^{-1} A_l] \\ &= \text{tr}[A_{kl}^{-1} A_l d_k A_k] - \text{tr}[A_{kl}^{-1} A_l A_k A_{kl}^{-1} \text{d}_k A_k] \\ &= (\text{vech}[(A_{kl}^{-1} A_l + A_l A_{kl}^{-1}) L_k] \\ &\quad - \text{vech}[(A_{kl}^{-1} A_l A_k A_{kl}^{-1} + A_{kl}^{-1} A_k A_l A_{kl}^{-1}) L_k])' \text{vech}[\text{d}L_k] \end{aligned} \quad (161)$$

For evaluation of the derivatives of the potential terms, we will refer again to the general form of the potential energy and the definitions of ρ and α. The general form of the differential is

$$d_k V_{kl} = \frac{V_{kl}}{S_{kl}} d_k S_{kl} - \frac{1}{\pi^{1/2}} S_{kl} \alpha^{-3/2} F_0\left(\frac{\rho}{\alpha}\right) d\alpha + \frac{2}{\pi^{1/2}} S_{kl} \alpha^{-1/2} F_0'\left(\frac{\rho}{\alpha}\right)\left(\frac{\alpha d_k \rho - \rho d_k \alpha}{\alpha^2}\right) \quad (162)$$

So in both cases it is only nessesary to find the derivatives of ρ and α. For $V_{kl} = ER_{kl}^{ij}$ we have

$$\begin{aligned} d_k \rho &= 2s' \bar{J}_{ij} d_k s = 2s' \bar{J}_{ij} \text{d}_k (\bar{A}_{kl}^{-1}) e + 2s' \bar{J}_{ij} \bar{A}_{kl}^{-1} d_k (\bar{A}_k) s_k \\ &= -2s' \bar{J}_{ij} \bar{A}_{kl}^{-1} d_k (\bar{A}_k) s + 2s' \bar{J}_{ij} \bar{A}_{kl}^{-1} d_k (\bar{A}_k) s_k \\ &= -2\text{tr}[(s \cdot s) J_{ij} A_{kl}^{-1} d_k (A_k)] + 2\text{tr}[(s_k \cdot s) J_{ij} A_{kl}^{-1} d_k (A_k)] \\ &= (2\text{vech}\ [(s_k \cdot s) J_{ij} A_{kl}^{-1} L_k] + 2\text{vech}\ [((s_k \cdot s) J_{ij} A_{kl}^{-1})' L_k] \\ &\quad - 2\text{vech}[((s \cdot s) J_{ij} A_{kl}^{-1} + A_{kl}^{-1} J_{ij} (s.s)) L_k])' \text{vech}[dL_k] \end{aligned} \quad (163)$$

and

$$\begin{aligned} d_k \alpha &= d_k \text{tr}[J_{ij} A_{kl}^{-1}] = -\text{tr}[J_{ij} A_{kl}^{-1} d_k (A_k) A_{kl}^{-1}] \\ &= -\text{tr}[A_{kl}^{-1} J_{ij} A_{kl}^{-1} d_k (A_k)] = -2\text{vech}\ [A_{kl}^{-1} J_{ij} A_{kl}^{-1} L_k]' \text{vech}[dL_k] \end{aligned} \quad (164)$$

So, for $V_{kl} = NA_{kl}^{it}$ we obtain

$$\begin{aligned} d_k \rho &= 2(s-t)' \bar{J}_{ii} d_k s = 2(s-t)' \bar{J}_{ii} \text{d}_k (\bar{A}_{kl}^{-1}) e + 2(s-t)' \bar{J}_{ii} \bar{A}_{kl}^{-1} d_k (\bar{A}_k) s_k \\ &= -2(s-t)' \bar{J}_{ii} \bar{A}_{kl}^{-1} d_k (\bar{A}_k) s + 2(s-t)' \bar{J}_{ii} \bar{A}_{kl}^{-1} d_k (\bar{A}_k) s_k \\ &= -2\text{tr}[(s \cdot (s-t)) J_{ii} A_{kl}^{-1} d_k (A_k)] + 2\text{tr}[(s_k \cdot (s-t)) J_{ii} A_{kl}^{-1} d_k (A_k)] \\ &= (2\text{vech}\ [(s_k \cdot (s-t)) J_{ii} A_{kl}^{-1} L_k] + 2\text{vech}\ [((s_k \cdot (s-t)) J_{ii} A_{kl}^{-1})' L_k] \\ &\quad - 2\text{vech}\ [(s \cdot (s-t)) J_{ii} A_{kl}^{-1} L_k] - 2\text{vech}\ [((s \cdot (s-t)) J_{ii} A_{kl}^{-1})' L_k])' \text{vech}[dL_k] \end{aligned} \quad (165)$$

and

$$\begin{aligned} d_k\alpha &= d_k\mathrm{tr}[J_{ii}A_{kl}^{-1}] = -\mathrm{tr}[J_{ii}A_{kl}^{-1}\mathrm{d}_k(A_k)A_{kl}^{-1}] \\ &= -\mathrm{tr}[A_{kl}^{-1}J_{ii}A_{kl}^{-1}d_k(A_k)] = -2\mathrm{vech}\ [A_{kl}^{-1}J_{ii}A_{kl}^{-1}L_k]'\mathrm{vech}\ [dL_k] \end{aligned} \tag{166}$$

which completes the gradient derivation.

5. *Non-Born–Oppenheimer Energy Gradients*

Again, these gradients differ only in the kinetic energy differential from the Born–Oppenheimer form:

$$\begin{aligned} d_{s_k}T_{kl} = \frac{T_{kl}}{S_{kl}}d_{s_k}S_{kl} &+ 4S_{kl}((s-s_k)\bar{A}_k\bar{M}\bar{A}_l\bar{A}_{kl}^{-1}\bar{A}_k \\ &+ (s-s_l)\bar{A}_k\bar{M}\bar{A}_l\bar{A}_{kl}^{-1}\bar{A}_k - (s-s_l)\bar{A}_k\bar{M}\bar{A}_l)'ds_k \end{aligned} \tag{167}$$

The L_k differential of the kinetic energy integral, with M instead of $\frac{1}{2}$, is

$$d_kT_{kl} = \frac{T_{kl}}{S_{kl}}d_kS_{kl} + S_{kl}(4d_k((s-s_k)'\bar{A}_k\bar{M}\bar{A}_l(s-s_l)) + 6d_k\mathrm{tr}[MA_lA_{kl}^{-1}A_k]) \tag{168}$$

Let us look at the first term in the parentheses:

$$\begin{aligned} d_k((s-s_k)'\bar{A}_k\bar{A}_l(s-s_l)) &= (-\mathrm{vech}[A_{kl}^{-1}A_kMA_l((s-s_l)\cdot s)L_k] \\ &- \mathrm{vech}[(A_{kl}^{-1}A_kMA_l((s-s_l)\cdot s))'L_k] \\ &+ \mathrm{vech}[A_{kl}^{-1}A_kMA_l((s-s_l)\cdot s_k)L_k] + \mathrm{vech}[(A_{kl}^{-1}A_kMA_l((s-s_l)\cdot s_k))'L_k] \\ &+ \mathrm{vech}[MA_l((s-s_l)\cdot(s-s_k))L_k] + \mathrm{vech}[(MA_l((s-s_l)\cdot(s-s_k)))'L_k] \\ &- \mathrm{vech}[(s\cdot(s-s_k))A_kMA_lA_{kl}^{-1}L_k] - \mathrm{vech}[((s\cdot(s-s_k))A_kMA_lA_{kl}^{-1})'L_k] \\ &+ \mathrm{vech}[(s_k\cdot(s-s_k))A_kMA_lA_{kl}^{-1}L_k] \\ &+ \mathrm{vech}[((s_k\cdot(s-s_k))A_kMA_lA_{kl}^{-1})'L_k])'\mathrm{vech}[dL_k] \end{aligned} \tag{169}$$

And finally let's look at the second term in the parentheses:

$$\begin{aligned} d_k\mathrm{tr}[MA_lA_{kl}^{-1}A_k] &= \mathrm{tr}[d_kA_kA_{kl}^{-1}MA_l] - \mathrm{tr}[A_kA_{kl}^{-1}\mathrm{d}_kA_kA_{kl}^{-1}MA_l] \\ &= \mathrm{tr}[A_{kl}^{-1}MA_ld_kA_k] - \mathrm{tr}[A_{kl}^{-1}MA_lA_kA_{kl}^{-1}\mathrm{d}_kA_k] \\ &= (2\mathrm{vech}[A_{kl}^{-1}MA_lL_k] - 2\mathrm{vech}[A_{kl}^{-1}MA_lA_kA_{kl}^{-1}L_k])'\mathrm{vech}[\mathrm{d}L_k] \end{aligned} \tag{170}$$

6. *Energy Gradients for the Electric Field Term*

The differential of the electric field term follows easily from the definition of s:

$$d_{s_k} FT^z_{kl} = \frac{FT^z_{kl}}{S_{kl}} d_{s_k} S_{kl} + S_{kl}(o' E_{ii} \bar{A}^{-1}_{kl} \bar{A}_k ds_k) \tag{171}$$

For the L_k differential, the general form is

$$d_k FT^z_{kl} = \frac{FT^z_{kl}}{S_{kl}} d_k S_{kl} + S_{kl} d(o' E_{ii} s) \tag{172}$$

The only new derivative needed is the leftmost term on the left-hand side:

$$\begin{aligned} d(o' E_{ii} s) &= o' E_{ii} ds \\ &= o' E_{ii} d(\bar{A}^{-1}_{kl} e) \\ &= -o' E_{ii} (\bar{A}^{-1}_{kl} d\bar{A}_k \bar{A}^{-1}_{kl} e - \bar{A}^{-1}_{kl} de) \\ &= -o' E_{ii} (\bar{A}^{-1}_{kl} d\bar{A}_k s - \bar{A}^{-1}_{kl} \mathrm{d}\bar{A}_k s_k) \end{aligned} \tag{173}$$

At this point it is useful to define two vectors: $v1 = E_{ii} o$ and $v2 = s_k - s$. Using these, the last equation becomes

$$\begin{aligned} d(o' E_{ii} s) &= \mathrm{tr}[(v2 \cdot v1) A^{-1}_{kl} dA_k] \\ &= \mathrm{tr}[(v2 \cdot v1) A^{-1}_{kl} (dL_k L'_k + L_k dL'_k)] \\ &= \mathrm{tr}[L'_k (v2 \cdot v1) A^{-1}_{kl} dL_k] + \mathrm{tr}[(v2 \cdot v1) A^{-1}_{kl} L_k dL'_k)] \\ &= \mathrm{tr}[L'_k (v2 \cdot v1) A^{-1}_{kl} dL_k] + \mathrm{tr}[L'_k A^{-1}_{kl} (v2 \cdot v1)' dL_k)] \\ &= (\mathrm{vech}[A^{-1}_{kl} (v2 \cdot v1)' L_k] + \mathrm{vech}[(v2 \cdot v1) A^{-1}_{kl} L_k])' \mathrm{vech}[dL_k] \end{aligned} \tag{174}$$

7. *Geometry Optimization Gradients*

When looking at Born-Oppenhiemer calculations, we find that additional gradients are needed in geometry optimization. The gradient of the energy with respect to nuclear coordinates is given by

$$\nabla_{r_N} E = \frac{1}{c' Sc} \left(\frac{\partial \mathrm{vech} H}{\partial r'_N} \right)' (\mathrm{vech}[2cc' - \mathrm{diag} cc']). \tag{175}$$

The derivatives needed are

$$\frac{\partial \mathrm{vech}\ H}{\partial r'_{N.}} = \frac{\partial \mathrm{vech}\ NR}{\partial r'_N} - \frac{\partial \mathrm{vech}\ NA}{\partial r'_N} \tag{176}$$

where NA stands for nuclear attraction energy and NR stands for nuclear repulsion energy. Thus we need to calculate the elements:

$$\frac{\partial NA_{kl}}{\partial r'_N} = \sum_{i=1}^{n} \sum_{t=1}^{N} \frac{\partial NA_{kl}^{it}}{\partial r'_N} \tag{177}$$

and

$$\frac{\partial NR_{kl}}{\partial r'_N} = \sum_{u>t} \frac{\partial NR_{kl}^{tu}}{\partial r'_N} \tag{178}$$

For the NA integral we cannot differentiate directly with respect to r_N, so we must differentiate with respect to t, since

$$\frac{\partial NA_{kl}^{it}}{\partial r'_{N_t}} = \frac{\partial NA_{kl}^{it}}{\partial t'} \tag{179}$$

where r_{N_t} is the three-element partition of r_N pertaining to the tth nucleus and it is understood that we include only the three nonzero elements of the vector $\partial NA_{kl}^{it}/\partial t'$ in the equality.

This is then easily differentiated using the chain rule:

$$dNA_{kl}^{it} = -\frac{4}{\pi^{1/2}} S_{kl} (\mathrm{tr}[J_{ii}A_{kl}])^{-3/2} F'_0 \left(\frac{(s-t)'\bar{J}_{ii}(s-t)}{\mathrm{tr}[J_{ii}A_{kl}]} \right) (s-t)'\bar{J}_{ii}\, dt \tag{180}$$

The elements of NR can be easily differentiated directly on r_N:

$$dNR_{kl}^{tu} = \frac{Z_t Z_u S_{kl}}{(r'_N \bar{J}_{tu} r_N)^{3/2}} r'_N \bar{J}_{tu} d(r_N) \tag{181}$$

8. *Expectation Values*

For non-Born–Oppenheimer calculations we have no molecular structure, per se, since all particles including nuclei are treated quantum mechanically. Thus the only information we can obtain about the structure of these molecules are the expectation values of distances between particles (r_{ij}) and powers of these distances. We present below the integrals needed in the calculation of these expectation values.

In order to find the r_{ij} integrals, we use the substitution

$$\frac{1}{r_{ij}} = \frac{2}{\pi^{1/2}} \int_0^\infty \exp(-u^2 r_{ij}^2)\, du \tag{182}$$

Thus for the integral we have

$$\begin{aligned}\langle g_k|r_{ij}|g_l\rangle &= r_{ij}^{kl} \\ r_{ij}^{kl} &= \langle g_k|\frac{r_{ij}^2}{r_{ij}}|g_l\rangle \\ r_{ij}^{kl} &= \frac{2}{(\pi)^{1/2}}\frac{\partial}{\partial a}\int_0^\infty du\frac{-1}{u^2}\langle g_k|\exp(-au^2r'\bar{J}_{ij}r)|g_l\rangle|_{a=1}\end{aligned} \tag{183}$$

Expanding the basis functions, we have

$$r_{ij}^{kl} = \frac{2}{(\pi)^{1/2}}\exp(-\gamma - s'\bar{A}_{kl}s)\frac{\partial}{\partial a}\int_{-\infty}^\infty dr\int_0^\infty du\,\frac{1}{u^2}\exp[-r'(\bar{A}_{kl}+au^2\bar{J}_{ij})r+2s'\bar{A}_{kl}r]|_{a=1} \tag{184}$$

Applying the formula

$$\int_{-\infty}^\infty \exp(-r'Mr + iz'r)\,dr = \pi^{n/2}|M|^{-1/2}\exp(-z'M^{-1}z/4) \tag{185}$$

where M is an $n \times n$ matrix, and r and z are vectors of length n, we find the integral over r:

$$\begin{aligned}r_{ij}^{kl} &= \frac{2}{(\pi)^{1/2}}\exp(-\gamma - s'\bar{A}_{kl}s)\frac{\partial}{\partial a}\int_0^\infty du\frac{\pi^{3n/2}}{u^2}|\bar{A}_{kl} + au^2\bar{J}_{ij}|^{-1/2} \\ &\times \exp[s'\bar{A}_{kl}(\bar{A}_{kl} + au^2\bar{J}_{ij})^{-1}\bar{A}_{kl}s + 2s'\bar{A}_{kl}r]|_{a=1}\end{aligned} \tag{186}$$

which, upon further simplification, becomes

$$r_{ij}^{kl} = S_{kl}\frac{2}{(\pi)^{1/2}}\frac{\partial}{\partial a}\int_0^\infty du\,\frac{1}{u^2}\left(1 + au^2\mathrm{tr}[J_{ij}A_{kl}^{-1}]\right)^{-3/2}\exp\left(\frac{-au^2s'\bar{J}_{ij}s}{1 + au^2\mathrm{tr}[J_{ij}A_{kl}^{-1}]}\right)\Bigg|_{a=1} \tag{187}$$

At this point we define $\alpha = \mathrm{tr}[J_{ij}A_{kl}^{-1}]$ and $\beta = s'\bar{J}_{ij}s$. Using these definitions and differentiating with respect to a, we have

$$r_{ij}^{kl} = S_{kl}\frac{2}{(\pi)^{1/2}}\int_0^\infty du\left[\frac{3\alpha}{2}(1+au^2\alpha)^{-5/2} + \beta(1+au^2\alpha)^{-7/2}\right]\exp\left(\frac{-au^2\beta}{1+au^2\alpha}\right)\Bigg|_{a=1} \tag{188}$$

These two integrals may be solved by clever substitution and use of the integrals:

$$\int_0^1 \exp(-bt^2)\,dt = \frac{1}{2}\left(\frac{\pi}{b}\right)^{1/2} \operatorname{erf}(b^{1/2}) \tag{189}$$

$$\int_0^1 t^2 \exp(-bt^2)\,dt = \frac{1}{4}\left(\frac{\pi}{b^3}\right)^{1/2} \operatorname{erf}(b^{1/2}) - \frac{1}{2b}\exp(-b) \tag{190}$$

and

$$\int_0^1 t^4 \exp(-bt^2)\,dt = \frac{3}{8}\left(\frac{\pi}{b^5}\right)^{1/2} \operatorname{erf}(b^{1/2}) - \frac{1}{2b}\exp(-b) - \frac{3}{4b^2}\exp(-b) \tag{191}$$

After simplification, we have

$$r_{ij}^{kl} = S_{kl}\left\{\left(\frac{1\alpha}{2\beta^{1/2}} - \beta^{1/2}\right)\operatorname{erf}\left[\left(\frac{\beta}{\alpha}\right)^{1/2}\right]\right\} - \left(\frac{\alpha}{\pi}\right)^{1/2}\exp\left(-\frac{\beta}{\alpha}\right) \tag{192}$$

The integrals for $(r^2)_{ij}^{kl}$ may be obtained as a subset of the above integral by setting it up as

$$(r^2)_{ij}^{kl} = \frac{\partial}{\partial u}\langle g_k|\exp(-ur'\bar{J}_{ij}r)|g_l\rangle|_{u=0} \tag{193}$$

With relative ease we obtain

$$(r^2)_{ij}^{kl} = S_{kl}\left(\frac{3}{2}\alpha - \beta\right) \tag{194}$$

9. *One-Particle Densities*

Here we will present the formulae needed for calculating the reduced one-particle density matrices from the floating correlated Gaussians used in this work. The first-order density matrix for wave function $\Psi(\mathbf{r}_1, \mathbf{r}_2, \ldots, \mathbf{r}_n)$ for particle 1 is defined as

$$P(\mathbf{r}_1, \mathbf{r}_1') = \int \Psi^*(\mathbf{r}_1, \mathbf{r}_2, \ldots, \mathbf{r}_n)\Psi(\mathbf{r}_1', \mathbf{r}_2, \ldots, \mathbf{r}_n)d\mathbf{r}_2 \cdots d\mathbf{r}_n \tag{195}$$

Since we are using the expansion

$$\Psi(\mathbf{r}) = \sum_{k=1}^{M} c_k \hat{Y}\phi_k(\mathbf{r}) \tag{196}$$

we need to only find the matrix elements:

$$P_{kl}(\mathbf{r}_1, \mathbf{r}'_1) = \int \phi_k(\mathbf{r}_1, \mathbf{r}_2, \ldots, \mathbf{r}_n)\phi_l(\mathbf{r}'_1, \mathbf{r}_2, \ldots, \mathbf{r}_n)d\mathbf{r}_2 \cdots d\mathbf{r}_n \tag{197}$$

for basis functions ϕ_k and ϕ_l.

We will follow the derivations presented in the work by Poshusta and Kinghorn [104] for the single-center Gaussians. The present work differs from theirs in that we use multicenter Gaussians to shift density away from the origin of the coordinate system. A definition used in the integral formulas is

$$\bar{\mathbf{M}} = \mathbf{M} \otimes \mathbf{I}_3$$

where $\mathbf{M}$ is a matrix and $\mathbf{I}_3$ is the 3×3 identity. As in the previous work [104] and for the convenience of the presentation, we define the following augmented coordinate vectors:

$$\mathbf{r}^+ = \begin{pmatrix} \mathbf{r}'_1 \\ \mathbf{r} \end{pmatrix} = \begin{pmatrix} \mathbf{u} \\ \mathbf{v} \end{pmatrix} \tag{198}$$

where

$$\mathbf{u} = \begin{pmatrix} \mathbf{r}'_1 \\ \mathbf{r}_1 \end{pmatrix}, \qquad \mathbf{v} = \begin{pmatrix} \mathbf{r}_2 \\ \mathbf{r}_3 \\ \cdots \\ \mathbf{r}_n \end{pmatrix} \tag{199}$$

These vectors allow us to separate the coordinates $\mathbf{r}_1$ and $\mathbf{r}'_1$ from the coordinates over which the integration is performed ($\mathbf{v}$) in the density calculation. Likewise, the vector of the shifts needs to be augmented to match the dimension of $\mathbf{r}^+$:

$$\mathbf{s}_k^+ = \begin{pmatrix} 0 \\ 0 \\ 0 \\ \mathbf{s}_k \end{pmatrix} \tag{200}$$

We will also need

$$\bar{\mathbf{A}}_k^+ = \begin{pmatrix} \bar{0} & \bar{0} \\ \bar{0} & \bar{\mathbf{A}}_k \end{pmatrix} \tag{201}$$

and

$$\bar{\prod} = \begin{pmatrix} \bar{0} & \bar{1} & \bar{0} \\ \bar{1} & \bar{0} & \bar{0} \\ \bar{0} & \bar{0} & \bar{\mathbf{I}}_{n-1} \end{pmatrix} \tag{202}$$

where $\bar{\mathbf{I}}_{n-1}$ is the $(n-1) \times (n-1)$ identity matrix. Using the above definitions, we can formally represent $\phi_k(\mathbf{r}_1, \mathbf{r}_2, \ldots, \mathbf{r}_n)$ and $\phi_k(\mathbf{r}'_1, \mathbf{r}_2, \ldots, \mathbf{r}_n)$ as

$$\phi_k(\mathbf{r}_1, \mathbf{r}_2, \cdots \mathbf{r}_n) = \exp\left[-(\mathbf{r}^+ - \mathbf{s}_k^+)'\bar{\mathbf{A}}_k^+(\mathbf{r}^+ - \mathbf{s}_k^+)\right] \tag{203}$$

and

$$\phi_l(\mathbf{r}'_1, \mathbf{r}_2, \cdots \mathbf{r}_n) = \exp\left[-\left(\mathbf{r}^+ - \bar{\prod}\mathbf{s}_l^+\right)'\left(\bar{\prod}\bar{\mathbf{A}}_l^+\bar{\prod}\right)\left(\mathbf{r}^+ - \bar{\prod}\mathbf{s}_l^+\right)\right] \tag{204}$$

The product of the Gaussians in the integrand of P_{kl} may then be written (using the abbreviations $\bar{\Pi}\bar{\mathbf{A}}_l^+\bar{\Pi} = \bar{\mathbf{A}}_l^{+p}$ and $\bar{\Pi}\mathbf{s}_l^+ = \mathbf{s}_l^{+p}$) as

$$\begin{aligned} &\phi_k(\mathbf{r}_1, \mathbf{r}_2, \ldots, \mathbf{r}_n)\phi_l(\mathbf{r}'_1, \mathbf{r}_2, \ldots, \mathbf{r}_n) \\ &\quad = \exp[-\mathbf{s}'_k\bar{\mathbf{A}}_k\mathbf{s}_k - \mathbf{s}'_l\bar{\mathbf{A}}_l\mathbf{s}_l - \mathbf{r}^{+\prime}(\bar{\mathbf{A}}_k^+ + \bar{\mathbf{A}}_l^{+p})\mathbf{r}^+ + 2\mathbf{r}^{+\prime}(\bar{\mathbf{A}}_k^+\mathbf{s}_k^+ + \bar{\mathbf{A}}_l^{+p}\mathbf{s}_l^{+p})] \end{aligned} \tag{205}$$

Using the definitions

$$\bar{\mathbf{Z}}^{kl} = \bar{\mathbf{A}}_k^+ + \bar{\mathbf{A}}_l^{+p} \tag{206}$$

and

$$\mathbf{z}^+ = \bar{\mathbf{A}}_k^+\mathbf{s}_k^+ + \bar{\mathbf{A}}_l^{+p}\mathbf{s}_l^{+p} \tag{207}$$

we can rewriting the above expression as

$$\phi_k(\mathbf{r}_1, \mathbf{r}_2, \ldots, \mathbf{r}_n)\phi_l(\mathbf{r}'_1, \mathbf{r}_2, \ldots, \mathbf{r}_n) = \exp[-\mathbf{s}'_k\bar{\mathbf{A}}_k\mathbf{s}_k - \mathbf{s}'_l\bar{\mathbf{A}}_l\mathbf{s}_l - \mathbf{r}^{+\prime}\bar{\mathbf{Z}}^{kl}\mathbf{r}^+ + 2\mathbf{r}^{+\prime}\mathbf{z}^+] \tag{208}$$

The last two terms in the exponent above can be expanded using the augmented matrices defined above:

$$\begin{aligned} &-\mathbf{r}^{+\prime}(\bar{\mathbf{A}}_k^+ + \bar{\mathbf{A}}_l^{+p})\mathbf{r}^+ + 2\mathbf{r}^{+\prime}(\bar{\mathbf{A}}_k^+\mathbf{s}_k^+ + \bar{\mathbf{A}}_l^{+p}\mathbf{s}_l^{+p}) \\ &= \exp\left[-(\mathbf{uv})\begin{pmatrix} \bar{\mathbf{Z}}_u^{kl} & \bar{\mathbf{Z}}_{uv}^{kl} \\ \bar{\mathbf{Z}}_{vu}^{kl} & \bar{\mathbf{Z}}_v^{kl} \end{pmatrix}\begin{pmatrix} \mathbf{u} \\ \mathbf{v} \end{pmatrix} + 2(\mathbf{uv})\begin{pmatrix} \mathbf{z}^{+u} \\ \mathbf{z}^{+v} \end{pmatrix}\right] \end{aligned} \tag{209}$$

where $\bar{\mathbf{Z}}_u^{kl}$ is a 6×6 matrix , $\bar{\mathbf{Z}}_v^{kl}$ is a $(n-1) \times (n-1)$ matrix, and $\mathbf{z}^{+u}$ is a vector of length 6, while $\mathbf{z}^{+v}$ is a vector of length $(3n-3)$. Expanding the above expression again we get

$$\exp[-\mathbf{u}'\bar{\mathbf{Z}}_u^{kl}\mathbf{u} - \mathbf{v}'\bar{\mathbf{Z}}_v^{kl}\mathbf{v} - 2\mathbf{u}'\bar{\mathbf{Z}}_{uv}^{kl}\mathbf{v} + 2\mathbf{u}'\mathbf{z}^{+u} + 2\mathbf{v}'\mathbf{z}^{+v}] \tag{210}$$

Now, after substituting the above expression into the original expression, we get

$$\begin{aligned}&\phi_k(\mathbf{r}_1,\mathbf{r}_2,\ldots,\mathbf{r}_n)\phi_l(\mathbf{r}_1',\mathbf{r}_2,\ldots,\mathbf{r}_n)\\&\quad=\exp[-\mathbf{s}_k'\bar{\mathbf{A}}_k\mathbf{s}_k-\mathbf{s}_l'\bar{\mathbf{A}}_l\mathbf{s}_l-\mathbf{u}'\bar{\mathbf{Z}}_u^{kl}\mathbf{u}+2\mathbf{u}'\mathbf{z}^{+u}]\times\exp[-\mathbf{v}'\bar{\mathbf{Z}}_v^{kl}\mathbf{v}-2\mathbf{u}'\bar{\mathbf{Z}}_{uv}^{kl}\mathbf{v}+2\mathbf{v}'\mathbf{z}^{+v}]\end{aligned} \tag{211}$$

We can now integrate over the variables in the $\mathbf{v}$ vector and we get

$$\begin{aligned}P_{kl}(\mathbf{r}_1,\mathbf{r}_1')=&\,\pi^{3(n-1)/2}|\bar{\mathbf{Z}}_v^{kl}|^{-3/2}\exp[-\mathbf{s}_k'\bar{\mathbf{A}}_k\mathbf{s}_k-\mathbf{s}_l'\bar{\mathbf{A}}_l\mathbf{s}_l-\mathbf{u}'\bar{\mathbf{Z}}_u^{kl}\mathbf{u}+2\mathbf{u}'\mathbf{z}^{+u}]\\&\times\exp[-\mathbf{u}'\bar{\mathbf{Z}}_{uv}^{kl}(\bar{\mathbf{Z}}_v^{kl})^{-1}\bar{\mathbf{Z}}_{vu}^{kl}\mathbf{u}-2\mathbf{u}'\bar{\mathbf{Z}}_{uv}^{kl}(\bar{\mathbf{Z}}_v^{kl})^{-1}\mathbf{z}^{+v}+\mathbf{z}^{+v\prime}(\bar{\mathbf{Z}}_v^{kl})^{-1}\mathbf{z}^{+v}]\end{aligned} \tag{212}$$

By grouping all the terms that do not include $\mathbf{u}$, we can define the following "augmented" overlap matrix element:

$$S_{kl}^{+} = \pi^{3(n-1)/2}|\bar{\mathbf{Z}}_v^{kl}|^{-3/2}\exp[-\mathbf{s}_k'\bar{\mathbf{A}}_k\mathbf{s}_k - \mathbf{s}_l'\bar{\mathbf{A}}_l\mathbf{s}_l + \mathbf{z}^{+v\prime}(\bar{\mathbf{Z}}_v^{kl})^{-1}\mathbf{z}^{+v}] \tag{213}$$

Using this quantity, we get

$$P_{kl}(\mathbf{r}_1,\mathbf{r}_1') = S_{kl}^{+} \times \exp[-\mathbf{u}'(\bar{\mathbf{Z}}_u^{kl} - \bar{\mathbf{Z}}_{uv}^{kl}(\bar{\mathbf{Z}}_v^{kl})^{-1}\bar{\mathbf{Z}}_{vu}^{kl})\mathbf{u} + 2\mathbf{u}'(\mathbf{z}^{+u} - \bar{\mathbf{Z}}_{uv}^{kl}(\bar{\mathbf{Z}}_v^{kl})^{-1}\mathbf{z}^{+v})] \tag{214}$$

Now we define two new matrices:

$$\bar{\mathbf{W}} = \bar{\mathbf{Z}}_u^{kl} - \bar{\mathbf{Z}}_{uv}^{kl}(\bar{\mathbf{Z}}_v^{kl})^{-1}\bar{\mathbf{Z}}_{vu}^{kl} \tag{215}$$

and

$$\mathbf{d} = \mathbf{z}^{+u} - \bar{\mathbf{Z}}_{uv}^{kl}(\bar{\mathbf{Z}}_v^{kl})^{-1}\mathbf{z}^{+v} \tag{216}$$

and we arrive at the following final expression for $P_{kl}(\mathbf{r}_1,\mathbf{r}_1')$:

$$P_{kl}(\mathbf{r}_1,\mathbf{r}_1') = S_{kl}^{+} \times \exp[-\mathbf{u}'\bar{\mathbf{W}}\mathbf{u} + 2\mathbf{u}'\mathbf{d}] \tag{217}$$

Thus the density matrix elements take on the familiar form of a Gaussian with a shifted center. We can now find the one-particle density matrix element by setting $\mathbf{r}_1 = \mathbf{r}'_1 \equiv \begin{pmatrix} x_1 \\ y_1 \\ z_1 \end{pmatrix}$:

$$D_{kl}(\mathbf{r_1}) = P_{kl}(\mathbf{r}_1, \mathbf{r}_1) = S^{+}_{kl} \times \exp\left[-(W_{11} + W_{22} + 2W_{12})(x_1^2 + y_1^2 + z_1^2) + 2\begin{pmatrix} \mathbf{r}_1 \\ \mathbf{r}_1 \end{pmatrix}' \mathbf{d}\right] \tag{218}$$

This last expression was used in the code to calculate the densities. As derived, the formulae apply to the density of pseudoparticle 1 (and all the particles identical to it). As mentioned, the density of other particles may be obtained by permuting the desired particle to position 1 in the basis functions.

B. Spherically Symmetric Molecules

In the standard language of chemistry we describe a molecular geometry with bond lengths and bond angles. In the true quantum-mechanical picture of a molecule, though, at best we can know average values of the distances between particles in our system. We will call these two viewpoints the chemical and the physical picture, respectively, following the designations of Monkhorst [55, 56]. It should be noted that we basically arrive at the chemical picture by making the adiabatic and BO approximations in the physical picture. In the chemical picture, we may solve the electronic problem for a fixed molecular geometry—that is, for fixed bond lengths—or we may optimize the nuclear geometry in the average field of the electrons and find fictitious equilibrium bond lengths. This bond length, for a linear molecule, corresponds to the very bottom of the potential energy curve; likewise, for polyatomic molecules, it corresponds to the very bottoms of potential energy surfaces (PES's). At no point does the molecule ever reside in this point in the PES. We know that for every normal mode of vibration of the molecule $E_{v_i} = h\nu\left(v_i + \frac{1}{2}\right)$ the molecule has to be at least in the $v_i = 0$ state, and this state lies above the troughs of the PES by $\frac{1}{2}h\nu_i$.

The above is a well-understood problem of the BO approximation, and the most accurate calculations of molecular properties takes this into account. A less well understood difference between the physical and chemical pictures is that, in the physical picture, the ground state of any molecule is spherically symmetric. This may be understood in the chemical picture by noting that when all degrees of freedom are taken into account, the total wave function contains the nuclear vibratrional and rotational wave functions as well as the electronic wave function:

$$\Psi(r_N, r_e) = \psi(r_e)\psi_{vib}(r_N)\psi_{rot}(r_N) \tag{219}$$

In the ground state, the nuclear rotation state is the $J = 0$ state, which is spherically symmetric. This total wave function may be thought of in analogy to the hydrogen atom as a radial part (which is the expansion in explicitly correlated Gaussians) multiplied by the rotational function, which for the ground state is simply a constant. This point may be understood by the classical picture of the molecule rotating about the center of mass. In this picture, the time-averaged wave function will be spherically symmetric, because the molecule over time has no preferred direction in space.

This presents a problem when we discuss the dipole moment of a polar heteronuclear diatomic molecule, *AX*, where *X* will be the more electronegative. In the chemical picture, it is quite common to say that in the ground state the molecule lies along some axis and that it has a definite dipole moment. In the physical picture, we say that the molecule has no measurable dipole moment in the ground state.

When we apply an electric field to the molecule, though, say along the Z axis, the molecule will begin to align itself with the field. We call this the rotational polarization of the molecule. In this case, though, when the Hamiltonian includes a field directed along the z axis, the rotational quantum number J is no longer a good quantum number, and we now have only eigenstates of $\hat{J}_z$. This molecule is no longer spherically symmetric, and thus we may measure the dipole moment. All that is necessary for the molecule to rotationally polarize is for the energy lowering due to the interaction with the field to be greater than a rotational excitation.

Another problem comes in examining the polarizability. In the physical picture, the spherically symmetric molecule, just like an atom, has isotropic polarizability. In the chemical picture, for a diatomic molecule we have two unique polarizabilities: (1) α_{zz} and α_{xx} in the internal coordinate system or (2) $\alpha_{ZZ} = \frac{1}{3}(\alpha_{xx} + \alpha_{xx} + \alpha_{zz})$ (isotropic polarizability) and $\Delta\alpha = \alpha_{zz} - \alpha_{xx}$ [polarizability anisotropy(PA)] in the laboratory coordinate system. The latter are the values that are measured in experiments. The isotropic polarizability of the physical picture is comparable to the BO-isotropic polarizability at very low field strengths but is comparable to α_{zz} at higher field strengths. We cannot extract the PA from the physical picture, because in this ideal model of dilute gas-phase molecules the polarizability is isotropic.

One possible solution to this problem may be had from examining the time development of a molecule on the electric field. Before the field is applied, the molecule is spherically symmetric, and no PA exists. As soon as the field is applied, the molecule will distort and lose the symmetry. Early in this distortion process, though, especially if the field is small, the molecule is still symmetric. If one can calculate the polarizability at this state, and then calculate the polarizability of the state when the molecule is fully aligned with the field, then these two values can give the PA.

C. Good Quantum Numbers and the Symmetry Properties of the Basis Functions

In the absence of an electric field, the non-BO Hamiltonian commutes with the square of the angular momentum operator, $[\hat{H}, \hat{J}^2] = 0$, and so the eigenfunctions of the Hamiltonian also have to be eigenfunctions of $\hat{J}^2$. This condition is met, for example, by functions such as

$$g_k = \prod_{i=1}^{n} x_i^{p_{ki}} y_i^{l_{ki}} z_i^{m_{ki}} \exp(-\alpha_i^k (r_i)^2) \prod_{j>i}^{n} \exp(-\beta_{ij}^k (r_i - r_j)^2) \tag{220}$$

In the presence of an electric field, this commutation is no longer true: $[\hat{H}, \hat{J}^2] \neq 0$, though if the field is applied along the z axis, then the Hamiltonian does commute with the z component of angular momentum: $[\hat{H}, \hat{J}_z] = 0$. In this case the eigenfunctions of the Hamiltonian must also be eigenfunctions of $\hat{J}_z$. This is met by functions such as those used in this work:

$$g_k = \prod_{i=1}^{n} \exp(-\alpha_i^k (r_i - R_i^k)^2) \prod_{j>i}^{n} \exp(-\beta_{ij}^k (r_i - r_j)^2) \tag{221}$$

so long as the only components of R_i^k that are nonzero are the z components. Thus, the wave functions used in this work are not rigorously correct for the field-free calculations, but the energy error should be only contamination by the first few rotational states and has, in practice, never been a problem.

D. The Finite Field Method

The response of the energy of a molecule to a static electric field along the z axis may be written as

$$E(f) = E_o - \frac{\partial E}{\partial f} f - \frac{1}{2!}\frac{\partial E^2}{\partial^2 f} f^2 - \frac{1}{3!}\frac{\partial E^3}{\partial^3 f} f^3 - \frac{1}{4!}\frac{\partial E^4}{\partial^4 f} f^4 \cdots \tag{222}$$

We define these derivatives as the usual response properties:

$$E(f) = E_o - \mu f - \frac{1}{2!}\alpha f^2 - \frac{1}{3!}\beta f^3 - \frac{1}{4!}\gamma f^4 \cdots \tag{223}$$

where α is called the polarizability, β is called the first hyperpolarizability, and γ is called the second hyperpolarizability. To calculate the response of a molecule to a static electric field only requires knowledge of these properties.

We use a very simple method for calculating the response of molecules to electric fields. We calculate the wave function and energy for molecules in electric fields of various strengths and plot this information. This plot is fitted to a polynomial, and the ith-order properties are extracted from the ith-order coefficients of the fit.

In the physical picture of a molecule, the non-BO energy at the field strength f is equal to the energy at the field $-f$, because as the direction of the field changes the the molecule orients itself with the field so as to have as low a potential energy as possible. Thus, for any ground-state spherically symmetric molecule the energy is an even function of the field, and if it is approximated by a polynomial, only even powers need to be used. This obviously results in a zero dipole, as well as zero-valued odd-ordered properties in general for any system.

An alternative approach is to apply stronger fields and only use energies calculated for positive field strengths in generating the polynomial fit. In this case the energy is a function of both odd and even powers in the polynomial fit. We will show that the dipole moments derived from our non-BO calculations with the procedure that uses only positive fields and polynomial fits with both even and odd powers match very well the experimental results. Thus in the present work we will show results obtained using interpolations with even- and odd-power polynomials. Methods other than the finite field method exist where the noise level in the numerical derivatives is smaller (such as the Romberg method), but such methods still do not allow calculation of odd-ordered properties in the non-BO model.

E. Vibrationally Corrected Electronic Values

The conventional approach used to describe the response of a molecule to a static electric field is either to perform pure electronic BO calculations or to perform calculations where the BO values are corrected for vibrational and rotational (thermal) motion of the nuclei. The vibrationally corrected polarizabilities usually do an excellent job of correcting the errors inherent in the pure electronic BO values. Bishop has written several excellent reviews on this topic [78–80].

F. Isotopomers of H_2

Please note that much of this material is reported in the recent article by Cafiero and Adamowicz [66].

There have been no previous direct non-BO studies of the response of H_2 and its isotopomers to electric fields. The ground-state dipole moment of HD has been determined experimentally by Nelson and Tabisz [81] to be 0.000345 a.u. There have been several theoretical studies of the dipole moment of HD, all within the BO approximation but including adiabatic corrections. The calculated values by Wolniewicz, 0.000329 [83], Ford and Browne, 0.000326 [82], and Thorson et al., 0.000334 [84], all agree well with the experimental value, although they are all about 5% too small. This is an extremely difficult experiment to carry out, and because all theoretical studies agree on the value, it may be that the experimental work is slightly inaccurate. There have been several studies of the vibrational corrections to the pure electronic BO values. Adamowicz [85] used numerical MCSCF electronic functions and numerical evaluation of the vibrational equations to obtain the following ground-state (ground electronic, $v = 0$, $J = 0$) values of α_{zz}: 6.623 (H_2), 6.569 (HD), and 6.509 (D_2) a.u. Also, he found the following for γ_{zzzz}: 1088(H_2), 1074(HD), and 1052(D_2) a.u. Bishop et al. [86] employed the sum over states method and found the static limit value of α_{zz} to be 6.397 a.u. In a different study, Bishop et al. [87] found γ_{zzzz} to be 1099 including electronic and vibrational contributions. There are no non-BO investigations of the values of β_{zzz}.

The wave functions used in this work were built from noncorrelated products of simple gaussians describing the electronic and nuclear wave functions. Initial guesses for the exponential coefficients (squares of the diagonal elements of L_k) and functional centers (s_k) were generated quasi-randomly. The exponential coefficients were randomly distributed around the value 4.0 with a variance of 1. Since the electric field in our calculations was applied along the z direction, the initial guesses for the functional centers were distributed randomly along the z axis around 1.4 bohr for the H-nucleus and one of the electrons, and around 0.0 bohrs for the second electron, with a variance of 0.015. The initial guess wave functions were optimized with respect to all of the elements of L_k and s_k using analytical gradients. The process of optimization builds the required correlation into the wave function by filling in the off-diagonal elements of L_k.

The correct permutational symmetry was implemented into the wave functions by projection onto irreducible representations of the total symmetry group S_2^e for heteronuclear species and $S_2^e \otimes S_2^H$ for homonuclear species, where e refers to electron exchange and H refers to nuclear exchange. The irreducible representations chosen were singlets in all cases.

The energy values obtained for the H_2 isotopomer series were all converged until the energy change with respect to variation was in the subpicohartree range, and the squared gradient norm for the total gradient vector was at most 10^{-14}. The energy curve was fitted with a fourth-order polynomial.

The values for the dipoles, polarizabilities, and hyperpolarizabilities of the H_2 series were obtained using (a) a 16-term basis with a fourfold symmetry projection for the homonuclear species and (b) a 32-term basis with a twofold symmetry projection for the heteronuclear species. These different expansion lengths were used so that when combined with the symmetry projections the resulting wave functions were of about the same quality, and the properties calculated would be comparable. A crude analysis shows that basis set size for an n particle system must scale as κ^n, where κ is a constant. In our previous work [64, 65] we used a 244-term wave function for the five-internal-particle system LiH to obtain experimental quality results. This gives a value of $\kappa^5 = 3^5 \approx 244$. Applying this analysis to HD, which has three internal particles,

TABLE XI
Values for the Zero-Field Energies, Dipole Moments (μ), Polarizabilities (α), Hyperpolarizabilities (β), and Second Hyperpolarizabilities (γ) for the Non-BO H_2 Isotopomer Series[a]

	$\langle H \rangle$	μ	α	(β)	γ
H_2	−1.153736345	1.00×10^{-8}	6.74	(0.0360)	1062
HD	−1.156234289	3.27×10^{-4}	6.67	(0.0306)	1038
HT	−1.157152576	4.37×10^{-4}	6.65	(0.0186)	1028
D_2	−1.159178760	9.00×10^{-9}	6.59	(0.0360)	1009
DT	−1.160312051	1.09×10^{-4}	6.56	(0.0312)	999
T_2	−1.161561149	9.00×10^{-9}	6.52	(0.0360)	989

[a]All values are in atomic units.

we get $\kappa^3 = 3^3 = 27$. Thus the basis set size of 32 terms should be sufficient to give results of similar quality to what was obtained in the LiH case. For H_2, where we have additional basis functions generated by the proton exchange symmetry projection, the basis may be smaller. The calculated values are presented in Table XI.

Allowing only positive and rather significant electric field values permits the calculation of the dipole moment of the heteronuclear species. Applying the same approach to homonuclear species (H_2, D_2, and T_2) should give the dipoles identically equal to zero. In our calculations, these actually come out to 10^{-8}. This small noise that entered our calculations was due in part to the previously mentioned fact that the zero-field wave function we use is not an eigenfunction of $\hat{J}^2$ as it should be. The level of noise introduced is negligible, because 10^{-8} is four orders of magnitude smaller than the size of the dipole moments for the heteronuclear species.

The dipole moment obtained for HD, 3.27×10^{-4} a.u., is very close to the values obtained by Wolniewicz (3.29×10^{-4}) [83] and by Ford and Brown (3.26×10^{-4}) [82]. All of these values are about 5% lower than the experimentally measured value. The ratio of the dipole moment of HD to those of HT and DT may be predicted based on electronegativity arguments to be 0.75 and 3.0. The values obtained here fulfill this prediction exactly.

We see the expected trend in polarizabilities and second hyperpolarizabilities going down the series according to total mass. H_2 has the highest polarizability, 6.74 a.u., because the light, very quantum-mechanical protons delocalize more easily in the electric field than the heavier D and T nuclei. Also, the electrons in H_2 are slightly farther away from the nuclei than in the heavier isotopomers and are more polarizable. Indeed when we add weight to the nuclei, as in HD, the polarizability goes down to 6.67 a.u. Another large jump in polarizability (6.65 a.u. to 6.59 a.u) comes between HT, the last isotopomer containing a proton, and D_2, where both nuclei are heavier. The second hyperpolarizability for H_2 agrees well with the previous values. Our value of 1062 a.u. is smaller than Bishop's 1099 and Adamowicz's 1088 a.u. The corresponding values for the other isotopomers are also smaller than the values by Adamowicz. The value of β for the homonuclear species should be identically zero. We obtain for all three homonuclear species a value of 0.0360 a.u. This very small value is consistent across all of these species, and it is larger than zero due to the very sensitive nature of this third-order property to small numerical inaccuracies. Although we show the values of β for all the heteronuclear isotopomers in Table XI (in parentheses), they are very small and too close to the numerical noise to be trusted.

G. LiH and LiD

There have been several recent attempts to find the nuclear corrections to the LiH dipole moment. Papadopoulos et al. [88] used the perturbation theory to calculate the corrections, and Tachikawa and Osamura [57] used the Dynamic Extended Molecular Orbital method to try to calculate the nonadiabatic result directly. Results for these methods are reported in Table XII. In all cases, the calculated values are outside of the range of the experimental results [89, 90], also reported in Table XII.

For LiH and LiD, 244-term non-BO wave functions were variationally optimized. The initial guess for the LiH non-BO wave function was built by multiplying a 244-term BO wave function expanded in a basis of explicitly correlated functions by Gaussians for the H nucleus centered at and around (in all three dimensions) a point separated from the origin by the equilibrium distance of 3.015 bohr along the direction of the electric field. Thus the centers

TABLE XII
Experimental (expt.) and Theoretical (calc.) Dipole Moments (μ) for LiH and LiD from the Literature[a]

m	μ
LiH	
[88], calc.	2.317
[57], calc.	2.389
[89], expt.	2.3145
[90], expt.	2.314 (0.001)
LiD	
[57], calc.	2.392
[90], expt.	2.309 (0.001)

[a]All values are in atomic units. Values given in parentheses are experimental uncertainty.

TABLE XIII
Expectation Values for Zero-Field Energies, Virial Coefficients (η), Dipole Moments (μ), and Static Polarizabilites (α) for Non-BO LiH/D for Various Expansion Lengths (m)[a]

m	$\langle H \rangle$	η	μ	α
LiH				
24	−8.0423294	1.000000	2.4047	24.42
64	−8.0592988	1.000000	2.3394	28.48
104	−8.0619267	1.000000	2.3261	29.41
144	−8.0629324	0.999999	2.3149	29.54
244	−8.0636331	0.999999	2.3140	29.57
Experimental			2.314 ± 0.001[90]	
LiD				
244	−8.0650331	1.000000	2.3088	
Experimental			2.309 ± 0.001[90]	

[a]All values are in atomic units.

corresponding to the hydrogen nucleus were scattered from about 2.9 to about 3.1 bohr. The lithium nucleus was, of course, placed at the origin of the internal coordinate system. The functional centers corresponding to the electrons were located primarily on the two nuclei, with two electrons at the origin (about 0.0 ± 0.001 bohr in all three directions) and two electrons near the H nucleus (about 3.05 ± 0.06 bohr) per basis set. This reflects the strong ionic character in the lithium–hydrogen bond. The LiD non-BO wave function was optimized starting from the converged LiH wave function. Wave functions of various smaller expansion lengths were optimized for LiH alone. Table XIII shows the convergence properties of the dipole moment for these basis sets. It can be seen that the calculated value of the dipole converges and reaches a value near that of experiment, 2.314 a.u., as larger basis sets are used. The reported results for all functions were converged to the point where the squared norm of the total gradient was at least on the order 10^{-8}, and the energy changed at most in the ninth decimal place. At this point the dipole moment was converged to seven decimal places, which is more than experimental accuracy. The total variational energy (also in Table XIII) for our 244-term wave function is −8.0636331 hartree. The most accurate non-BO energy calculated before was that of Scheu et al. [120] equal to −8.0661557 hartree. Thus our energy value for this basis set is in error by only about 0.0025 hartree. Despite this small error, we deem the obtained convergence of the dipole moment value with the basis set size quite satisfactory, and it is highly unlikely that further enlargement of the basis can change this value by an amount close to the uncertainty level of the experiment where the LiH/LiD dipole moments were measured.

The optimized basis functions show strong correlation between the nuclei, between the nuclei and the electrons, and between the electrons. The centers describing the H/D nucleus show a spread of values from about 2.9 bohr to 3.1 bohr, with the expectation value for the internuclear distance being 3.063 bohr for LiH and 3.052 bohr for LiD. These values are in good agreement with the value 3.061 bohr for LiH obtained by Scheu et al. [120] in their non-BO calculations and are, as expected, longer than the BO 'equilibrium' value. Our values are believed to be much more accurate than those of Ref. 57, which are 3.119937 bohr and 3.104819 bohr for LiH and LiD, respectively.

The value of the dipole moment of LiH obtained in this work, 2.3140 a.u., is essentially identical to the experimental value, 2.314 ± 0.001 [90]. Our calculations simulate experiment more closely than any previous calculations. The results also provide validation of the perturbation approach of Ref. 88, since the perturbation result, 2.317 a.u., is very close to our value. At the same time, our results are much more accurate than those of Ref. 57, the only other "direct" calculation of the LiH dipole moment. The value of the dipole moment of LiD, 2.3088, is also of good accuracy, compared to the experimental result, 2.309 ± 0.001 [90]. Again, our result is much more accurate than that of Ref. 57.

There have been a few recent studies of the corrections due to nuclear motion to the electronic diagonal polarizability (α_{zz}) of LiH. Bishop et al. [92] calculated vibrational and rotational contributions to the polarizability. They found for the ground state ($v = 0$, the state studied here) that the vibrational contribution is 0.923 a.u. Papadopoulos et al. [88] use the perturbation method to find a corrected value of 28.93 a.u. including a vibrational component of 1.7 a.u. Jonsson et al. [91] used cubic response functions to find a corrected value for α_{zz} of 28.26 a.u., including a vibrational contribution of 1.37 a.u. In all cases, the vibrational contribution is approximately 3% of the total polarizability.

The results for the non-BO diagonal polarizability are shown in Table XIII. Our best—and, as it seems, well-converged—value of α, 29.57 a.u., calculated with a 244-term wave function, is slightly larger than the previously obtained "corrected" electronic values, 28.93 and 28.26 a.u. [88,91]. It is believed that the non-BO correction to the polarizability will be positive and on the order of less than 1 a.u. [92], but it is not possible to say if the difference between the value obtained in this work and the previous values for polarizability are due to this effect or to other effects, such as the basis set incompleteness in the BO calculations. An effective way of testing this would be to perform BO calculations of the electronic and vibrational components of polarizability using an extended, well-optimized set of explicitly correlated Gaussian functions. This type of calculation is outside of our current research interests and is quite expensive. It may become a possibility in the future. As such, we would like the polarizability value of 29.57 a.u. obtained in this work to serve as a standard for non-BO polarizability of LiH.

VII. THE USE OF SHIFTED GAUSSIANS IN NON-BO CALCULATIONS ON POLYATOMIC MOLECULES

We have shown that in order to calculate a non-Born–Oppenheimer wave function we must use basis functions that are eigenvalues of the total angular momentum. We have used such functions in the calculation of atoms and diatomic molecules and have shown that we are working on extending these basis sets to triatomic molecules. We may also perform non-Born–Oppenheimer calculations of reasonable accuracy using basis functions which are not rigorous eigenfunctions of angular momentum but which are complete sets and may approximate such eigenfunctions.

Non-Born–Oppenheimer wave functions calculated in this way look more like their Born–Oppenheimer counterparts in the smaller basis set limits, and thus a good starting guess for these may be taken from Born–Oppenheimer calculations in the same basis. Thus we calculate the electronic part first (this requires much fewer basis functions than does a full non-Born–Oppenhimer calculation) and then form the total basis function by multiplying each electronic portion by a guess for the nuclear portion:

$$\psi_T(r_{total}) = \psi_N(r_{nuc})\psi_e(r_{elec}) \tag{224}$$

We will present below a short description of some Born–Oppenheimer calculations we have done on this basis, followed by examples of triatomic non-Born–Oppenheimer calculations on this basis.

A. Born–Oppenheimer Calculations in a Basis of Explicitly Correlated Gaussians

FSECGs have been used in BO calculations for some time, but always with quite small systems. It has been said [94] that the bottleneck for the application of ECGs to larger systems ($n > 4$) is the large amount of time spent in optimization of the wave functions. We have removed a large part of the bottleneck in these calculations by implementing analytical gradients in the optimization and parallelizing the entire code. The use of the analytical gradients speeds up optimization, because the optimization routine needs to make only one function call to calculate the energy and the total gradient vector, rather than several function calls to calculate finite-difference gradients. The analytical gradients are also more accurate and lead to faster optimization paths.

B. Test Calculations on H_3 and H_3^+

Please note that much of this work may be found in a previous article [70].

A new upper bound for the BO energy of the ground state of H_3^+ was recently obtained by Cencek et al. [48] using explicitly correlated Gaussians. Below we

TABLE XIV
Expectation Values of the Hamiltonian, $\langle H \rangle$, Virial Coefficients, η, and Squared Gradient Norms, $\| grad \|^2$, For the Ground State of H_3^+ $(R1 = R2 = R3 = 1.65)$ for the 75-Term Wave Function

	Cencek et al. [48]	This Work
$\langle H \rangle$	-1.343834724	-1.343834853
η	0.9999977	0.9999978
$\| grad \|^2$	$O(10^{-8})$	$O(10^{-12})$

show how our calculations compare against those. The Young projection operator used was for the singlet state, and the D_{3h} point group symmetry was ensured by constraining the molecule to the xy plane and applying the operator:

$$\hat{P} = \hat{1} + \hat{C}_3 + \hat{C}_3^2 + \hat{\sigma}_1 + \hat{\sigma}_2 + \hat{\sigma}_3 \tag{225}$$

We did expect to gain a slight lowering of the energy due to our use of analytical gradients in our optimization as opposed to a numerical optimization used by Cencek et al., and we did. Table XIV shows the energy obtained by Cencek et al. for the 75-term wave function, as well as the squared gradient norm—that is, the sum of the squares of the gradients with respect to all the variational parameters. As can be seen, the wave function obtained by Cencek et al. [48] had considerable "room" for further optimization: The gradients with respect to most variational parameters were indeed quite small, but the gradients with respect to the off-diagonal exponential coefficients were somewhat larger and thus made the squared norm of the gradient overall larger. After several cycles of our unconstrained optimization in which we reoptimized all variational parameters, the energy for this basis set was lowered and the gradients with respect to all parameters, and thus the squared norm of the gradient, became much smaller. This final wave function is more tightly converged than that obtained before.

The formulas derived in Cencek et al. [49] involve a simplifying approximation of only including one correlation term per basis function; that is, only two electrons are correlated per function. In this work, all functions include correlation among all electrons. For two electrons there is no difference in the two forms of the basis, but for three or more electrons the formulas derived here should prove more efficient; that is, they should converge faster and with a smaller basis set size.

We tested a 76-term wave function for the system H_3, including permutational and point group symmetry. The initial guess for the nonlinear parameters in the ECGs were generated randomly using Matlab. The Young

TABLE XV
Expectation Values of the Hamiltonian, $\langle H \rangle$, Virial Coefficients, η, and Squared Gradient Norms, $\| grad \|^2$, for the Ground State of H_3 ($R1 = R2 = 1.75$) for the 76-Term wave Function (This Work) and the 100-Term Wave Function (Cencek and Rychlewski) [49]

	100-term [49]	76-term
$\langle H \rangle$	−1.658491	−1.658565
η	—	1.00012
$\| grad \|^2$	—	$O(10^{-13})$

projection operator used was a doublet, and the $D_{\infty h}$ point group symmetry was ensured by constraining the molecule to the z axis and applying the operator:

$$\hat{P} = \hat{1} - \hat{\sigma_h} \tag{226}$$

The energy obtained after optimization for the 76-term function was found to be lower than the energy obtained by Cencek and Rychlewski [49] for the 100-term wave function, illustrating the power of including correlation among all the electrons in each basis term (see Table XV).

C. Geometry Optimization

Please note that much of this work may be found in a previous article [69].

In the course of this research, we introduced the first geometry optimization via analytical gradients to be used in very-high-accuracy electronic structure calculations with FSECGs [69]. In this method we simultaneously optimize the nonlinear parameter of the basis functions and the molecular structure parameters. Simultaneous optimization of both types of parameter provides a unique path to the high accuracy of the calculation. The explicitly correlated Gaussians are a particularly interesting basis to use in geometry optimization since, by monitoring the magnitude of correlation coefficients (off-diagonal elements in the L_k matrix), we can see the dynamic correlations of pairs of electrons increasing or decreasing as the molecule undergoes changes in nuclear geometry. A simple example of this is found in the results on small hydrogen clusters presented here.

Sample calculations were carried out on H_2, H_3^+, and H_3. Geometry optimizations were carried out in internal coordinates. The projection operators used in the expansion (4) represented a singlet state for H_2 and H_3^+ and a doublet state for H_3. Starting geometries that were used are given in Table XVI. The initial wave functions were centered at the nuclei. For all of the initial functions the correlation parameters were set to zero (that is, the matrices A_k were

TABLE XVI
Starting Geometries (in parentheses), Energies, $\langle \hat{H} \rangle$, Virial Coefficients, η, Squared Gradient Norms, $\| grad \|^2$, and Optimized Geometries for the Hydrogen Clusters[a]

	$\langle \hat{H} \rangle$	η	$\| grad \|^2$	Geometry
H_2(R12 = 2.000)	−1.173092	1.000000	10^{-13}	R12 = 1.399
H_3^+	−1.334711	0.999999	10^{-13}	R12 = 1.650
(R12 = 1.400,				R13 = 1.650
R13 = R23 = 1.220)				R23 = 1.649
H_3(global minimum)	−1.673468	0.999999	10^{-13}	R12 = 1.400
(R12 = R23 = 1.750)				R23 = 6.442
H_3(saddle point)	−1.655734	1.000000	10^{-13}	R12 = 1.757
(R12 = R23 = 1.750)				R23 = 1.757

[a]All reported values are in atomic units.

diagonal). Expansion lengths used were $m = 16$ for H_2 and H_3^+ and $m = 64$ for H_3.

The nonlinear optimization is always sensitive to the initial guess of the parameters (nonlinear and geometrical). For each system, several initial guesses for the parameters were used, and we report the results of the calculations which produced the lowest energies. In all cases, equilibrium structures were found; additionally, the saddle point for H_3 was found using a constrained optimization. This point is defined in the work by Liu [98] as the minimum under the conditions of linearity and σ_h symmetry. We thus reduced the optimization to a single variable—that is, the bond length. It is interesting to note that for the H_3 saddle point the correlation coefficients (off-diagonal elements of A_k) were all of about the same order of magnitude (10^{-2}), while for the equilibrium point, where H_2 and H are separated, the magnitudes of the correlation coefficients between the electrons of H_2 were much larger (10^{-2}) than those between electrons on different subunits (10^{-5}). This reflects the expected result that there is strong correlation of electrons within the localized chemical bond in H_2 and much weaker correlation of the electrons that are spatially separated. In the case of the H_3 equilibrium structure, the optimization settled into the van der Waals' well corresponding to linear H_3 with the $H_2 \cdots H$ distance equal to 6.442 bohr. This structure is in good agreement with the potential surface minima for this system determined by Truhlar and Horowitz [99], Wu et al. [100], and Tang and Toennies [101]. Final energies, virial coefficients, squared gradient norms, and geometries are given in Table XVI.

We compared the van der Waals region obtained with our method to that which may be obtained with a fairly high-end standard method, CCSD(T)/cc-pVTZ, as implemented in *Gaussian* 98 [102]. The CCSD(T) energy for this system, though not variational, should provide a reasonable estimate of the variational energy in this basis, because the single and double excitations are much more important than the triple excitations. The CCSD(T) energy was -1.6721738 hartree, about 1.3 millihartree higher than our result. The geometry obtained with CCSD(T) was $R12 = 1.403$ bohr and $R23 = 6.478$ bohr. The expected value of $R12$ is 1.401 bohr, which is slightly closer to our value of 1.400 bohr. The literature value of $R23$ is between 6.44 and 6.48 bohr. The relatively newer result of Tang and Toennies [101] predicts a value of 6.46 bohr, roughly between the values obtained using our method and CCSD(T).

D. Extension to Non-Born–Oppenheimer

We show here how we may take the information obtained above and use it as a starting point for non-Born–Oppenheimer calculations. The five-particle systems of non-Born–Oppenheimer H_3^+ and its isotopomers were transformed via separation of the center-of-mass Hamiltonian to four-pseudoparticle systems as described above. The resulting total position vector is

$$r = \begin{pmatrix} r_1 \\ r_2 \\ r_3 \\ r_4 \end{pmatrix} \tag{227}$$

For the illustrative calculations shown here, the spin-free wave functions, Ψ, for the H_3^+ isotopomers were obtained as 50-term expansions in a basis of FSECG's $g_k(r)$:

$$\Psi(r) = \sum_{k=1}^{50} c_k \hat{Y} g_k(r) \tag{228}$$

where $\hat{Y}$ is the total Young operator described above.

The FSECG basis function for four particles is

$$g_k(r) = \exp\{-(r - s_k)'[(L_k L_k') \otimes I_3](r - s_k)\} \tag{229}$$

where s_k is a 12 vector of 1"shifts" that are variational parameters, and L_k is an upper triangular 4×4 matrix of variational exponential parameters.

The ground-state energy values and the wave functions for the considered systems were variationally optimized:

$$E = \min \frac{\langle \Psi | \hat{H} | \Psi \rangle}{\langle \Psi | \Psi \rangle} \tag{230}$$

Each wave function was optimized with respect to the parameters L_k, s_k, and c_k. This lead to $\frac{1}{2}n(n+1) + 3n + 1$ variational parameters per basis function (23 for the H_3^+ systems). We began with an initial guess for the wave function built from the above Born–Oppenheimer calculations and then use analytical gradients in a truncated Newton-type algorithm to find the lowest value for the energy. The initial guesses for the nuclear portions of the basis functions were randomly generated with all the s_k centers lying in the xy plane. For small basis sets such as those used here, it has been found that placing functional centers off of the xy plane makes a negligible difference in the energy and structural expectation values. Because the basis set is increased and the planar optimization space becomes saturated, the placement of functions off of the xy plane would become more important, with the basis set limit featuring a full three-dimensional spherical symmetry for the ground state.

For the homonuclear (HON) species, the permutation–symmetry operator had the following form: $\hat{Y} = \hat{Y}_N(S_3) \otimes \hat{Y}_e(S_2)$, where $\hat{Y}_N(S_3)$ is a Young operator for the third-order symmetric group which permutes the nuclear coordinates and $\hat{Y}_e(S_2)$ is a Young operator for the second-order symmetric group which permutes the electronic coordinates. For the fermionic nuclei (H and T, spin = 1/2) the Young operators corresponded to doublet-type representations, while for the bosonic D nuclei we use operators that correspond to the totally symmetric representation. In all cases the electronic operator corresponded to a singlet representation.

For the heteronuclear (HEN) A_2B-type species, the symmetry operator was $\hat{Y} = \hat{Y}_N(S_2) \otimes \hat{Y}_e(S_2)$. In this case the nuclear operator was a singlet for H pairs and symmetric for D pairs. Finally for the HEN HDT-type isotopomer we had $\hat{Y} = \hat{Y}_e(S_2)$. Again in all cases the electron operators represented a singlet. For discussion of the construction of the operators, see, for example, the excellent work of Pauncz [73].

It is obvious that the projection operators for the different species have different numbers of terms in them. The HON species have 12 terms ($3! \times 2!$) while the A_2B-type species have four terms, and the HDT^+ isotopomer has only two terms. This results in different sizes of the spin-projected basis sets, and for this reason the properties obtained in this work are not precisely comparable between the A_3, A_2B, and ABC systems, although a very good idea of the trends may be obtained from the data in Table XVI. While all of the above are given in terms of the original particles, it should be noted that the permutations used in the internal particle basis functions are "pseudo"-permutations induced by the permutations on real particles.

The geometrical parameters reported in Table XVI include the distances from the particle at the origin of coordinates to the other two nuclei (r_1 and r_2) as well the distance between the two "loose" nuclei (r_{12}); also included are the squares of all of these distances. For the HON species, the quantity actually calculated is $\langle r_1 + r_2 + r_{12} \rangle$. The value reported in the table is this number divided by three. For the HEN A_2B species we calculate $\langle r_1 + r_2 \rangle$ and $\langle r_{12} \rangle$

TABLE XVII
Values for the Nonadiabatic Energies, Virial Coefficients (η), and Bond Lengths of DH_2^+ for Various Expansion Lengths[a]

m	$\langle H \rangle$	η	r_{ij}
50	−1.316992613	0.999999	$r_{DH} = 1.735$
			$r_{HH} = 1.746$
80	−1.318112939	0.999996	$r_{DH} = 1.731$
			$r_{HH} = 1.741$
118	−1.318845090	0.999999	$r_{DH} = 1.731$
			$r_{HH} = 1.740$
489	−1.321226255	0.999742	$r_{DH} = 1.724$
			$r_{HH} = 1.734$

[a]All values are in atomic units.

separately. Finally, for the HDT^+ isotopomer, we find each distance separately. The same patterns were used for the squares of the distances. Since the wave functions used were nonadiabatic and include both electronic and nuclear coordinates, the bond lengths calculated are not the usual equilibrium bond lengths, r_e (i.e., the very bottom of potential wells), obtained in conventional BO calculations, but rather the r_0 bond lengths (i.e., bond lengths that are more comparable to bond lengths obtained in the BO calculations by averaging the internuclear distances over the ground vibrational state of the system).

For one isotopomer, DH_2^+, two additional larger basis sets were optimized ($m = 80$, 118, 489) in order to determine the convergence of the structure parameters. This data are presented in Table XVII. As may be seen, the energies are not yet converged. Because there are no references for the fully nonadiabatic energy of DH_2^+, we may estimate it by adding the zero-point vibrational energy to the best Born–Oppenheimer energy obtained so far. These data may be found in the work by Jaquet et al. [103], and the value is −1.326672 hartree. Although our energies for DH_2^+ are above this value, the geometry is well known to converge before the energy. The difference in bond length for the DH bond decreases by 0.003 bohr, going from 50 basis functions to 80, and then decreases by less than 0.001 bohr, going from 80 to 118 basis functions. Finally, taking a very large step to 489 basis functions, the bond lengths are seen to decrease by less than 0.01 bohr. It may be noted also that this basis set produces an energy much closer to the estimated energy than the smaller basis sets, and so the geometry may be that much more reliable.

Assuming that any additional increase in basis set size will cause a decrease on the order 0.005 bohr or less, we may say that the geometries of the systems considered here obtained with 50 basis functions are fairly well optimized. The isotopic differences across the isotopomers seem to be consistant as well. It will be expected that bond lengths obtained from the HON species will be slightly more accurate than those obtained from the HEN species due to the difference in the spin-projected basis set size. Likewise, the HDT^+ isotopomer will have the least accurate bond lengths.

E. Discussion

The commonly accepted equilibrium bond length, r_e, for H_3^+ is 1.650 bohr (see, for example, Cencek et al. [48] and references therein). This number corresponds to the very center of the potential energy well. In actual fact, molecules in their ground states reside above this trough in the zero-point vibrational state. The more anharmonic the well is, however, the more the bond length will be shifted to higher values. H_3^+ is a particularly floppy molecule, and so the r_0 bond length would be expected to be significantly larger than the r_e bond length. This is in fact the case, since the value we obtained in our nonadiabatic calculation is 1.748 bohr.

As the nuclei get heavier, the displacement of the molecule from its equilibrium structure in the zero-point vibration becomes smaller and the r_0 bond lengths should approach the r_e bond lengths. This is in fact the case, as we see from Table XVIII. T_3^+ is about 0.04 bohr closer to the expected well than H_3^+, which is about 0.1 bohr from the expected value. Another phenomenon seen in the data is that the uncertainty in the nuclear position, calculated as $\delta = (|\langle r\rangle^2 - \langle r^2\rangle|)^{1/2}$, gets progressively smaller for each HON isotopomer, going from 0.238 to 0.194 to 0.179 as we move from H_3^+ down to T_3^+.

An interesting point in the data presented here is that for the HON species and the A_2B HEN species it is impossible to determine from the r data alone if the molecule is linear or planar triangular. Even if the expectation values of the

TABLE XVIII
Values for the Nonadiabatic Expectation Values of the Ground-State Energies, Virial Coefficients (η), Interparticular Distances, and Squares of Interparticular Distances, for Some Isotopomers of H_3^{+a}

	$\langle H\rangle$	η	r_{ij}	r_{ij}^2
H_3^+	−1.314383574	0.999999	$r_{HH} = 1.748$	$r_{HH}^2 = 3.112$
D_3^+	−1.322718305	0.999999	$r_{DD} = 1.720$	$r_{DD}^2 = 2.996$
T_3^+	−1.326427799	0.999997	$r_{TT} = 1.707$	$r_{TT}^2 = 2.946$
DH_2^+	−1.316992613	0.999999	$r_{DH} = 1.735$	$r_{DH}^2 = 3.059$
			$r_{HH} = 1.746$	$r_{HH}^2 = 3.105$
HD_2^+	−1.319779541	0.999999	$r_{HD} = 1.734$	$r_{HD}^2 = 3.056$
			$r_{DD} = 1.719$	$r_{DD}^2 = 2.997$
HDT^+	−1.320907124	0.999999	$r_{TD} = 1.715$	$r_{DT}^2 = 2.976$
			$r_{TH} = 1.729$	$r_{DD}^2 = 3.034$
			$r_{HD} = 1.734$	$r_{DD}^2 = 3.053$

[a]All values are calculated for an optimized 50-term explicitly correlated Gaussian basis set and are in atomic units.

angles were calculated, we would find that due to indistinguishability of the nuclei the angles would come out equal in either configuration. The only way to predict from the nonadiabatic calculations the actual structure of the molecules and determine whether they are linear or planar triangular is to consider the average bond lengths calculated for the HDT^+ isotopomer. In HDT^+ the indistinguishability plays no role, and in this system only we find that the molecular geometry is a near-equilateral triangle, and not linear. Thus, just as in experiment, "isotopic substitution" is necessary to extract information about the molecular structure from the nonadiabatic calculations.

VIII. FUTURE WORK

At this stage we are at the very beginning of development, implementation, and application of methods for quantum-mechanical calculations of molecular systems without assuming the Born–Oppenheimer approximation. So far we have done several calculations of ground and excited states of small diatomic molecules, extending them beyond two-electron systems and some preliminary calculations on triatomic systems. In the non-BO works, we have used three different correlated Gaussian basis sets. The simplest one without r_{ij} premultipliers ($\phi_k = \exp[-\mathbf{r}'(A_k \otimes I_3)\mathbf{r}]$) was used in atomic calculations; the basis with premultipliers in the form of powers of r_1 ($\phi_k = r_1^{m_k} \exp[-\mathbf{r}'(A_k \otimes I_3)\mathbf{r}]$) was used in calculations for diatomic systems; and Gaussians with shifted centers ($\phi_k = \exp\left[-(\mathbf{r}-\mathbf{s})'(A_k \otimes I_3)(\mathbf{r}-\mathbf{s})\right]$) were used in non-BO calculations of diatomic molecules in the static electric field. The latter basis was also used in non-BO calculations of ground states of some simple triatomic systems (H_3, H_3^+ and their isotopomers).

At present our effort concentrates on the development of methods for non-BO calculations of excited states of molecules with three nuclei. Our aim is to match in such calculations the accuracy we have been getting for the diatomic systems. This development will open to us the possibility of studying highly vibrationally excited, charged, and neutral clusters of hydrogen and its isotopomers. H_3 and H_3^+ are among the most interesting cases in this cathegory. At present we consider two different approaches for such calculations. The first is based on implementation of correlated Gaussian basis functions with preexponential multipliers consisting of products of all three internuclear distances (r_1, r_2, and r_{12}) raised to even powers ($\phi_k = r_1^{m_k} r_2^{n_k} r_{12}^{l_k} \exp[-\mathbf{r}'(A_k \otimes I_3)\mathbf{r}]$). Such a basis should very effectively describe the coupled motions of electrons and nuclei in systems with three heavy, repelling particles. In the second approach we will use the correlated Gaussians with shifted centers ($\phi_k = \exp\left[-(\mathbf{r}-\mathbf{s})'(A_k \otimes I_3)(\mathbf{r}-\mathbf{s})\right]$). Since these functions are not rotationally invariant eigenfunctions, we will need to implement a projection procedure for separating functional manifolds corresponding to different quantum numbers of the square of the total angular momentum operator $\hat{J}^2$. This may be done by including a penalty term in the Hamiltonian matrix ($a \times |\langle\phi_k|\hat{J}^2 - J(J-1)|\phi_k\rangle|$; a being a positive penalty factor) which will elevate the energies of functions corresponding to the rotation eigenvalues different from the one considered in the calculation (J) to higher energies effectively separating them away.

Another development that we will undertake in the near future is development of algorithms for non-BO calculations of molecules with π-electrons (the CH radical is an example of such a system). We also contemplate development of methods for describing systems where only the light nuclei (apart from electrons) are treated as quantum particles, and the other heavier nuclei are described either classically or by using a low-level approximation. This development would move us closer to cosidering the quantum dynamics of such reactions as inter- and intramolecular proton transfer.

Finally, the development of new algorithms in atomic and molecular non-BO calculations must be carried out in parallel with the development of the computer technology, particularly in parallel with the advances in parallel computing. An option that can be considered involves the so-called meta-computing. Meta-computing on a grid of distributed computing platforms connected via high-speed networks can revolutionize computational research by enabling hybrid computations that integrate multiple systems distributed over wide geographical locations [46]. The non-BO method developed in our work is well-suited for implementation on parallel computational platforms. In our laboratory we use parallel "Beowulf" clusters based on commercially available PC components and connected via a fast Ethernet switch. The clusters use a Unix (Linux) operational system, and the software parallelism is facilitated by MPI. For our non-BO calculations we found the Beowulf clusters to be very cost-effective since the calculations can be easily distributed over a network of processors and executed in parallel with quite little interprocessor communication and without the need to share a common operational memory. Utilization of massively parallel systems has given us momentum to proceed with the development of the non-BO method which we hope the computers of the future will allow to apply to larger systems more central to chemistry. Hence, while the applications presented in this chapter concern very small systems, the emphasis in the development we have carried out is placed on creating a general method that is applicable to molecular systems with an arbitrary number of electrons and nuclei.

References

1. M. Born and J. P. Oppenheimer, *Ann. Phys.* **84**, 457 (1927).
2. J. M. Cobes and R. Seiler, *Quantum Dynamics of Molecules*, R. G. Woolley, ed., Plenum Press, New York, 1980, p. 435.

3. M. Klein, A. Martinez, R. Seiler, and X. P. Wang, *Commun. Math. Phys.* **143**, 607 (1992).
4. A. Carington and R. A. Kennedy, *Gas Phase Ion Chemistry*, Vol. 3, M. T. Bowers, ed., Academic Press, New York, p. 393.
5. D. R. Yarkony, *J. Phys. Chem. A* **105**, 6277 (2001).
6. A. Toniolo, M. Ben-Nun, and T. J. Martinez, *J. Phys. Chem. A* **106**, 4679 (2002).
7. R. S. Friedman, T. Podzielinski, L. S. Cederbaum, et al., *J. Phys. Chem. A* **106**, 4320 (2002).
8. S. Mahapatra, *J. Chem. Phys.* **116**, 8817 (2002).
9. M. Brouard, P. O'Keeffe, and C. Vallance, *J. Phys. Chem. A* **106**, 3629 (2002).
10. M. Baer, A. M. Mebel, and R. Englman, *Chem. Phys. Lett.* **354** 243 (2002).
11. S. Matsika and D. R. Yarkony, *J. Phys. Chem. A* **106**, 2580 (2002).
12. S. Matsika and D. R. Yarkony, *J. Chem. Phys.* **116**, 2825 (2002).
13. S. Matsika and D. R. Yarkony, *J. Chem. Phys.* **115**, 5066 (2001).
14. S. Matsika and D. R. Yarkony, *J. Chem. Phys.* **115**, 2038 (2001).
15. N. Iordanova and S. Hammes-Schiffer, *J. Am. Chem. Soc.* **124**, 4848 (2002).
16. P. K. Agarwal, S. R. Billeter, and S. Hammes-Schiffer, *J. Phys. Chem. B* **106**, 3283 (2002).
17. P. K. Agarwal, S. R. Billeter, P. T. R. Rajagopalan, et al., *Proc. Natl. Acad. Sci. USA* **99**, 2794 (2002).
18. S. Hammes-Schiffer, *CHEMPHYSCHEM* **3**, 33 (2002).
19. S. R. Billeter, S. P. Webb, P. K. Agarwal, et al., *J. Am. Chem. Soc.* **123**, 11262 (2001).
20. S. Hammes-Schiffer and S. R. Billeter, *Int. Rev. Phys. Chem.* **20**, 591 (2001).
21. M. N. Kobrak and S. Hammes-Schiffer, *J. Phys. Chem. B* **105**, 10435 (2001).
22. I. Rostov and S. Hammes-Schiffer, *J. Chem. Phys.* **115**, 285 (2001).
23. T. Iordanov, S. R. Billeter, S. P. Webb, et al., *Chem. Phys. Lett.* **338**, 389 (2001).
24. N. Iordanova, H. Decornez, and S. Hammes-Schiffer, *J. Am. Chem. Soc.* **123**, 3723 (2001).
25. S. Hammes-Schiffer, *Accounts Chem. Res.* **34**, 273 (2001).
26. S. R. Billeter, S. P. Webb, T. Iordanov, P. K. Agarwal, and S. Hammes-Schiffer, *J. Chem. Phys.* **114**, 6925 (2001).
27. A. M. Frolov and V. H. Smith, *J. Chem. Phys.* **115**, 1187 (2001).
28. F. E. Harris, A. M. Frolov, and V. H. Smith, *J. Chem. Phys.* **119**, 8833 (2003).
29. A. M. Frolov and V. H. Smith, *J. Chem. Phys.* **119**, 3130 (2003).
30. A. M. Frolov, *J. Phys. B* **35**, L331 (2002).
31. A. M. Frolov, *Phys. Rev. A* **67**, 064501 (2003).
32. A. M. Frolov, *Phys. Rev. A* **69**, 062507 (2004).
33. A. M. Frolov, *J. Phys. B* **37**, 853 (2004).
34. A. M. Frolov, *Phys. Rev. A* **69**, 022505 (2004).
35. L. Adamowicz and A. J. Sadlej, *J. Chem. Phys.* **67**, 4398 (1977).
36. L. Adamowicz and A. J. Sadlej, *Chem. Phys. Lett.* **48**, 305 (1977).
37. L. Adamowicz, *J. Quant. Chem.* **13**, 265 (1978).
38. L. Adamowicz, *Acta Phys. Pol.* **A53**, 471 (1978).
39. L. Adamowicz and A. J. Sadlej, *J. Chem. Phys.* **69**, 3992 (1978).
40. L. Adamowicz and A. J. Sadlej, *Acta Phys. Pol. A* **54**, 73 (1978).
41. L. Adamowicz and A. J. Sadlej, *Chem. Phys. Lett.* **53**, 377 (1978).

42. K. Szalewicz, L. Adamowicz, and A. J. Sadlej, *Chem. Phys. Lett.* **61**, 548 (1979).
43. D. B. Kinghorn and L. Adamowicz, *J. Chem. Phys.* **106**, 4589 (1997).
44. K. R. Lykke, K. K. Murray, and W. C. Lineberger, *Phys. Rev. A* **43**, 6104 (1991).
45. W. F. Drake, *Nucl. Instr. Meth. Phys. Res. B* **31**, 7 (1988).
46. I. Forster and C. Kesselman, *The Grid: Blueprint for a New Computing Infrastructure*, Morgan-Kaufmann, San Francisco, 1999.
47. S. F. Boys, *Proc. R. Soc. London*, Ser. A **258**, 402 (1960).
48. W. Cencek, J. Komasa, and J. Rychlewski, *Chem. Phys. Lett.* **246**, 417 (1995).
49. W. Cencek and J. Rychlewski, *J. Chem. Phys.* **98**, 1252 (1993).
50. W. Cencek, J. Komasa, and J. Rychlewski, *Chem. Phys. Lett.* **304**, 293 (1999).
51. J. Komasa, *J. Chem. Phys.* **112**, 7075 (2000)
52. R. D. Poshusta, *Int. J. Quant. Chem.* **8**, 27 (1978).
53. Y. Suzuki and K. Varga, *Stochastic Variational Approach to Quantum Mechanical Few-Body Problems*, Springer-Verlag, Berlin, 1998.
54. N. C. Handy and A. M. Lee, *Chem. Phys. Lett.* **252**, 425 (1996).
55. H. J. Monkhorst, *Phys. Rev. A* **36**, 1544 (1987).
56. H. J. Monkhorst, *Int. J. Quantum Chem.* **72**, 281 (1999).
57. M. Tachikawa, K. Mori, K. Suzuki, and K. Iguchi, *Int. J. Quantum Chem.* **70**, 491 (1998).
58. H. Nakai, K. Sodeyama, and M. Hoshino, *Chem. Phys. Lett.* **345**, 118 (2001).
59. Y. Shigeta, H. Nagao, K. Nishikawa, and K. Yamaguchi, *J. Chem. Phys.* **111**, 6171 (1999).
60. D. B. Kinghorn and L. Adamowicz, *J. Chem. Phys.* **113**, 1203 (2000).
61. P. Kozlowski and L. Adamowicz, *J. Comp. Chem.* **13**, 100 (1992).
62. P. Kozlowski and L. Adamowicz, *J. Chem. Phys.* **96**, 9013 (1992).
63. D. B. Kinghorn and R. D. Poshusta, *Phys. Rev. A* **47**, 3671 (1996).
64. M. Cafiero and L. Adamowicz, *Phys. Rev Lett.* **88**, 33002 (2002).
65. M. Cafiero and L. Adamowicz, *J. Chem. Phys.* **116**, 5557 (2002).
66. M. Cafiero and L. Adamowicz, *Phys. Rev Lett.* in press.
67. V. I. Korobov, *Phys. Rev. A* **63**, 044501 (2001).
68. D. M. Bishop and S. A. Solunac, *Phys. Rev. Lett.* **55**, 1986 (1985).
69. M. Cafiero and L. Adamowicz, *Chem. Phys. Lett.* **335**, 404 (2001).
70. M. Cafiero and L. Adamowicz, *Int. J. Quantum Chem.* **82**, 151 (2001).
71. G. Schatz and M. Ratner, *Quantum Mechanics in Chemistry*, Dover, Mineola, NY, 2002.
72. F. A. Cotton, *Chemical Applications of Group Theory*, Wiley, New York, 1990.
73. R. Pauncz, *Spin Eigenfunctions*, Plenum, New York, 1979
74. M. Hamermesh, *Group Theory and its Application to Physical Problems*, Dover, Mineola, NY, 1989.
75. K. Singer, *Proc. R. Soc. London*, Ser. A **258**, 412 (1960).
76. S. G. Nash, *SIAM J. Num. Anal.* **21**, 770 (1984).
77. M. Snir, S. Otto, S. Huss-Lederman, D. Walker, and J. Dongarra, *MPI—The Complete Reference*, Vol. 1, The MIT Press, Cambridge, MA, 1998.
78. D. M. Bishop, *Rev. Mod. Phys.* **62**, 343 (1990).
79. D. M. Bishop, *Adv. Quant. Chem.* **25**, 1 (1994).

80. D. M. Bishop, *Adv. Chem. Phys.* **104**, 1 (1998).
81. J. B. Nelson and G. C. Tabisz, *Phys. Rev. A* **28**, 2157 (1983).
82. A. L. Ford and J. C. Browne, *Phys. Rev. A* **16**, 1992 (1977).
83. L. Wolniewicz, *Can. J. Phys.* **54**, 672 (1976).
84. W. R. Thorson, J. H. Choi, and S. K. Knudson, *Phys. Rev. A* **31**, 34 (1985).
85. L. Adamowicz, *Mol. Phys.* **65**, 1047, (1988).
86. D. M. Bishop, J. Pipin, and S. M. Cybulski, *Phys. Rev. A* **43**, 4845 (1991).
87. D. M. Bishop and B. Lam, *Chem. Phys. Lett.* **134**, 283 (1986).
88. M. G. Papadopoulos, A. Willetts, N. C. Handy, and A. E. Underhill, *Mol. Phys.* **88**, 1063 (1996).
89. F. J. Lovas, and E. Tiemann, *J. Phys. Chem. Ref. Data* **3**, 609 (1974).
90. L. Wharton, L. P. Gold, and W. Klemperer, *J. Chem. Phys.* **37**, 2149 (1962).
91. D. Jonsson, P. Norman, and H. Agren, *J. Chem. Phys* **105**, 6401 (1996).
92. D. M. Bishop, B. Lam, and S. T. Epstein, *J. Chem. Phys* **88**, 337 (1988).
93. P. Kozlowski and L. Adamowicz, *J. Comp. Chem* **13**, 602 (1992).
94. J. Rychlewski, *Adv. Quantum Chem.* **31**, 173 (1998).
95. D. W. Gilmore, P. M. Kozlowski, D. B. Kinghorn, and L. Adamowicz, *Int. J. Quant. Chem.* **63**, 991 (1997).
96. P. Pulay, *Mol. Phys.* **17**, 197 (1969).
97. H. B. Schlegel, *Theor. Chem. Acc.* **103**, 294 (2000).
98. B. Liu, *J. Chem. Phys.* **58**, 1925 (1973).
99. D. G. Truhlar and C. J. Horowitz, *J. Chem. Phys.* **68**, 2466 (1978).
100. Y.-S. M. Wu, A. Kuppermann, and J. B. Anderson *Phys. Chem. Chem. Phys.* **1**, 929 (1999).
101. K. T. Tang and J. P. Toennies *Chem. Phys. Lett.*, **151**, 301 (1988).
102. Gaussian 98 (Revision A.7), M. J. Frisch, et al., Gaussian, Inc., Pittsburgh, PA, 1998.
103. R. Jaquet, W. Cencek, W. Kutzelnigg, and J. Rychlewski, *J. Chem. Phys.* **108**, 2837 (1998).
104. R. Poshusta and D. Kinghorn, *Int. J. Quantum Chem.* **60**, 213 (1996).
105. J. R. Magnus and H. Neudecker, *Matrix Differential Calculus with Applications in Statistics and Econometrics*, Wiley, Chichester, 1988.
106. W. Heitler and F. London, *Z. Phys.* **44**, 455 (1927).
107. L. Wolniewicz, private communication.
108. L. Wolniewicz, *J. Chem. Phys.* **103**, 1792 (1995).
109. I. Ben-Itzhak, E. Wells, K. D. Carnes, V. Krishnamurthi, O. L. Weaver, and B. D. Esry, *Phys. Rev. Lett.* **85**, 58 (2000).
110. A. Carrington, I. R. McNab, C. A. Montgomerie–Leach, and R. A. Kennedy, *Mol. Phys.* **72**, 735 (1991).
111. R. E. Moss and I. A. Sadler, *Mol. Phys.* **61**, 905 (1987).
112. R. E. Moss, *Mol. Phys.* **78**, 371 (1993).
113. B. D. Esry and H. R. Sadeghpour, *Phys. Rev. A* **60**, 3604 (1999).
114. R. E. Moss and L. Valenzano, *Mol. Phys.* **100**, 649 (2002).
115. M. Cafiero, S. Bubin, and L. Adamowicz, *Phys. Chem. Chem. Phys.* **5**, 1491 (2003).
116. D. B. Kinghorn, *Int. J. Quantum Chem.* **57**, 141 (1996).
117. D. B. Kinghorn and R. D. Poshusta, *Int. J. Quantum Chem.* **62**, 223 (1997).

118. D. B. Kinghorn and L. Adamowicz, *J. Chem. Phys.* **110**, 7166 (1999).
119. D. B. Kinghorn and L. Adamowicz, *Phys. Rev. Lett.* **83**, 2541 (1999).
120. C. E. Scheu, D. B. Kinghorn, and L. Adamowicz, *J. Chem. Phys.* **114**, 3393 (2001).
121. S. Bubin and L. Adamowicz, *J. Chem. Phys.* **118**, 3079 (2003).
122. S. Bubin and L. Adamowicz, *J. Chem. Phys.* **120**, 6051 (2004).
123. S. Bubin and L. Adamowicz, *J. Chem. Phys.* **121**, 6249 (2004).
124. S. Bubin, E. Bednarz, and L. Adamowicz, submitted for publication.
125. I. Dabrowski, *Can. J. Phys.* **62**, 1639 (1984).
126. H. W. Sarkas, J. H. Hendricks, S. T. Arnold, and K. H. Bowen, *J. Chem. Phys.* **100**, 1884 (1994).
127. D. Chang, K. Reimann, G. Surratt, G. Gellene, P. Lin, and R. Lucchese, *J. Chem. Phys.* **117**, 5757 (2002).
128. A. M. Frolov and V. H. Smith, *Phys. Rev. A* **55**, 2662 (1997).
129. J. Usukura, K. Varga, and Y. Suzuki, *Phys. Rev. A* **58**, 1918 (1998).
130. Z.-C. Yan and Y. K. Ho, *Phys. Rev. A* **60**, 5098 (1999).
131. M. Mella, G. Morosi, D. Bressanini, and S. Elli, *J. Chem. Phys.* **113**, 6154 (2000).
132. M. Mella, S. Chiesa, and G. Morosi, *J. Chem. Phys.* **116**, 2852 (2002).
133. K. Strasburger and H. Chojnacki, *J. Chem. Phys.* **108**, 3218 (1998).
134. K. Strasburger, *J. Chem. Phys.* **111**, 10555 (1999).
135. K. Strasburger, *J. Chem. Phys.* **114**, 615 (2001).
136. J. Mitroy and G. G. Ryzhikh, *J. Phys. B* **33**, 3495 (2000).
137. W. Cencek and J. Rychlewski, *Chem. Phys. Lett.* **320**, 549 (2000).
138. Z.-C. Yan and G. W. F. Drake, *Phys. Rev. A* **61**, 022504 (2000).
139. X. Li and J. Paldus, *J. Chem. Phys.* **118**, 2470 (2003).
140. W. C. Stwalley and W. T. Zemke, *J. Phys. Chem. Ref. Data* **22**, 87 (1993).
141. Z.-C. Yan and Y. K. Ho, *Phys. Rev. A* **60**, 5098 (1999).

AUTHOR INDEX

Numbers in parentheses are reference numbers and indicate that the author's work is referred to although his name is not mentioned in the text. Numbers in *italic* show the pages on which the complete references are listed.

Advances in Chemical Physics, Volume 131, edited by Stuart A. Rice

SUBJECT INDEX

Advances in Chemical Physics, Volume 131, edited by Stuart A. Rice